Advances in Experimental Medicine and Biology

Volume 859

More information about this series at http://www.springer.com/series/5584

Marco Canepari • Dejan Zecevic
Olivier Bernus
Editors

Membrane Potential Imaging in the Nervous System and Heart

Springer

Editors
Marco Canepari
Laboratoire Interdisciplinaire de Physique
Grenoble, France

Olivier Bernus
L'Institut de rythmologie et modélisation cardiaque LIRYC
Université de Bordeaux
Bordeaux, France

Dejan Zecevic
Department of Cellular and Molecular Physiology
Yale University School of Medicine
New Haven, CT, USA

ISSN 0065-2598 ISSN 2214-8019 (electronic)
Advances in Experimental Medicine and Biology
ISBN 978-3-319-37144-3 ISBN 978-3-319-17641-3 (eBook)
DOI 10.1007/978-3-319-17641-3

Springer Cham Heidelberg New York Dordrecht London

Softcover reprint of the hardcover 1st edition 2015

Printed on acid-free paper

Springer International Publishing AG Switzerland is part of Springer Science+Business Media (www.springer.com)

Preface

This book is a new edition of a previously published manual entitled *Membrane Potential Imaging in the Nervous System: Methods and Applications*. A significant advance in this new version is the broadening of the subject to include optical techniques for recording electrical activity of the heart in addition to the nervous system. Development of optical biotechnologies gives scientists the ability to probe more deeply into the functions of the nervous system and the heart. One of the most significant developments since the publication of the previous edition has been the concentrated effort to engineer genetically encoded protein probes of membrane potential that can be selectively targeted and expressed in specific cell types. Several laboratories have recently described a number of novel fluorescence protein voltage sensors. This new edition includes an updated critical review on the current status of these developments in terms of signal size, time course, and *in vivo* function.

The year 2015 is the UNESCO International Year of Light. In this respect, the publication of this book is timely since the optical readout of the nervous activity and of the activity of the heart is an important part of the "photon revolution," bringing about predictions that the photon will progressively replace the electron for probing excitable tissue. The imaging approach provides a method for monitoring electrical signaling with submicron and sub-millisecond resolution. Two main advantages over conventional electrical measurements make this experimental technique both unique and valuable: (a) the ability to record voltage transients simultaneously at hundreds or thousands of sites in the nervous system and heart and (b) the ability to monitor signals from structures that are too small for electrode measurements (e.g., terminal dendritic branches, axons, and dendritic spines). This recording technique represents a complementary component in a rapidly developing avant-garde strategy to implement a relatively non-invasive all-optical investigation of living neural circuits using light-activated ion channels, and the recent development of novel imaging technologies such as depth-resolved optical imaging of cardiac electrical activity. *Membrane Potential Imaging in the Nervous System and Heart* reviews the current status of optical techniques to monitor electrical signaling in the excitable cells. Special emphasis is given to limitations of these techniques as well as new developments and future directions.

The knowledge upon which voltage-imaging technique is based was generated over several decades and described in a wide variety of publications. This book provides a single comprehensive source of information on different types of voltage-imaging techniques, including overviews, methodological details, examples of experimental measurements, and future developments. The book is structured into five parts, each containing several chapters written by experts and major contributors to particular topics. All of the chapters have been updated to include the most recent developments. The volume starts with a historical perspective and fundamental principles of membrane potential imaging and covers the measurement of membrane potential signals from dendrites and axons of individual neurons, measurements of the activity of many neurons with single cell resolution, monitoring of population signals from the nervous system, optical mapping of cardiac electrophysiological properties and arrhythmias, concluding with the overview of new approaches to voltage imaging. The book is targeted at all scientists interested in this mature but also rapidly expanding imaging approach.

Grenoble, France — Marco Canepari
Bordeaux, France — Olivier Bernus
New Haven, CT, USA — Dejan Zecevic

Contents

About the Editors

Marco Canepari is first class INSERM researcher (CR1) working at the *Laboratoire Interdisciplinaire de Physique* in Grenoble. He graduated in Physics at the University of Genoa and received his PhD in Biophysics from the International School for Advanced Studies in Trieste. He worked at the National Institute for Medical Research in London, at Yale University, and at the University of Basel. Dr. Canepari is an expert on several optical techniques applied to neurophysiology. He and Dr. Zecevic collaborated for a number of years using voltage-imaging and calcium imaging approaches to study mechanisms underlying synaptic plasticity.

Olivier Bernus is Professor and Scientific Director at the Electrophysiology and Heart Modeling Institute LIRYC at the University of Bordeaux, France. He graduated in Mathematical Physics at the University of Ghent, Belgium, and obtained his PhD in Physics from the same university. He subsequently worked as a postdoctoral fellow at SUNY Upstate Medical University, USA, and as a tenure track research fellow at the University of Leeds, UK. Dr. Bernus is an expert in cardiac optical imaging and is currently working on the development of novel techniques for three-dimensional depth-resolved optical imaging of cardiac electrical activity.

Dejan Zecevic is Senior Research Scientist at the Department of Cellular and Molecular Physiology, Yale University School of Medicine. He received his PhD in Biophysics from the University of Belgrade, Serbia, and was trained in the laboratory of Dr. Lawrence Cohen, who initiated the field of voltage-sensitive dye recording. Dr. Zecevic is the pioneer of intracellular voltage-sensitive dye imaging technique, a unique and a cutting-edge technology for monitoring the membrane potential signaling in axons, dendrites, and dendritic spines of individual nerve cells.

Part I
Principles of Membrane Potential-Imaging

Chapter 1
Historical Overview and General Methods of Membrane Potential Imaging

Oliver Braubach, Lawrence B. Cohen, and Yunsook Choi

Abstract Voltage imaging was first conceived in the late 1960s and efforts to find better organic voltage sensitive dyes began in the 1970s and continue until today. At the beginning it was difficult to measure an action potential signal from a squid giant axon in a single trial. Now it is possible to measure the action potential in an individual spine. Other chapters will discuss advances in voltage imaging technology and applications in a variety of biological preparations. The development of genetically encoded voltage sensors has started. A genetically encoded sensor could provide cell type specific expression and voltage recording (see Chap. 20).

Optimizing the signal-to-noise ratio of an optical recording requires attention to several aspects of the recording apparatus. These include the light source, the optics and the recording device. All three have improved substantially in recent years. Arc lamp, LED, and laser sources are now stable, more powerful, and less expensive. Cameras for recording activity have frames rates above 1 kHz and quantum efficiencies near 1.0 although they remain expensive. The sources of noise in optical recordings are well understood. Both the apparatus and the noise sources are discussed in this chapter.

Keywords History • Richard Keynes • David Hill • Ichiji Tasaki • Dora Rosental • David Gilbert • Shot noise • Signal-to-noise ratio • Out-of-focus blurring • Scattering

O. Braubach • Y. Choi
Center for Functional Connectomics, Korea Institute of Science and Technology (KIST), Seoul 136-791, Korea

L.B. Cohen (✉)
Center for Functional Connectomics, Korea Institute of Science and Technology (KIST), Seoul 136-791, Korea

Department of Cellular and Molecular Physiology, Yale University School of Medicine, New Haven, CT 06520, USA
e-mail: lawrence.b.cohen@hotmail.com

M. Canepari et al. (eds.), *Membrane Potential Imaging in the Nervous System and Heart*, Advances in Experimental Medicine and Biology 859,
DOI 10.1007/978-3-319-17641-3_1

1 Motivation

Several factors make optical methods interesting to physiologists and neuroscientists. First, is the possibility of making simultaneous measurements from multiple sites. Because many different neurons in a nervous system are active during each behavior, it is obviously important to be able to monitor activity from many cells simultaneously. Similarly, different parts of a neuron may perform different functions and again it would be essential to make simultaneous observations over the whole structure of the neuron. Finally, many different brain regions are simultaneously active during behaviors and recording of population signals from these regions would be useful.

Second, optical measurement can be made with very high time resolution. Some voltage sensitive dyes are known to respond to changes in membrane potential extremely rapidly, with time constants of less than 1 μs. Other indicators of activity, such as hemodynamic changes, monitored as intrinsic signals, can be much slower, with latencies and durations in the range of seconds, but even so these recordings reflect real-time measurements of events in the brain. Third, optical measurements are in some sense non-invasive; recordings can be made from cells and processes too small for electrode measurements; these include T-tubules, presynaptic terminals, and dendritic spines. On the other hand, optical measurements using extrinsic dyes and proteins can be invasive if they cause pharmacological or photodynamic effects. The degree of this kind of invasiveness varies from relatively innocuous, to measurements where pharmacology and photodynamic damage are important factors, a subject discussed in detail in later chapters.

The above three factors are clearly attractive because the use of voltage imaging has grown dramatically since its beginnings in the 1970s. Nonetheless it is important to note that optical methods cannot be used to record from deep in a tissue, an important capability which other, slower, imaging methods such as fMRI and 2-deoxyglucose do provide.

Three kinds of optical signals have been extensively used to monitor activity; indicators of membrane potential, intracellular ion concentrations, and intrinsic signals. This book chapter is an historical and technical introduction of voltage imaging.

2 History

2.1 *Early Milestones in Imaging Activity*

The first optical signals detected during nerve activity were light scattering changes that accompany trains of stimuli to nerves (Hill and Keynes 1949; Hill 1950). During and following the stimuli there is first an increase in scattering followed by a slower decrease lasting many tens of seconds (Fig. 1.1a). On the top left is a

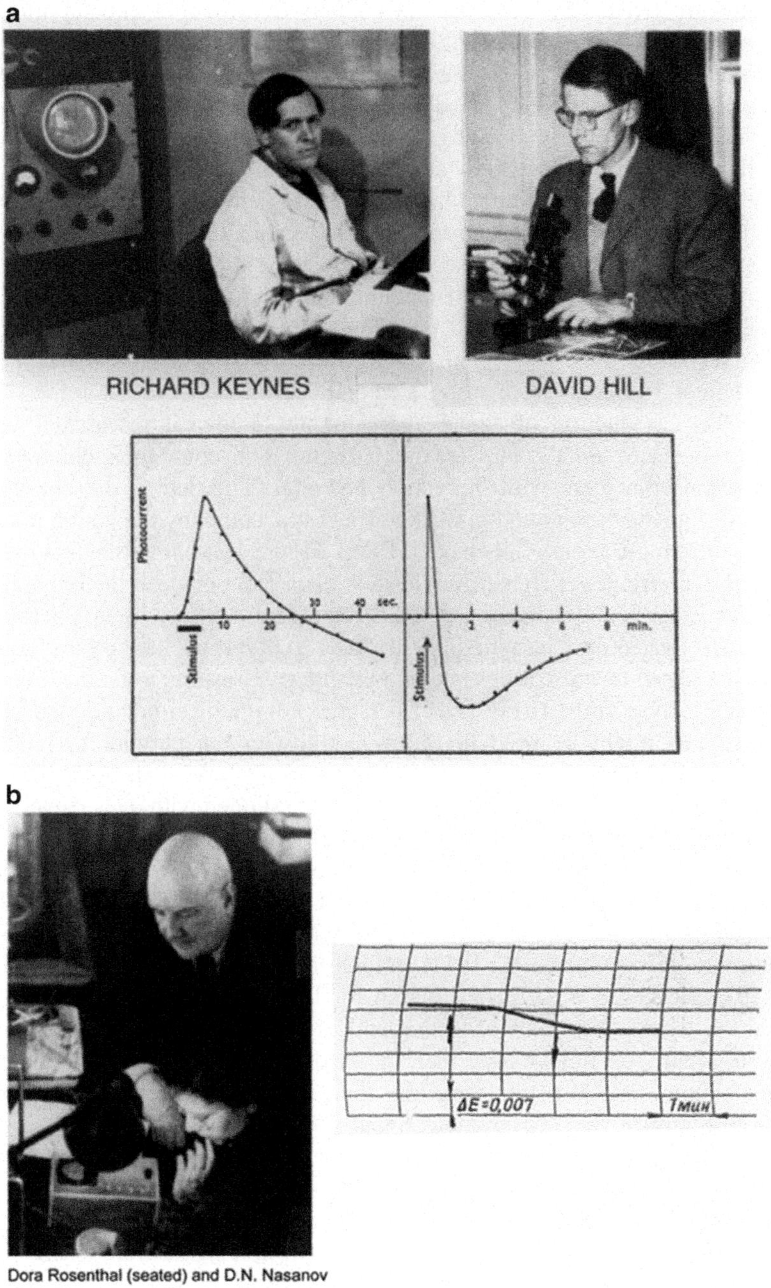

Fig. 1.1 (**a**) *Top*. Photographs of Richard Keynes and David Hill. *Bottom panel*. A *Carcinus maenas* leg nerve was stimulated 250 times during the bar labeled stimulus. The stimulus resulted in an increase in scattering that lasted about 20 s. This was followed by a scattering decrease which persisted for several minutes. From Hill and Keynes (1949). (**b**) A photograph of D.N. Nasonov and Dora Rosenthal taken in 1956 together with a recording of the change in Direct Turquoise absorption as a result of stimulating a crab axon 50 times per second for 2 min (6,000 stimuli). The optical recording is from Levin et al. (1968)

photograph of Richard Keynes (together with an oscilloscope) taken at the time of those measurements; on the top right is a photograph of David Hill taken about 20 years later. In the bottom panel is the recording of light scattering from a whole crab leg nerve as a function of time during and following a train of 250 stimuli. The light scattering response is shown on two time bases.

Starting at about the same time and continuing into the 1960s, scientists at the Institute of Cytology in Leningrad began measuring changes in the optical properties of dye stained nerves (Nasonov and Suzdal'skaia 1957; Vereninov et al. 1962; Levin et al. 1968). Again they used relatively long stimulus trains and recorded signals with time courses of tens of seconds. The photograph in Fig. 1.1b includes two of the scientists: D. N. Nasonov (standing) and Dora Rosenthal (seated). On the right is a recording from Levin et al. (1968) illustrating the change in absorption of the dye Direct Turquoise from an isolated crab axon.

With the introduction of signal averaging to biological research it became possible to measure signals that were much smaller. One could then look for signals that occurred coincident with the action potential. This led to the discovery of changes in light scattering and birefringence that accompany the action potentials in nerves and single axons (Cohen et al. 1968). Figure 1.2a shows the first measurement of the birefringence signal from a squid giant axon made at the Laboratory of the Marine Biological Association in 1967. The signal was very small; the fractional change, $\Delta I/I$, was only 1 part in 100,000. And, 20,000 trials had to be averaged to achieve the signal-to-noise ratio shown in the figure. Brushing aside the tiny size of the signals, David Gilbert (Fig. 1.2b; personal communication) pointed out that optical signals might be used to follow activity in the nervous system. Soon thereafter Ichiji Tasaki (Fig. 1.2b) and collaborators (Tasaki et al. 1968) discovered fast changes in the fluorescence of axons that were stained with dyes. Several years later changes in dye absorption (Ross et al. 1974) and birefringence (Ross et al. 1977) were also found. Almost all of these dye signals were shown to depend on changes in membrane potential (Davila et al. 1974; Cohen et al. 1974; Salzberg and Bezanilla 1983) a conclusion that is now widely accepted. However, there was early disagreement (Conti et al. 1971) and Ichiji was always unconvinced.

In 1971 we began a deliberate search for larger optical signals by screening many dyes. After a few years of testing and ~1,000 dyes we had obtained signals that were about 100 times larger than the intrinsic signals described above. One hundred times larger doesn't mean that the signals are now large. Care is still needed to optimize the measurement signal-to-noise ratio. The second section of this chapter together with other chapters provide information about the needed optimization. Because of this improvement in signal size, we were able to record optical signals from a smaller membrane area and with higher time resolution. The top panel of Fig. 1.3 illustrates a single trial absorption signal from a single action potential in a 50 μm diameter barnacle neuron made with the merocyanine dye whose structure is illustrated. The five adults on the right hand side of the photograph at the bottom were in large part responsible for the dye screening. Other important contributors during those early years (not shown) were Vicencio Davila, Jeff Wang, Ravinder Gupta and Alan Waggoner.

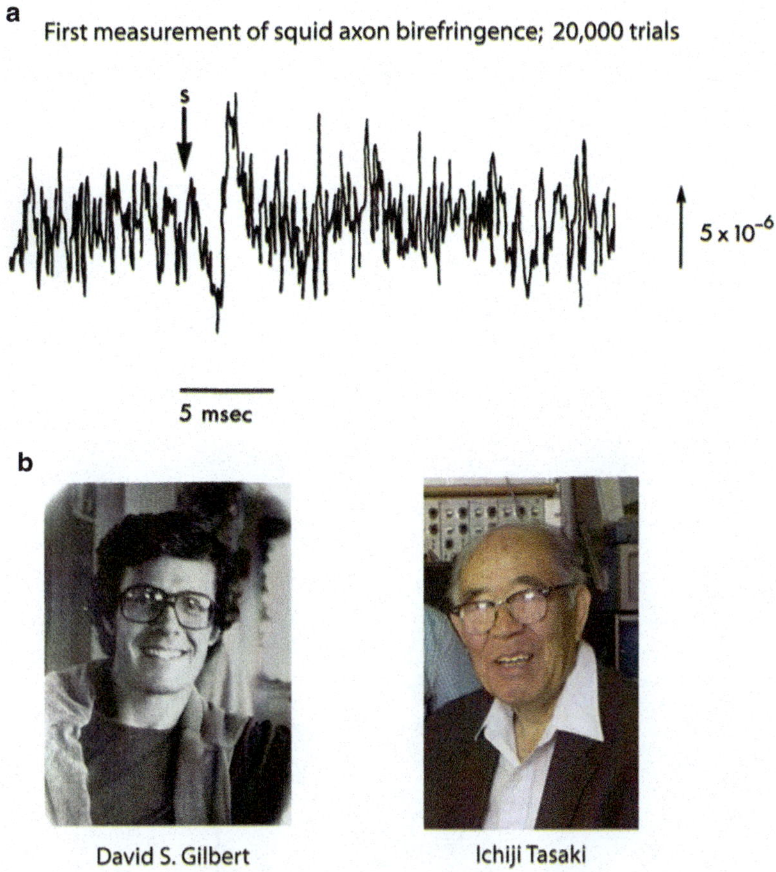

Fig. 1.2 (**a**) The first recording of an optical signal (birefringence) from an individual axon on a time scale appropriate for measuring the time course of the signal during the action potential. 20,000 trials were averaged. In this and subsequent figures the size of the *vertical line* represents the stated value of the fractional change in intensity ($\Delta I/I$ or $\Delta F/F$). (L.B. Cohen and R.D. Keynes, unpublished result). (**b**) Photographs of David Gilbert and Ichiji Tasaki

2.2 *Ion Sensitive Dyes and Their Comparison with Voltage Sensitive Dyes*

The first attempt to use organic optical indicators of intracellular ion concentration were measurements of calcium with the dye murexide (Jobsis and O'Connor 1966). Ironically, those results are now thought to be an artifact (Maylie et al. 1987). Ashley and Ridgway (1970) later introduced the use of the photoprotein aequorin. Following the suggestion of John Cooper (personal communication), the Arsenazo class of calcium dyes was introduced (Brown et al. 1975). More recently, Roger Tsien and collaborators have developed a large number of widely used ion indicator dyes based on the chemical structure of the chelator EGTA (e.g. Tsien 1989).

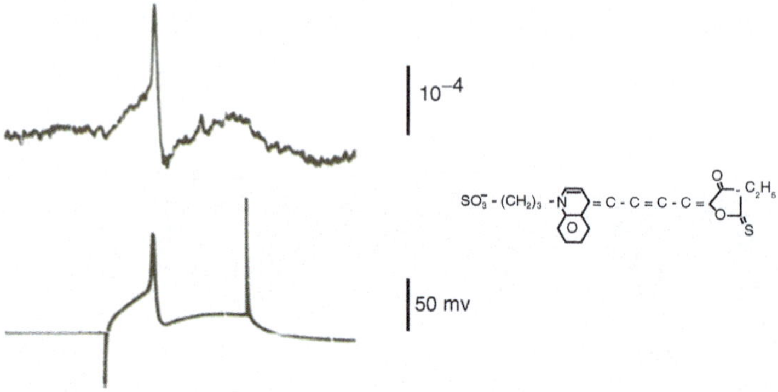

Absorption signals from a single neuron in a barnacle ganglion.

Fig. 1.3 On the *top* is an absorption change measured during a single action potential from a cell body in a barnacle supraesophageal ganglion. The structure of the absorption dye, synthesized by Jeff Wang and Alan Waggoner, is illustrated at the *top*. In the photograph are, from the *left*, Naoko Kamino, Kyoko Kamino, Irit Grinvald, Eran Grinvald, Amiram Grinvald, Larry Cohen, Kohtaro Kamino, Kaeko Kamino, Brian Salzberg and Bill Ross. The photograph, made by Sarah Lesher, is from 1976. The optical recording is from Salzberg et al. (1977)

Figure 1.4 illustrates two time course comparisons between voltage, calcium, and intrinsic signals. The top pair compares voltage and calcium signals from the same location on a turtle olfactory bulb. The signals are in response to a shock to the olfactory nerve. The voltage sensitive dye signal rises faster and falls faster than the calcium signal. Some of the slowness of the calcium signal is thought to result

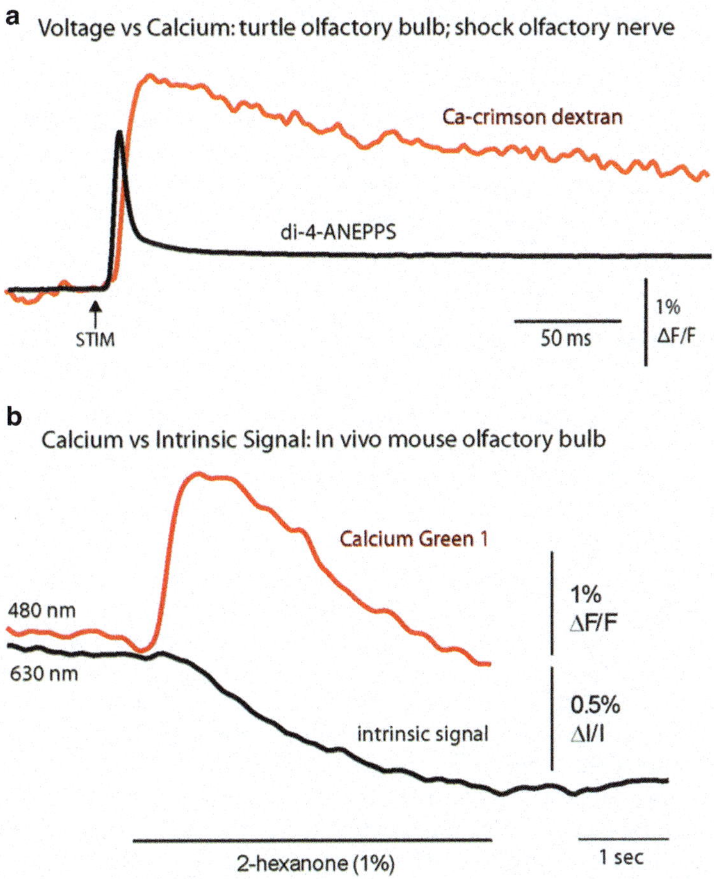

Fig. 1.4 Comparisons of voltage, calcium, and intrinsic imaging signals. (**a**) A comparison of the time course of a voltage and a calcium dye signal from the same location near the merger of the olfactory nerve and the olfactory bulb in an *in vitro* turtle preparation. The voltage signal is faster (M. Wachowiak and L. B. Cohen, unpublished). (**b**) A comparison of the time course of the calcium signal and the intrinsic imaging signal in a mouse olfactory bulb glomerulus in response to the odorant, 2-hexanone. The intrinsic imaging signal is substantially slower than the calcium signal. (Modified from Wachowiak and Cohen 2003)

from the dye response itself and some from the time taken by the cells to bring their calcium concentration back to its original level. The bottom pair of traces compares a calcium signal and an intrinsic signal (Blasdel and Salama 1986; Grinvald et al. 1986) from the same glomerulus in a mouse olfactory bulb. Here the stimulus was an odorant presented to the nose. The calcium signal reaches a peak in less than 0.5 s while the intrinsic signal doesn't reach its peak until about 5 s after the stimulus onset.

2.3 *Multi-site Measurements of Brain Activity*

The first multi-site measurements of brain activity were made by Schuette et al. (1974) who used an image intensifier and camera tube to monitor changes in NADH fluorescence from cortex during epileptic seizures. This was followed by methods using several individual silicon photodiodes (Salzberg et al. 1977), and later by a variety of photodiode arrays, camera tubes, CCD and CMOS cameras.

3 Measurement Principles

Even though the changes in membrane potential during action potentials and synaptic activity are small in mV, they are giant in volts/cm because the membrane is very thin. Thus, 100 mV across 3 nm is 300,000 V/cm, a voltage gradient long known to be able to alter the spectral properties of merocyanine dyes (Labhart 1963; Bucher et al. 1969). This direct effect of voltage on dye spectra, called electrochromism, is one mechanism thought to give rise to signals from organic voltage sensitive dyes (Loew et al. 1985). Other mechanisms with spectral supporting evidence are voltage sensitive shifts in monomer-dimer equilibria (Waggoner and Grinvald 1977) and dye rotation in the electric field (Conti 1975; Fromherz et al. 1991). Because the electric field change is so large, it is perhaps not surprising that we were able to measure signals in ~500 of the ~2,000 dyes that were tested on squid giant axons.

Action potential propagation and synaptic currents also give rise to extracellular and intracellular currents which means that there will also be voltage gradients in these locations. However these extracellular currents flow over distances of mm (rather than nm) and thus the voltage gradients are smaller by six orders of magnitude. It is for this reason that the voltage sensitive dye signals are presumed to arise solely from changes in dye molecules embedded in or directly adjacent to the membrane.

4 The Definition of a Voltage Sensitive Dye

The voltage-sensitive dye signals described in this book are "fast" signals (Cohen and Salzberg 1978) that are presumed to arise from membrane-bound dye. Figure 1.5, top, illustrates the kind of result that is used to define a voltage sensitive dye. In a giant axon from a squid, these optical signals are fast, following membrane potential with a time constant of <10 μs (Loew et al. 1985). The dye signals have a time course that is qualitatively different from the time course of the ionic currents or the membrane permeability changes. With almost all of the tested dyes the signal size is linearly related to the size of the change in potential (e.g. Gupta et al. 1981).

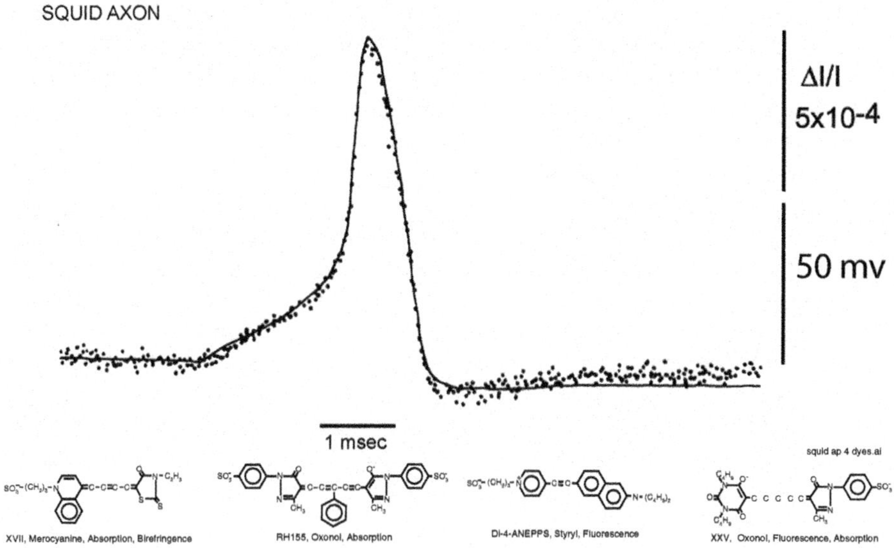

Fig. 1.5 *Top*. Changes in absorption (*dots*) of a giant axon stained with a merocyanine dye, XVII, (*bottom left* dye) during a membrane action potential (*smooth trace*) recorded simultaneously. The change in absorption and the action potential had the same time course. The response time constant of the light measuring system was 35 μs; 32 sweeps were averaged. *Bottom*. Examples of four different chromophores that have been used to monitor membrane potential. The merocyanine dye, XVII (WW375), and the oxonol dye, RH155, were commercially available as NK2495 and NK3041 from Nippon Kankoh-Shikiso Kenkyusho Co. Ltd., Okayama, Japan. The oxonol, XXV (WW781) and styryl, di-4-ANEPPS, are available commercially as dye R-1114 and D-1199 from Invitrogen (Molecular Probes), Junction City, OR

Thus, these dyes provide a direct, fast, and linear measure of the change in membrane potential of the stained membranes.

A direct comparison of membrane potential and voltage-sensitive dye signal like that illustrated in Fig. 1.5 cannot often be made. Several of the following chapters discuss the important issues involved in interpreting less direct comparisons.

Several organic voltage-sensitive dyes have been used to monitor changes in membrane potential in a variety of preparations. Figure 1.5, bottom, illustrates four different chromophores (the merocyanine dye, XVII, was used for the measurement illustrated in Fig. 1.5, top). The merocyanine dye, XVII (WW375), and the oxonol dye, RH155 were commercially available as NK2495 and NK3041 from Nippon Kankoh-Shikiso Kenkyusho Co. Ltd., Okayama, Japan. The oxonol, XXV (WW781) and styryl, di-4-ANEPPS, are commercially available as dye R-1114 and D-1199 from Invitrogen (Molecular Probes), Junction City, OR. For each of the four chromophores approximately 100 analogues have been synthesized in an attempt to optimize the signal-to-noise ratio. (This screening was made possible by synthetic efforts of three laboratories: Jeff Wang, Ravender Gupta and Alan Waggoner then at Amherst College; Rina Hildesheim and Amiram Grinvald at the Weizmann Institute; and Joe Wuskell and Leslie Loew at the University of

Connecticut Health Center.) For each of the chromophores, there were 10 or 20 dyes that gave approximately the same signal size on squid axons (Gupta et al. 1981). However, dyes that had nearly identical signal size on squid axons often had very different responses on other preparations, and thus tens of dyes usually have to be tested to obtain the largest possible signal. A common failure was that the dye did not penetrate through connective tissue or along intercellular spaces to the membrane of interest.

The following rules-of-thumb seem to be useful. First, each of the chromophores is available with a fixed charge, which is either a quaternary nitrogen (positive) or a sulfonate (negative). Generally the positive dyes have given larger signals when dye is applied either extracellularly or intracellularly in vertebrate preparations. Second, each chromophore is available with carbon chains of several lengths. The more hydrophilic dyes (methyl or ethyl) work best if the dye has to penetrate through a compact tissue (vertebrate brain) or be transported to distant branches of a dendritic tree. More hydrophobic dyes (hexyl or octyl) work best in tissue culture; in tissue culture the less hydrophobic dyes unbind rapidly.

5 Preliminary Considerations

The optical signals are small in many of the examples of voltage sensitive dye measurements given in the chapters in this book. Some examples involve fractional intensity changes, $\Delta I/I$, that are as low as 10^{-4}. In order to measure these signals, the noise in the measurements had to be an even smaller fraction of the resting intensity. In the sections that follow, some of the considerations necessary to achieve such a low noise are outlined.

5.1 Signal Type

Sometimes it is possible to decide in advance which kind of optical signal will give the best signal-to-noise ratio but, in other situations, an experimental comparison is necessary. The choice of signal type often depends on the optical characteristics of the preparation. Extrinsic birefringence signals are relatively large in preparations that, like axons, have a cylindrical shape and radial optic axis (Ross et al. 1977). However, in preparations with spherical symmetry (e.g., cell soma), the birefringence signals in adjacent quadrants will cancel (Boyle and Cohen 1980).

Thick preparations (e.g. mammalian cortex) also dictate the choice of signal. In this circumstance transmitted light measurements are not easy (a subcortical implantation of a light guide would be necessary), and the small size of the absorption signals that are detected in reflected light (Ross et al. 1977; Orbach and Cohen 1983) meant that fluorescence would be optimal (Orbach et al. 1985). Another factor that affects the choice of absorption or fluorescence is that the signal-to-noise ratio in fluorescence is more strongly degraded by dye bound to extraneous material.

Fluorescence signals have most often been used in monitoring activity from tissue-cultured neurons. Both kinds of signals have been used in measurements from ganglia and brain slices. Fluorescence has always been used in measurements from intact vertebrate brains.

5.2 Amplitude of the Voltage Change

Activity signals are often presented as a fractional intensity change, ΔI/I. The signals give information about the time course of the potential change but no direct information about the absolute magnitude. However, in some instances, approximate estimations can be obtained. For example, the size of the optical signal in response to a sensory stimulus can be compared to the size of the signal in response to an epileptic event (Orbach et al. 1985). Or in single neuron experiments long lasting hyperpolarizing steps can be induced in the soma after blocking ionic currents and this signal can be used for calibration. Another approach is the use of ratiometric measurements at two independent wavelengths (Gross et al. 1994). However, one must know the fraction of the fluorescence that results from dye bound to active membranes *versus* dye bound to non-active membranes. This requirement is almost never met.

6 Noise

6.1 Shot Noise

The limit of accuracy with which light can be measured is set by the shot noise arising from the statistical nature of photon emission and detection. Fluctuations in the number of photons emitted per unit time occur, and if an ideal light source (e.g. a tungsten filament) emits an average of 10^{16} photons/ms, the root-mean-square (RMS) deviation in the number emitted is the square root of this number or 10^{8} photons/ms. Because of shot noise, the signal-to-noise ratio is directly proportional to the square root of the number of measured photons and inversely proportional to the square root of the bandwidth of the photodetection system (Braddick 1960; Malmstadt et al. 1974). The basis for the square root dependence on intensity is illustrated in Fig. 1.6. In the top plot, the result of using a random number table to distribute 20 photons into 20 time windows is shown. In the bottom plot the same procedure was used to distribute 200 photons into the same 20 bins. Relative to the average light level there is more noise in the top trace (20 photons) than in the bottom trace (200 photons). The improvement from A to B is similar to that

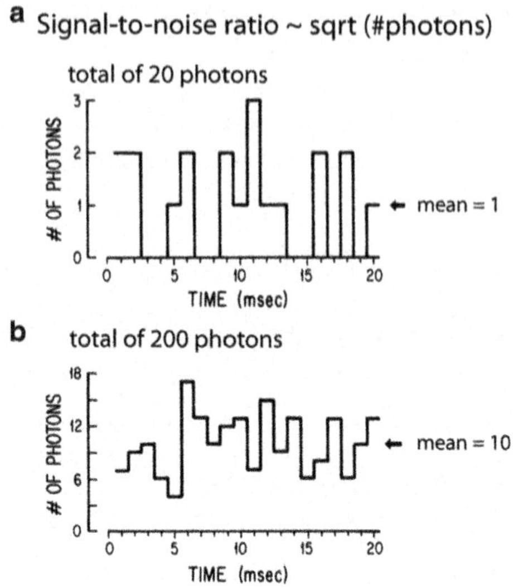

Fig. 1.6 Plots of the results of using a table of random numbers to distribute 20 photons (*top*, **a**) or 200 photons (*bottom*, **b**) into 20 time bins. The result illustrates the fact that when more photons are measured the signal-to-noise ratio is improved. The improvement is close to that expected from the known square root relationship between the signal-to-noise ratio and the measured intensity. Redrawn from Wu and Cohen (1993)

expected from the square-root relationship. This square-root relationship is indicated by the straight red line in Fig. 1.7 which plots the light intensity divided by the noise in the measurement versus the light intensity. In a shot-noise limited measurement, improvement in the signal-to-noise ratio can only be obtained by (1) increasing the illumination intensity, (2) improving the light-gathering efficiency of the measuring system, or (3) reducing the bandwidth.

6.2 *The Optimum Signal to Noise Ratio*

Because only a small fraction of the 10^{16} photons/ms emitted by a tungsten filament source will be measured, a signal to noise ratio of 10^{8} (see above) cannot be achieved. A partial listing of the light losses follows. A 0.9 NA lamp collector lens would collect 10 % of the light emitted by a point source. Only 20 % of that light is in the visible wavelength range; the remainder is infrared (heat). Limiting the incident wavelengths to those which elicit a signal means that only 10 % of the visible light is used. Thus, the light reaching the preparation might typically be reduced to 10^{13} photons/ms. If the light-collecting system that forms the image has high efficiency e.g., in an absorption measurement, about 10^{13} photons/ms will reach the image plane. (In a fluorescence measurement there will be much less light measured because (1) only a fraction of the incident photons are absorbed by the fluorophores, (2) only a fraction of the absorbed photons reappear as emitted photons, and (3) only a fraction of the emitted photons are collected by the objective.) If the camera has a quantum efficiency of 1.0, then, in

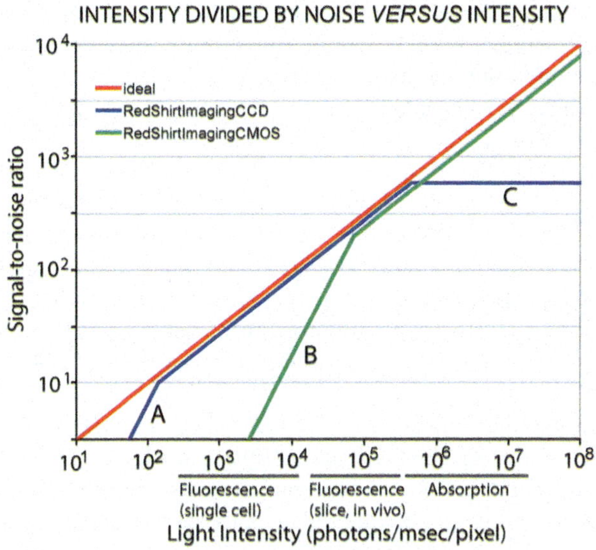

Fig. 1.7 The ratio of light intensity divided by the noise in the measurement as a function of light intensity in photons/ms/pixel. The theoretical optimum signal-to-noise ratio (*red line*) is the shot-noise limit. Two camera systems are shown, a cooled, back-illuminated, 2 kHz frame rate, 80 × 80 pixel CCD camera (*blue lines*) and a CMOS camera (*green lines*). The CMOS camera provides an optimal signal-to-noise ratio at higher intensities while the CCD camera is better at lower intensities. The approximate light intensity per detector in fluorescence measurements from a single neuron, fluorescence measurements from bath application of dye to a slice or *in vivo* preparation, and in absorption measurements from a ganglion or a slice is indicated along the x axis. The signal-to-noise ratio for the CMOS camera falls away at low intensities because of dark noise (*B*). The lower dark noise of the cooled CCD allows it to function at the shot-noise limit at lower intensities until read noise dominates (*A*). The CCD camera saturates at intensities above 5×10^6 photons/ms/0.2 % of the object plane (*C*)

absorption, a total of 10^{13} photoelectrons/ms will be measured. With a camera of 10,000 pixels, there will be 10^9 photoelectrons/ms/pixel. The r.m.s. shot noise will be $10^{4.5}$ photoelectrons/ms/pixel; thus the very best that can be expected is a noise that is $10^{-4.5}$ of the resting light (a signal-to-noise ratio of 90 dB). The extra light losses in a fluorescence measurement will further reduce the maximum obtainable signal-to-noise ratio.

One way to compare the performance of different camera systems and to understand their deviations from optimal (shot-noise limited) is to determine the light intensity divided by the noise in the measurement and plot that versus the number of photons reaching the each pixel of the camera. The straight red line in Fig. 1.7 is the plot for an ideal camera. At high light intensities this ratio is large and thus small changes in intensity can be detected. For example, at 10^8 photons/ms a fractional intensity change of 0.1 % can be measured with a signal-to-noise ratio of 10. On the other hand, at low intensities the ratio of intensity divided by noise is small and only large signals can be detected. For example, at 10^4 photons/ms the same fractional change of 0.1 % can be measured with a signal-to-noise ratio of only 1 and that only after averaging 100 trials.

In addition, Fig. 1.7 compares the performance of two particular cameras, a cooled 80 × 80 pixel CCD camera (blue line), and a 128 × 128 pixel CMOS camera (green line). The CMOS camera approaches the shot-noise limitation over the range of intensities from 5×10^4 to 10^8 photons/ms/pixel. This is the range of intensities obtained in absorption measurements and fluorescence measurements on bulk-stained *in vitro* slices and intact brains. On the other hand, the cooled CCD camera approaches the shot noise limit over the range of intensities from 2×10^2 to 5×10^5 photons/ms/pixel. This is the range of intensities obtained from fluorescence experiments on processes of individual cells and neuron cell bodies.

In the discussion that follows we will indicate the aspects of the measurements and the characteristics of the two camera systems which cause them to deviate from the shot noise ideal. The two camera systems we have chosen to illustrate in Fig. 1.7 have relatively good dark noise and saturation characteristics; other cameras would be dark noise limited at higher light intensities and would saturate at lower intensities. At present there is no single camera which can cover the entire intensity range of interest to physiologists with close to shot noise limited performance.

6.3 Extraneous Noise

A second type of noise, termed extraneous or technical noise, is more apparent at high light intensities where the sensitivity of the measurement is high because the fractional shot noise and dark noise are low. One type of extraneous noise is caused by fluctuations in the output of the light source (see below). Two other sources of extraneous noise are vibrations and movement of the preparation. A number of precautions for reducing vibrational noise have been described (Salzberg et al. 1977, London et al. 1987). The pneumatic isolation mounts on many vibration isolation tables are more efficient in reducing vertical vibrations than in reducing horizontal movements. An improvement is air-filled soft rubber tubes (Newport Corp, Irvine, CA). For more severe vibration problems, Minus K Technology (Inglewood, CA) sells vibration isolation tables with very low resonant frequencies. For even tougher situations the Halcyonics Micro 60 (Menlo Park, CA) is an active (piezoelectric drivers) isolator and can defeat airborne vibrations as well as those transmitted from the floor (Brian Salzberg, personal communication). The increasing improvement comes at increasing cost. Nevertheless it has been difficult to reduce vibrational noise to less than 3×10^{-5} of the total light. With this amount of vibrational noise, increases in measured intensity beyond 10^9 photons/ms would not improve the signal-to-noise ratio.

6.4 Dark Noise/Read Noise

At 1 kHz frame rates the read noise is likely to be larger than the dark noise in CCD and CMOS cameras. The read noise will degrade the signal-to-noise ratio at very low light levels (<100 photons/pixel/frame for the RedShirtImaging CCD camera). The read noise of cooled CCD cameras can be substantially lower than that of large pixel CMOS cameras (Table 1.2). The larger read noise in the CMOS camera accounts for the fact that segment B in Fig. 1.7 is substantially to the right of segment A.

6.5 Preparation Movement

Preparation movement is often the limiting noise factor in wide-field *in vivo* measurements. The movement artifacts *in vivo* usually consist of irregular movements of the entire animal (less of a problem in anesthetized preparations) as well as of heart beat- and breathing-related movements of the brain. The heart beat pulsations are larger in regions with a high density of blood vessels. The heart beat- and breathing-related artifacts significantly increase their amplitudes when the skull and the *dura matter* are removed. The stability of recordings also depends on the diameter of the craniotomy. Thus, openings larger than 1 mm in diameter are often accompanied by larger movement artifacts. This noise can be reduced by covering the skull opening with 2 % agarose and a glass coverslip (Svoboda et al. 1997), by keeping the temperature of the brain surface stable (with a precision of 0.1 °C (Garaschuk et al. 2006)), and by subtraction techniques (Orbach et al. 1985).

7 Light Sources

Three kinds of sources have been used. Tungsten filament lamps are a stable source, but their intensity is relatively low, particularly at wavelengths less than 480 nm. Arc lamps are somewhat less stable but can provide higher intensity. Laser illumination can provide even more intense illumination. Newly available LED sources are bright, stable, and convenient.

7.1 Tungsten Filament Lamps

It is not difficult to provide a power supply stable enough so that the output of the bulb fluctuates by less than 1 part in 10^5. In absorption measurements, where the fractional changes in intensity are relatively small, only tungsten filament sources have been used. On the other hand, fluorescence measurements often have larger fractional changes that will better tolerate light sources with small amounts of systematic noise, and the measured intensities are lower, making improvements in signal-to-noise ratio from brighter sources attractive.

7.2 Arc Lamps

Cairn Research Ltd (Faversham, UK) provides xenon power supplies, lamp housings, and arc lamps with noise that is in the range of 1 part in 10^4. A 150 W lamp yielded 2–3 times more light at 520 ± 45 nm than a tungsten filament bulb. The extra intensity is especially useful where the light intensity is low and the measurement is shot noise/dark noise limited (e.g. for fluorescence measurements from processes of single neurons).

7.3 Light Emitting Diodes (LEDs)

Salzberg et al. (2005) reported that high power light-emitting diodes (LED's) exhibit low-frequency noise characteristics that are clearly superior to those of tungsten halogen lamps. This holds, *a fortiori*, for arc lamps, which are generally quite poor. Also, their non-coherence is important when measuring absorption or scattering change, where the speckle from lasers is a problem (less of a problem when measuring fluorescence). In general, it is preferable to choose a single wavelength LED if this is possible, rather than filtering the output of a white LED. The former have more power in the spectral range of interest. Prizmatix, Ltd., (Givat-Shmuel, Israel) provides LED illumination systems.

7.4 Lasers

Laser illumination can provide very high illumination intensity but avoiding photodynamic damage requires careful attention to optimizing the trade-off between intensity and photodamage. The interference from speckle noise can be eliminated by reducing the beam coherence (Dejan Zecevic and Thomas Knopfel, personal communication).

8 Optics

8.1 *Numerical Aperture*

The need to maximize the number of measured photons has been a dominant factor in the choice of optical components. In epifluorescence, both the excitation light and the emitted light pass through the objective, and the intensity reaching the photodetector is proportional to the fourth power of numerical aperture (Inoue 1986). Clearly, numerical aperture is an important consideration in the choice of lenses. Direct comparison of the intensity reaching the image plane has shown that the light collecting efficiency of an objective is not completely determined by the stated magnification and numerical aperture. Differences of a factor of five between lenses of the same specification were found. We presume that these differences depend on the number of lenses, the coatings, and absorbances of glasses and cements. We recommend empirical tests of several lenses for efficiency.

8.2 *Depth of Focus*

Salzberg et al. (1977) determined the effective depth of focus for a 0.4 NA objective lens by recording an optical signal from a neuron when it was in focus and then moving the neuron out of focus by various distances. They found that the neuron had to be moved 300 μm out of focus to cause a 50 % reduction in signal size. Using 0.5 NA optics, 100 μm out of focus led to a reduction of 50 % (Kleinfeld and Delaney 1996).

8.3 *Light Scattering and Out-of-Focus Light*

Light scattering and out-of-focus light can limit the spatial resolution of an optical measurement. These are very large effects in measurements from mammalian brains; with wide-field microscopy individual cells cannot be imaged unless they are within 50 μm of the surface. Figure 1.8 illustrates the results of measurements carried out with a 500 μm slice of salamander cortex. The top section indicates that when no tissue is present, essentially all of the light (750 nm) from a small spot falls on one detector. The bottom section illustrates the result when the olfactory bulb slice is present. The light from the small spot is spread to about 200 μm. Mammalian cortex appears to scatter more than the salamander olfactory bulb. Clearly, light scattering causes considerable blurring of signals in vertebrate brain preparations.

The second source of blurring is signal from regions that are out of focus. For example, if the active region is a cylinder (a column) perpendicular to the plane of focus, and the objective is focused at the middle of the cylinder, then the light from

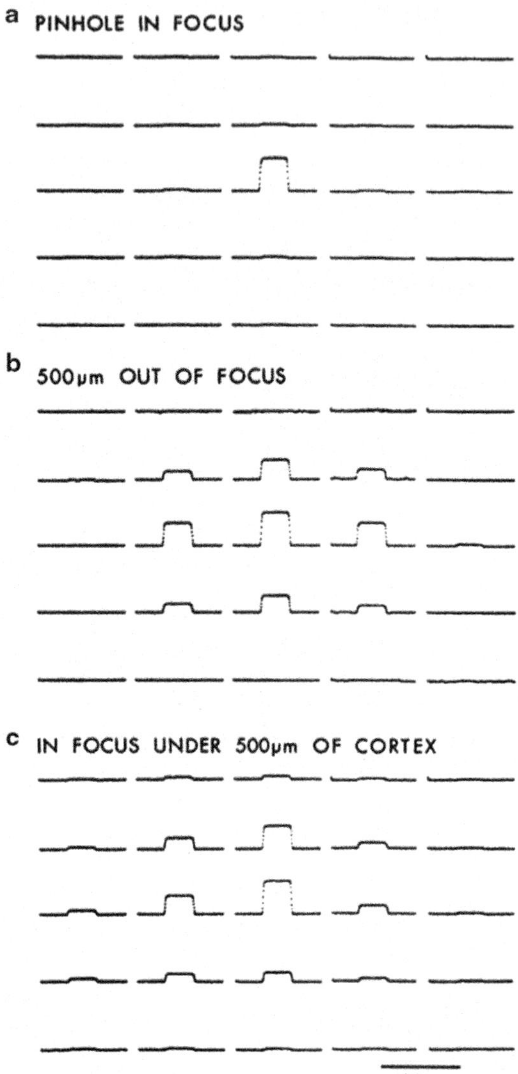

Fig. 1.8 Effects of focus and scattering on the distribution of light from a point source onto a photodetector array. (**a**) A 40 μm pinhole in aluminum foil covered with saline was illuminated with light at 750 nm. The pinhole was in focus. More than 90 % of the light fell on one detector. (**b**) The stage was moved downward by 500 μm. Light from the out-of-focus pinhole was now seen on several detectors. (**c**) The pinhole was in focus but covered by a 500 μm slice of salamander cortex. Again the light from the pinhole was spread over several detectors. A 10 × 0.4 NA objective was used. Kohler illumination was used. The recording gains were adjusted so the largest signal in each of the three trials would be approximately the same size in the figure. Redrawn from Orbach and Cohen (1983)

the focal plane will have the correct diameter at the image plane. However, the light from the regions above and below are out of focus and will have a diameter that is too large. The middle section of Fig. 1.8 illustrates the effect of moving the small spot of light 500 μm out of focus. The light from the small spot is spread to about 200 μm. Thus, in preparations with considerable scattering or with out-of-focus signals, the actual spatial resolution will often be limited by the preparation and not by the number of pixels in the camera.

8.4 2-Photon Microscopes

The effect of scattering and out of focus light are greatly reduced in 2-photon microscopy. However this comes with a greatly reduced fluorescence intensity. The generation of fluorescent photons in the 2-photon microscope is not efficient. We directly compared the signals from Calcium Green-1 in the mouse olfactory bulb using 2-photon and ordinary microscopy. In this comparison the number of photons contributing to the intensity measurement in the 2-photon microscope was about 1,000 times smaller than the number measured with the conventional wide-field microscope and a CCD camera. As a result the signal-to-noise ratio in the CCD recording is much larger even though the spatial resolution was much poorer (Baker et al. 2005). The factors that contribute to the relatively small number of photons in the 2-photon measurement are: (1) The incident light interacts with many fewer dye molecules because only a thin section receives high intensity illumination and (2) The presently available dyes have a low 2-photon cross section which results in a low optical efficiency. This low efficiency cannot be overcome by increasing the incident intensity because higher intensity will heat the preparation. On the other hand, the advantages of 2-photon microscopy are clear; rejection of scattered light and very shallow depth of focus results in much better x-y and z-axis resolution. The two kinds of imaging systems are optimal for different niches in the parameter space of imaging.

Table 1.1 compares the number of photons in three different kinds of measurements. A very large range of measured intensities is obtained in full frame (10,000–100,000 pixels) optical recordings from the nervous system. This range is affected both by the preparation and by the imaging method. Applications of 2-photon microscopy on voltage imaging are discussed in detail in another chapter.

8.5 Random-Access Fluorescence Microscopy

Bullen et al. (1997) have used acousto-optic deflectors to construct a random scanning microscope and were able to measure signals from parts of cultured hippocampal neurons. Relatively large signal-to-noise ratios were obtained using voltage-sensitive dyes. This method can have the advantage that only a small proportion of the preparation is illuminated, thereby reducing the photodynamic damage from the very bright laser light source. However, this method will probably be restricted to preparations such as cultured neurons where there is relatively little light scattering. Random-access fluorescence microscopy is discussed in detail in another chapter.

Table 1.1 Photons/ms/pixel

Object	Imaging method	Photons/ms/per pixel
Cell body	2-Photon fluorescence	1
Distal dendrite	Wide field fluorescence, intracellular dye, arc lamp	1,000
Brain slice	Wide field absorption	1,000,000

9 Cameras

9.1 Quantum Efficiency

Because the signal-to-noise ratio in a shot noise limited measurement is proportional to the square root of the number of photons converted into photoelectrons (see above), quantum efficiency is important. Silicon photodiodes have quantum efficiencies approaching the ideal (1.0) at wavelengths where most dyes absorb or emit light (500–800 nm). In contrast, only specially chosen vacuum photocathode devices (phototubes, photomultipliers, or image intensifiers) have a quantum efficiency as high as 0.5 and this only at certain wavelengths. Thus, in shot-noise limited situations, a silicon diode will have a larger signal-to-noise ratio. Quantum efficiencies near 1.0 in CCD and CMOS cameras are only obtained with "back illuminated" camera chips. Front illuminated chips have quantum efficiencies closer to 0.5. Photographic film has a much lower quantum efficiency, 0.01, (Shaw 1979) and thus has not been used for the kinds of measurements discussed in this book.

9.2 Imaging Devices

Perhaps the most important considerations in choosing an imaging system are the requirements for spatial and temporal resolution. Increases in either temporal or spatial resolution reduce the signal-to-noise ratio. Our discussion considers systems that have frame rates at 1 kHz or faster. In most of these systems, the camera has been placed in the objective image plane of a microscope. However, Tank and Ahmed (1985) suggested a scheme by which a hexagonal close-packed array of optical fibers is positioned in the image plane, and individual photodiodes are connected to the other end of the optical fibers. NeuroPlex-III, a 464 pixel photodiode array camera (WuTech Instruments, Gaithersburg, MD) is based on this scheme.

9.3 Silicon Diode Imagers: Parallel Readout Arrays

Diode arrays with 256–1,020 elements are used in several laboratories (e.g. Hirota et al. 1995). These arrays are designed for parallel readout; each detector followed by its own amplifier whose output can be digitized at frame rates of >1 kHz. While the need to provide a separate amplifier for each diode element limits the number of pixels in parallel read-out systems, it contributes to the very large (10^5) dynamic range that can be achieved.

9.4 CCD Cameras

By using a serial readout, the number of amplifiers is greatly reduced. In addition, it is easier to cool CCD (charge-coupled device) chips to reduce the dark noise. However, because of saturation, presently available CCD cameras are not optimal for the higher intensities available in some neurobiological experiments (Fig. 1.7). The high intensity limit of the CCD camera is set by the light intensity which fills the electron wells on the CCD chip. The well depth of commercially available CCD chips is less than 10^6 e−. This accounts for the bending over of the CCD camera performance at segment C in Fig. 1.7. A dynamic range of even 10^3 is not easily achieved with a CCD camera. A CCD camera will not be optimal for measurements of absorption (Fig. 1.7). The incident light intensity would have to be reduced with a consequent decrease in signal-to-noise ratio. On the other hand, CCD cameras are close to ideal for measurements from individual neurons stained with internally injected dyes.

9.5 CMOS Cameras

SciMedia (Taito-ku, Tokyo) and RedShirtImaging now market CMOS cameras with a well depth of 10^6 e− to 10^8 e− that are close to ideal for the range of intensities between 10^5 and 10^8 photons/msec/pixel. Table 1.2 compares a CCD camera and the CMOS cameras.

9.6 EM-CCD Cameras

These cameras have on chip multiplication and should lead to better signal to noise performance at very low light levels. However, the multiplication process adds

Table 1.2 Characteristics of three fast CCD and CMOS camera systems (as reported by the manufacturer)

	Frame rate (Hz) Full frame	Well size (×1,000 e-)	Read noise (electrons)	Back illum.	Bits a-to-d	Pixels
RedShirtImaging NeuroCCD-SMQ[a]	2,000	200	20	Yes	14	80 × 80
RedShirtImaging NeuroCMOS-SM[a]	2,500, 10,000	100,000	100	No	14	128 × 128
SciMedia MiCAM Ultima[b] CMOS	10,000	1,000,000	300	No	14	100 × 100

[a]www.redshirtimaging.com
[b]www.scimedia.com

noise (a factor of 1.4) and some existing chips are even noisier than expected from the factor of 1.4. If an ordinary CCD has a read noise of 10 e−, then an ideal EM-CCD camera will have a better signal-to-noise ratio only at light levels less than 100 photons/pixel/frame. The light level achieved in the neurobiological and cardiac experiments is almost always greater than 1,000 photons/pixel/ms. Thus CCD or CMOS cameras will have a better signal-to-noise ratio than an EM-CCD.

10 Organic vs. Genetically Encoded Voltage Sensors

Organic voltage and calcium sensitive dyes stain all the cell types in the preparation. In some situations this is a positive feature but in others it is a drawback. In the vertebrate CNS the uniform staining makes it painfully difficult to determine which cell types are responsible for which components of a population signal. Similarly in a 2-photon measurement of calcium signals from cell bodies in the CNS there is no trivial way to determine which of the many cell types present are responding. In principle, a protein sensor of membrane potential or calcium would get around this problem. A variety of tools are available to obtain cell type specific expression of a protein sensor. There has been remarkable recent progress in developing protein sensors of membrane potential with larger signals a faster kinetics (see Chap. 20).

Voltage sensitive dyes were first introduced in the 1970s. Looking back, it is clear that there has been remarkable progress in the development and utilization of these tools over the past 45 years. On the other hand, from the point of view of participants in these developments, progress has seemed to be painfully slow. We should take heart from the fact that the development of the microscope continues apace after more than 400 years of effort. Voltage recording methods have already been applied to a wide variety of neurobiological and cardiac problems that range from dendritic diversity in single neurons to activity maps that cover large areas of mammalian cortex or the heart. It seems likely that new applications and tools will continue to be developed.

Acknowledgements The authors are indebted to their collaborators Bradley Baker, Vicencio Davila, Amiram Grinvald, Kohtaro Kamino, Ying-wan Lam, Leslie Loew, Bill Ross, Brian Salzberg, Alan Waggoner, Matt Wachowiak, Jian-young Wu, and Michal Zochowski for numerous discussions about optical methods.

References

Ashley CC, Ridgway EB (1970) On the relationship between membrane potential, calcium transient and tension in single barnacle muscle fibres. J Physiol 209:105–130

Baker B, Kosmidis E, Vucinic D, Falk CX, Cohen LB, Djurisic M, Zecevic D (2005) Imaging brain activity with voltage- and calcium-sensitive dyes. Cell Mol Neurobiol 25:245–282

Blasdel GG, Salama G (1986) Voltage-sensitive dyes reveal a modular organization in monkey striate cortex. Nature 321:579–585

Boyle MB, Cohen LB (1980) Birefringence signals that monitor membrane potential in cell bodies of molluscan neurons. Fed Proc 39:2130

Braddick HJJ (1960) Photoelectric photometry. Rep Prog Phys 23:154–175

Brown JE, Cohen LB, De Weer P, Pinto LH, Ross WN, Salzberg BM (1975) Rapid changes in intracellular free calcium concentration. Detection by metallochromic indicator dyes in squid giant axon. Biophys J 15:1155–1160

Bucher H, Wiegand J, Snavely BB, Beck KH, Kuhn H (1969) Electric field induced changes in the optical absorption of a merocyanine dye. Chem Phys Lett 3:508–511

Bullen A, Patel SS, Saggau P (1997) High-speed, random-access fluorescence microscopy: I. High resolution optical recording with voltage-sensitive dyes and ion indicators. Biophys J 73:477–491

Cohen LB, Keynes RD, Hille B (1968) Light scattering and birefringence changes during nerve activity. Nature 218:438–441

Cohen LB, Salzberg BM, Davila HV, Ross WN, Landowne D, Waggoner AS, Wang CH (1974) Changes in axon fluorescence during activity: molecular probes of membrane potential. J Membr Biol 19:1–36

Cohen LB, Salzberg BM (1978) Optical measurement of membrane potential. Rev Physiol Biochem Pharmacol 83:35–88

Conti F (1975) Fluorescent probes in nerve membranes. Annu Rev Biophys Bioeng 4:287–310

Conti F, Tasaki I, Wanke E (1971) Fluorescence signals in ANS-stained squid axons during voltage clamp. Biophys J 8:58–70

Davila HV, Cohen LB, Salzberg BM, Shrivastav BB (1974) Changes in ANS and TNS fluorescence in giant axons from *Loligo*. J Membr Biol 15:29–46

Fromherz P, Dambacher KH, Ephardt H, Lambacher A, Muller CO, Neigl R, Schaden H, Schenk O, Vetter T (1991) Fluorescent dyes as probes of voltage transients in neuron membranes: progress report. Ber Bunsenges Phys Chem 95:1333–1345

Garaschuk O, Milos RI, Grienberger C, Marandi N, Adelsberger H, Konnerth A (2006) Optical monitoring of brain function in vivo: from neurons to networks. Pflugers Arch 453:385–396

Grinvald A, Lieke E, Frostig RD, Gilbert CD, Wiesel TN (1986) Functional architecture of cortex revealed by optical imaging of intrinsic signals. Nature 324:361–364

Gross E, Bedlack RS, Loew LM (1994) Dual-wavelength ratiometric fluorescence measurements of the membrane dipole potential. Biophys J 67:208–216

Gupta RK, Salzberg BM, Grinvald A, Cohen LB, Kamino K, Lesher S, Boyle MB, Waggoner AS, Wang CH (1981) Improvements in optical methods for measuring rapid changes in membrane potential. J Membr Biol 58:123–137

Hill DK (1950) The effect of stimulation on the opacitiy of a crustacean nerve trunk and its relation to fibre diameter. J Physiol 111:283–303

Hill DK, Keynes RD (1949) Opacity changes in stimulated nerve. J Physiol 108:278–281

Hirota A, Sato K, Momose-Sato Y, Sakai T, Kamino K (1995) A new simultaneous 1020-site optical recording system for monitoring neural activity using voltage-sensitive dyes. J Neurosci Methods 56:187–194

Inoue S (1986) Video microscopy. Plenum, New York

Jobsis FF, O'Connor MJ (1966) Calcium release and reabsorption in the sartorius muscle of the toad. Biochem Biophys Res Commun 25:246–252

Kleinfeld D, Delaney KR (1996) Distributed representation of vibrissa movement in the upper layers of somatosensory cortex revealed with voltage-sensitive dyes. J Comp Neurol 375:89–108

Labhart H (1963) Bestimmung von moleküleigenschaften aus elektrooptischen effekten. Tetrahedron 19(Suppl 2):223–241

Levin SV, Rosenthal DL, Komissarchik YY (1968) Structural changes in the axon membrane on excitation. Biofizika 13:180–182

Loew LM, Cohen LB, Salzberg BM, Obaid AL, Bezanilla F (1985) Charge-shift probes of membrane potential. Characterization of aminostyrylpyridinium dyes on the squid giant axon. Biophys J 47:71–77
London JA, Zecevic D, Cohen LB (1987) Simultaneous optical recording of activity from many neurons during feeding in *Navanax*. J Neurosci 7:649–661
Malmstadt HV, Enke CG, Crouch SR, Harlick G (1974) Electronic measurements for scientists. Benjamin, Menlo Park, CA
Maylie J, Irving M, Sizto NL, Boyarsky G, Chandler WK (1987) Calcium signals recorded from cut frog twitch fibers containing tetramethylmurexide. J Gen Physiol 89:145–176
Nasonov DN, Suzdal'skaia IP (1957) Changes in the cytoplasm of myelinated nerve fibers during excitation. Fiziol Zh SSSR 43:664–672
Orbach HS, Cohen LB (1983) Optical monitoring of activity from many areas of the *in vitro* and *in vivo* salamander olfactory bulb: a new method for studying functional organization in the vertebrate central nervous system. J Neurosci 3:2251–2262
Orbach HS, Cohen LB, Grinvald A (1985) Optical mapping of electrical activity in rat somatosensory and visual cortex. J Neurosci 5:1886–1895
Ross WN, Salzberg BM, Cohen LB, Davila HV (1974) A large change in dye absorption during the action potential. Biophys J 14:983–986
Ross WN, Salzberg BM, Cohen LB, Grinvald A, Davila HV, Waggoner AS, Wang CH (1977) Changes in absorption, fluorescence, dichroism, and birefringence in stained giant axons : optical measurement of membrane potential. J Membr Biol 3:141–183
Salzberg BM, Bezanilla F (1983) An optical determination of the series resistance in Loligo. J Gen Physiol 82:807–817
Salzberg BM, Grinvald A, Cohen LB, Davila HV, Ross WN (1977) Optical recording of neuronal activity in an invertebrate central nervous system: simultaneous monitoring of several neurons. J Neurophysiol 40:1281–1291
Salzberg BM, Kosterin PV, Muschol M, Obaid AL, Rumyantsev SL, Bilenko Y, Shur MS (2005) An Ultra-stable non-coherent light source for optical measurements in neuroscience and cell physiology. J Neurosci Methods 141:165–169
Schuette WH, Whitehouse WC, Lewis DV, O'Connor M, VanBuren JM (1974) A television fluorimeter for monitoring oxidative metabolism in intact tissue. Med Instrum 8:331–333
Shaw R (1979) Photographic detectors. Appl Opt Optical Eng 7:121–154
Svoboda K, Denk W, Kleinfeld D, Tank DW (1997) In vivo dendritic calcium dynamics in neocortical pyramidal neurons. Nature 385:161–165
Tank D, Ahmed Z (1985) Multiple-site monitoring of activity in cultured neurons. Biophys J 47:476A
Tasaki I, Watanabe A, Sandlin R, Carnay L (1968) Changes in fluorescence, turbidity, and birefringence associated with nerve excitation. Proc Natl Acad Sci U S A 61:883–888
Tsien RY (1989) Fluorescent probes of cell signaling. Annu Rev Neurosci 12:227–253
Vereninov AA, Nikolsky NN, Rosenthal DL (1962) Neutral red sorption by the giant axon of Sepia at excitation. Tsitologiya 4:666–668
Wachowiak M, Cohen LB (2003) Correspondence between odorant-evoked patterns of receptor neuron input and intrinsic optical signals in the mouse olfactory bulb. J Neurophysiol 89:1623–1639
Waggoner AS, Grinvald A (1977) Mechanisms of rapid optical changes of potential sensitive dyes. Ann N Y Acad Sci 303:217–241
Wu JY, Cohen LB (1993) Fast multisite optical measurements of membrane potential. In: Mason WT (ed) Fluorescent and luminescent probes for biological activity. Academic Press, London, pp 389–404

Chapter 2
Design and Use of Organic Voltage Sensitive Dyes

Leslie M. Loew

Abstract The chemistry and the physics of voltage sensitive dyes (VSDs) should be understood and appreciated as a prerequisite for their optimal application to problems in neuroscience cardiology. This chapter provides a basic understanding of the properties of the large variety of available organic VSDs. The mechanisms by which the dyes respond to voltage guides the best set up of the optics for recording or imaging electrophysiological activity. The physical and chemical properties of the dyes can be tuned to optimize delivery to and staining of the cells in different experimental preparations. The aim of this chapter is to arm the experimentalists who use the dyes with enough information and data to be able to intelligently choose the best dye for their specific requirements.

Keywords Fluorescence • Optics • Membrane potential • Voltage sensor • Electrochromism • Chemistry • Spectroscopy

1 Introduction and Overview of Optical Imaging and Recording Applications

The pioneering work of Lawrence Cohen in the mid 1970s led to the establishment of optical methods as a way to measure the electrical activity of large populations of cells either through a microscope or in bulk suspension where traditional micro-electrode methods are not applicable. Cohen's laboratory used an extensive and comprehensive assay of commercially available dyes on the voltage clamped squid giant axon to screen for dyes with large optical responses to membrane potential changes (Cohen et al. 1974). This resulted in the identification of the merocyanine class of dyes as effective candidates for further refinement via alteration of both the chromophore and side chains (Gupta et al. 1981; Ross et al. 1977).

L.M. Loew (✉)
Department of Cell Biology, R. D. Berlin Center for Cell Analysis and Modeling, University of Connecticut Health Center, Farmington, CT 06030-6406, USA
e-mail: les@volt.uchc.edu

M. Canepari et al. (eds.), *Membrane Potential Imaging in the Nervous System and Heart*, Advances in Experimental Medicine and Biology 859,
DOI 10.1007/978-3-319-17641-3_2

This laboratory joined the effort to develop potentiometric dyes by applying rational design methods based on molecular orbital calculations of the dye chromophores and characterization of their binding and orientations in membranes (Loew et al. 1978, 1979a). Several important general-purpose dyes have emerged from this effort including di-5-ASP (Loew et al. 1979b), di-4-ANEPPS (Fluhler et al. 1985; Loew et al. 1992), di-8-ANEPPS (Bedlack et al. 1992; Loew 1994), and di-4-ANEPPDHQ (Fisher et al. 2008; Obaid et al. 2004). All of these dyes provide rapid absorbance and fluorescence responses to membrane potential and are therefore capable of recording action potentials. They have chromophores that are in the general structural category called hemicyanine or styryl dyes. The characteristics of this class of chromophores will be discussed in detail in the next section. In addition, we have developed TMRM, TMRE (Ehrenberg et al. 1988), which are slow responding dyes that are capable of measuring smaller changes in plasma membrane potential or mitochondrial potential via confocal imaging (Farkas et al. 1989; Loew 1993), but the primary focus of this chapter, in keeping with the theme of the book, is on fast dyes that are designed to image electrical activity in excitable cells and tissue. All of the above-mentioned dyes and several others from this lab are now available in the Molecular Probes/Invitrogen catalog; some are also distributed by smaller companies.

The dyes have been of great utility to neuroscientists interested in mapping patterns of electrical activity in complex neuronal preparations with numerous examples spanning the past 20 years (Djurisic et al. 2003; Grinvald and Hildesheim 2004; Wu et al. 1998). Interestingly, our styryl dyes have been of increasing utility when microinjected in single cells (Acker et al. 2011; Antic et al. 2000; Antic 2003; Canepari et al. 2007, 2008, 2010; Djurisic and Zecevic 2005; Foust et al. 2010; Holthoff et al. 2010; Kampa and Stuart 2006; Milojkovic et al. 2005; Nishiyama et al. 2008; Palmer and Stuart 2009; Popovic et al. 2014; Stuart and Palmer 2006; Zecevic 1996; Zhou et al. 2007, 2008) to follow patterns of electrical activity along very thin dendrites and dendritic spines where direct electrical recording is precluded.

The dyes developed in our laboratory, primarily di-4-ANEPPS, have also been successfully applied for optical mapping of electrical activity in studies of cardiac physiology and pathophysiology (Efimov et al. 2004; Loew 2001). But the need to monitor cardiac activity in the presence of blood demanded a new set of VSDs that had red-shifted spectra to avoid overlap with the spectrum of hemoglobin. We therefore synthesized voltage-sensitive dyes (VSDs) with new hemicyanine chromophores that have absorbance and emission further toward the red end of the visible spectrum (Kee et al. 2008; Matiukas et al. 2006, 2007; Wuskell et al. 2006; Yan et al. 2008; Zhou et al. 2007). These dyes are often called "blue dyes" because by absorbing long wavelength red light they appear blue. The longer wavelength absorbance and emission properties of these dyes make them useful for experiments where absorbance by endogenous chromophores, notably hemoglobin, needs to be avoided or when voltage imaging is done in conjunction with imaging of other indicators. We have collaborated with several of these cardiac labs to monitor activity of calcium and voltage together in single myocytes (Warren et al. 2010), whole hearts (Lee et al. 2011, 2012c) or live animals (Lee et al. 2012a). Long wavelength also permits recording from deeper within tissue because light scattering, which limits the depth of optical penetration, is dependent on inverse fourth power of the wavelength. This enabled studies of deep electrical signaling in the human sinoatrial node and atrioventricular junction (Fedorov et al. 2010, 2011). There are several chapters in this volume that provide in depth treatments of voltage imaging in cardiology research.

Most recently, our laboratory has developed a series of over 20 new fluorinated hemicyanine dyes (Yan et al. 2012). There were two motivations for the development of these new VSDs. Firstly, fluorination is known to impart better photostability to any chromophore. Secondly, strategic placement of one or more fluorines on different parts of the chromophore shifts the spectrum to either the blue or red. Using this strategy, we produced VSDs with excitation wavelengths ranging from 410 to 680 nm and emission ranging from 530 to 850 nm. This wide choice of available spectral characteristics allows an experiment to be designed around the required optical properties of the biological preparation and/or the appropriate light source for excitation. For example, the fluorinated dye di-2-AN(F)EPPTEA was perfectly tuned to the highest available wavelength in a Ti-sapphire femtosecond laser, 1,060 nm, allowing 2-photon single voxel recording of activity in single dendritic spines in a brain slice (Acker et al. 2011). Other fluorinated dye were used for 2-photon excitation all the way out to 1,350 nm and could be used in simultaneous 2-photon excitation of calcium indicators (Yan et al. 2012).

There are several other labs that have been actively working on synthesizing new potentiometric dyes. Rina Hildesheim has been working for about 25 years in the laboratory of Amiram Grinvald and has synthesized hundreds of dyes. The latest dyes to emerge from these efforts are RH1691 and RH1692, oxonols that absorb at around 630 nm (Derdikman et al. 2003; Shoham et al. 1999; Slovin et al. 2002). They have been successfully used for *in vivo* studies on awake animals. The relative response ($\Delta F/F$) to electrical activity in mammalian brains for these dyes is ca. 10^{-3}, but it is hard to compare this to other dyes as no controlled voltage clamp studies have been reported on isolated cells or membranes. The dyes have become available through Optical Imaging, Inc., a company founded by Dr. Grinvald. The laboratory of Peter Fromherz has long been interested in the photophysical properties of the hemicyanine chromophores (also called “styryl” dyes) that have been developed in this lab. Recently, they have developed a new chromophore series, the “ANINE”s that are hemicyanines enclosed in a completely rigid annelated ring framework. ANINE-6 has been particularly promising (Kuhn and Fromherz 2003; Kuhn et al. 2004) and a new version, ANINE-6P, has improved solubility so that it may be more readily used in neuroscience applications (Fromherz et al. 2008). They are both excited at shorter wavelength than di-4-ANEPPS. Roger Tsien and his colleagues have proposed VSDs based on intramolecular electron transfer, which might be modulated by the presence of an electric field (Miller et al. 2012). The Tsien lab has also advocated 2-component Fluorescence Resonance Energy Transfer (FRET) systems in which one fluorescent molecule is anchored to one side of the membrane and a negatively charged quencher can flip from one side to the other in a voltage-dependent manner (Cacciatore et al. 1999; Gonzalez and Tsien 1995, 1997). The idea of a 2-component voltage sensor was further improved by the introduction of dipicrylamine as a non-fluorescent quencher (Bradley et al. 2009; Fink et al. 2012).

An alternative to extrinsically applied VSDs for voltage sensing is Genetically Encoded Voltage Indicators (GEVIs), which can be introduced into cells tissues or organisms through molecular biology. The first attempts at developing GEVIs simply inserted variants of green fluorescent protein into the gating regions of voltage-dependent ion channels (Ataka and Pieribone 2002; Guerrero et al. 2002; Sakai et al. 2001; Siegel and Isacoff 1997). However, the initially reported constructs had problems of being too slow, too insensitive or not properly inserted into the plasma membrane (Baker et al. 2007). Over the last several years, several groups have screened large libraries of constructs based on voltage dependent phosphatases or opsins derived from microorganisms, producing GEVIs with sensitivities and speeds approaching those of organic VSDs (Akemann et al. 2010; Gong et al. 2014; Hochbaum et al. 2014; Jin et al. 2012; Kralj et al. 2011; St-Pierre et al. 2014; Tsutsui et al. 2008). It is likely that GEVIs will continue to be improved and that they will permit the mapping of activity in specific cellular subpopulations in transgenic mice. The prospects for this exciting technology are reviewed elsewhere in this volume. The rest of this chapter will detail the mechanisms and uses of VSDs, as these also have great prospects for continued improvement and diverse application.

2 Mechanisms of Dye Optical Responses to Changes in Membrane Potential

It is important to appreciate how VSDs work as voltage sensors in order to apply them intelligently to different experimental situations. The dyes all respond by some change in their spectral properties in response to a change in membrane potential, so a prerequisite to understanding the possible mechanisms by which they can respond, is an understanding of the physical chemistry underlying the absorption and emission of a photon.

The core of any dye is the chromophore—the portion of the molecule that actually interacts with the light. The absorption of a photon is possible when the difference in energy between the ground state of the chromophore and an excited state matches the energy of the photon, which is given by h/λ, Planck's constant divided by the wavelength of the light. The range of wavelengths that can excite the molecule is broad because of the existence of vibrational sublevels within the chromophore's electronic states and, importantly, a range of possible interactions of the chromophore with its molecular environment. Once the molecule is promoted to an excited electronic state, it immediately readjusts the position of its atoms to the lowest energy through vibrational relaxation processes. But it stays in this vibrationally relaxed, electronically excited, state for some time, usually a few nanoseconds, before losing its energy either by emitting a photon (fluorescence) or through heat to regenerate the ground state. The fact that the excited state has had a chance to relax its structure before it emits a photon, possibly adopting a conformation that is most stable for the state's distribution of electrons and reorganizing the surrounding solvent molecules to the most stable configuration, invariably produces a longer wavelength (lower energy) fluorescence emission spectrum compared to the band of wavelengths that is used to excite a given chromophore. It is also possible for the excited state to undergo a chemical change which would destroy the chromophore; this is generally known as bleaching. It is possible to engineer dyes so that an electric field can either interact with the chromophore's electron distribution directly or cause the dye to change its environment. For such dyes, a change in membrane voltage can alter the absorbance or fluorescence spectrum. Good VSDs have chromophores that are bright (efficient excitation and emission), are highly sensitive to environment, and are photostable (i.e. resistant to photobleaching).

As briefly summarized in the previous section, there are several established mechanisms by which dyes can respond to action potentials because of a change in their molecular environment: ON-OFF, reorientation and FRET. The ON-OFF, reorientation and FRET mechanisms all involve a change in the location of a charged dye as a result of the changing membrane voltage (Fig. 2.1). For the ON-OFF mechanism, the dye moves from the aqueous extracellular medium to the cell membrane; because of their environmental sensitivity, the dyes will typically display a substantial increase in fluorescence upon association with the membrane and this is how the voltage change becomes transduced to a fluorescence change. Dyes with cyanine and oxonol chromophores often utilize this mechanism (Waggoner et al. 1977). A problem with this mechanism is that although the sensitivity is large, the response time of the system is often too slow to record action potentials. In the reorientation mechanism, the changing electric field within the membrane causes a membrane-bound dye to flip from an orientation perpendicular to the cell surface to a an orientation where its long axis is parallel to the surface. Again therefore, the molecular environment of the dye is changed and this produces a change in the spectral properties; moreover, there is also a change in the average orientation of the dye molecules with respect to the propagation direction of the exciting light and this can also produce a change in the efficiency of light absorption. Merocyanine dyes have been shown to often utilize this mechanism (Dragsten and Webb 1978). This mechanism can be very fast, but the sensitivity can be low and quite variable from preparation to preparation. In the FRET mechanism, a donor fluorophore is anchored to the outer surface of the cell membrane and transfers its energy to nearby acceptor chromophore which then emit fluorescence at longer wavelength. If the acceptor is a negatively charged membrane-permeant dye, it will redistribute to the inner surface of the bilayer when the membrane depolarizes, thus reducing FRET and reducing the long wavelength emission. This idea was demonstrated for a coumarin-conjugated lipid donor and a permeant oxonol acceptor (Cacciatore et al. 1999; Gonzalez and Tsien 1997). The sensitivity can be high for this mechanism, but the requirement for the application of two dyes

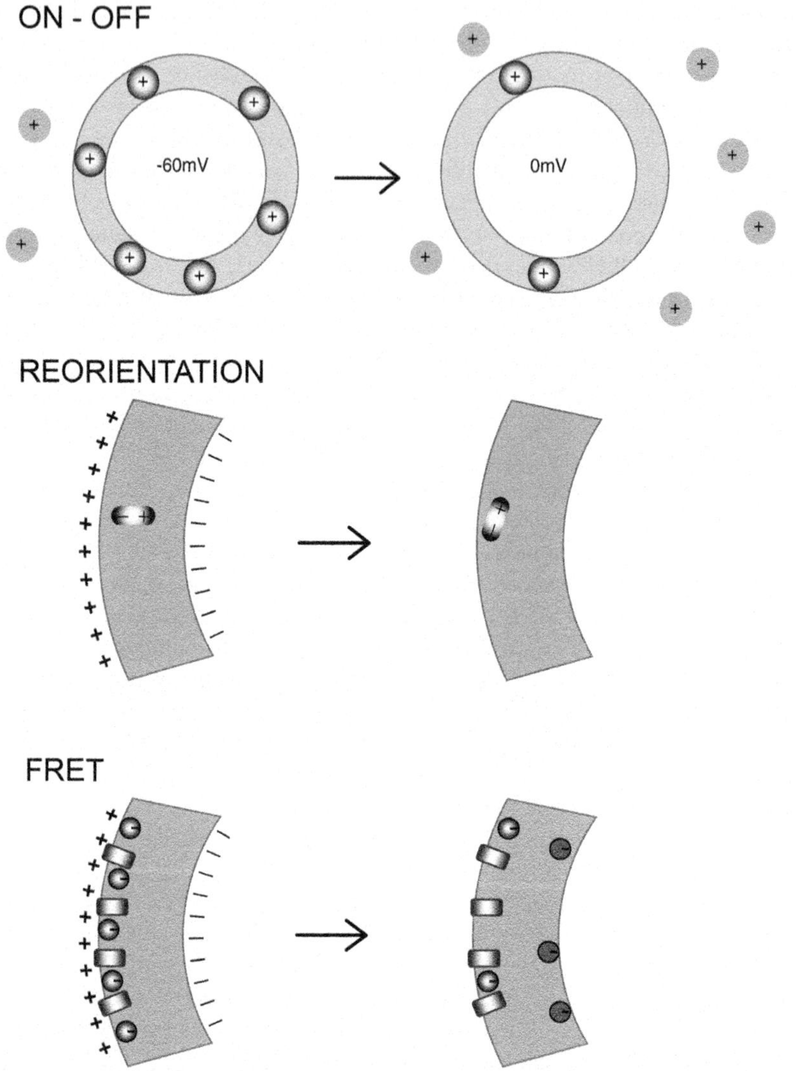

Fig. 2.1 Mechanisms used by VSDs that can change their location in response to membrane depolarization

at relatively high concentration has impeded its widespread adoption. In all three of the cases illustrated in Fig. 2.1, the voltage change has to tip the equilibrium balance between two states of the dye molecule resulting in the movement of the dye to a new environment. Since the intrinsic properties of membranes are themselves quite variable and can therefore affect the equilibrium and/or the kinetics of dye relocation, a dye that is sensitive to potential in one preparation or set of experimental conditions may be completely ineffective in another.

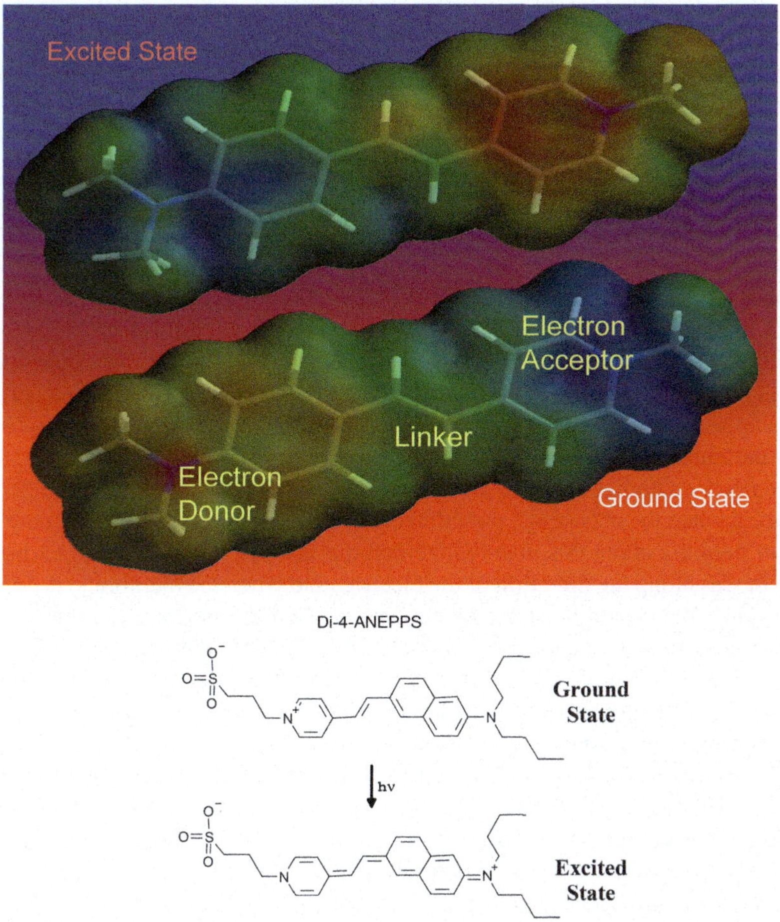

Fig. 2.2 Electrochromic mechanism of voltage sensitivity. The *top* shows how the electrons, and therefore the charge distribution, shifts upon excitation of a typical electrochromic dye. These images were generated from molecular orbital calculations where low electron density (i.e. regions of positive charge) are represented by *bluer* shades and high electron density (i.e. negative charge) is represented by *redder* colors. The *lower* portion of the figure shows resonance structures for the ground and excited states of one of the most widely used VSDs, di-4-ANEPPS. In this chromophore, the donor moiety is an aminonaphthyl group, the linker is a simple double bond and the acceptor is a pyridinium moiety

It would be preferable to have a mechanism that produces dyes with sensitivities that are more robust and reliable from one experimental situation to another. In pursuit of this goal, we have concentrated on producing dyes with chromophores that interact directly with the membrane electric field by an electrochromic mechanism (often referred to as a molecular Stark effect). The idea is explained in Fig. 2.2, which uses the results of molecular orbital calculations and qualitative resonance structures to demonstrate how a dye which has an electron rich π-system on one end and an electron deficient π-system on the other can switch its electron distribution upon absorption of a photon to produce the excited state. di-4-ANEPPS, the most popular of the VSDs to emerge from our work, provides a specific example. The chromophore of this dye changes its electron configuration upon excitation such that the charge shifts from the pyridinium nitrogen in the ground state to the amino nitrogen in the excited state. This behavior is predicted from molecular orbital calculations (Loew et al. 1978) such as the results shown at the top of Fig. 2.2. A polar group is appended to one side of the chromophore in the form of a propylsulfonate moiety and two hydrocarbon chains are included at the opposite end. This, together with the intrinsic amphiphilicity of the chromophore, serves to anchor it in the membrane bilayer in an orientation that is approximately perpendicular to the surface. This orients the direction of the excitation-induced charge motion parallel to the electric field vector within the membrane, as depicted in the upper diagram of Fig. 2.3, where the dye is shown as if having been inserted from the outside of the cell. As a consequence, the ground and excited states are differentially stabilized by the intramembrane electric field, causing a shift in the spectrum when the membrane potential changes (bottom of Fig. 2.3). An equivalent way of thinking about this is to realize that the excitation induced charge displacement moves along the direction of the electric field when the membrane is polarized and against the direction of the electric field when the membrane is depolarized. Experiments from voltage-clamped bilayer membranes (Fluhler et al. 1985; Loew et al. 1979a; Loew and Simpson 1981) have provided evidence that di-4-ANEPPS and related hemicyanine dyes do respond to membrane potential via an electrochromic mechanism; indeed, these bilayer experiments can even be used to estimate the magnitude of the excitation induced charge redistribution (Loew and Simpson 1981). Although electrochromic dyes such as di-4-ANEPPS can produce unpredictable complications (discussed further in Sect. 4, below) we and others have been able to successfully utilize them in a very wide variety of experimental preparations (Loew et al. 1992). Dyes designed to utilize an electrochromic mechanism come closest to being universally applicable and calibrateable VSDs.

3 Measuring Voltage Changes with VSDs

A number of modalities are available for measuring voltage dependent optical changes from excitable cells and tissues. The most common is fluorescence. This is advantageous because of the high contrast and low background intrinsic to this technique and the wide availability of fluorescence and confocal microscopes. Other chapters in this book will ably describe the considerations required for optimizing the speed and sensitivity of the instrumentation. In this section we will describe how the properties of the VSDs should be considered, focusing on fluorescence, but then briefly discussing absorbance and non-linear optical imaging of membrane potential.

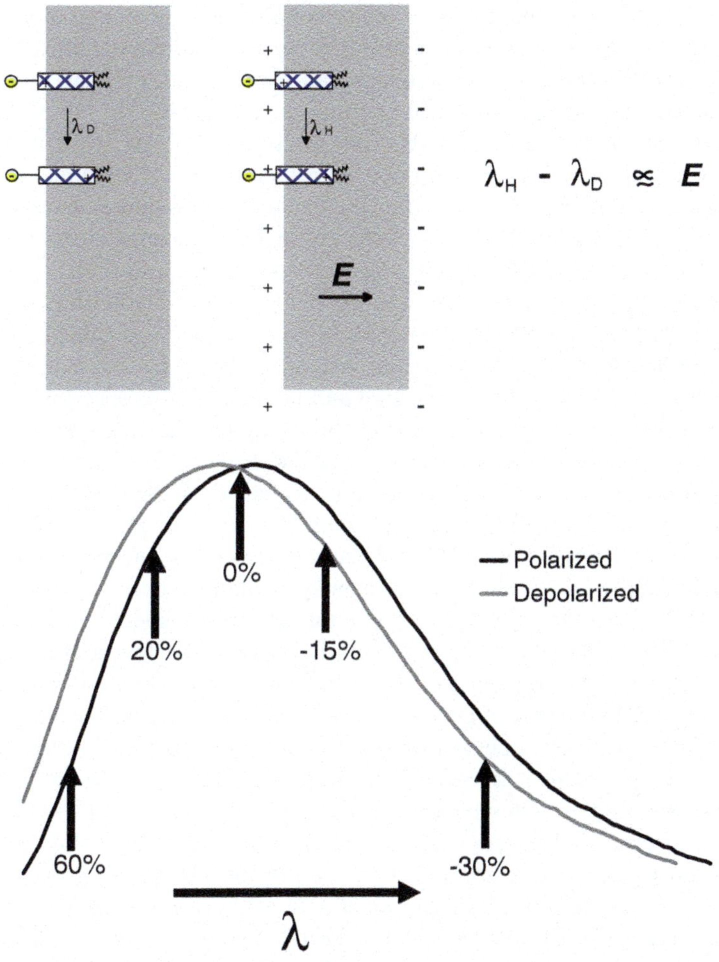

Fig. 2.3 The shift in the spectrum of an electrochromic dye upon depolarization. The upper panel shows how the electric field in the membrane, **E**, can perturb the spectrum of an electrochromic chromophore, represented by the stippled rectangles embedded in the outer surface of a membrane. The spectra shown in the *lower half* of this figure can represent absorbance, excitation or emission spectra. The relative change in fluorescence, ΔF/F is shown as a percentage change for different choices of detection wavelengths

The voltage-dependent shift in the excitation or emission spectrum to be expected from an electrochromic dye is depicted in Fig. 2.3, where the dye is assumed to be bound to the extracellular surface of the membrane and undergo charge redistribution in its excited state in the direction shown in Fig. 2.2. Under these circumstances, the spectra shift to shorter wavelength upon membrane depolarization. The magnitude of the spectral shift will be linearly related to the change in potential. In most experiments, a fixed band of wavelengths is continuously monitored, rather than attempting to scan the entire spectrum. Fig. 2.3 shows that the choice of wavelength can have a drastic effect on the observed sensitivity of the dye response. Contrary to what might be an experimentalist's first impulse, the wavelengths of maximal absorbance or emission are actually the worst choices for achieving optimal sensitivity. The best sensitivity is achieved at the wings of the spectra and will show changes in opposite directions at the high and low wavelength wings. Furthermore, measuring the change in the fluorescence relative to total fluorescence signal, ΔF/F, biases the optimal wavelengths even further to the extreme edges of the excitation and emission spectra because F, the denominator, deceases steadily at the edges. An approximately linear relationship between the voltage change and the fluorescence or absorbance change is obtained at the wings because the local slope of the spectrum is approximately constant, at least until it starts flattening out toward the very high and low ends of the spectrum.

Another important consideration in choosing wavelengths, which mitigates against going too far out to the edges of the spectrum, is the signal to noise ratio (S:N). Essentially, the detection system must collect a sufficient number of photons per measurement so as to be able to determine the small modulation by the voltage with statistical reliability. For fluorescence measurements, the number of photons can be boosted by simply utilizing an intense excitation source; it should then be possible to pick wavelengths at the edge of the absorbance spectrum, where ΔF/F is maximal, without sacrificing signal. This is because the low absorbance at the spectral edge can be compensated by a higher intensity light source. This will permit a sufficient number of excitation events to occur in a sufficiently short time to be able to record action potentials. But a key consideration that prohibits applying this argument with respect to the edge of the *emission* spectrum is the limitation imposed by photobleaching. By detecting only a narrow band at the emission spectrum edge, most of the emitted photons will be lost, effectively wasting most of the excitation events. It would then be more probable that the dye will bleach before collecting sufficient photons for a good signal to noise ratio. Therefore, the best compromise between ΔF/F and S:N is to use an intense narrow band excitation source at the edge of the VSD absorbance spectrum, while collecting the emitted light through a filter that will pass all wavelengths longer than the emission spectrum maximum.

We and others have also taken advantage of the opposite changes in the voltage-dependence optical signals at the low and high wavelength wings of the spectra (Fig. 2.3) to implement a dual wavelength ratio detection scheme for measuring membrane potential with VSDs (Beach et al. 1996; Bullen et al. 1997; Knisley et al. 2000; Montana et al. 1989). The idea is to collect light from both wings of the spectrum and calculate the ratio of their fluorescence intensities as a measure of the voltage change. The primary advantage is that the sensitivity of the ratio is approximately equal to the additive absolute sensitivities at each wavelength. A second advantage is that the dual wavelength ratio can be, in principle, be used to report the absolute value of the intramembrane electric field. A single wavelength fluorescence measurement can only report a relative change in potential because the intensity depends on the level of staining, which can be quite variable even along the membrane of a single neuron. The ratio effectively normalizes away any differential staining levels because the fluorescence intensity at both wavelengths will be proportional to dye density but oppositely responsive to intramembrane electric field. We introduced the idea of dual excitation wavelength imaging (Montana et al. 1989) and showed that it could be used to map the membrane potential along a neuron induced by an external electric field (Bedlack et al. 1992). Importantly, because the dye reports on the local electric field, any variations in dipole potential or surface potential may also produce variations in the ratio (Bedlack et al. 1994; Gross et al. 1994; Xu and Loew 2003); therefore, any spatial variations in ratio must be interpreted with caution and my not reflect a variation in transmembrane potential. On the other hand, fast temporal variations in the ratio are likely to reflect changes in transmembrane potential because the changes in lipid composition that would be required for changes in other sources of intramembrane electric field would be slow on the timescale of action potentials. A dual emission wavelength measurement is preferred for rapid spatial mapping of electrical activity because two fixed detection paths can be employed (Bullen et al. 1997; Bullen and Saggau 1999) rather than mechanically switching between two excitation wavelength filters. In general, dual wavelength ratiometric measurements require more complex instrumentation and analysis than simply monitoring ΔF/F, so the adoption of this approach has been limited to experiments that require extracting a voltage dependent signal from other confounding variables.

A commonly used alternative to fluorescence for optically monitoring electrical activity, is to record or image the light transmitted through a specimen stained with a VSD (Glover et al. 2008; Habib et al. 2013; Sasaki et al. 2002). A change in the transmittance of a stained preparation simply reflects the change in the absorbance spectrum of the dye. Of course, the transmitted light corresponds to the light that actually does not get absorbed by the dye. The amount of light that is absorbed by a single dye-stained membrane depends on the size of the cell, the surface density of the dye molecules on the membrane and the extinction coefficient of the dye at the chosen wavelength. But it can be estimated that this will never exceed 1 part in 10^3 for even the largest cell. The modulation of the transmitted light signal by an action potential would therefore never be much greater than ~1 part in 10^4. A comparable estimate for the modulation of fluorescence is one part in ten, because the entire light signal emanates from the stained membrane. So if ΔT/T is intrinsically so much lower than ΔF/F, why would anyone prefer to use transmitted light for optical electrophysiological recording? Indeed, the technique is almost never used for recording from single cells. In experiments where the collective activity of a large population of cells in a tissue needs to be imaged, the sensitivity of the technique is increased because more of the incident light is absorbed. Furthermore, the S:N in a transmitted light detection is not limited by photon statistics so simpler and less noisy photodetectors can be used.

Another approach that is just beginning to emerge for VSDs is 2-photon-excited fluorescence (2PF) (Acker and Loew 2013; Acker et al. 2011; Fisher et al. 2008; Kuhn et al. 2008; Loew et al. 2002; Yan et al. 2012). There is really no difference in principle in the way that 1-photon and 2-photon excited fluorescence measures the dye response, except, of course, that the 2-photon modality requires a femtosecond pulsed laser operating at twice the wavelength and a microscope configuration that is optimized for the purpose (Denk et al. 1990). The major advantage of 2-photon excited fluorescence is the ability to probe deep inside a specimen with high 3D resolution. The disadvantage, in addition to complex instrumentation, is that the laser scanning required to obtain a full image of the specimen is too slow for recording of action potential activity; therefore, line scans or special spatial sampling protocols are used to record optical signals from a small number of sites. A later chapter in this volume is devoted to 2-photon microscopy of VSDs.

Second harmonic generation (SHG) is another non-linear optical process (Campagnola and Loew 2003; Millard et al. 2003a) that can take place in a microscope coupled to an ultrafast laser. As in the case of two-photon excitation the probability of SHG is proportional to the square of the incident light intensity. While 2PF involves the near-simultaneous absorption of two photons to excite a fluorophore, followed by relaxation and non-coherent emission, SHG is a nearly instantaneous process in which two photons are converted into a single photon of twice the energy, emitted coherently. Furthermore, SHG is confined to loci lacking a center of symmetry; this constraint is readily satisfied at cellular membranes in which SHG-active constituents are unevenly distributed between the two leaflets of the lipid bilayer. The first examination of the non-linear optical properties of VSDs was in a collaboration between this lab and Aaron Lewis' lab, where we analyzed SHG from a monolayer of di-4-ANEPPS in a Langmuir-Blodgett trough (Huang et al. 1988). Subsequently (Ben-Oren et al. 1996; Bouevitch et al. 1993; Campagnola et al. 1999) we were able to show that the SHG signal from electrochromic dye stained membranes is sensitive to membrane potential. These experiments were continued (Clark et al. 2000; Millard et al. 2003b, 2004, 2005a, b; Teisseyre et al. 2007) in an attempt to characterize and optimize the SHG response of our dyes to membrane potential. They indicated that the mechanism could not be explained as a direct electrooptic effect because the kinetics of the response for some of the dyes was slow, although these were not always the conclusions reached by others (Jiang et al. 2007; Pons et al. 2003). Whatever the mechanism, the laboratories of Rafael Yuste, Kenneth Eisenthal and Watt Webb (Araya et al. 2006, 2007; Dombeck et al. 2004, 2005; Nuriya et al. 2006) were able to demonstrate the measurement of action potentials using second harmonic imaging microscopy. These studies used the hemicyanine dye FM-464 applied to the interior of the cell through a patch pipette. Because the SHG can only come from dye that is non-centrosymmetrically distributed on the spatial scale of the wavelength, the light emanates only form the internally stained plasma membrane and no signal appears from internally stained organelles with highly convoluted membranes such as the endoplasmic reticulum and mitochondria. Because of this elimination of background from these internal electrically inactive membranes, the voltage sensitivity of the SHG signal can approach 20 %/ action potential - much higher than is practically possible with fluorescence. However, these measurements had very low S:N and required extensive signal averaging because of the low intensity of the SHG. Therefore, it is not clear how generally useful this modality will ultimately be until new dyes are developed with much stronger SHG efficiencies. I present a detailed treatment of the topic of SHG imaging of membrane potential in Chap. 19 of this volume.

4 Choosing the Best Dye for an Experiment

Many of the hemicyanine dyes synthesized in our and other labs are capable of sensing membrane potential changes with high sensitivity. But different experimental designs require different spectral properties, solubility properties and membrane binding affinities. Furthermore, competing processes can sometimes diminish the sensitivity of a given dye in a particular experimental preparation. So there is a need for a large repertoire of dyes to meet these varying needs and this section will consider how to make these choices. We focus on dyes developed in this laboratory as examples, but the same principles can be applied to the dyes developed by others.

A prerequisite for choosing a dye is simply being able to appreciate the structural variety of VSDs and establishing some nomenclature that permits us to refer to them in a recognizable way. As discussed in Sect. 2, the hemicyanine chromophores are all characterized by one end that is electron rich in the ground state, also called an electron donor moiety, and another end that is electron deficient, an electron acceptor moiety (Fig. 2.2). When these groups are linked with a π-electron linker, it is possible for the electrons to readily shift in response to excitation. Chromophores with such electronic structures are commonly called "push-pull" chromophores because of the complementary tendencies of the electrons at the two ends. Fig. 2.4 lists all of the donor, linker and acceptor moieties that we have employed in our lab to develop push-pull chromophores. We have synthesized approximately 300 dyes utilizing over 50 of these push-pull hemicyanine chromophores. The R1 groups on the donor ends are usually hydrocarbon chains that anchor the dye to the hydrophobic interior of the membrane. The R2 groups on the acceptor ends are usually hydrophilic groups that protrude into the aqueous medium adjacent to the membrane and help to maintain the orientation of the chromophore perpendicular to the membrane surface (cf. Fig. 2.3). The laboratories of Amiram Grinvald and Alan Waggoner have also developed large numbers of VSDs using some of these groups. Table 2.1 provides the spectral characteristics of most of the chromophores constructed from these parts in ethanol, water and when bound to lipid vesicle membranes. An important attribute of almost all of these dyes is that they have strong fluorescence quantum efficiencies in the membrane bound forms, but typically at least 2 orders of magnitude lower fluorescence in aqueous solution. This means that the background signal from unbound dye can usually be completely neglected even if the preparation is allowed to remain exposed to the aqueous staining solution. For the reasons discussed in Sect. 3, the optimal wavelengths for voltage sensitivity are typically ~50 nm above the maximum wavelengths reported for the lipid spectral in Table 2.1.

Fig. 2.4 The components that have been used to construct hemicyanine push-pull chromophores. A combinatorially large repertoire of chromophores have been constructed by mixing the D-L-A moieties. The abbreviations for some of the more common components are shown alongside the number scheme that was employed to designate the chromophores in Table 2.1. VSDs are then realized from the chromophores by appending hydrophobic tails, R1, and hydrophilic head groups, R2

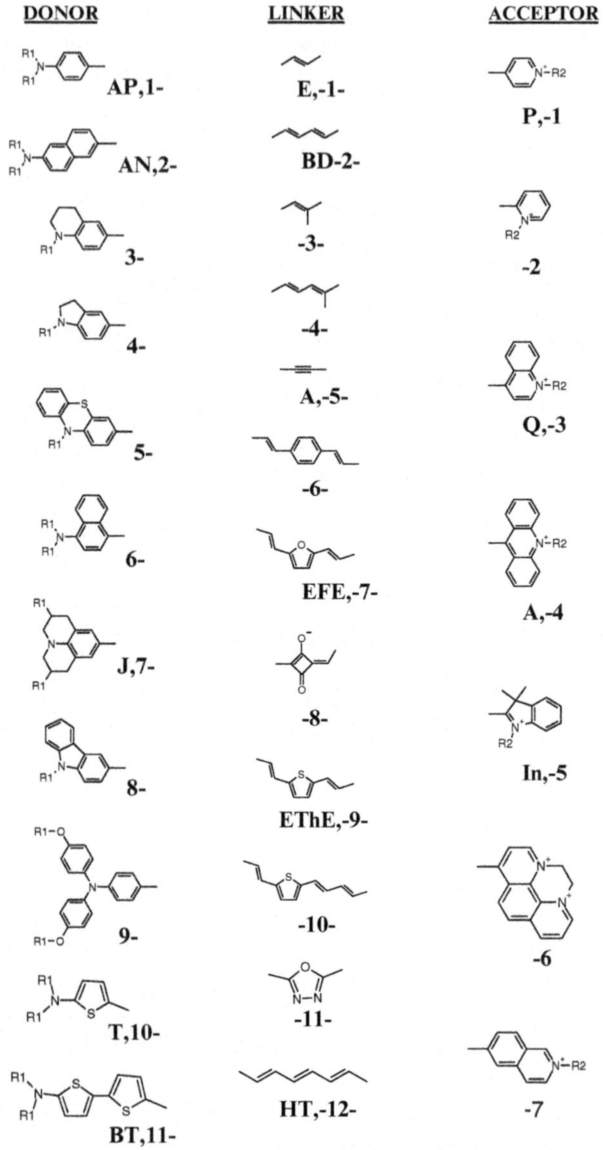

The systematic chemical names for these compounds are extremely cumbersome because of their complexity. Therefore, the field has adopted different styles of abbreviation. Most commonly, a dye is identified by the initials of the chemist who synthesized it followed by a number. For example JPW-3080 was synthesized by Joseph P. Wuskell and its synthesis was first recorded on page 080 of his third notebook. Similarly, RH-160 is dye number 160 prepared by Rena Hildesheim in Amiram Grinvald's lab. Of course these abbreviations don't convey much

Table 2.1 Wavelength ranges of absorbance and emission maxima for styryl dye chromophores

Chromophore	Number	λ_{ABS} (EtOH)	λ_{EM} (EtOH)	λ_{ABS} (lip)	λ_{EM} (lip)
[1-1-1]	18	490 ± 9	617 ± 8	468 ± 7	600 ± 10
[1-1-2]	2	470	610 ± 10	441 ± 2	590 ± 30
[1-1-3]	3	564 ± 5	684 ± 7	533 ± 5	680 ± 70
[1-1-4]	3	564 ± 5	684 ± 7	533 ± 5	680 ± 70
[1-1-5]	1	567	599	562	601
[1-1-6]	2	590 ± 1	679	578	662
[1-2-1]	5	510 ± 10	680 ± 50	468 ± 7	620 ± 20
[1-2-3]	2	560 ± 30	675 ± 7	540 ± 20	
[1-2-4]	3	700 ± 20	800 ± 50	611 ± 10	771 ± 1
[1-2-5]	1	650	700	638	
[1-3-1]	1	475	610	450	584
[1-4-1]	1	485	715	463	690
[1-5-4]	2	630 ± 10	720 ± 40	580 ± 10	660 ± 10
[1-6-1]	1	482		461	670
[1-7-3]	3	640 ± 20	890 ± 70	580 ± 10	740 ± 30
[1-8-3]	1	607		579	646
[1-8-5]	1	626	639	634	647
[1-9-3]	4	590 ± 20	800 ± 100	548 ± 5	720 ± 50
[1-9-5]	2	690 ± 30	870	640 ± 10	757
[1-10-3]	1	626	664	616	650
[1-11-1]	1	440	641	420	
[1-11-4]	1	391	692	404	552
[2-1-1]	47	500 ± 10	710 ± 10	470 ± 10	640 ± 20
[2-1-7]	1	486		464	627
[2-2-3]	1	584	812	526	702
[2-2-6]	1	609			
[3-1-3]	3	580 ± 6	689 ± 1	556 ± 5	680
[3-1-4]	1	681	770	658	
[4-1-3]	1	608	833	589	760
[4-1-4]	3	705 ± 7	833	612 ± 1	716
[4-1-6]	1	610		610	
[4-7-3]	1	592	873	597	715
[5-1-3]	1	514	620	500	714
[6-1-1]	1	474	656	432	630
[6-1-3]	2	520 ± 20	714 ± 9	480 ± 20	671 ± 8
[6-1-5]	1	616	636	620	643
[7-1-1]	1	523	640		
[7-1-3]	3	620 ± 10	720 ± 10	579 ± 1	673 ± 4
[7-1-4]	4	726 ± 10	810 ± 40	678 ± 7	770 ± 60
[7-1-5]	1	588	612	594	618
[7-1-6]	1	600	708	618	
[7-2-4]	1	825	880	674	790

(continued)

Table 2.1 (continued)

Chromophore	Number	λ_{ABS} (EtOH)	λ_{EM} (EtOH)	λ_{ABS} (lip)	λ_{EM} (lip)
[**7-8-3**]	1	663	687	668	680
[**7-9-3**]	2	620 ± 30		559	
[**8-1-1**]	1	440	588	421	566
[**8-1-3**]	1	500	648	486	640
[**9-1-1**]	1	472	672	457	622
[**9-1-2**]	2	540 ± 20	760 ± 40	530 ± 30	681 ± 8
[**9-1-3**]	2	540 ± 20	760 ± 40	530 ± 30	681 ± 8

information about the properties of the dyes and we have attempted to formulate more descriptive abbreviations for the more common hemicyanine VSDs. The general formulation we have adopted uses the following scheme:

di-n-DLAH

This provides a designation for their structural components as follows: alkyl chain lengths (n) - π-Donor – Linker(s) - π-Acceptor – hydrophilic Head group (Wuskell et al. 2006). A pair of alkyl chains, R1, is appended to the amino terminus of most of the dyes (right side of each structure in Fig. 2.4). The number of carbons in these chains is indicated by n, in the 'Di-n' portion of the naming scheme. The π-donor moiety, D, can be aminophenyl (AP, 1- in Fig. 2.4), aminonaphthyl (AN, 2- in Fig. 2.4), etc. The linker, L, is ethene (E, -1- in Fig. 2.4) diene (D, -2-), ethene-furan-ethene (EFE, -7-), etc. The acceptors, A, can be pyridinium (P, -1), quinolinium (Q, -3) indolenium (In, -5), etc. Finally, we have dyes with various head groups, H, the most common of which are shown in Fig. 2.5. Thus, di-4-ANEPPS in Fig. 2.2 has a pair of butyl groups (n = 4) attached to the amino group of an aminonaphthyl donor (D = AN), which is linked via an ethene linker (L = E) to a pyridinium (A = P) acceptor, which is appended with a propylsulfonate (H = PS) head group.

The chromophore determines the optical properties of the VSD (Table 2.1) and also its intrinsic voltage sensitivity. We screen every new dye using a voltage clamped hemispherical bilayer apparatus (Fluhler et al. 1985; Loew and Simpson 1981; Wuskell et al. 2006). From these experiments, as well as experience in other preparations including squid axon and lobster nerve (Loew et al. 1992; Wuskell et al. 2006), the chromophores with the best intrinsic voltage sensitivity are (DLA): APEP, ANEP, ANEQ, ANBDQ, APEThEQ, ANEThEQ, ANBDIn, APEThEIn and ANEThEIn. These chromophores span a range of over 200 nm in optimal excitation and emission wavelengths. In particular, the long wavelength VSDs have become increasingly important in applications involving multiple optical indicators (e.g. voltage and calcium), optical manipulation (e.g. optogenetic tools combined with voltage imaging), deeper tissue penetration or avoidance of interference from intrinsic chromophores such as hemoglobin in blood (Fedorov et al. 2010, 2011; Kee et al. 2008; Lee et al. 2011, 2012a, b, c; Matiukas et al. 2006, 2007; Tsuda et al. 2012; Zhou et al. 2007, 2008).

As mentioned above, we have recently further extended the wavelength range of the hemicyanine VSDs by strategic placement of fluorine substituents on the donor,

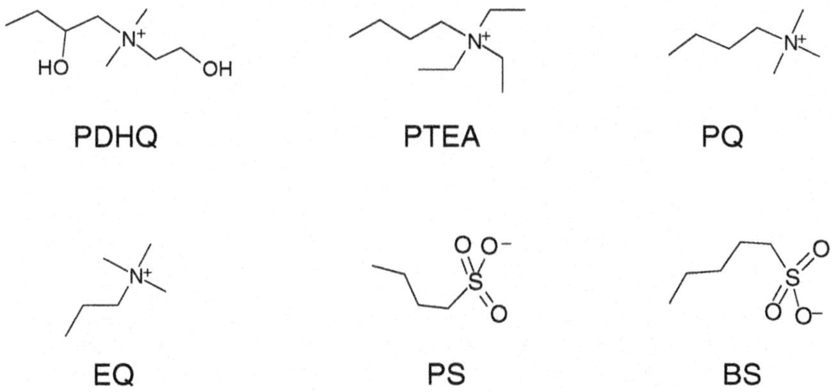

Fig. 2.5 Structures and abbreviations of commonly utilized polar head groups corresponding to R2 in Fig. 2.4

linker or acceptor moieties (Table 2.2) (Yan et al. 2012). The position of the fluorine substituents is coded into the name of the chromophore by simply appending a parenthetical (F), (F2) or (CF3) after the corresponding structural component that has been modified. Placing the fluorines on the donor lowers the excitation and emission wavelengths, while placement on the linker or acceptor produces a red shift in the spectra. In this way, we have produced a set of VSDs that span the entire visible spectrum for excitation and extend out to the near infrared in their emission spectra. A dye from Table 2.2 can be chosen to optimally pair with available 1-photon or 2-photon excitation light source and to minimize spectral interference with other fluorescent probes. For example, Fig. 2.6 show how di-2-AN(F)EPPTEA was paired with 2-photon excitation at 1,060 nm and a calcium indicator to simultaneously measure the electrical and calcium signals evoked by back propagating action potentials in single dendritic spines in a brain slice (Yan et al. 2012). In addition, fluorination generally imparts greater photostability to dye chromophores and this was the case for the VSDs in Table 2.2.

The choice of alkyl chains provides direct control of the solubility and membrane binding characteristics. Short alkyl chains impart good water solubility at the expense of strong membrane binding; this would be appropriate for staining a thick tissue where the dye needs to penetrate many cell layers. However, VSDs with greater solubility can also be more readily washed out of a preparation. We have found that longer R1 alkyl chains increase the strength of binding of the dyes to the membrane and also slow the rate of internalization. Internalization is detrimental to the potentiometric response because as the dye equalizes on either side of the bilayer the voltage-dependent spectral shifts from the two leaflets will cancel each other (cf. Fig. 2.2). However, long side chains decrease the solubility of the dye, making staining slow and inefficient and impractical for thick tissue. This problem can be ameliorated through the use of vehicles such as Pluronic F127 (Lojewska and Loew 1987) or γ-cyclodextrin (Wuskell et al. 2006), which form complexes with the dyes to keep them disbursed in solution.

Table 2.2 Fluorinated voltage sensitive dyes[a]

Name	Structure	Membrane Abs/Em maxima (nm)	Ethanol solution
Di-n-AN(CF3)EPPTEA n = 2, 3, 4	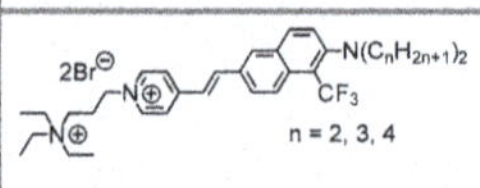	n = 2: 400/589 n = 3: 411/595 n = 4: 418/597	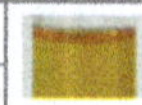
Di-4-AN(CF3)E(F)PPTEA	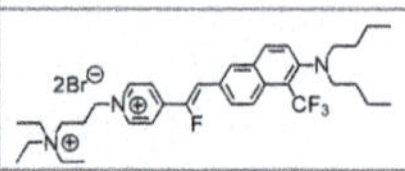	422/590	
Di-4-AN(CF3)EP(F2)PTEA	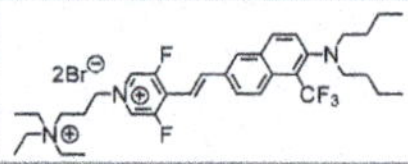	436/608	
Di-n-AN(F)EPPTEA n = 2, 4	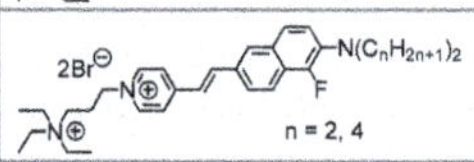	n = 2: 453/616 n = 4: 444/610	
Di-4-AN(F)E(F)PPTEA	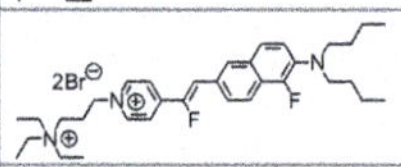	464/643	
Di-4-AN(F)EP(F)PTEA	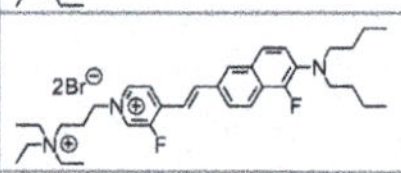	465/632	
Di-4-AN(F)E(F)P(F)PTEA	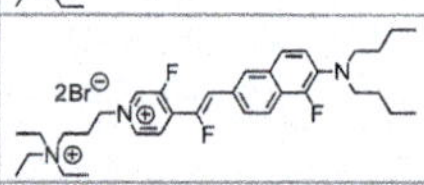	470/651	
Di-4-ANEPPTEA	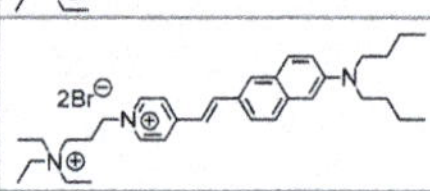	476/616	
Di-n-ANE(F)PPTEA n = 2, 4	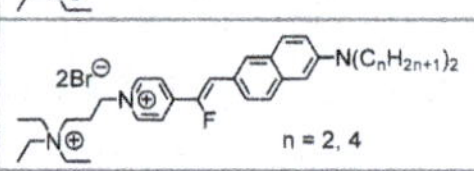	n = 2: 495/642 n = 4: 488/636	
Di-n-ANEP(F)PTEA n = 2, 4	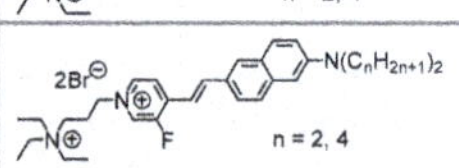	n = 2: 489/632 n = 4: 495/632	
Di-4-ANE(F)P(F)PTEA	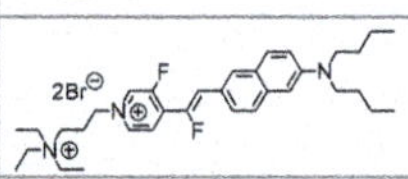	500/660	
Di-n-ANEP(F2)PTEA n = 2, 4	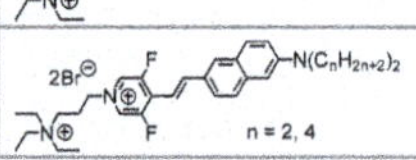	n = 2: 509/648 n = 4: 511/646	
Di-2-ANEQPTEA	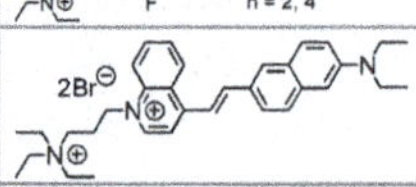	534/664	
Di-4-ANEQ(F)PTEA	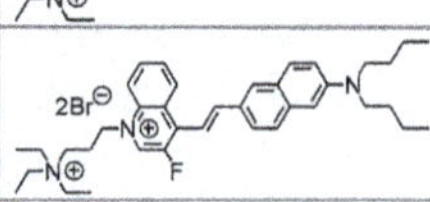	547/686	
Di-2-ANBDQPTEA	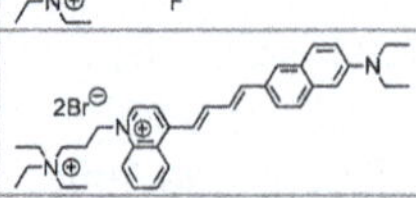	542/702	
Di-2-ANBDQ(F)PTEA	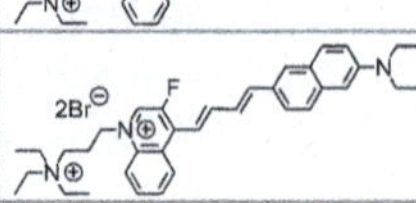	553/743	

[a]Table 1 in Yan et al. 2012. Reprinted by permission of the Proceedings of the National Academy of Sciences

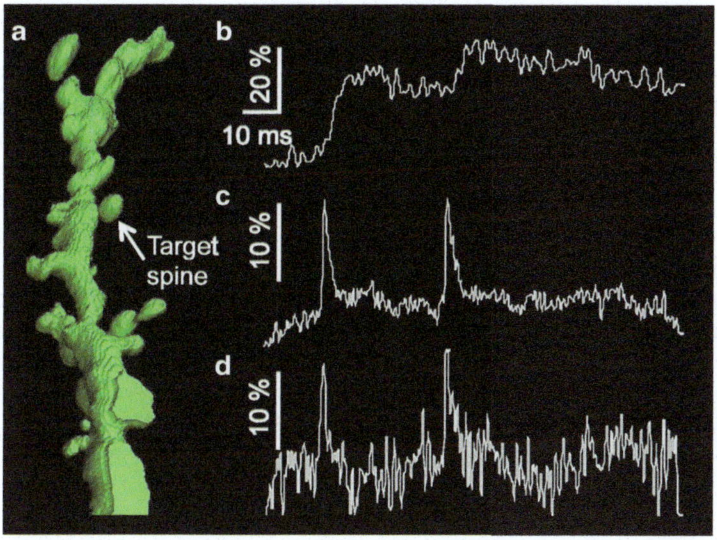

Fig. 2.6 Simultaneous voltage and calcium imaging in a single dendritic spine. (**a**) Reconstruction of the segment of proximal basal dendrite studied from a mouse acute cortical brain slice indicating the targeted spine. (**b**) Calcium Green 1 Dextran fluorescence in response to two somatically initiated bAPs. (**c**) Simultaneously recorded di-2-AN(F)EPPTEA voltage-sensitive dye (VSD) fluorescence transients. Both dyes are excited with the same 2-photon laser tuned to 1,020 nm. Waveforms are averages of ten trials, with bleaching subtracted. (**d**) Single-sweep VSD fluorescence. The vertical scale bars in **b**, **c** and **d** correspond to $\Delta F/F$. (Figure 1 from Yan et al. 2012; reprinted with kind permission from the Proceedings of the National Academy of Sciences)

With respect to R2, the negatively charged sulfonate head groups (PS and BS in Fig. 2.5) are attractive because they provide internal counter ions for the positive charge of the chromophore, resulting in an overall neutral VSD. These are less likely to interact and possibly interfere with other cellular components. The positively charged head groups in Fig. 2.5 provide some additional water solubility and can therefore partially offset the solubility lowering effect of moderately sized R1 side chains. Indeed, di-3-ANEPPDHQ and di-4-ANEPPDHQ have proven to be excellent general purpose dyes for staining of multicellular preparations (Obaid et al. 2004). Similarly, several of the new long wavelength dyes with positive head groups can be used for labeling for recording of large neuronal populations in brain slices (Kee et al. 2008). An important finding for the doubly positively charged VSDs with the positive head groups is that, given enough time, they can become internalized and will be retained in the cell interior because of the polarized resting potential. This has led to applications where long R1 side chain dyes with the PQ head group are applied in an ethanol solution to cause the dyes to precipitate out and remain localized in a confined anatomical region. They then can become internalized and be transported over a period of hours by retrograde flow or diffusion to remote regions of the neurons that were originally stained. This strategy was first employed with di-8-ANEPPQ and di-12-ANEPEQ where the dye was applied to chick spinal cord to ultimately label and record from remote neuronal cell bodies, dendrites and axons (Tsau et al. 1996; Wenner et al. 1996). When the positively charged head groups are combined with R1 chains of just one or two carbons in length, the resulting VSDs are extremely water soluble, often at a solubility of several mg/ml. This is important for applications where the dyes must be applied internally to a single cell through a patch pipette. Such experiments are designed to allow imaging of membrane potential from different regions of a single neuron in a brain slice or in vivo (Acker et al. 2011; Antic et al. 1999; Antic and Zecevic 1995; Antic 2003; Canepari et al. 2007; Milojkovic et al. 2005; Palmer and Stuart 2009; Stuart and Palmer 2006; Yan et al. 2012; Zhou et al. 2007, 2008). For example, Palmer and Stuart have been able to image membrane potential from dendritic spines with di-1-ANEPEQ (Palmer and Stuart 2009) and Acker et al. have been able to measure back propagating action potentials in individual dendritic spines with di-2-AN(F)EPPTEA (Acker et al. 2011).

5 How Can Better Dyes Expand the Scope of Possible Experiments?

One major area for improvement is dye photostability. Obviously, more photostable dyes will permit longer duration recording experiments; but as importantly, they will permit high excitation intensities with a resultant higher emitted photon flux. This is important because the signal to noise ratio (S:N) for recording fast electrical signals is limited by the stochastic shot noise of the number of photons that can be collected—the S:N is proportional to (photon count)$^{1/2}$. Higher dye brightness (the product of extinction coefficient and fluorescence quantum efficiency) will also help in the same way. The fluorinated VSDs in Table 2.2 are an advance in this direction.

Greater sensitivity of the optical signal to voltage changes would expand the range of applications for VSDs. We have produced dyes that can give ten-fold or more fluorescence changes for a 60 mV change in potential (Ehrenberg et al. 1988; Loew et al. 1983), but these dyes depend on a voltage-dependent redistribution across the membrane that renders the response too slow to follow an action potential. Fast dyes necessarily have much smaller voltage sensitivities. The best of our fast dyes, when excited at the spectral edge, can produce relative fluorescence changes of ~40 %/100 mV. Frankly, we don't anticipate much improvement in this sensitivity for electrochromic dyes, although improved instrumentation can further stretch the range of experiments accessible to the technology. Other voltage dependent mechanisms may be devised that retain the speed of electrochromism but also produce greater sensitivity. Recent attempts at electron transfer (Miller et al. 2012) and FRET- based mechanisms (Bradley et al. 2009; Fink et al. 2012) are steps in this direction for fluorescent VSDs.

On the other hand, SHG represents a relatively unexplored optical modality that could produce much greater sensitivity than fluorescence and this is the rationale for our continued interest in developing this technique despite the S:N limitation. Our reasoning is that by resolving the mystery of how SHG responds to potential, we should be able to use this understanding to significantly improve the already high sensitivity by optimizing the design of the dyes and the optics (see Chap. 19 in this volume).

Dyes, light sources and optical modalities are required to permit deeper penetration into thick tissue or in vivo imaging. Longer wavelength dyes will allow for deeper penetration because their spectra are removed from interference by endogenous chromophores and because light scattering diminishes as the inverse of the fourth power of wavelength. Non-linear optical modalities like 2-photon excited fluorescence and SHG will allow deeper optical sectioning if they can be implemented with longer wavelength ultrafast pulsed lasers.

One major area for future chemistry is to devise methods for more specific delivery of the dyes to individual, preferably multiple, cells in a complex preparation. Indeed, this is one of the main motivations for the intense activity in trying to devise a genetically encodable fluorescent protein voltage sensor. But by combining engineered organic dyes, photochemistry, and new optical technologies, it should be possible for chemists to devise general solutions to this challenge as well.

Acknowledgment I am indebted to the many talented chemists, microscopists and neuroscientists who have collaborated with me and who have carried out much of the research summarized in this chapter. Most notably, I wish to acknowledge my long term collaborators Larry Cohen, Aaron Lewis, Mei-de Wei and Joe Wuskell. The more recent work in my lab has benefited from collaborations with Ping Yan, Corey Acker and Srdjan Antic. This work was supported by NIH EB001963.

References

Acker CD, Loew LM (2013) Characterization of voltage-sensitive dyes in living cells using two-photon excitation. Methods Mol Biol 995:147–160

Acker CD, Yan P, Loew LM (2011) Single-voxel recording of voltage transients in dendritic spines. Biophys J 101:L11–L13

Akemann W, Mutoh H, Perron A, Rossier J, Knopfel T (2010) Imaging brain electric signals with genetically targeted voltage-sensitive fluorescent proteins. Nat Methods 7:643–649

Antic S, Major G, Zecevic D (1999) Fast optical recordings of membrane potential changes from dendrites of pyramidal neurons. J Neurophysiol 82:1615–1621

Antic S, Wuskell JP, Loew L, Zecevic D (2000) Functional profile of the giant metacerebral neuron of Helix aspersa: temporal and spatial dynamics of electrical activity in situ. J Physiol 527:55–69

Antic S, Zecevic D (1995) Optical signals from neurons with internally applied voltage-sensitive dyes. J Neurosci 15:1392–1405

Antic SD (2003) Action potentials in basal and oblique dendrites of rat neocortical pyramidal neurons. J Physiol (Lond) 550:35–50

Araya R, Jiang J, Eisenthal KB, Yuste R (2006) The spine neck filters membrane potentials. Proc Natl Acad Sci U S A 103:17961–17966

Araya R, Nikolenko V, Eisenthal KB, Yuste R (2007) Sodium channels amplify spine potentials. Proc Natl Acad Sci U S A 104:12347–12352

Ataka K, Pieribone VA (2002) A genetically targetable fluorescent probe of channel gating with rapid kinetics. Biophys J 82:509–516

Baker BJ, Lee H, Pieribone VA, Cohen LB, Isacoff EY, Knopfel T, Kosmidis EK (2007) Three fluorescent protein voltage sensors exhibit low plasma membrane expression in mammalian cells. J Neurosci Methods 161:32–38

Beach JM, McGahren ED, Xia J, Duling BR (1996) Ratiometric measurement of endothelial cell depolarization in arterioles with a potential sensitive dye. Am J Physiol 270:H2216–H2227

Bedlack RS, Wei M-d, Fox SH, Gross E, Loew LM (1994) Distinct electric potentials in soma and neurite membranes. Neuron 13:1187–1193

Bedlack RS, Wei M-d, Loew LM (1992) Localized membrane depolarizations and localized intracellular calcium influx during electric field-guided neurite growth. Neuron 9:393–403

Ben-Oren I, Peleg G, Lewis A, Minke B, Loew LM (1996) Infrared nonlinear optical measurements of membrane potential in photoreceptor cells. Biophys J 71:1616–1620

Bouevitch O, Lewis A, Pinevsky I, Wuskell JP, Loew LM (1993) Probing membrane potential with non-linear optics. Biophys J 65:672–679

Bradley J, Luo R, Otis TS, DiGregorio DA (2009) Submillisecond optical reporting of membrane potential in situ using a neuronal tracer dye. J Neurosci 29:9197–9209

Bullen A, Patel SS, Saggau P (1997) High speed random access fluorescence microscopy: I. High resolution optical recording with voltage-sensitive dyes and ion indicators. Biophys J 73:477–491

Bullen A, Saggau P (1999) High-speed, random-access fluorescence microscopy: II. Fast quantitative measurements with voltage-sensitive dyes [in process citation]. Biophys J 76:2272–2287

Cacciatore TW, Brodfuehrer PD, Gonzalez JE, Jiang T, Adams SR, Tsien RY, Kristan WB Jr, Kleinfeld D (1999) Identification of neural circuits by imaging coherent electrical activity with FRET-based dyes. Neuron 23:449–459

Campagnola P, Loew LM (2003) Second-harmonic imaging microscopy for visualizing biomolecular arrays in cells, tissues and organisms. Nat Biotechnol 21:1356–1360

Campagnola PJ, Wei M-d, Lewis A, Loew LM (1999) High resolution optical imaging of live cells by second harmonic generation. Biophys J 77:3341–3349

Canepari M, Djurisic M, Zecevic D (2007) Dendritic signals from rat hippocampal CA1 pyramidal neurons during coincident pre- and post-synaptic activity: a combined voltage- and calcium-imaging study. J Physiol 580:463–484

Canepari M, Vogt K, Zecevic D (2008) Combining voltage and calcium imaging from neuronal dendrites. Cell Mol Neurobiol 28(8):1079–1093

Canepari M, Willadt S, Zecevic D, Vogt KE (2010) Imaging inhibitory synaptic potentials using voltage sensitive dyes. Biophys J 98:2032–2040

Clark HA, Campagnola PJ, Wuskell JP, Lewis A, Loew LM (2000) Second harmonic generation properties of fluorescent polymer encapsulated gold nanoparticles. J Am Chem Soc 122:10234–10235

Cohen LB, Salzberg BM, Davila HV, Ross WN, Landowne D, Waggoner AS, Wang CH (1974) Changes in axon fluorescence during activity: molecular probes of membrane potential. J Membr Biol 19:1–36

Denk W, Strickler JH, Webb WW (1990) Two-photon laser scanning fluorescence microscopy. Science 248:73–76

Derdikman D, Hildesheim R, Ahissar E, Arieli A, Grinvald A (2003) Imaging spatiotemporal dynamics of surround inhibition in the barrels somatosensory cortex. J Neurosci 23:3100–3105

Djurisic M, Zecevic D (2005) Imaging of spiking and subthreshold activity of mitral cells with voltage-sensitive dyes. Ann N Y Acad Sci 1048:92–102

Djurisic M, Zochowski M, Wachowiak M, Falk CX, Cohen LB, Zecevic D (2003) Optical monitoring of neural activity using voltage-sensitive dyes. Methods Enzymol 361:423–451

Dombeck DA, Blanchard-Desce M, Webb WW (2004) Optical recording of action potentials with second-harmonic generation microscopy. J Neurosci 24:999–1003

Dombeck DA, Sacconi L, Blanchard-Desce M, Webb WW (2005) Optical recording of fast neuronal membrane potential transients in acute mammalian brain slices by second-harmonic generation microscopy. J Neurophysiol 94:3628–3636

Dragsten PR, Webb WW (1978) Mechanism of the membrane potential sensitivity of the fluorescent membrane probe merocyanine 540. Biochemistry 17:5228–5240

Efimov IR, Nikolski VP, Salama G (2004) Optical imaging of the heart. Circ Res 95:21–33

Ehrenberg B, Montana V, Wei M-d, Wuskell JP, Loew LM (1988) Membrane potential can be determined in individual cells from the Nernstian distribution of cationic dyes. Biophys J 53:785–794

Farkas DL, Wei M, Febbroriello P, Carson JH, Loew LM (1989) Simultaneous imaging of cell and mitochondrial membrane potential. Biophys J 56:1053–1069

Fedorov VV, Ambrosi CM, Kostecki G, Hucker WJ, Glukhov AV, Wuskell JP, Loew LM, Moazami N, Efimov IR (2011) Anatomic localization and autonomic modulation of atrioventricular junctional rhythm in failing human hearts. Circ Arrhythm Electrophysiol 4:515–525

Fedorov VV, Glukhov AV, Chang R, Kostecki G, Aferol H, Hucker WJ, Wuskell JP, Loew LM, Schuessler RB, Moazami N, Efimov IR (2010) Optical mapping of the isolated coronary-perfused human sinus node. J Am Coll Cardiol 56:1386–1394

Fink AE, Bender KJ, Trussell LO, Otis TS, DiGregorio DA (2012) Two-photon compatibility and single-voxel, single-trial detection of subthreshold neuronal activity by a two-component optical voltage sensor. PLoS One 7, e41434

Fisher JAN, Barchi JR, Welle CG, Kim G-H, Kosterin P, Obaid AL, Yodh AG, Contreras D, Salzberg BM (2008) Two-photon excitation of potentiometric probes enables optical recording of action potentials from mammalian nerve terminals in Situ. J Neurophysiol 99:1545–1553

Fluhler E, Burnham VG, Loew LM (1985) Spectra, membrane binding and potentiometric responses of new charge shift probes. Biochemistry 24:5749–5755

Foust A, Popovic M, Zecevic D, McCormick DA (2010) Action potentials initiate in the axon initial segment and propagate through axon collaterals reliably in cerebellar purkinje neurons. J Neurosci 30:6891–6902

Fromherz P, Hubener G, Kuhn B, Hinner MJ (2008) ANNINE-6plus, a voltage-sensitive dye with good solubility, strong membrane binding and high sensitivity. Eur Biophys J 37:509–514

Glover JC, Sato K, Momose-Sato Y (2008) Using voltage-sensitive dye recording to image the functional development of neuronal circuits in vertebrate embryos. Dev Neurobiol 68:804–816

Gong Y, Wagner MJ, Zhong Li J, Schnitzer MJ (2014) Imaging neural spiking in brain tissue using FRET-opsin protein voltage sensors. Nat Commun 5:3674

Gonzalez JE, Tsien RY (1995) Voltage sensing by fluorescence resonance energy transfer in single cells. Biophys J 69:1272–1280

Gonzalez JE, Tsien RY (1997) Improved indicators of membrane potential that use fluorescence resonance energy transfer. Chem Biol 4:269–277

Grinvald A, Hildesheim R (2004) VSDI: a new era in functional imaging of cortical dynamics. Nat Rev Neurosci 5:874–885

Gross E, Bedlack RS, Loew LM (1994) Dual-wavelength ratiometric measurement of the membrane dipole potential. Biophys J 67:208–216

Guerrero G, Siegel MS, Roska B, Loots E, Isacoff EY (2002) Tuning FlaSh: redesign of the dynamics, voltage range, and color of the genetically encoded optical sensor of membrane potential. Biophys J 83:3607–3618

Gupta RK, Salzberg BM, Grinvald A, Cohen LB, Kamino K, Lesher S, Boyle MB, Waggoner AS, Wang C (1981) Improvements in optical methods for measuring rapid changes in membrane potential. J Membr Biol 58:123–137

Habib ERMS, Komuro R, Yan P, Hayashi S, Inaji M, Momose-Sato Y, Loew LM, Sato K (2013) Evaluation of voltage-sensitive fluorescence dyes for monitoring neuronal activity in the embryonic central nervous system. J Membr Biol 246:679–688

Hochbaum DR, Zhao Y, Farhi SL, Klapoetke N, Werley CA, Kapoor V, Zou P, Kralj JM, Maclaurin D, Smedemark-Margulies N, Saulnier JL, Boulting GL, Straub C, Cho YK, Melkonian M, Wong GK, Harrison DJ, Murthy VN, Sabatini BL, Boyden ES, Campbell RE, Cohen AE (2014) All-optical electrophysiology in mammalian neurons using engineered microbial rhodopsins. Nat Methods 11(8):825–833

Holthoff K, Zecevic D, Konnerth A (2010) Rapid time course of action potentials in spines and remote dendrites of mouse visual cortex neurons. J Physiol 588:1085–1096

Huang JY, Lewis A, Loew LM (1988) Non-linear optical properties of potential sensitive styryl dyes. Biophys J 53:665–670

Jiang J, Eisenthal KB, Yuste R (2007) Second harmonic generation in neurons: electro-optic mechanism of membrane potential sensitivity. Biophys J 93:L26–L28

Jin L, Han Z, Platisa J, Wooltorton JRA, Cohen LB, Pieribone VA (2012) Single action potentials and subthreshold electrical events imaged in neurons with a fluorescent protein voltage probe. Neuron 75:779–785

Kampa BM, Stuart GJ (2006) Calcium spikes in basal dendrites of layer 5 pyramidal neurons during action potential bursts. J Neurosci 26:7424–7432

Kee MZ, Wuskell JP, Loew LM, Augustine GJ, Sekino Y (2008) Imaging activity of neuronal populations with new long-wavelength voltage-sensitive dyes. Brain Cell Biol 36:57–72

Knisley SB, Justice RK, Kong W, Johnson PL (2000) Ratiometry of transmembrane voltage-sensitive fluorescent dye emission in hearts. Am J Physiol Heart Circ Physiol 279:H1421–H1433

Kralj JM, Hochbaum DR, Douglass AD, Cohen AE (2011) Electrical spiking in escherichia coli probed with a fluorescent voltage-indicating protein. Science 333:345–348

Kuhn B, Denk W, Bruno RM (2008) In vivo two-photon voltage-sensitive dye imaging reveals top-down control of cortical layers 1 and 2 during wakefulness. Proc Natl Acad Sci U S A 105:7588–7593

Kuhn B, Fromherz P (2003) Anellated hemicyanine dyes in a neuron membrane: molecular Stark effect and optical voltage recording. J Phys Chem B 107:7903–7913

Kuhn B, Fromherz P, Denk W (2004) High sensitivity of stark-shift voltage-sensing dyes by one- or two-photon excitation near the red spectral edge. Biophys J 87:631–639

Lee P, Bollensdorff C, Quinn TA, Wuskell JP, Loew LM, Kohl P (2011) Single-sensor system for spatially-resolved, continuous and multi-parametric optical mapping of cardiac tissue. Heart Rhythm 8:1482–1491

Lee P, Taghavi F, Yan P, Ewart P, Ashley EA, Loew LM, Kohl P, Bollensdorff C, Woods CE (2012a) In situ optical mapping of voltage and calcium in the heart. PLoS One 7, e42562

Lee P, Wang K, Woods CE, Yan P, Kohl P, Ewart P, Loew LM, Terrar DA, Bollensdorff C (2012b) Cardiac electrophysiological imaging systems scalable for high-throughput drug testing. Pflugers Arch 464:645–656

Lee P, Yan P, Ewart P, Kohl P, Loew LM, Bollensdorff C (2012c) Simultaneous measurement and modulation of multiple physiological parameters in the isolated heart using optical techniques. Pflugers Arch 464:403–414

Loew LM (1993) Confocal microscopy of potentiometric fluorescent dyes. Methods Cell Biol 38:194–209

Loew LM (1994) Voltage sensitive dyes and imaging neuronal activity. Neuroprotocols 5:72–79

Loew LM (2001) Mechanisms and principles of voltage sensitive fluorescence. In: Rosenbaum DS, Jalife J (eds) Optical mapping of cardiac excitation and arrhythmias. Futura Publishing, Armonk, NY, pp 33–46

Loew LM, Bonneville GW, Surow J (1978) Charge shift optical probes of membrane potential. Theory. Biochemistry 17:4065–4071

Loew LM, Campagnola P, Lewis A, Wuskell JP (2002) Confocal and nonlinear optical imaging of potentiometric dyes. Methods Cell Biol 70:429–452
Loew LM, Cohen LB, Dix J, Fluhler EN, Montana V, Salama G, Wu J-Y (1992) A naphthyl analog of the aminostyryl pyridinium class of potentiometric membrane dyes shows consistent sensitivity in a variety of tissue, cell, and model membrane preparations. J Membr Biol 130:1–10
Loew LM, Rosenberg I, Bridge M, Gitler C (1983) Diffusion potential cascade. Convenient detection of transferable membrane pores. Biochemistry 22:837–844
Loew LM, Scully S, Simpson L, Waggoner AS (1979a) Evidence for a charge-shift electrochromic mechanism in a probe of membrane potential. Nature 281:497–499
Loew LM, Simpson L (1981) Charge shift probes of membrane potential. A probable electrochromic mechanism for ASP probes on a hemispherical lipid bilayer. Biophys J 34:353–365
Loew LM, Simpson L, Hassner A, Alexanian V (1979b) An unexpected blue shift caused by differential solvation of a chromophore oriented in a lipid bilayer. J Am Chem Soc 101:5439–5440
Lojewska Z, Loew LM (1987) Insertion of amphiphilic molecules into membranes is catalyzed by a high molecular weight non-ionic surfactant. Biochim Biophys Acta 899:104–112
Matiukas A, Mitrea BG, Pertsov AM, Wuskell JP, Wei MD, Watras J, Millard AC, Loew LM (2006) New near-infrared optical probes of cardiac electrical activity. Am J Physiol Heart Circ Physiol 290:H2633–H2643
Matiukas A, Mitrea BG, Qin M, Pertsov AM, Shvedko AG, Warren MD, Zaitsev AV, Wuskell JP, Wei M-d, Watras J, Loew LM (2007) Near infrared voltage sensitive fluorescent dyes optimized for optical mapping in blood-perfused myocardium. Heart Rhythm 4:1441–1451
Millard AC, Campagnola PJ, Mohler W, Lewis A, Loew LM (2003a) Second harmonic imaging microscopy. In: Marriott G, Parker I (eds) Methods in enzymology, vol. 361B. Academic, San Diego, pp 47–69
Millard AC, Jin L, Lewis A, Loew LM (2003b) Direct measurement of the voltage sensitivity of second-harmonic generation from a membrane dye in patch-clamped cells. Opt Lett 28:1221–1223
Millard AC, Jin L, Wei M-d, Wuskell JP, Lewis A, Loew LM (2004) Sensitivity of second harmonic generation from styryl dyes to trans-membrane potential. Biophys J 86:1169–1176
Millard AC, Jin L, Wuskell JP, Boudreau DM, Lewis A, Loew LM (2005a) Wavelength- and time-dependence of potentiometric non-linear optical signals from styryl dyes. J Membr Biol 208:103–111
Millard AC, Lewis A, Loew LM (2005b) Second harmonic imaging of membrane potential. In: Yuste R, Konnerth A (eds) Imaging in neuroscience and development. Cold Spring Harbor Laboratory Press, Cold Spring Harbor, pp 463–474
Miller EW, Lin JY, Frady EP, Steinbach PA, Kristan WB, Tsien RY (2012) Optically monitoring voltage in neurons by photo-induced electron transfer through molecular wires. Proc Natl Acad Sci U S A 109:2114–2119
Milojkovic BA, Wuskell JP, Loew LM, Antic SD (2005) Initiation of sodium spikelets in basal dendrites of neocortical pyramidal neurons. J Membr Biol 208:155–169
Montana V, Farkas DL, Loew LM (1989) Dual wavelength ratiometric fluorescence measurements of membrane potential. Biochemistry 28:4536–4539
Nishiyama M, von Schimmelmann MJ, Togashi K, Findley WM, Hong K (2008) Membrane potential shifts caused by diffusible guidance signals direct growth-cone turning. Nat Neurosci 11:762–771
Nuriya M, Jiang J, Nemet B, Eisenthal KB, Yuste R (2006) Imaging membrane potential in dendritic spines. Proc Natl Acad Sci U S A 103:786–790
Obaid AL, Loew LM, Wuskell JP, Salzberg BM (2004) Novel naphthylstyryl-pyridinium potentiometric dyes offer advantages for neural network analysis. J Neurosci Methods 134:179–190

Palmer LM, Stuart GJ (2009) Membrane potential changes in dendritic spines during action potentials and synaptic input. J Neurosci 29:6897–6903
Pons T, Moreaux L, Mongin O, Blanchard-Desce M, Mertz J (2003) Mechanisms of membrane potential sensing with second-harmonic generation microscopy. J Biomed Opt 8:428–431
Popovic MA, Gao X, Carnevale NT, Zecevic D (2014) Cortical dendritic spine heads are not electrically isolated by the spine neck from membrane potential signals in parent dendrites. Cereb Cortex 24:385–395
Ross WN, Salzberg BM, Cohen LB, Grinvald A, Davila HV, Waggoner AS, Wang CH (1977) Changes in absorption, fluorescence, dichroism, and birefringence in stained giant axons: optical measurement of membrane potential. J Membr Biol 33:141–183
Sakai R, Repunte-Canonigo V, Raj CD, Knopfel T (2001) Design and characterization of a DNA-encoded, voltage-sensitive fluorescent protein. Eur J Neurosci 13:2314–2318
Sasaki S, Yazawa I, Miyakawa N, Mochida H, Shinomiya K, Kamino K, Momose-Sato Y, Sato K (2002) Optical imaging of intrinsic signals induced by peripheral nerve stimulation in the in vivo rat spinal cord. Neuroimage 17:1240–1255
Shoham D, Glaser DE, Arieli A, Kenet T, Wijnbergen C, Toledo Y, Hildesheim R, Grinvald A (1999) Imaging cortical dynamics at high spatial and temporal resolution with novel blue voltage-sensitive dyes. Neuron 24:791–802
Siegel MS, Isacoff EYIN (1997) A genetically encoded optical probe of membrane voltage. Neuron 19:735–741
Slovin H, Arieli A, Hildesheim R, Grinvald A (2002) Long-term voltage-sensitive dye imaging reveals cortical dynamics in behaving monkeys. J Neurophysiol 88:3421–3438
St-Pierre F, Marshall JD, Yang Y, Gong Y, Schnitzer MJ, Lin MZ (2014) High-fidelity optical reporting of neuronal electrical activity with an ultrafast fluorescent voltage sensor. Nat Neurosci 17(6):884–889
Stuart GJ, Palmer LM (2006) Imaging membrane potential in dendrites and axons of single neurons. Pflugers Arch 453:403–410
Teisseyre TZ, Millard AC, Yan P, Wuskell JP, Wei M-d, Lewis A, Loew LM (2007) Non-linear optical potentiometric dyes optimized for imaging with 1064 nm light. J Biomed Opt 12:044001
Tsau Y, Wenner P, O'Donovan MJ, Cohen LB, Loew LM, Wuskell JP (1996) Dye screening and signal-to-noise ratio for retrogradely transported voltage-sensitive dyes. J Neurosci Methods 170:121–129
Tsuda S, Kee MZL, Cunha C, Kim J, Yan P, Loew LM, Augustine GJ (2012) Probing the function of neuronal populations: combining micromirror-based optogenetic photostimulation with voltage-sensitive dye imaging. Neurosci Res 75(1):76–81
Tsutsui H, Karasawa S, Okamura Y, Miyawaki A (2008) Improving membrane voltage measurements using FRET with new fluorescent proteins. Nat Methods 5:683–685
Waggoner AS, Wang CH, Tolles RL (1977) Mechanism of potential-dependent light absorption changes of lipid bilayer membranes in the presence of cyanine and oxonol dyes. J Membr Biol 33:109–140
Warren M, Spitzer KW, Steadman BW, Rees TD, Venable PW, Taylor T, Shibayama J, Yan P, Wuskell JP, Loew LM, Zaitsev AV (2010) High precision recording of the action potential in isolated cardiomyocytes using the near-infrared fluorescent dye di-4-ANBDQBS. Am J Physiol Heart Circ Physiol 299:H1271–H1281
Wenner P, Tsau Y, Cohen LB, O'Donovan MJ, Dan Y (1996) Voltage-sensitive dye recording using retrogradely transported dye in the chicken spinal cord: staining and signal characteristics. J Neurosci Methods 170:111–120
Wu J-Y, Lam Y-W, Falk C, Cohen LB, Fang J, Loew L, Prechtl JC, Kleinfeld D, Tsau Y (1998) Voltage-sensitive dyes for monitoring multi-neuronal activity in the intact CNS. Histochem J 30:169–187

Wuskell JP, Boudreau D, Wei MD, Jin L, Engl R, Chebolu R, Bullen A, Hoffacker KD, Kerimo J, Cohen LB, Zochowski MR, Loew LM (2006) Synthesis, spectra, delivery and potentiometric responses of new styryl dyes with extended spectral ranges. J Neurosci Methods 151:200–215

Xu C, Loew LM (2003) The effect of asymmetric surface potentials on the intramembrane electric field measured with voltage-sensitive dyes. Biophys J 84:2768–2780

Yan P, Acker CD, Zhou WL, Lee P, Bollensdorff C, Negrean A, Lotti J, Sacconi L, Antic SD, Kohl P, Mansvelder HD, Pavone FS, Loew LM (2012) A palette of fluorinated voltage sensitive hemicyanine dyes. Proc Natl Acad Sci U S A 109:20443–20448

Yan P, Xie A, Wei M-d, Loew LM (2008) Amino(oligo)thiophene-based environmentally sensitive biomembrane chromophores. J Org Chem 73:6587–6594

Zecevic D (1996) Multiple spike-initiation zones in single neurons revealed by voltage-sensitive dyes. Nature 381:322–325

Zhou W-L, Yan P, Wuskell JP, Loew LM, Antic SD (2007) Intracellular long wavelength voltage-sensitive dyes for studying the dynamics of action potentials in axons and thin dendrites. J Neurosci Methods 164:225–239

Zhou WL, Yan P, Wuskell JP, Loew LM, Antic SD (2008) Dynamics of action potential backpropagation in basal dendrites of prefrontal cortical pyramidal neurons. Eur J Neurosci 27:923–936

Part II
Membrane Potential Signals with Single Cell Resolution

Chapter 3
Imaging Submillisecond Membrane Potential Changes from Individual Regions of Single Axons, Dendrites and Spines

Marko Popovic, Kaspar Vogt, Knut Holthoff, Arthur Konnerth, Brian M. Salzberg, Amiram Grinvald, Srdjan D. Antic, Marco Canepari, and Dejan Zecevic

Abstract A central question in neuronal network analysis is how the interaction between individual neurons produces behavior and behavioral modifications. This task depends critically on how exactly signals are integrated by individual nerve cells functioning as complex operational units. Regional electrical properties of branching neuronal processes which determine the input-output function of any

M. Popovic • D. Zecevic
Department of Cellular and Molecular Physiology, Yale University School of Medicine, New Haven, CT 06520, USA

K. Vogt
Division of Pharmacology and Neurobiology, Biozentrum – University of Basel, Basel, Switzerland

K. Holthoff
Hans Berger Klinik für Neurologie, Friedrich-Schiller-Universität Jena, Jena, Germany

A. Konnerth
Center for Integrated Protein Science and Institute of Neuroscience, Technical University Munich, Munich, Germany

B.M. Salzberg (✉)
Department of Neuroscience, University of Pennsylvania School of Medicine, Philadelphia, PA, USA

Department of Physiology, University of Pennsylvania School of Medicine, Philadelphia, PA, USA
e-mail: bmsalzbe@mail.med.upenn.edu

A. Grinvald
Department of Neurobiology, Weizmann Institute of Science, PO Box 26, Rehovot 76100, Israel

S.D. Antic
Department of Neuroscience, U. Conn. Health Center, Farmington, CT 06030, USA

M. Canepari
Laboratoire Interdisciplinare de Physique, (CNRS Université Joseph Fourier UMR 5588), Grenoble, France

M. Canepari et al. (eds.), *Membrane Potential Imaging in the Nervous System and Heart*, Advances in Experimental Medicine and Biology 859,
DOI 10.1007/978-3-319-17641-3_3

neuron are extraordinarily complex, dynamic, and, in the general case, impossible to predict in the absence of detailed measurements. To obtain such a measurement one would, ideally, like to be able to monitor, at multiple sites, subthreshold events as they travel from the sites of origin (synaptic contacts on distal dendrites) and summate at particular locations to influence action potential initiation. It became possible recently to carry out this type of measurement using high-resolution multisite recording of membrane potential changes with intracellular voltage-sensitive dyes. This chapter reviews the development and foundation of the method of voltage-sensitive dye recording from individual neurons. Presently, this approach allows monitoring membrane potential transients from all parts of the dendritic tree as well as from axon collaterals and individual dendritic spines.

Keywords Fluorescence • Glutamate uncaging • Voltage-imaging • Potentiometric probes • Voltage-sensitive dyes • Neurons • Axons • Dendrites • Dendritic spines

1 Introduction

Understanding the biophysical properties and functional organization of single neurons and how they process information is fundamental to understanding how the brain works. Because the primary function of any nerve cell is to process electrical signals (i.e. membrane potential (V_m) transients), usually from multiple sources, there is a need for detailed spatiotemporal analysis of electrical events in thin axonal and dendritic processes. This requirement resulted in the development of new measurement techniques that allow monitoring of the electrical activity of different parts of the same cell simultaneously. A major experimental advance in this field, which also underscored the importance of such measurements, was achieved by the development of the multiple patch-electrode recording method that made possible simultaneous monitoring of voltage transients from two or more dendritic locations on a single neuron (Stuart et al. 1993; Stuart and Sakmann 1994). More recently, methods were developed for patch-pipette recording of electrical signals from cut ends of axons of layer 5 pyramidal cells in the cerebral cortex (Shu et al. 2006) as well as from presynaptic axon terminals of the giant synapses of mossy fiber-CA3 connections (Geiger et al. 2006). These techniques, however, are still limited in their capacity for assessing spatiotemporal patterns of signal initiation and propagation in complex dendritic and axonal processes. Moreover, many subcellular structures including small diameter terminal dendritic branches as well as dendritic spines and most axon terminals and axon collaterals are not accessible to electrodes. To overcome these limitations, it was highly desirable to complement the patch-electrode approach with technologies that permit extensive parallel recordings from all parts of a neuron with adequate spatial and temporal resolution. An adequate temporal resolution in recording neuronal action potential (AP) and synaptic potential signals is in the sub-millisecond range, as determined by the duration of different phases of the AP and SP waveforms. An adequate spatial resolution is on the order of 1 μm, as determined by the dimensions of neuronal terminal processes and dendritic spines. This spatio-temporal resolution can now be realized using optical recording of V_m changes with organic voltage-sensitive dyes (V_m-imaging) (Cohen and Salzberg 1978). The sensitivity of this measurement technique has recently reached a level that permits single trial optical recordings of V_m transients from all parts of a neuron, including axon terminals and collaterals, terminal dendritic branches, and individual dendritic spines. Whenever the experimental design allows signal averaging, a relatively small number of trials (4–9) will result in two or threefold improvements in the signal-to-noise ratio (S/N).

2 Determinants of V_m-Imaging Sensitivity

The central figure-of-merit in V_m-imaging is the recording sensitivity expressed as the S/N. The rules governing the sensitivity of light intensity measurements are well understood and may be put in a nutshell with an expression:

$$S/N \propto (\Delta F/F)\sqrt{\Phi} \tag{3.1}$$

where $\Delta F/F$ is the fractional fluorescence signal per unit change in V_m and Φ is the number of detected photons per unit time (fluorescence intensity expressed as photon flux). This relationship is valid under the shot noise limited conditions fulfilled in most modern fluorescence measurements. Shot noise limited conditions imply that the noise arising from the statistical nature of the emission and detection of photons (shot noise) is the dominant source of noise in the recording system, while the following sources of noise are negligible: (a) the noise in the incident light intensity arising from an unstable light source (50 or 60 Hz and harmonic ripple noise and/or arc-wander noise); (b) noise caused by mechanical vibration of the image projected on the photodetector; (c) electrical noise in the amplifier circuits (dark-noise), and (d) read noise of the CCD camera (see also Chap. 1). Under these conditions, one way to increase the S/N for a given dye is to increase Φ by increasing either the incident light intensity or the detection efficiency, or both. An additional possibility is to increase $\Delta F/F$ per unit change in V_m (sensitivity of the dye). Thus, close attention must be paid to these two parameters.

The detected fluorescence photon flux Φ is a function of several parameters. These include: (a) the excitation light intensity; (b) the extinction coefficient of the dye absorption; (c) the quantum yield of a given dye (the efficiency of the fluorescence process defined as the ratio of the number of photons emitted to the number of photons absorbed); (d) the overall light throughput of the optical elements in the light path, including objectives, mirrors, auxiliary lenses and optical filters; (e) the quantum efficiency of the recording device (the ratio between electrons generated to the number of photons absorbed at the photoreactive surface). In addition to these factors that are usually constant for a given experiment, the photon flux will also depend on three additional variables: (f) the number of membrane-bound dye molecules in the light path which will be a function of the amount of staining and the membrane surface area projected onto individual pixels of the recording device; (g) the fraction of the fluorescent dye bound to the external membrane that changes potential and (h) the time interval over which photons are collected for each data point as determined by the imaging frame rate. Thus, the photon flux and S/N will decrease as the spatial and temporal resolutions are increased.

The voltage sensitivity of the dye, expressed as ΔF/F, depends on its chemical structure as described in Chap. 2. In addition, for a given dye, the voltage sensitivity is a function of the excitation wavelength, as demonstrated early in the development of organic probes (Cohen et al. 1974; Loew 1982). The sensitivity increases at extreme wings of the absorption spectrum for charge-shift voltage-sensitive probes. Also, the optimal sensitivity will be reached when all excitation occurs for the wavelength with the best response. In other words, monochromatic illumination at the appropriate wavelength (as opposed to the conventional use of a bandpass interference filter to select a range of wavelengths) will result in optimal sensitivity (Kuhn et al. 2004). These considerations argue that a laser emitting monochromatic light at an appropriate wavelength will be the optimal excitation light source. It should be noted that the advantages of laser illumination are more pronounced in fluorescent voltage-sensitive dye measurements since the loss of coherence introduced by fluorescence emission eliminates the speckle noise that plagues transmission (absorption) measurements using laser sources (Daintry 1984).

3 Foundation of Presently Available Recording Sensitivity

The feasibility of multiple site optical recording from individual nerve cells was initially demonstrated using invertebrate ganglia and monolayer neuronal culture and extracellular application of the voltage-sensitive dye (Salzberg et al. 1977, Grinvald et al. 1981). It was subsequently shown that sufficient sensitivity of recording with subcellular spatial resolution from intact parts of the nervous system can only be achieved if nerve cells are labeled selectively by the intracellular application of the membrane impermeant probe. This section reviews the limits to the sensitivity of recording using extracellular and intracellular application of the voltage-sensitive dyes.

3.1 Extracellular Application of Dyes

3.1.1 Dissociated Neurons in Culture

In 1981 Grinvald, Ross, and Farber demonstrated that voltage-sensitive dyes and multi-site optical measurements can be employed successfully to determine conduction velocity, space constants and regional variations in the electrical properties

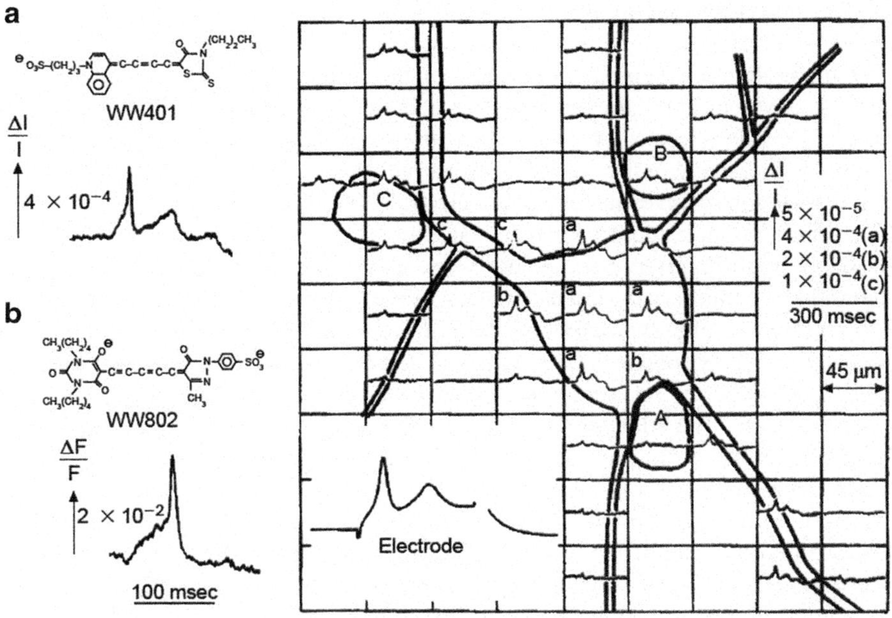

Fig. 3.1 Transmission (**a**) and fluorescence (**b**) signals recorded with a single photodiode from extracellularly stained neuroblastoma cells in culture. Multiple site recording using a photodiode array is shown on the *right*. Adapted from Grinvald et al. (1981)

of neuronal processes in measurements from dissociated neurons in monolayer (Grinvald et al. 1981) culture. Both absorption and fluorescence measurements were used in these experiments and the individual dissociated neurons in a culture dish were stained from the *extracellular* side by bath application of the dye (Grinvald and Farber 1981). Typical optical recordings obtained from absorption and fluorescence measurements using the most sensitive probes available at the time are shown in Fig. 3.1. It has also been possible, using the same approach, to study synaptic interactions between several interconnected neurons in culture (Parsons et al. 1989, 1991), and the cell-to-cell propagation of the AP in patterned growth cardiac myocytes forming two-dimensional hearts in culture (Rohr and Salzberg 1994). The monolayer neuronal culture is a low opacity system especially convenient for the extracellular selective staining of the outside cellular membrane as well as for both absorption and fluorescence measurements. However, primary cultures are networks of neurons that grow under artificial conditions and are substantially different from intact neuronal networks. Therefore, a number of important questions in cellular neurophysiology can only be studied in intact or semi-intact preparations.

3.1.2 Invertebrate Ganglia

Extending the same approach to *in situ* conditions has proven difficult and, hence, slow to develop. In *in situ* conditions, extracellular application of fluorescent dyes cannot provide single cell resolution because of the large background fluorescence from the dye bound indiscriminately to all membranes in the preparation. Thus, in the first attempt to investigate regional electrical properties of individual neurons *in situ*, in the barnacle supraesophageal ganglion, (Ross and Krauthamer 1984) transmission (absorption) measurements were used and, as in the experiments on neurons in culture, bath applied voltage-sensitive dyes to record optical signals from processes of individual nerve cells. Because the S/N of these measurements was not nearly as good as in recordings from monolayer culture, extensive averaging (approximately 300 trials) was required to obtain signals from processes of single neurons. The relatively small signal size which requires extensive averaging limits the utility of this approach for studying synaptic interactions and plasticity. Also, when many neurons are active in a densely packed neuropile it is difficult to determine the source of the signal if all the cells and processes are stained by extracellular dye application (Konnerth et al. 1987). Thus, the prospect for using measurement of voltage-sensitive dye absorption or fluorescence with extracellular staining is limited, and no further experiments requiring subcellular resolution have been reported using this type of staining. The limits to the sensitivity in these measurements were determined primarily by: (a) the relatively low voltage sensitivity of the available dyes; and, (b) the large background light intensity inherent to both absorption and fluorescence measurements with extracellular application of the dye, which translates into low fractional change in light intensity ($\Delta I/I$) related to V_m transients (Waggoner and Grinvald 1977).

3.2 *Intracellular Application of Dyes*

3.2.1 Invertebrate Ganglia

A different approach to optical analysis of electrical events in the processes of individual nerve cells is to stain particular neurons *in situ* selectively by intracellular application of an impermeant fluorescent voltage-sensitive dye. This approach is based on pioneering measurements, carried out on the giant axon of the squid, which demonstrated that optical signals may be obtained when the dye is applied from the inside (Davila et al. 1974; Cohen et al. 1974; Salzberg 1978; Gupta et al. 1981). In these experiments, optical signals were obtained from the large membrane surface area of the giant axon. Additional efforts were required to demonstrate that the same approach is feasible at the spatial scale of normal size neurons and their processes. The initial experiments using intracellular application of the voltage-sensitive dyes were carried out by Obaid, Shimizu and Salzberg (Obaid et al. 1982) and by Grinvald, Salzberg, Lev-Ram and Hildesheim (Grinvald et al. 1987) on leech neurons. The results demonstrated the essential advantages of using intracellular application of fluorescent potentiometric probes. Fluorescence measurements are more effective than absorption measurements when measuring from a small membrane area (Waggoner and Grinvald 1977), particularly in situations where the image of the object (e.g., thin process) is much smaller than the size of the photodetector picture element (pixel) (Grinvald et al. 1982; Cohen and Lesher 1986). When transmitted light is used, only a small fraction of the total light captured by individual pixels will be modulated by the signal from the neuronal process (dendrite or axon) projected onto that pixel (ΔI). Most of the light will be projected directly and will only contribute to the resting light intensity (I). Thus, the fractional signal (ΔI/I) will be very small. On the other hand, in fluorescence measurements, practically all of the light projected onto individual pixels will come from the object (if auto fluorescence is negligible) regardless of the fraction of the pixel surface area covered by the image of the object. In the experiments on leech neurons (Grinvald et al. 1987), fluorescence measurements were used to record APs and synaptic potential signals from processes of selectively stained single neurons (Fig. 3.2). As predicted, the S/N was substantially improved relative to prior absorption measurements, but it was still too low to be of practical value—the available sensitivity was insufficient for multiple site optical recording of complex electrical interactions at the level of thin neuronal processes. Due to the low sensitivity of the measurements, elaborate correction procedures and extensive temporal averaging were necessary to improve the S/N. Additionally, spatial resolution was sacrificed and the signals reported were recorded from one location with a single photodiode. Moreover, the signals originated from a relatively large region of neuropile containing many processes, albeit from a single identified neuron (Fig. 3.2c). Despite low sensitivity, these experiments clearly showed the advantages of selective staining of individual neurons by intracellular application of the dye. The limits to the sensitivity of these measurements were determined primarily by: (a) the relatively low voltage sensitivity of the available dyes (amino-phenyl styryl dyes, e.g. RH437 and RH461; fractional change in fluorescence intensity per AP, in intracellular application, of the order of 0.01–0.1 %); (b) the choice of the excitation light bandwidth, and (c) the relatively low intensity of the incident light that could be obtained from a 100 W mercury arc lamp.

At the time, the reported sensitivity was the result of a modest screening effort suggesting that better signals might be obtained by: (a) synthesizing and screening new molecules for higher sensitivity; (b) increasing the concentration of the dye to increase the fluorescence intensity; (c) using an excitation light source capable of providing higher excitation light intensity and better stability and (d) using detector devices with lower dark-noise and adequate spatial and temporal resolution. Following this rationale, the first substantial improvement in the S/N was obtained by finding an intracellular voltage-sensitive dye with sensitivity in intracellular application two orders of magnitude higher than what was previously available. The amino-naphthalene styryl dye JPW1114 synthesized by J. P. Wuskell and L. M. Loew at the University of Connecticut Health Center had the sensitivity expressed as ΔF/F per AP, in intracellular application, of the order of 10 %

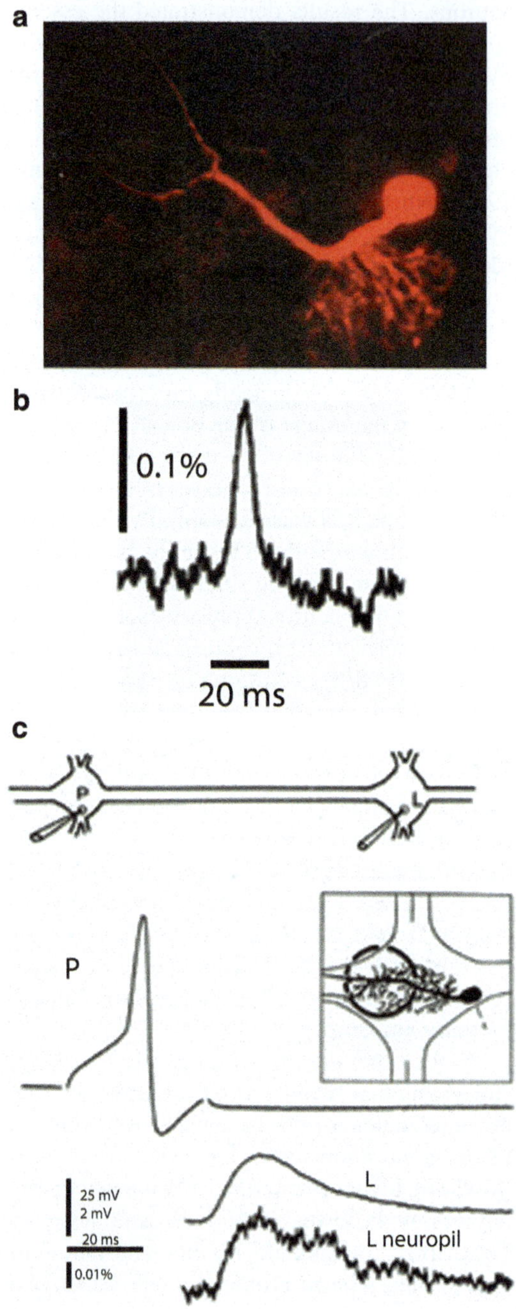

Fig. 3.2 Single photodiode optical recording from a neuron selectively stained by intracellular application of a voltage-sensitive dye. (**a**) Leech motoneuron following a 20 min iontophoretic injection with the styryl dye, RH-461. The soma, main process and its arborization within the neuropile are all clearly visible. (**b**) Optical recording of the AP in 175 μm long axonal segment of a P-sensory neuron in the anterior root. 12 trials were averaged. (**c**) Optical recording of the excitatory postsynaptic potential in the arborization of the L-motorneuron. The experimental arrangement for interganglionic stimulation and recording shown on *top*. Microelectrode recordings of the presynaptic AP in the P sensory neuron and the postsynaptic potential in the soma of the L motor neuron are shown. The region of the neuropile that was monitored is shown in the *inset*. The optical recording of the post synaptic potential from the neuropile in the adjacent ganglion is shown at the *bottom*. Adapted from Grinvald et al. (1987)

(Antić and Zecević 1995). The higher sensitivity of this dye translates directly into higher S/N. At this stage, the sensitivity was also improved by using a more powerful excitation light source (Osram, XBO 250 W OFR) powered by a low-noise power supply (Model 1700; Opti-Quip, Highland Mills, NY). The dramatic improvement in sensitivity and the use of 12 × 12 or 24 × 24 element photodiode detector arrays (Centronix, Inc., Newbury Park, CA) allowed simultaneous multiple site monitoring of electrical signals from processes of invertebrate neurons in isolated ganglia (Antic et al. 2000; Antić and Zecević 1995; Zecević 1996). Figure 3.3 shows a diagram of the optical arrangement for epi-fluorescence measurement and a typical multiple site recording from axo-dendritic processes of an individual neuron from the land snail *Helix aspersa*. This type of recording permitted the analysis of the pattern of initiation and propagation of spikes in neuronal processes and provided direct information about the location and the number of spike trigger zones in a particular nerve cell. Also, it was possible to monitor directly spike propagation failure at axonal branch points (Antic et al. 2000). In all experiments on invertebrate neurons, individual nerve cells in isolated ganglia were observed using wide-field transmission microscopy and selectively stained intracellularly with the voltage-sensitive dye by iontophoresis from beveled sharp electrodes (Obaid et al. 1982; Grinvald et al. 1987; Antić and Zecević 1995).

The substantial improvement in the sensitivity of optical recording was accompanied by new demands for increased spatial and temporal resolution. Increased resolution, in turn, tended to diminish the S/N. The invertebrate neuron studies described above required spatial and temporal resolution that could only be achieved by using substantial temporal averaging (from 4 to 100 trials) to improve the S/N. In addition, the spatial resolution was limited by the available S/N to about 20–50 μm long sections of axo-dendritic processes. Clearly, further improvements were needed in order to reduce the temporal averaging and increase spatial resolution. The limits to the sensitivity in these measurements were determined primarily by: (a) the sensitivity of the available voltage dyes; (b) the choice of the excitation bandwidth; (c) the maximum excitation light intensity that could be obtained from a 250 W Xenon arc-lamp; and (d) the dark noise of the photodiode array which was the dominant noise at the fluorescence intensities recorded from individual neurons stained by intracellular application of the dye.

3.2.2 Vertebrate Brain Slices

It was of considerable interest to apply V_m-imaging techniques to the dendrites of vertebrate CNS neurons in brain slices. This possibility was first demonstrated with work on layer V neocortical pyramidal neurons (Antic et al. 1999) that established a basic protocol for intracellular labeling of individual nerve cells in slices and described the method for simultaneous optical recording of electrical signals from multiple sites on apical, oblique, and basal dendrites. In these experiments, individual pyramidal neurons in slices were observed using infrared dark-field video microscopy and stained selectively by intracellular application

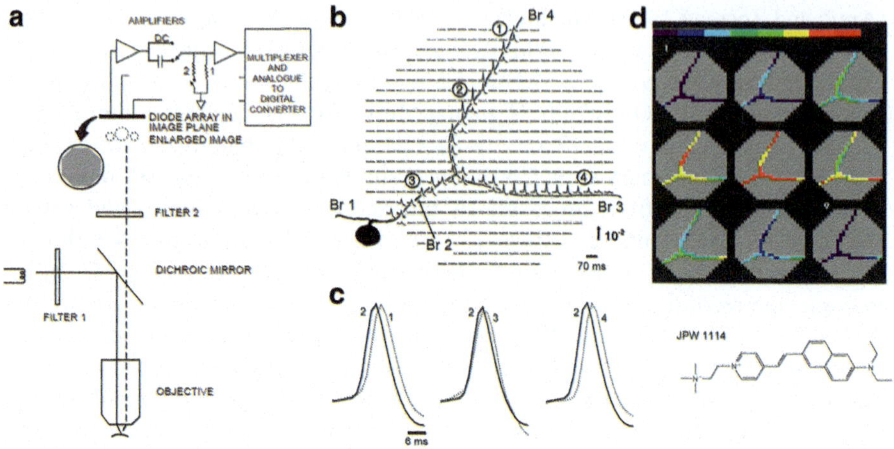

Fig. 3.3 Multiple site optical recording with a photodiode array from a neuron selectively stained by intracellular application of a voltage-sensitive dye. (**a**) Schematic drawing of the optical apparatus for multiple site epi-fluorescence measurements. The photodiode array was placed at the plane where the objective makes a real, magnified and inverted image. Epi-illumination was provided via a 10×, 0.4 NA, long working distance objective. The preparation was illuminated with the green portion (interference filter 520 ± 45 nm) of the output of a 100 W xenon arc lamp. Fluorescence emission from the preparation was selected by means of a dichroic mirror and an appropriate barrier filter (Schott RG610). The output of each detector was amplified individually, filtered, multiplexed, digitized and stored in a computer also used to analyze and display the data. (**b**) Optical recordings of AP signals from elements of a photodiode array positioned over the image of the axonal arborization of a *Helix* metacerebral cell from the left cerebral ganglion. The neuron was injected with the voltage-sensitive dye JPW1114 (structural formula shown in *lower right panel*). Spikes were evoked by transmembrane current pulses delivered through the recording microelectrode in the soma. The cell was depolarized from the resting V_m of −64 mV. Each optical trace represents 70 ms of recording centered on the peak of the spike. The recordings are shown divided by the resting light intensity. The *arrow* indicates the direction and the stated value of the relative fluorescence change (ΔF/F). Each diode received light from a 50 × 50 μm area in the object plane. A total of 90 trials were averaged to improve the S/N from distal processes. Optical signals were found in the regions of the array that corresponded closely to the geometry of the cell. Data from the detectors that do not receive light from the cell were deleted to improve clarity and allow color-coded display, shown in (**d**), based on a relative scale applied separately to each trace. (**c**) AP signals from four different locations indicated in (**b**) scaled to the same height, are compared on an expanded time scale to determine the site of the origin of the AP and the direction of propagation. (**d**) Color-coded representation of the data shown in (**b**) indicating the size and location of the primary spike trigger zone and the pattern of spike propagation. Consecutive frames represent data points that are 1.6 ms apart. Color scale is in relative units with the peak of the AP shown in *red*. Adapted from Zecević (1996)

of the voltage-sensitive dye by diffusion from a patch electrode in a whole-cell configuration. The major problem in injecting vertebrate neurons from patch pipettes was leakage of the dye from the electrode into the extracellular medium before the electrode is attached to the neuron. Patching requires pressure to be applied to the electrode during electrode positioning and micromanipulation through the tissue. This pressure ejects solution from the electrode. To avoid extracellular deposition of the dye that binds to the slice and produces large background fluorescence, the tip of the electrode was filled with dye-free solution, and the electrode was backfilled with dye solution. The amount of dye-free solution in the electrode tip and the applied pressure has to be adjusted empirically to ensure that no dye leaks from the electrode before the seal is formed. Usually, no pressure is applied before the electrode enters the slice. Low pressure (10–20 mBar) is used during electrode positioning in the slice, and the final pressure of 50–100 mBar, necessary for cleaning the surface membrane of the cell, is applied immediately (1 s) before the seal formation. Using this method, it was possible to load neurons routinely without increasing the background fluorescence of the surrounding tissue. At the end of a ~1 h diffusion period, the patch pipette was detached and the injected dye was allowed to spread for an additional 2 h before the start of optical recording. Within this time, dye would reach the apical tuft branches in layer I, ~1 mm from the soma. The results of these experiments demonstrated: (a) that loading vertebrate neurons with a voltage-sensitive dye using patch electrodes was possible without contamination of the extracellular environment; (b) that brain slices did not show significant autofluorescence at the excitation/emission wavelengths used (in contrast with the experience of Grinvald et al. (1987)); (c) that pharmacological effects of the dye were completely reversible; (d) that the level of photodynamic damage was low enough to permit meaningful measurements and could be reduced further; and (e) that the sensitivity of optical recording with a 464 element photodiode array was comparable to that reported for invertebrate neurons. At this level of sensitivity, however, modest temporal averaging in recording AP signals was still required, and spatial resolution was limited by the available S/N to approximately 20 μm long sections of dendritic processes (Antic et al. 1999). The limits to the S/N in these initial measurements from vertebrate neurons were determined by the same factors as described above for invertebrate preparations.

3.2.3 Improvement in Sensitivity with CCD Cameras

The second significant improvement in recording sensitivity in measurements from individual nerve cells (after the synthesis of a dye far more suitable for intracellular application) was the introduction of a cooled, back-illuminated CCD camera in place of the 464-element diode array. Fluorescence measurements from individual neurons have a lower range of intensities (with correspondingly lower absolute amplitude of the shot noise) so that the dominant noise, in recordings with a 464-diode array, was the relatively large dark noise of the amplifier-photodiode circuit. Thus, the dark noise of the photodiode array limited the detection system sensitivity, rendering it significantly lower than the theoretically attainable shot noise limit. It was then realized that in this range of illumination intensities (from 5×10^3 to 5×10^6 photons/ms) the CCD camera can approach the ideal (shot noise limited) sensitivity (Wu et al. 1999) (see also Chap. 1). The back illuminated CCD camera, because it is cooled and can have about 1,000 times smaller pixel surface area and proportionally smaller dark currents compared to available diode arrays, exhibits substantially lower dark noise. Test measurements performed at low light level (Wu, Cohen and Zecevic, unpublished observations), established that a particular cooled CCD camera (FastOne, Pixel Vision Inc., Beaverton, OR) improved the sensitivity by a factor of approximately 20 (Fig. 3.4a). The upper trace in Fig. 3.4a shows the recording of a step change in light intensity adjusted in amplitude to be at the limit of resolution (S/N = 1) for the photodiode array.

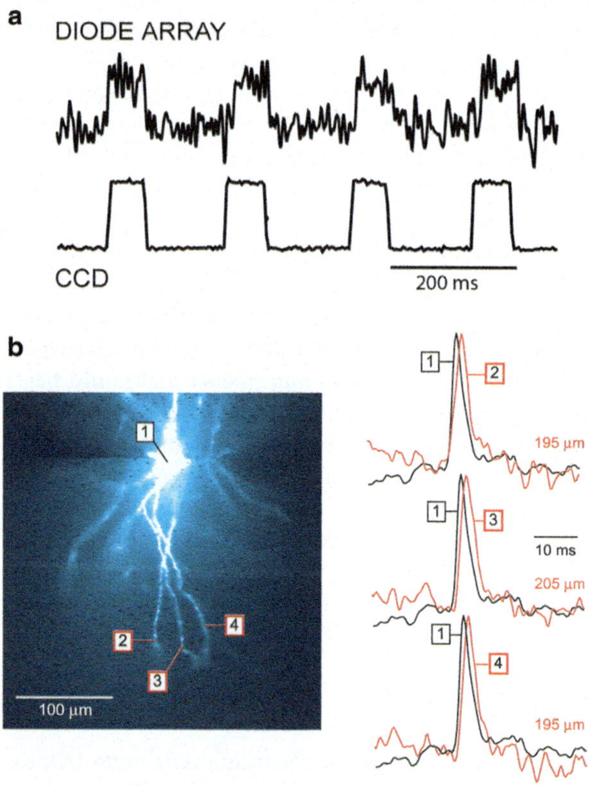

Fig. 3.4 Optical detectors for voltage-sensitive dye recordings. (**a**) The performance of a 464-element photodiode array and a cooled back-illuminated CCD camera are compared directly using the same preparation. The measurements were done at a relatively low resting light level, comparable to the resting fluorescence intensity in recordings from processes of selectively stained individual neurons. Upper trace: recording of a step change in light intensity adjusted in amplitude to be at the limit of resolution (S/N $\approx$ 1) for the photodiode array. *Lower trace*: the recording of the same signals using the high-speed, cooled CCD camera. The improvement in the S/N ratio by a factor of approximately 20 is attributable to the large reduction in dark noise. (**b**) AP signals in distal segments of basal dendrites of a pyramidal neuron obtained with a cooled CCD camera. A composite fluorescence image of a layer V pyramidal neuron in recording position shown on the *left*. Optical signals from distal dendritic segments at distances larger than 180 mm, as indicated, shown on the *right*. Each trace is the product of temporal (four trials) and spatial (6–9 pixels) averaging. Adapted from Antic (2003)

The lower trace is the recording of the same steps using the high-speed, cooled CCD camera. This result is a demonstration of the remarkable improvement in S/N resulting from the large reduction in dark noise.

The increase in sensitivity with the introduction of advanced CCD cameras allowed, for the first time, recording of the spatial and temporal dynamics of electrical events in thin, distal dendritic processes that cannot be probed with conventional patch-electrode techniques. The optical approach made possible simultaneous measurements of the V_m transients in basal and oblique dendrites of pyramidal neurons during single APs and trains of APs (Fig. 3.4b; (Antic 2003)), as well as the analysis of synaptic potential initiation and spread in terminal dendrites of the glomerular tuft of individual mitral cells in the olfactory bulb slice (Djurisic et al. 2004).

The optical measurements from basal dendrites (Antic 2003) provided unique evidence that, in contrast to apical dendrites, basal and oblique dendritic processes impose modest amplitude and time course modulation on the backpropagating AP (bAP) and are robustly invaded by the somatic spike even when somatic firing rates reach 40 Hz (Fig. 3.4b). A compartmental model incorporating AP peak latencies and half-widths obtained from optical measurements indicated that the specific intracellular resistance (R_i) is less than 100 Ω cm. The data obtained simultaneously from multiple dendritic sites and subsequently linked to a numerical simulation, revealed that all synaptic locations along basal and oblique dendrites, situated within 200 μm of the soma, experience strong and near-simultaneous (latency <1 ms) voltage transients during somatic firing (Antic 2003; Zhou et al. 2008). The continuation of these studies (Acker and Antic 2009) demonstrated that dendritic multi-site voltage sensitive dye recordings can be combined with pharmacological manipulations of membrane excitability to test and optimize multicompartmental numerical simulations. For example, the bAPs in basal dendrites have been monitored before and after blocking the voltage gated sodium channels with tetrodotoxin (TTX), or blocking A-type potassium channels with 4-Aminopyridine, in order to determine the contribution of each membrane conductance to the dendritic AP signal (Acker and Antic 2009).

Experiments on mitral cells (Djurisic et al. 2004) showed that optical data can be used to measure the amplitude and shape of subthreshold excitatory postsynaptic potentials (EPSPs) evoked by olfactory nerve stimulation at the site of origin (glomerular tuft) and to determine its attenuation along the entire length of the primary dendrite. In addition, direct evidence was provided for the number, location, and stability of spike trigger zones, the excitability of terminal dendritic branches, and the pattern and nature of spike initiation and propagation in the primary and secondary dendrites. In a subsequent study (Djurisic et al. 2008), V_m-imaging was used to analyze the electrical properties and the functional organization of the terminal dendritic tuft, which was not accessible to standard electrode recordings (for single site patch-electrode recordings from small dendritic branches see (Nevian et al. 2007; Larkum et al. 2009)). Optical recording provided direct evidence that the dendritic tuft functions as a single electrical compartment for subthreshold signals within the range of amplitudes detectable by voltage-sensitive dye recording.

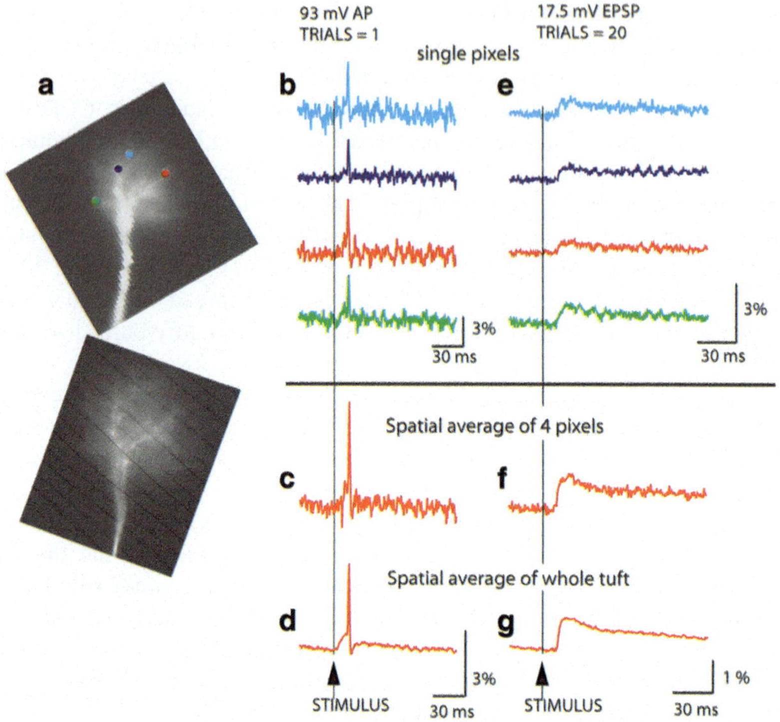

Fig. 3.5 Sensitivity of voltage imaging with low noise CCD camera in recordings from the apical dendritic tuft of a mitral cell of the olfactory bulb. (**a**) Low resolution (*top*) and high resolution (*bottom*) fluorescence images of the terminal dendritic tuft in recording position. (**b**) Recordings of AP signals from four individual pixels receiving light from four 3.7 × 3.7 μm locations on the dendritic tuft. (**c**) The S/N ratio with spatial averaging of four pixels marked in (**b**). (**d**) The S/N ratio with spatial averaging of the entire tuft area. (**e**) An evoked EPSP recorded from four individual locations on the dendritic tuft; 20 trials averaged. (**f**) Spatial average of signals shown in (**e**). (**g**) Spatial average of the entire tuft area. Adapted from Djurisic et al. (2004)

Figure 3.5 illustrates the sensitivity of multiple site optical recordings of AP related signals and subthreshold, EPSP signals from mitral cell tuft dendrites in wide-field epi-fluorescence mode. With this approach, it was possible to record AP signals at a frame rate of 2 kHz in single trial recordings from ~4 μm long sections of terminal dendritic branches less than 1 μm in diameter. As shown in the figure, the modest S/N obtained in single-trial, single-pixel measurements could be further improved by spatial and temporal averaging. A representative optical recording of EPSPs from the site of origin on glomerular dendritic branches is shown in Fig. 3.6. These experiments demonstrated that, at the spatial scale of dendritic processes, 3–5 mV synaptic potentials can be resolved optically in single-trial recordings. The relatively weak S/N in measurements of small, subthreshold V_m transients could be improved by modest temporal averaging (4–9 trials).

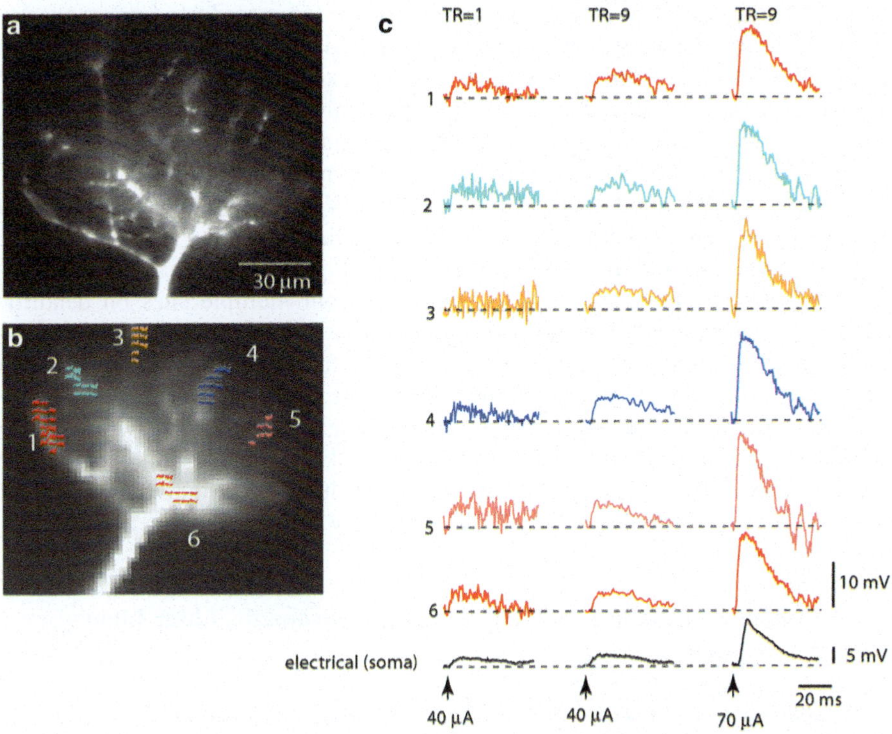

Fig. 3.6 Spatial distribution of subthreshold EPSP signals is uniform in thin terminal dendritic branches of a mitral cell dendritic tuft. (**a**, **b**) Fluorescence image of a mitral cell tuft obtained with a conventional high-resolution CCD (**a**) and with a fast, low-resolution (80 × 80 pixels) CCD used for voltage imaging (**b**). Six color-coded recording locations indicated on the image in (**b**) correspond to traces shown in (**c**). (**c**) Trace display of V_m signals. Traces are spatial averages of colored pixels. *Bottom black traces*, electrical recordings from soma. *Left traces*: Single-trial recordings of the small EPSP signal are at the limit of amplitude resolution. *Middle traces:* Nine averages of the small EPSP signals shown on the *left* improved the S/N ratio. *Right traces:* Nine averages of the EPSP four times larger in amplitude. Adapted from Djurisic et al. (2008)

The same methodology based on the use of high-speed, low read noise CCD camera (FastOne, Pixel Vision Inc., Beaverton, Oregon or NeuroCCD-SM, RedShirtImaging LLC, Decatur, Georgia), has been applied to the analysis of dendritic signals from rat hippocampal CA1 pyramidal neurons during coincident pre- and post-synaptic activity using combined voltage- and calcium-imaging from neurons labeled with two indicators (see Chap. 4 and (Canepari et al. 2007)). The combined imaging technique was also applied to the analysis of dendritic V_m and Ca^{2+} signals associated with parallel fiber and climbing fiber stimulation in cerebellar Purkinje neurons (Chap. 4 and (Canepari and Vogt 2008)), as well as to the study of the correlation between dendritic electrical transients (plateau potentials) and dendritic calcium signals during suprathreshold glutamatergic synaptic input (Milojkovic et al. 2007). In addition, using exactly the same recording apparatus and protocols, similar spatial resolution was reported in recordings from axons and basal dendrites of layer 5 pyramidal neurons (Palmer and Stuart 2006; Kampa et al. 2006). However, in these studies, a significantly lower range of excitation light intensities was utilized, reducing the S/N significantly. The loss in sensitivity was compensated by extensive temporal averaging, which is only possible at reduced light intensity because of reduced photodynamic damage.

The studies described above documented that V_m-imaging is quite efficient at the spatial scale of dendritic branches. Clearly, the sensitivity was adequate to monitor suprathreshold, regenerative AP signals (Figs. 3.3, 3.4, and 3.5) as well as subthreshold, synaptic input signals (Fig. 3.6) from multiple sites on dendritic processes. The available S/N, however, was not adequate to allow multisite recordings from thin axons and axon collaterals at high frame rates (5–10 kHz) without extensive averaging (>100 trials). Such high frame rates are required for monitoring fast axonal APs at physiological temperature. In addition, it was still not possible to monitor V_m signals at the higher optical magnification and finer spatial scale necessary for resolving electrical events at the level of individual dendritic spines. Clearly, further improvements in sensitivity were needed.

3.2.4 Improvements in Sensitivity with Monochromatic, Laser Light Excitation

A possible approach toward achieving high speed multiple site optical recordings of membrane voltage from axons and axonal collaterals as well as from individual dendritic spines in brain slices is wide-field epi-fluorescence microscopy applied at high optical magnification. Several alternative optical approaches to V_m-imaging are also available; these will be discussed in Sect. 3.2.5 below.

The limited sensitivity of available epi-fluorescence V_m-imaging methods required extensive averaging (>100 trials) in recording fast axonal AP signals because these recordings necessitate high frame rates (10 kHz) (Palmer and Stuart 2006). Similar low sensitivity is obtained in recording signals from individual dendritic spines because an order of magnitude higher optical magnification is required to resolve spines (Palmer and Stuart 2009). Because extensive averaging (>100 trials) makes most physiologically relevant experiments very difficult or impossible, a substantial improvement, by a factor of 10–100, in the sensitivity was desirable.

Within the constraints imposed by Eq. (3.1), one way to increase the S/N with a given voltage-sensitive dye is to increase the photon flux (Φ) by increasing the excitation light intensity. Another possibility is to increase the fractional fluorescence change per unit change in V_m (sensitivity of the dye) by choosing the optimal excitation wavelength. Following this rationale, Holthoff et al. (2010) used one of the most sensitive voltage-sensitive dyes in terms of S/N (JPW3028; (Djurisic et al. 2004; Zhou et al. 2008; Acker and Antic 2009)), and improved both the excitation (and, hence, emission) light intensity and the relative fluorescence change in response to V_m change by utilizing a laser as an illumination

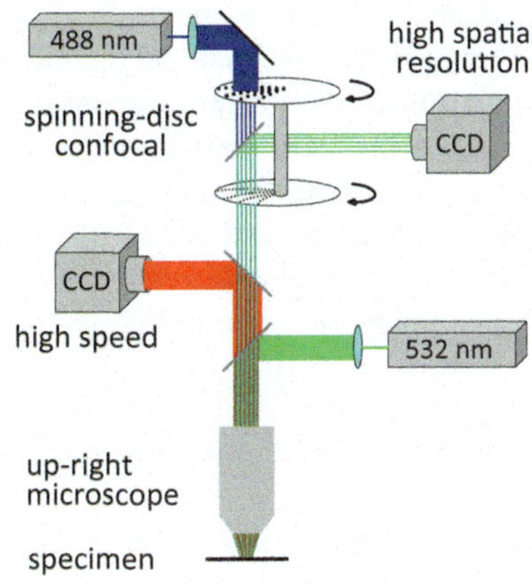

Fig. 3.7 Experimental setup for voltage-imaging from dendritic spines in cortical brain slices. The voltage-sensitive dye was excited using a 532 nm solid-state laser in wide-field illumination mode. Emission light was recorded with a high-speed CCD-camera. A spinning-disc confocal scanner with 488 nm laser was utilized for morphological reconstruction. Adapted from Holthoff et al. (2010)

source in wide-field epi-fluorescence microscopy mode (Holthoff et al. 2010). In measurements from layer 5 pyramidal neurons in rat visual cortex slices, the light from a frequency-doubled 200 mW diode-pumped Nd:YVO_4 continuous wave (CW) laser emitting at 532 nm (Excelsior 532 single mode; Newport-Spectra-Physics, Mountain View, CA) was directed to a quartz optical fiber (TILL Photonics GmbH, Gräfelfing, Germany) coupled to the microscope via a single-port epi-fluorescence condenser (TILL Photonics) designed to overfill the back aperture of the objective. In this way, approximately uniform illumination of the object plane was attained. Figure 3.7 depicts the experimental set-up schematically. A stationary stage upright microscope (Model BX51WI Olympus Inc, Japan) equipped with three camera ports was used in these experiments. One camera port had a standard, high spatial resolution CCD camera for infrared DIC video-microscopy (CCD-300-RC, Dage-MTI, Michigan City, IN). The second camera port had a fast data acquisition camera (up to 20,000 frames/s) with relatively low spatial resolution (80×80 pixels) but outstanding dynamic range (14 bits) and exceptionally low read noise (NeuroCCD-SM, RedShirtImaging LLC, Decatur, GA). The third camera port had a CCD camera with high spatial resolution ($1{,}392 \times 1{,}024$ pixels; Pixelfly-qe, PCO Imaging, Kelheim, Germany) mounted on a Yokogawa spinning-disc confocal scanner used to collect z-stacks of confocal images for detailed morphological reconstruction of the stained cell. The fractional amplitude noise in low-noise solid-state lasers (<0.2 %) did not interfere with recording sensitivity because it was below the typical fractional shot-noise in fluorescence voltage-sensitive dye recordings (Iwasato et al. 2000; Matsukawa et al. 2003; Zhou et al. 2008). The excitation light was reflected to the preparation by a dichroic mirror having a central wavelength of 560 nm, and the fluorescence emission was passed through a 600 nm barrier filter (a Schott RG600). Imaging was performed with a low-noise high-speed CCD camera.

With this apparatus, a series of control experiments were designed to test three critical methodological parameters: (a) improvement in the S/N; (b) the extent of photodynamic damage at the required excitation light intensity; (c) the limits to the spatial resolution as determined by light scattering in wide-field epi-fluorescence measurements. Robust and unambiguously favorable results were obtained regarding all three parameters (see below).Following these initial measurements, further experiments were carried out to characterize bAP signals in dendrites and spines.

As described above (Fig. 3.7) imaging was performed with a high-speed CCD camera. Neurons with easily discernible dendrites were loaded from a patch-pipette with the voltage-sensitive dye JPW3028. Figure 3.8b shows a composite fluorescence image of a pyramidal neuron labelled with the voltage-sensitive dye.

Improved V_m-Imaging from Dendritic Branches

In the initial series of experiments, the extent of propagation and the time course of bAP signals were monitored from dendrites at relatively low magnification (the full CCD frame (80 × 80 pixels) corresponded to a 300 × 300 μm area in the object plane). At this magnification the dendritic spines could not be resolved even in principle because their image was smaller than individual pixels. Before optical measurements, the dye-loaded pyramidal neurons were re-patched with a dye-free pipette (Fig. 3.8b). Spikes were elicited by depolarizing current pulses delivered from a patch electrode in the soma and bAPs recorded optically from various dendritic compartments covering the entire dendritic tree (Fig. 3.8a–c). The recordings from different areas were obtained by sequentially repositioning the field of view. As indicated in the figure, at all dendritic sites, including the terminal apical tuft (Fig. 3.8a) as well as the oblique (Fig. 3.8c) and the basal dendrites (Fig. 3.8b), bAPs could be monitored with unprecedented sensitivity by the voltage probe JPW3028 at the laser excitation wavelength of 532 nm in single-trial measurements. The combined effect of an increase in light intensity and the use of a near optimal excitation wavelength was a dramatic improvement in the sensitivity of voltage imaging by a factor that varied from 10 to 50 in recording from different sites on neuronal processes. The single-trial signals (no averaging) from some of the individual pixels (Fig. 3.8c, location 5) yielded fractional fluorescence changes of about 60 % recorded with the S/N of approximately 50. The sensitivity of these measurements is the largest ever obtained from individual nerve cells in slices. Clearly, the amplitude of ΔF/F signals corresponding to the back-propagating spikes varied widely at different locations. This variability corroborates the earlier findings (Djurisic et al. 2004; Canepari et al. 2007) that the sensitivity of the optical measurements from different dendritic regions is not uniform. The sensitivity varied because the ratio of the dye bound to the plasma membrane that changes potential to the dye bound to intracellular membranes varies between different regions on the dendritic arbor as a function of many factors that cannot be determined accurately, including the surface to volume ratio and the amount of intracellular membranes (see Chap. 4 and (Canepari et al. 2008) for more detailed explanation on dye signal calibration). A small number of dendritic regions exhibited extremely large fractional changes (e.g. location 5 in Fig. 3.8c). This result suggest that the intrinsic voltage sensitivity of the dye is very high closely approaching the theoretical sensitivity limit for charge-shift probes excited by a

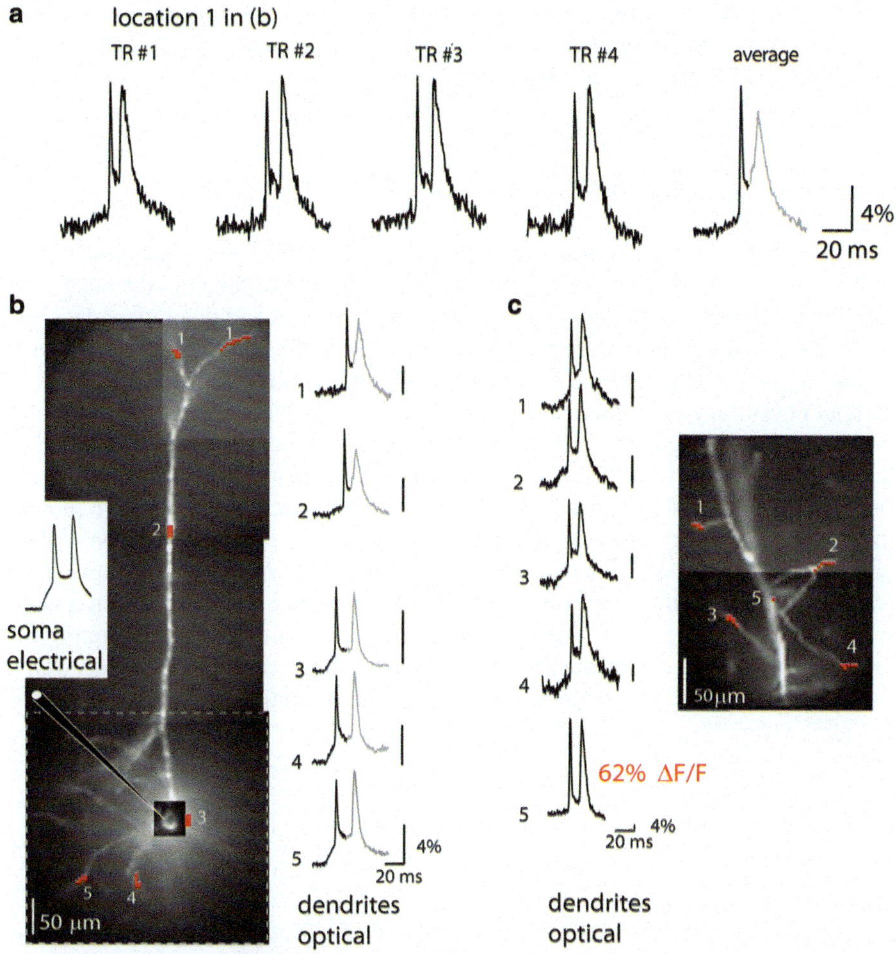

Fig. 3.8 Dendritic bAP signals in apical, basal and oblique dendrites. (**a**) The bAP signals from the apical terminal tuft 700 μm from the soma (spatial average from location 1 in **b**). Single-trial recordings and a temporal average of four trials illustrate the sensitivity of recording (S/N ~ 15 and 30 respectively). (**b**) A composite fluorescence image of a pyramidal neuron. The size of the full frame imaging region is outlined by dashed-line rectangle. Traces on the right are temporal averages of four trials and spatial averages from locations 1 to 5 indicated on the image. The peak of the first bAP was used as a reference point for averaging. Consequently, the second spike shown in *gray* did not average coherently because of temporal jitter. The electrode recording from the soma is on the *left*. (**c**) A composite fluorescence image of a dendritic region from another neuron with several oblique dendrites in focus for voltage-imaging. Single-trial recordings on the *left* are spatial averages from locations 1 to 4 indicated on the image. Adapted from Holthoff et al. (2010)

wavelength close to the extreme red edge of the absorption spectrum (Kuhn et al. 2004). The simplest explanation for the large optical signals recorded from a small subset of dendritic regions is that the fluorescence from these sites was largely dominated by the dye bound to the plasma membrane, possibly because of a favorable surface-to-volume ratio and the absence of internal membranes at these locations. Apart from these sites, the optical signals related to bAP from most of dendritic regions were in the range of 10–20 %.

Improved V_m-Imaging from Axons

Another series of experiments was carried out at the same optical magnification to utilize the improved sensitivity of the voltage-imaging technique in the analysis of electrical events in individual axons. Recent experimental and theoretical data (Shu et al. 2007; Bennett and Muschol 2009) indicate that the functional capabilities of axons are much more diverse than traditionally thought. Consequently, the field of axonal physiology is rapidly expanding. At the same time, the traditional tools for analyzing axonal signaling were limited to single site electrode recordings from the cut end of axons. Clearly, new tools to gain a better understanding of signal integration in axonal processes are critical. The first successful attempt to monitor optically AP signals from multiple sites on the axon of a pyramidal neuron (Palmer and Stuart 2006) clearly demonstrated advantages of V_m-imaging by providing direct evidence on the approximate location of the spike initiation site in a vertebrate nerve cell (Fig. 3.9). The sensitivity of these recordings was limited by the amount of light that can be obtained from a 100 W tungsten-halogen bulb as well as by the choice of less favorable wavelengths of the excitation light. Thus, extensive

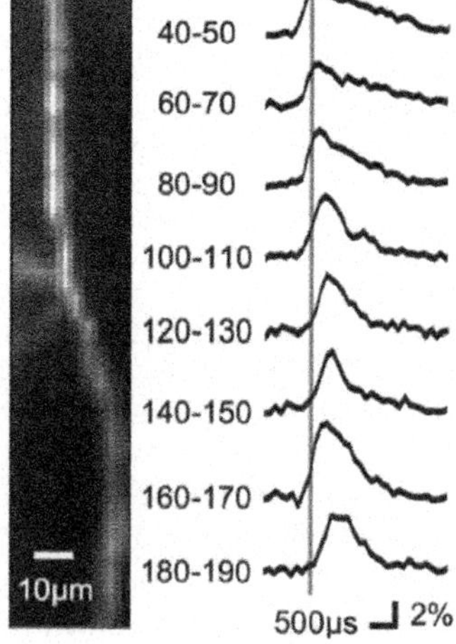

Fig. 3.9 High-speed recording of AP signals from axons of a cortical pyramidal neuron. High-magnification image of the axon of a layer 5 neocortical pyramidal neuron filled with voltage-sensitive dye. Average fluorescence change of 130 individually aligned APs recorded from indicated axonal locations at a frame rate of 10 kHz. APs were evoked by somatic current injection

signal averaging (>100 trials) was required to obtain useful information. This amount of averaging makes experiments difficult and prone to errors. A significant improvement in sensitivity brought about by laser excitation facilitated further studies. By using those improvements in imaging technique that provided exceptional temporal resolution, it was possible to analyze the individual Purkinje neurons and resolve the region of spike initiation as well as follow spike propagation into axon collaterals for each action potential initiated in single trials (Fig. 3.10). The results of this study demonstrated that the axon initial segment is a critical decision point in Purkinje cell processing and that the properties of axon branch points are adjusted to maintain faithful transmission (Foust et al. 2010). In another study focused on layer V pyramidal neurons (Popovic et al. 2011), the authors utilized the same refinements of dye-based measurements to elucidate the behavior of cortical pyramidal cell axons and their small, branched collaterals. Figure 3.11 illustrates how multiple site optical recording was used to directly determine the location and length of the spike trigger zone (TZ) as defined in functional terms. The results show that in mature axons of mouse cortical layer 5 pyramidal cells, action potentials initiate in a region ~20 μm in length centered in the axon initial segment between 20 and 40 μm from the soma. From this region, the AP depolarizing wave invades initial nodes of Ranvier within a fraction of a millisecond and propagates in a saltatory fashion into axonal collaterals without failure at all physiologically relevant frequencies. Optical signal processing that was utilized to generate the kind of results described above is illustrated in Fig. 3.12. More detailed explanation of the limits to spatiotemporal resolution is given below.

How accurate are optical measurements of fast electrical events? The accuracy of these measurements is a function of the available sensitivity of V_m imaging which is, in turn, a function of both spatial and temporal resolution. Thus, these parameters have to be considered in every particular study. For example, it is known that the initiation of the AP in the sub-region of the axonal initial segment (AIS) and its propagation to the first node of Ranvier is accomplished in less than 100 μs in layer-5 pyramidal neurons (Palmer and Stuart 2006) and in Purkinje cells (Foust et al. 2010). It is also known from electrode measurements that accurate recording of the fastest frequency components of the axonal AP (maximum dV/dt of 1,100–1,200 V/s at 32–37 °C) requires sampling rates of 20–80 kHz. Further increase in the sampling rate gave similar results (Yu et al. 2008). While this range of sampling frequencies are difficult to realize in optical recordings because of limited sensitivity, it was possible to effectively apply a frame rate of 10 kHz (Foust et al. 2010; Popovic et al. 2011). These studies demonstrated that the filtering effect of insufficiently high sampling rate had a very small effect on determining soma–axon delays used to establish the location and length of the spike TZ. In other words, it was possible to assume that the AP waveform reconstructed by cubic spline interpolation from the data obtained at a sampling rate of 10 kHz is an adequate approximation of the true AP shape for the purpose of measuring soma-axon delays. Thus, the reconstructed waveform was treated as an analog signal that has a known value at all times and the continuous cubic spline interpolation curve, re-sampled at 100 kHz, was used to characterize timing information with a temporal resolution of 10 μs.

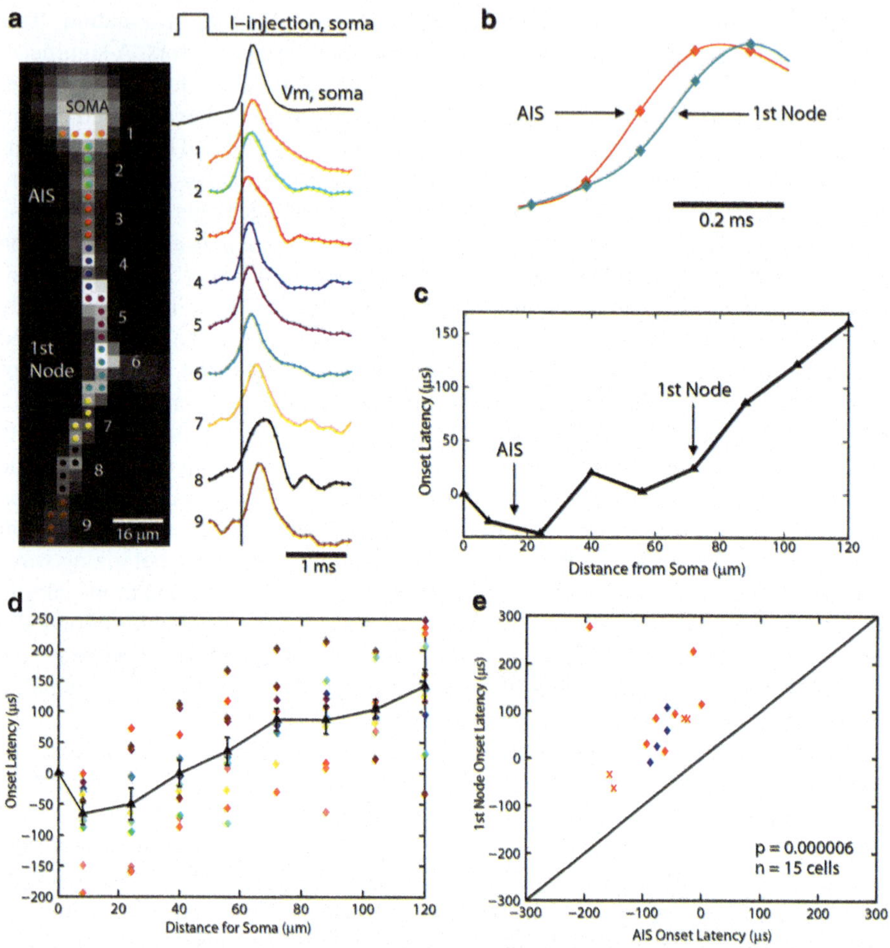

Fig. 3.10 APs initiate in the Purkinje cell axon initial segment. (**a**) Voltage-sensitive dye fluorescence image of the soma–axon region in recording position shown on the *left*. Traces on the *right*, Single-trial recordings of AP signals, initiated by somatic current injection, from nine locations indicated on the image. Each trace is a spatial average of four to five same-color pixels. Signals are scaled to the same peak height. The *symbols* represent actual data points; the *solid lines* are reconstructed signals using cubic spline interpolation. The *vertical line* denotes the time at half-maximum amplitude at the edge of the soma (location 1). Note that the AIS segment traces (*green* and *red*) initiate the action potential first. (**b**) AP recordings from the AIS and the first node of Ranvier scaled to the same height and compared on an expanded timescale. (**c**) The time to half-maximum amplitude subtracted from the time at half-max for the edge of the soma ("onset latency") as a function of distance from the soma. (**d**) Group data (n = 15) illustrating the level of consistency in the spatial localization of spike onset and the spike propagation velocity across cells. Error bars indicate SEM. (**e**) Group data statistics. The onset latency is lowest in the axon initial segment, and in all cells preceded the spike in the first node of Ranvier (p = 0.000006; n = 15). The *diagonal line* represents equity. The *diamonds* represent spikes initiated with a current pulse. The data obtained from spontaneous spikes are indicated by *crosses*. The data in *red* was obtained at 35 °C, whereas the data in *blue* was obtained at 23 °C

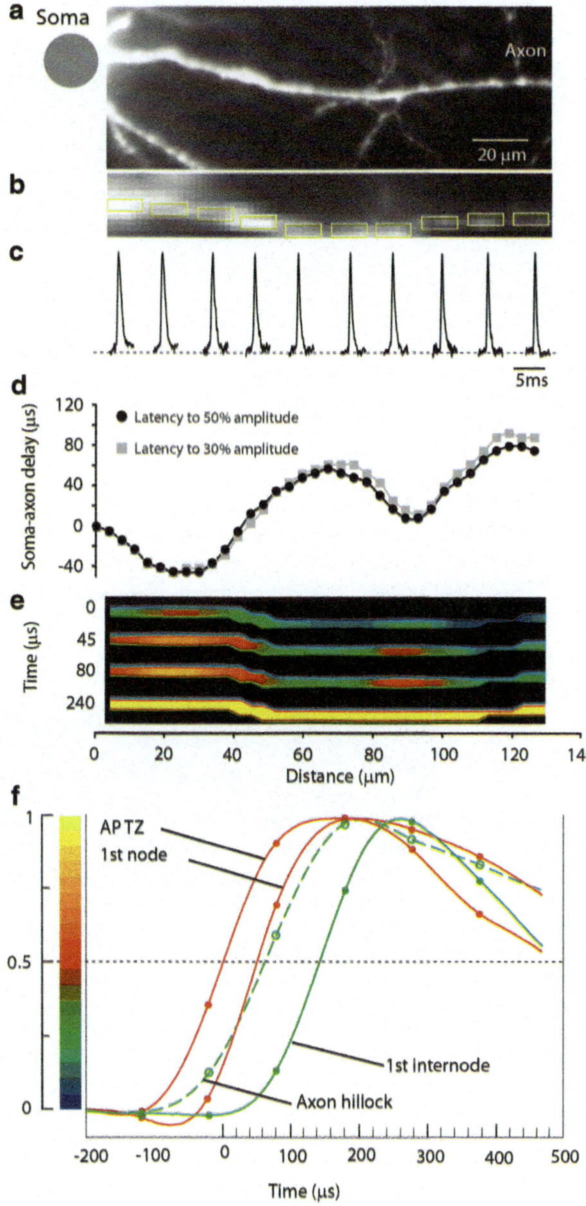

Fig. 3.11 Measurement of the spatial distribution of membrane potential as a function of time along the proximal axon during AP initiation. (**a**) High resolution confocal image of the axon in recording position. (**b**) Low spatial resolution fluorescence image of the axon obtained by CCD used for V_m imaging. (**c**) AP signals from ten locations indicated by *yellow rectangles*, each 10 μm in length. (**d**) Soma–axon latency to 30 % (*grey*) and 50 % (*black*) AP amplitude as a function of distance from the cell body. The first minimum identifies the location and length of the spike TZ. (**e**) Time sequence of frames showing spatial profile of color coded relative V_m amplitude in the axon at four characteristic time points: 0 μs—AP initiation at TZ; 45 and 80 μs—invasion of the first node; 240 μs—peak depolarization. (**f**) Comparison of AP signals from four characteristic locations on an expanded time scale. The measured data points and cubic spline interpolation curves are shown. *Red traces*—TZ and first node; *green dashed trace*—axon hillock; *green trace*—first internodal region. Membrane potential color scale shown on *left*

The necessary spatial resolution is dictated by the anticipated length of the spike TZ. The average length of the AIS has been determined to be ~40–50 μm (Palmer and Stuart 2006). The length of the spike TZ defined in functional terms has been unknown because there have been no direct recordings of this parameter. A plausible prediction for the length of the TZ based on structural studies as well as

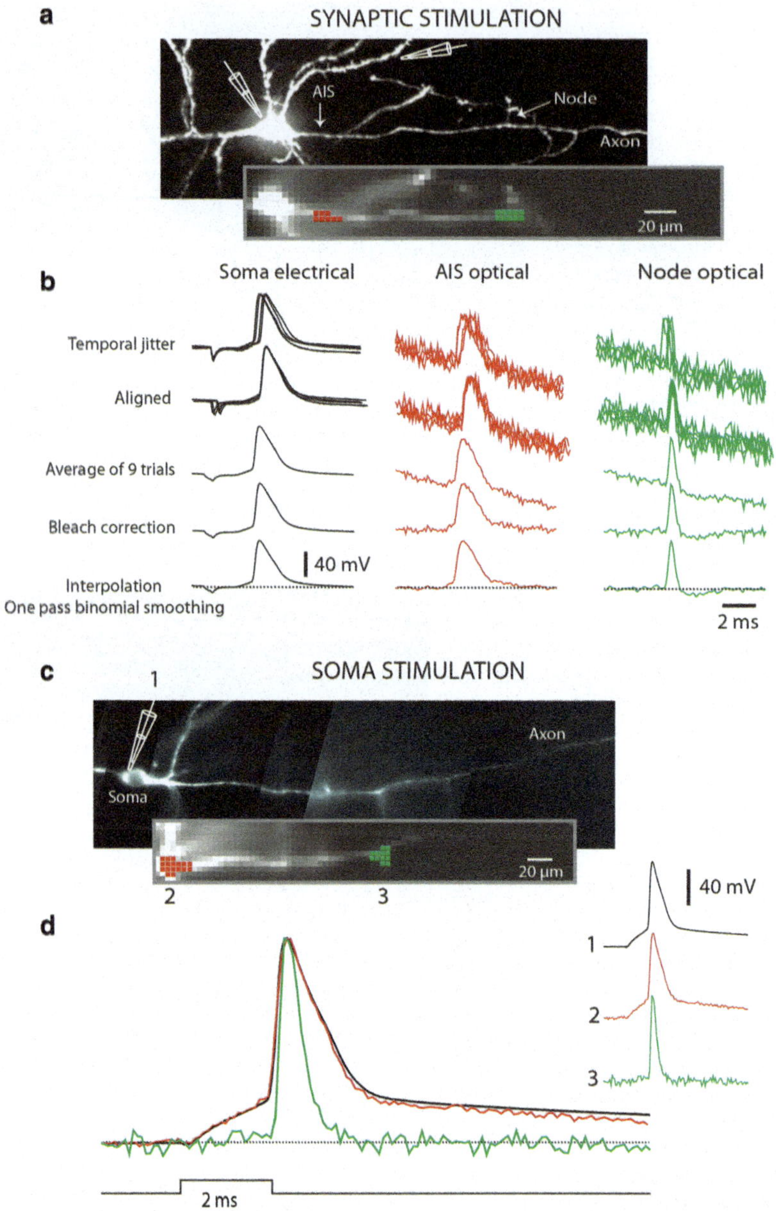

Fig. 3.12 Signal processing. (**a**) Synaptic stimulation. *Upper image*, high resolution confocal image of a stained neuron with axon in recording position. Recording electrode attached to soma and stimulating electrode next to basal dendrite shown schematically. *Lower image*, low spatial resolution fluorescence image of the axon obtained by CCD used for V_m imaging. (**b**) electrode recordings from soma, optical recordings from spike TZ (*red*), and from node of Ranvier (*green*). *Top traces*: row data from nine trials showing temporal jitter in AP initiation following synaptic activation. *Second row of traces*: temporally aligned signals. *Third row of traces*: averaged signal. *Fourth row of traces*: bleach correction. *Bottom traces*: cubic spline interpolation with one pass of

from modelling is in the range of 10–40 μm (Hu et al. 2009). In the experiment shown in Fig. 3.11 the optical magnification was adjusted so that 1 pixel of the CCD used for V_m imaging received light from a surface area of ~4 × 4 μm in the object plane. Next, one has to consider the sensitivity (amplitude resolution) of the optical measurement at the spatiotemporal resolution described above (10 μs; 4 μm). First, it is necessary to take into account relevant V_m changes during AP initiation that are likely to occur at the temporal scale on the order of 10 μs. The steepest slope (dV/dt) of the AP upstroke measured optically at 32–34 °C, was ~400–550 mV/ms in the soma and ~550–700 mV/ms in the axon (Popovic et al. 2011). The slope of the axonal AP upstroke signal indicates that the membrane potential will change by ~5–7 mV in 10 μs during a maximum rate of depolarization. Thus, a meaningful optical readout of V_m transients with a temporal resolution of 10 μs requires that the sensitivity of optical recording (expressed as the S/N) is sufficient to reliably resolve V_m changes of this magnitude. Figure 3.13 addresses this question. In the measurement shown in Fig. 3.13c, AP signals were recorded optically from an axonal region 4 μm in length. The trace on the left is a temporal average of optical data from four trials characterized by a S/N of 7. The middle trace is a temporal average of 16 trials characterized by an increase in sensitivity by a factor of 2 (S/N = 14). The noise in the signal can cause an error in the measurement of the time of the transition to a given signal amplitude. Figure 3.13c illustrates that this error was negligible. The slow noise of relatively small amplitude recorded at the resting membrane potential did not have a significant effect on amplitude measurement during the most rapidly changing phase of the signal. Thus, the measurements of AP initiation in the axon were limited to the spatial resolution of 4 μm, temporal resolution of 10 μs and amplitude resolution of 4 mV. It was necessary to employ modest temporal averaging (4–25 trials) to attain this sensitivity in different experiments, depending on the level of intracellular staining and on axon location relative to the surface of the slice. It is noteworthy that temporal averaging has strict limitations. Because the S/N increases with the square root of the number of trials averaged, it is possible to improve the sensitivity by a factor of 2–5 (by averaging 4–25 trials) without significant difficulties. Further improvement in the sensitivity, however by a factor of only 2, would require extensive averaging (>100 trials), which is not easily realizable because of the preparation run-down with time and because repetitive exposure to high-intensity excitation light results in photodynamic damage.

The same recording technology can be used to analyze the spatial pattern of AP propagation as revealed by monitoring transmembrane potential over longer

Fig. 3.12 (continued) temporal smoothing. (**c**) Somatic stimulation. *Upper image*, high resolution confocal image of another neuron with axon in recording position. *Lower image*, low spatial resolution fluorescence image of the axon obtained by CCD used for V_m imaging. *Traces on right*: AP transients from three locations: 1-electrode recording from soma; 2-optical recording from axon hillock; 3-optical recording from the first node of Ranvier. *Bottom traces*: superimposed signal from same three locations

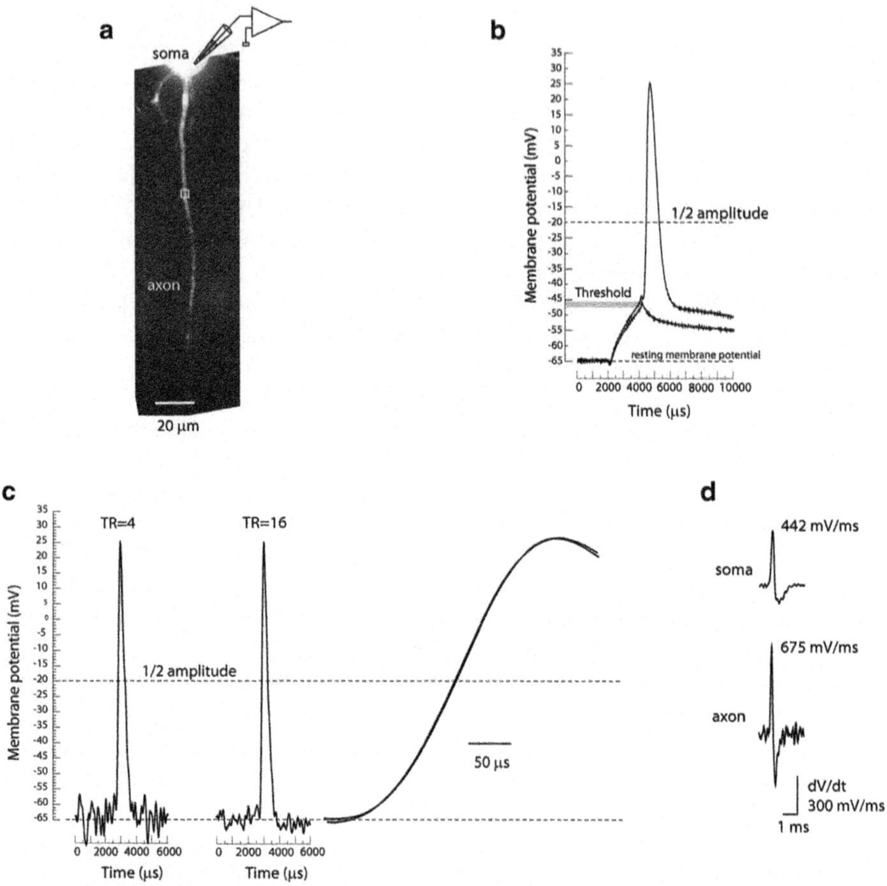

Fig. 3.13 Spatiotemporal resolution and recording sensitivity. (**a**) Fluorescence confocal image of a dye loaded pyramidal neuron with proximal axon in recording position. (**b**) The size and shape of the AP recorded from the soma of layer 5 pyramidal neuron at 34 °C. (**c**) Optical recording of AP related signal from axonal region 4 μm in length indicated by *white rectangle* in (**a**). *Left trace*, average of four trials. Middle trace, average of 16 trials. *Right trace*, superimposed upstroke of the two AP signals on expanded time scale. (**d**) First derivative (d*V*/d*t*) of the somatic and axonal AP

sections of individual myelinated axons. Previously, this information was not available for any neuron. A representative experiment is illustrated in Fig. 3.14. The spatial plot of the soma–axon latency along an axonal section of approximately 300 μm clearly identified the position of the spike TZ and putative nodes of Ranvier, all characterized by localized reduction in soma–axon latency typical for saltatory conduction (Fig. 3.14a). It is likely that the internodal regions, invaded by the depolarizing wave with a delay, were myelinated. The spatial plot of AP latency provides a functional readout for the position of the nodes of Ranvier. Figure 3.14b shows the initiation and the propagation of an AP as revealed by a time sequence of the colour-coded spatial maps of the depolarizing AP wave. The time sequence of

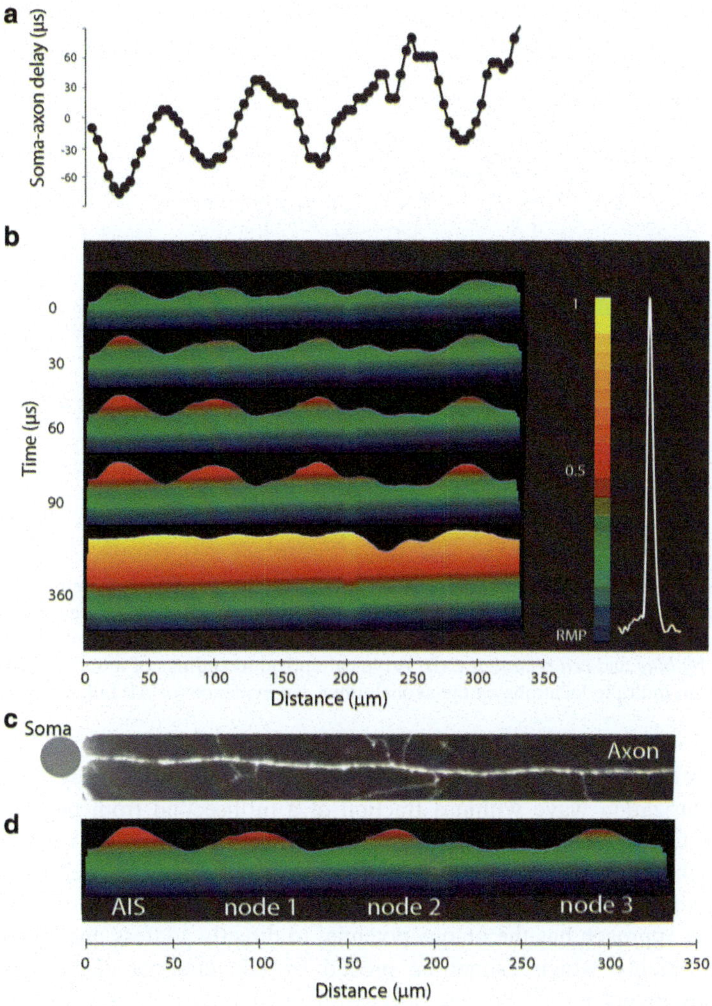

Fig. 3.14 Spatial pattern of AP initiation and propagation in an individual axon. (**a**) Soma–axon latency to 50 % AP amplitude as a function of distance from the soma. (**b**) Time sequence of frames showing spatial profile of colour coded relative V_m amplitude in the axon at five characteristic points in time: 0 μs—AP initiation at TZ; 30, 60 and 90 μs—invasion of the nodes; 360 μs—the peak depolarization. Membrane potential color scale shown on *right*. (**c, d**) Alignment of image of the axon, showing axonal collaterals, and membrane potential changes at spike initiation. Note that each point of issuance of an axon collateral appears functionally to be a node of Ranvier. *Grey dot* denotes the location of the soma

spatial maps shows the colour coded representation of the relative V_m amplitude on the y-axes (omitted for clarity) at five characteristic points in time—spike initiation at time 0, the gradual invasion of the three putative nodes of Ranvier during the following 90 μs, and the peak depolarization at 360 μs. It is noteworthy

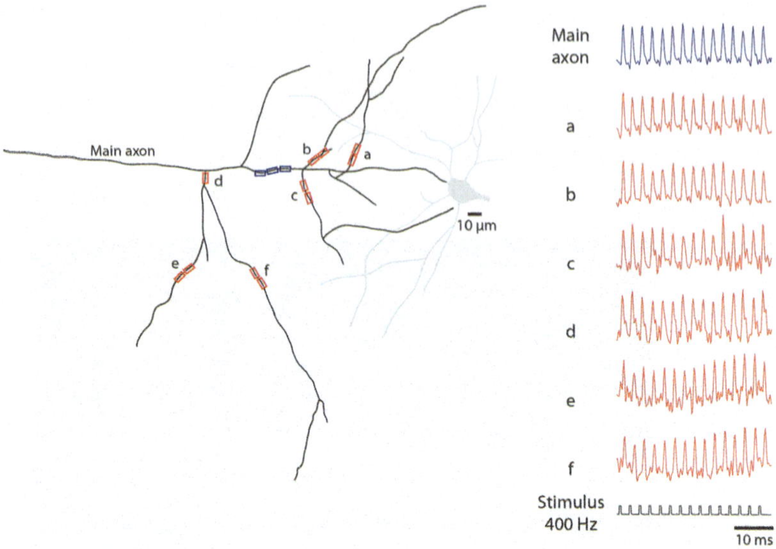

Fig. 3.15 Reliability of AP propagation in the axonal arbor. (**a**) A schematic representation of the axonal arbour reconstructed from a confocal image of a L5 pyramidal neuron loaded with the voltage-sensitive dye. Seven recording locations on the main axon and several collaterals are indicated by *blue* and *red rectangles*. (**b**) A typical optical recordings of a train of APs evoked at 400 Hz from multiple locations on the axonal arbor. No evidence for AP failures was found

that the entire section of the axon, about 300 µm in length, is invaded by the peak of the AP depolarization wave within a fraction of a millisecond from the time the V_m threshold for excitation was reached. Figure 3.14c, d illustrates the clear correlation between the positions of the putative nodes of Ranvier determined by functional imaging with anatomical data showing axonal regions that give rise to collaterals. This same approach has the unique potential to directly address questions of action potential fidelity which cannot be probed by conventional electrophysiological techniques. Figure 3.15 illustrates a successful propagation of a train of 15 APs evoked at the frequency of 400 Hz. In this work (Popovic et al. 2011) optical recordings of a total of 1,770 APs from the main axon and collaterals revealed that the propagation of APs was reliable for all frequencies of firing supported by the spike TZ. These measurements established that the superior sensitivity of wide field laser-excitation epi-fluorescence recordings permit V_m-imaging from the axonal arbor, including small diameter axon collaterals, with high temporal and spatial resolution.

Improved V_m-Imaging of bAPs from Dendritic Spines

The high sensitivity of recording from dendrites and axons at relatively low optical magnification described above indicated that it should be possible to increase the optical magnification by a factor of 10 and monitor V_m-transients from individual dendritic spines and, potentially, from presynaptic axonal varicosities/boutons. Recording electrical events from individual dendritic spines is important for several reasons. Spines play a critical role in the input-output transform carried out by an individual neuron. They receive most of the excitatory synapses in many brain regions and serve as calcium compartments, which appear to be necessary for input-specific synaptic plasticity (Yuste and Denk 1995; Sabatini et al. 2002; Noguchi et al. 2005). In the last decade, a number of investigators, (e.g. Winfried Denk; Rafael Yuste; Karel Svoboda; Bernardo Sabatini; Haruo Kasai) revisited previously well-articulated theoretical questions and ideas (Rall 1974; Koch and Zador 1993) and analyzed open problems in spine physiology with the aid of novel experimental technologies. The methodology has been developed to measure Ca^{2+} signals from individual spines with great precision using two-photon microscopy, and many aspects of spine physiology related to Ca^{2+} signaling have been illuminated (Sala and Segal 2014).

The electrical behavior of spines, however, is less well understood and controversial. For a long time, the electrical role of spines had to be considered on purely theoretical grounds because it was technically impossible to measure V_m-signals from individual spines. Theoretical work (Rall 1974; Segev and Rall 1998; Jack et al. 1975) and, more recently, several experimental studies (Bloodgood and Sabatini 2005; Noguchi et al. 2005; Araya and Jiang 2006; Araya et al. 2006; Harnett et al. 2012; Gulledge et al. 2012) provided indirect evidence that the electrical characteristics of dendritic spines might have important implications for integrative function and for the plastic properties of neurons. Other studies based on diffusion measurements (Svoboda et al. 1996; Tønnesen et al. 2014) as well as on multi-compartmental modeling (Koch and Zador 1993) indicated that spines may not play a significant electrical role. This question is still unresolved because it has never been possible to document directly the electrical behavior of dendritic spines owing to the limited sensitivity of the available measurement techniques at the requisite spatial resolution. Thus, a critical challenge, both conceptually and technically, was to develop an approach for the direct analysis of V_m signals from individual dendritic spines. The improvement in sensitivity described above made this type of recording possible (Holthoff et al. 2010) as illustrated in Fig. 3.16. In these measurements a magnified image of a spiny dendrite of a stained neuron was projected onto a CCD camera and fluorescence intensity changes were monitored from multiple locations. The bAP related signals from dendritic spines were clearly resolved in single trial measurements and the S/N was further improved by averaging a small number of trials (4–9). One concern with respect to these measurements was that scattered light from the parent dendrite, in wide-field fluorescence measurements, might contaminate signals from individual spines. To address this issue, Holthoff et al. (2010) compared dendritic signals from individual spines with recordings from analogous regions without spines and found that the signals from regions without spines were smaller than the noise in the measurement (Fig. 3.16a, b, locations 2 and 3). Thus, the interference from scattered light was insignificant in the superficial layers of the slice. Another concern was the amount of photodynamic damage caused by the high intensity excitation light. In several experiments, the very first single-trial optical

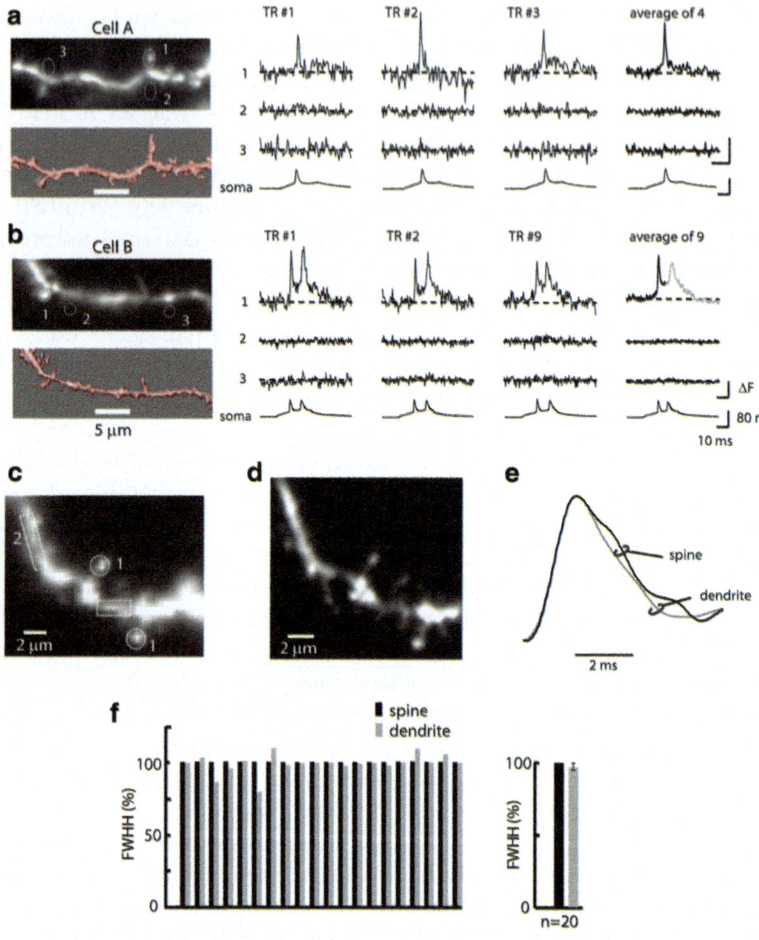

Fig. 3.16 bAP signals from individual dendritic spines and parent dendrites. (**a**, **b**) *Left panels*: *Upper micrographs*—fluorescence images of dendritic spiny branches obtained with the CCD camera for voltage-imaging. *Lower micrographs*—anatomical reconstructions obtained from deconvoluted stacks of spinning-disk confocal images. *Right panels*: Fluorescence intensity traces from locations 1 to 3 outlined on CCD images. Single-trial recordings and temporal averages of 4 and 9 trials are shown. *Bottom traces*: electrode recordings from the soma. The bAP signals are absent from regions without spines (locations 2 and 3). (**c**) A fluorescence image of a section of a spiny dendrite. (**d**) Anatomical reconstruction from a stack of confocal images. (**e**) Comparison of bAP waveform in the spines and parent dendrite. (**f**) Summary results: comparison of bAP duration in the spines and parent dendrites. The individual results of waveform comparison from 20 neurons are shown together with the average values ± SEM. Adapted from Holthoff et al. (2010)

recording of the bAP (control signal) was compared with the last bAP signal at the end of the experiment. The results indicated that, in the range tested (up to 25 recording trials) the first and the last bAP signal were identical, indicating that the photodynamic damage was not significant.

From the type of data shown in Fig. 3.16, it was possible, for the first time, to determine and compare the time course of the bAPs signals in spines and parent dendrites (Holthoff et al. 2010).The waveforms of the bAP signals were reconstructed from the data obtained at 2 kHz frame rate using cubic spline interpolation and compared on an expanded time scale. Figure 3.16e shows that both the upstroke and the downstroke of the AP in spines and dendrites closely overlapped. The summary result from 20 different neurons (Fig. 3.16f) showed that signals from spines and dendrites did not differ significantly. Thus, these results demonstrate that bAPs in spines have a rapid time-course that is very similar to that of spikes recorded in the parent dendrite. The rapid time-course of the bAP in the spines may be a critical determinant for the precise regulation of spike timing-dependent synaptic plasticity within a very narrow time window (Caporale and Dan 2008).

Improved V_m-Imaging of Subthreshold EPSP from Dendritic Spines

On the conceptual level, a key open question regarding the electrical role of dendritic spines is whether the hypothetical electrical isolation of synapses on spine heads caused by a narrow spine neck provides specific functions which are not available to synapses on dendrites. Several such functions have been proposed based on computational modeling and indirect evidence: (1) Spines reduce location-dependent variability of local EPSPs and, thus, standardize and enhance synaptic activation of voltage-gated channels. (2) Changes in structure and electrical resistance of the spine neck under hypothesized activity control mediate the induction of synaptic plasticity underlying learning and memory formation. (3) Electrical properties of spines promote either linear or nonlinear (depending on evidence) dendritic integration and associated forms of plasticity, thus fundamentally enhancing the computational capabilities of neurons. (4) Spines have the capacity to act as discrete electrogenic compartments that amplify synaptic potentials by activation of voltage-sensitive channels. These functions, if directly demonstrated, would define the electrical role of spines. However, a direct demonstration of the electrical behaviour of dendritic spines would require recording of unitary subthreshold synaptic responses at the spatial scale of individual spines with adequate sensitivity and temporal resolution to allow quantitative analysis. The central problem was the sensitivity of optical recordings in terms of the signal-to-noise ratio (S/N) at the required spatiotemporal resolution. The best existing sensitivity which allowed optical monitoring of bAP from dendritic spines, as shown above, was insufficient because EPSP signals would be 5–10-fold smaller in amplitude. The required improvement in sensitivity was recently accomplished by: (a) further increase in the excitation intensity from a laser at the wavelength that has the best signal; (b) minimizing photodynamic damage by restricting the excitation light to a small area (18 μm × 18 μm) around the spine of interest and by briefly lowering oxygen concentration in the extracellular solution during optical recording. The sensitivity of optical recording under these conditions is illustrated in Fig. 3.17. A subthreshold EPSP was evoked by stimulating one synapse on a

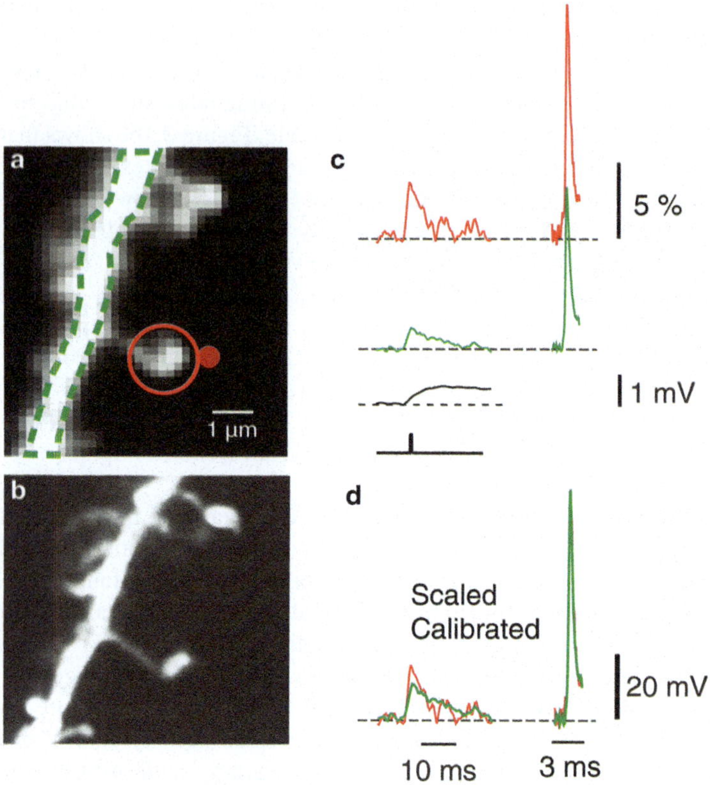

Fig. 3.17 Recording subthreshold electrical events from individual dendritic spines. (**a**) Fluorescence images of a small section of a spiny basal dendritic branch obtained with the CCD camera for voltage-imaging. (**b**) High-resolution mage of the same area from a z-stack of confocal images. (**c**) Traces on *left*: EPSP recordings from spine head (*red*) and parent dendrite (*green*). Average of eight trials. *Bottom black traces*: Somatic electrode recording and the uncaging command pulse. Traces on *right*: bAP signals from same locations. Average of four trials. (**d**) *Left traces*: superimposed EPSP signals from spine head and parent dendrite calibrated in terms of membrane potential. *Right traces*: bAP signals corrected for recording sensitivity difference

dendritic spine using 2-photon uncaging of glutamate while optical signals were acquired at a frame rate of 2 kHz from a small segment of a basal dendrite of a layer 5 pyramidal neuron. The uncaging was followed by a bAP evoked by a depolarizing current pulse delivered from a somatic patch-electrode. Both the subthreshold signal and the action potential signal can be clearly resolved in optical recordings from the spine head and the parent dendrite with modest signal averaging; the EPSP and bAP recordings are average of 8 and 4 trials respectively. Calibration of optical signals on an absolute scale (in mV) was obtained by normalizing the subthreshold signal to an optical signal from a bAP which has a known declining amplitude along basal dendrites (Nevian et al. 2007).These measurements established that voltage-sensitive dye recording based on wide-field laser-excitation fluorescence microscopy permits monitoring of subthreshold electrical signal integration in individual dendritic spines and parent dendrites in acute brain slice preparations.

V_m-Imaging of Inhibitory Synaptic Potentials

In parallel to the improved ability to resolve EPSPs from dendritic spines, laser light excitation also allows discriminating inhibitory postsynaptic potentials (IPSPs) from larger dendritic segments. The possibility of obtaining this type of measurement was demonstrated by Canepari et al. (2010) who monitored IPSPs in the apical dendrites of CA1 hippocampal pyramidal neurons (Canepari et al. 2010). In contrast to EPSPs, IPSPs are mediated by $GABA_A$ or glycine receptors and Cl^- ions which have negative reversal potential under physiological conditions resulting in a smaller driving force at rest. Thus, IPSPs are typically smaller in amplitude compared to EPSPs and their discrimination requires averaging fluorescence over dendritic segments >10 μm. Nevertheless IPSPs of 1 mV or smaller can be optically measured as shown in the example of Fig. 3.18 (Canepari et al. 2010). Beside the possibility of resolving this important synaptic signal locally and with a relatively good spatial resolution, the optical measurement of IPSPs offers another advantage compared to the electrode measurement. Indeed, the polarity and the size of the IPSPs depend on the intracellular Cl^- concentration which is perturbed by the dialyzing effect of the invasive electrode recording. In the same pilot study

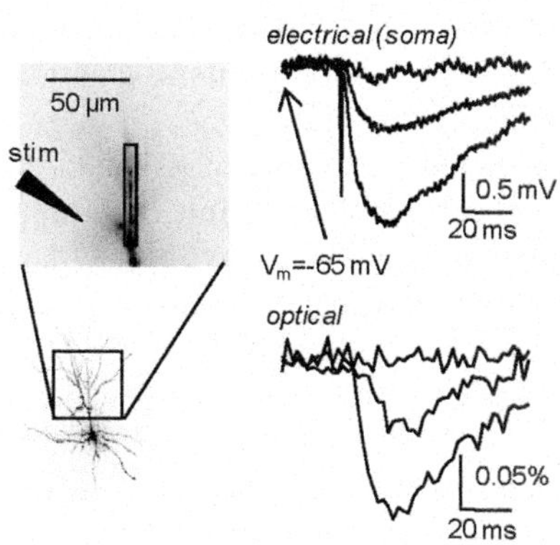

Fig. 3.18 Optical recording of dendritic IPSP. The reconstruction of a CA1 hippocampal pyramidal neuron and the dendrite portion in the recording position are shown on the *left*. Optical recordings are spatial average from dendritic region indicated by the *rectangle*. The position of the stimulating electrode is illustrated. Electrical recordings from the soma (*upper traces*) and optical recordings from the dendrite (*lower traces*) following one extracellular stimulus at three different intensities are shown on the *right*; the weakest intensity failed to evoke an IPSP; the intermediate intensity evoked a unitary IPSP of less than 1 mV; the strongest intensity evoked a larger IPSP of ~2 mV. Recordings are averages of 16 trials obtained in the presence of glutamate receptors antagonists. Adapted from Canepari et al. (2010)

Canepari et al. (2010) showed that it is possible to image IPSPs under physiological Cl^- concentration without concomitant electrode recording. Therefore, voltage imaging offers a tool to investigate IPSPs during the evolution of Cl^- homeostasis occurring in early postnatal development (Cherubini et al. 1991).

3.2.5 Alternative Approaches

A possible way to further improve V_m-imaging from individual neurons is by synthesizing new organic probes with characteristics designed specifically for a particular application. In recent studies, Leslie Loew, Srdjan Antic, and their collaborators introduced a new series of long-wavelength voltage-sensitive dyes for intracellular application (Wuskell et al. 2006; Zhou et al. 2008). These new voltage probes called blue dyes (Shoham et al. 1999) extended the range of excitation wavelengths to near 700 nm, with emission reaching 800–900 nm. Longer wavelength dyes permit deeper penetration by the excitation light into the nervous tissue which could be of considerable importance in optical measurements for both brain slice and *in vivo* preparations (see Chap. 2). These dyes also offer new possibilities for the design of combined recordings and optogenetic stimulations with multiple probes (Canepari et al. 2008; Willadt et al. 2014) (see also Chap. 4). The new probes seem to be characterized by somewhat higher voltage sensitivity in terms of the fractional fluorescence change in response to a unit change in V_m, when applied intracellularly (but see Fig. 3.8c and the discussion of dye sensitivity in Sect. 3.2.4). The sensitivity of recording from individual neurons in terms of the S/N, however, has not, as yet, been fully exploited with the application of blue dyes (Zhou et al. 2007, 2008). Additional sensitivity comparisons remain to be carried out at similar excitation light intensities and for each particular preparation.

Both the new long-wavelength voltage-sensitive dyes and the more conventional probes excited by the green portion of the visible spectrum were used, together with stationary, small spot laser illumination, to investigate the dynamics of bAPs in the terminal sections of basal dendrites (dendritic tips) of prefrontal cortical pyramidal neurons, a region that has never been probed for electrical signals by any method (Zhou et al. 2008). In this recording mode, the excitation light was provided from laser sources emitting at 532, 633, or 658 nm. The laser beam was directed into one end of a 0.2 mm diameter optical fiber (light guide) with a collimator attached to the other end of the fiber which was coupled to the epifluorescence port of a microscope. In this way, a stationary spot of laser light (25–50 μm in diameter) was projected onto the object plane. For recordings of voltage-sensitive dye signals, a selected region of the neuronal process of interest was moved into the laser spot using the X–Y microscope positioning stage. The sensitivity of these measurements in recordings from 25 to 50 μm long sections on dendritic processes allowed optical monitoring of AP signals from the tips of basal dendrites, as well as from individual axons, with minimal temporal averaging. Representative AP related signals obtained in these studies are shown in Fig. 3.19. The results indicated that in short

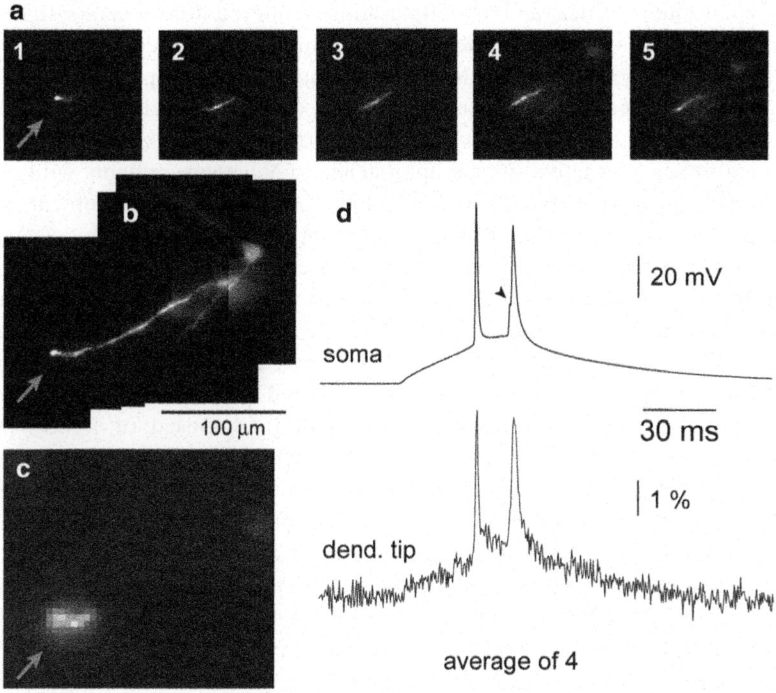

Fig. 3.19 AP signals from basal dendrites of a pyramidal neuron from the prefrontal cortex slice obtained with laser spot illumination.(**a**) Five images (1–5) of different dendritic segments illuminated by a motionless laser spot (~50 μm in diameter). (**b**) A composite image of the whole basal dendrite (from images 1 to 5 shown in **a**). (**c**) Enlargement of a section from image 1. *Arrow* marks the tip of the dendrite. (**d**) Simultaneous whole-cell electrode recording from soma and optical recording from the tip of the dendrite of evoked APs. The optical signal is the spatial average from eight neighboring pixels averaged over four trials. The peak of the first AP was used as a reference point for averaging. Adapted from Zhou et al. (2008)

(<150 μm) and medium (150–200 μm) length basal dendrites APs backpropagated with modest changes in AP half-width or AP rise-time. The lack of substantial changes in AP shape is inconsistent with the AP-failure model based mainly on electrode recordings (Nevian et al. 2007). Large access resistance, up to 200 MΩ, and incomplete electrode capacitance compensation associated with high-resistance patch pipettes (100 MΩ) manufactured to patch thin (submicron diameter) dendritic branches (Nevian et al. 2007) may affect both the amplitude and half-width of fast sodium APs (see Fig. 9 in Zhou et al. 2008). Thus, the information from optical recordings (Zhou et al. 2008) calls into question the commonly held view that spikes are severely attenuated in both apical and basal dendrites and fail to propagate to the most distal dendritic regions (see Vetter et al. (2001) for references).

The disadvantage of the small spot laser illumination approach is that it sacrifices spatial resolution by recording from a single site on the neuron. On the other hand, this method has the advantage that it improves the sensitivity by reducing the background fluorescence and by eliminating scattered fluorescence light from neighboring regions which degrades spatial resolution in the wide-field illumination mode. Additionally, small spot illumination reduces possible photodynamic damage and allows longer recording periods because only a relatively small region on a neuron is exposed to high intensity light at any one time.

Several other alternative optical approaches to V_m measurements with organic dyes, including fluorescence resonance energy transfer (FRET), confocal fluorescence microscopy, second harmonic generation (SHG) microscopy as well as imaging methods with genetically encoded membrane protein V_m sensors, are currently being explored. At present, however, these methods are less efficient and seem less likely to achieve adequate response time and sensitivity at high frame rates and single-spine resolution. Only two-photon excitation of voltage-sensitive dye fluorescence (Chap. 17) has, thus far, exhibited the requisite sensitivity when tested on mouse neurohypophysis, a preparation composed of tightly packed excitable axon terminal membranes and, hence, uniquely convenient for V_m-imaging. More recently this technique has been successfully applied to detect back-propagating action potentials in single dendritic spines (Acker et al. 2011) as well as spikes in single axonal boutons (Rowan et al. 2014; see also Chap. 17).

Another important approach to multiple site voltage-imaging from individual neurons is the use of hybrid voltage-sensors consisting of fluorescent proteins and voltage-dependent quenchers. This idea, based on Fluorescence Resonance Energy Transfer (FRET), was introduced by González and Tsien (1995) using fixed fluorescent donors and translocating oxonol acceptors (see Chap. 6). The two component system was subsequently improved (Chanda et al. 2005) to include a genetically encoded component—farnesylated enhanced GFP (eGFP-F)—and a synthetic voltage sensing molecule, dipicrylamine (DPA).The sensitivity of this approach in terms of $\Delta F/F$ per unit change in membrane potential was further optimized by substituting eGFP-F component for a fluorescent neuronal tracer dye DiO (Bradley et al. 2009; Fink et al. 2012) or a cerulean fluorescent protein (Wang et al. 2010). The hybrid voltage sensor (hVOS) which includes a fluorescence protein has an advantage that it can be genetically targeted to a specific cell type (Wang et al. 2010). The genetic targeting of hVOS has been demonstrated in transgenic mice (Wang et al. 2010) but the specificity of expression was insufficient for single cell resolution. The use of hVOS which includes DiO applied to individual neurons by electroporation is limited to the soma and proximal axonal and dendritic processes because DiO is lipophilic and its diffusion within a neuron is too slow (Fink et al. 2012). The FRET response time constant tends to be inversely proportional to the hydrophobicity of the acceptor (González and Tsien 1995, 1997) or the donor dye (Bradley et al. 2009); the highly lipophilic dyes that yield fast response time constants and high sensitivity to V_m changes are more difficult or impossible to load into cell membranes at locations distal to the dye injection site. Thus, while voltage-sensitive probes based on FRET demonstrated relatively high sensitivity in recordings from proximal neuronal processes (Fink et al. 2012), they have not been successfully employed in imaging V_m signals from distal processes of individual neurons (see Chap. 6 for more detailed description of the FRET imaging approach).

The application of confocal fluorescence microscopy for measuring signals from voltage-sensitive dyes had been limited by another type of difficulty. The superior spatial resolution of confocal microscopy technique has not been effectively utilized in V_m-imaging from dendritic spines and other very small structures owing to insufficient S/N. This is mainly because of the high fractional shot noise related to the small number of photons available for collection (Kuhn et al. 2004; Dombeck et al. 2006; Kerr and Denk 2008); (see also Chap. 17). A study by Palmer and Stuart (2009) demonstrated that, in spite of the technical difficulties to which we have already alluded, V_m transients from presynaptic terminals and individual spines could be monitored using voltage-sensitive dyes and the confocal fluorescence microscopy mode with the aid of extensive temporal averaging. The fluorescence signals related to V_m transients were detected from individual dendritic spines using confocal microscopy (Palmer and Stuart 2009). Optical signals from spines and parent dendrites were monitored with Olympus FV300 confocal microscope with a 60× objective (Olympus; NA 0.9) using an open pinhole to maximize light intensity. The voltage-sensitive dye JPW3028 was excited using a 543 nm laser (HeNe; Melles Griot). An important advantage of this approach is its excellent spatial resolution that allowed visualization of individual dendritic spines below the surface of a brain slice as well as monitoring of V_m signals clearly isolated to individual spines and neighboring parent dendrites (Fig. 3.20). There are, however, three important disadvantages of confocal V_m-imaging. The first disadvantage is related to the relatively low photon flux that can be achieved in this microscopy mode. Due to low light intensity, the AP signal size was approximately ten times

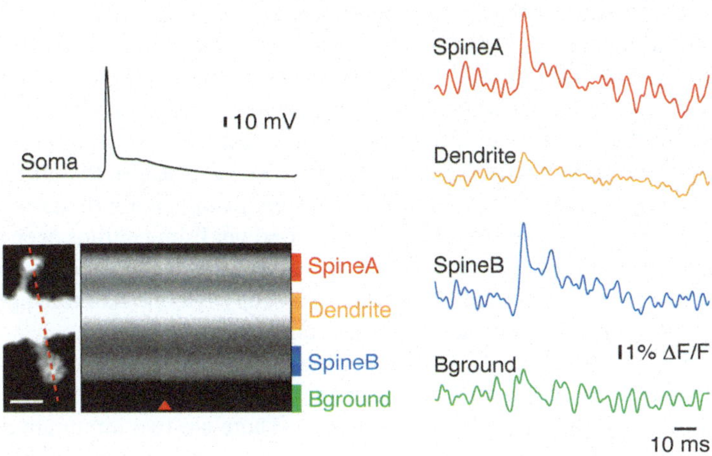

Fig. 3.20 Imaging V_m in dendritic spines using confocal line scan microscopy mode. (**a**) A fluorescence image of a basal dendrite with two dendritic spines. Scale bar = 1 μm. *Dashed red line* indicates a laser line-scan traversing two spines and the resulting average fluorescence change in response to ~100 bAPs (*right*). *Red arrow head* indicates timing of stimulus, and colored bars represent four labeled regions of interest. (**b**) Fluorescence traces (*colored*) in response to somatically evoked bAPs (*black*) for the regions of interest shown in (**a**). Adapted from Palmer and Stuart (2009)

smaller than the shot noise. Therefore, extensive averaging (>100 trials) as well as adding averaged results to create so called super-averages, were required to extract useful information. The necessity for extensive averaging makes repeated recordings under different conditions difficult or impossible. In addition, signal averaging limits this approach to studies of relatively simple phenomena while many aspects of dendritic signal integration are too complex to be analyzed by extensive averaging. The second disadvantage of confocal measurements is that the recording speed is restricted to approximately 800 Hz by the available S/N. This sampling rate is insufficient for correct reconstruction of AP and EPSP signal size and shape according to the Nyquist–Shannon sampling theorem (Roberts 2004). The theorem defines the Nyquist rate, the minimum sampling rate required to avoid distortion of the analog signal, to be equal to twice the highest frequency contained within the signal. The upstroke of the AP waveform (threshold to peak) at room temperature is completed in 300–600 μs corresponding to frequencies of 1.6–3 kHz and, thus, requiring optical recording frame rates in the range of 3–6 kHz. At physiological temperature this requirement is even more stringent. It is helpful that the general shapes of the AP and EPSP waveforms are well defined from electrical measurement, so that aliasing, a certain type of signal distortion, can be safely excluded. This fact allows a modest relaxation of the strict Nyquist rate rule. Nevertheless, the data obtained with sampling rates significantly lower than the Nyquist rate must be regarded as an approximation because of the significant distortion of both the waveform and the amplitude of the signal. Finally, the third important limitation of laser-scanning confocal V_m-imaging is that, in applications where the goal is to monitor electrical events at many locations on an individual neuron, a line scan imaging mode cannot replace a true multiple site optical recording. Notwithstanding these limitations, the results described in Palmer and Stuart (2009) argue convincingly that bAPs invade dendritic spines without significant voltage loss. Additionally, the measurements of EPSP signals at different V_m levels showed that voltage-activated channels do not significantly boost the voltage response in dendritic spines during synaptic input. This work also emphasizes the crucial importance of adequate spatial resolution in recordings from small structures within an opaque brain slice tissue. If the recording is carried out from neurons deep in a slice, the required resolution can only be provided by microscopy modes that reject scattered and out of focus light (confocal microscopy, second harmonic generation (SHG) microscopy, and two-photon fluorescence microscopy).

The same limitation in sensitivity applies to SHG imaging of V_m signals, a nonlinear optical technique that generates a similar or lower photon flux compared to two-photon imaging (Dombeck et al. 2005). There are two important additional restrictions to SHG imaging. First, SHG is, essentially, a non-linear scattering phenomenon and, as such, is predominantly in the forward direction (see Chaps. 17 and 19), limiting this approach to thin preparations and precluding epi-illumination. Secondly, SHG requires a non-centro-symmetrical radiating (scattering) source. If the molecular distribution of the probe in a focal volume is partially symmetrical, with molecules oriented in opposing directions, they will produce SHG signals with opposing phases. This leads to a partial destructive interference of SHG signals, reducing the signal size by a factor that is unknown in the general case (Moreaux et al. 2000). Thus, the same V_m transient in different dendritic compartments may produce optical signals of dramatically different amplitudes, preventing correct calibration of the optical response in terms of V_m from measurements at any one site. The same difficulty in calibrating optical signals on an absolute scale exist in V_m-imaging using intracellular voltage-sensitive dyes, as described above, albeit for a different reason. The first attempt to analyze electrical events in individual spines (Nuriya et al. 2006) was based on recording SHG signals. The results of this work provided excellent impetus for making V_m-imaging with single spine resolution possible. The same results, however, are a good illustration of the current methodological difficulties. The low sensitivity of these measurements required extensive averaging (>100 trials) but still resulted in an insufficient S/N, precluding quantitative analysis of the signal size and shape. It is noteworthy that signal averaging has strict limitations and a further improvement in the S/N by a factor of only 2 would require averaging more than 1,000 trials, which is not feasible in most experiments. Thus, no further application of this approach to monitoring electrical events from individual spines has been reported.

Finally, the most modern approach to V_m-imaging is a methodology focused on genetically encoded protein V_m sensors (see Chap. 20 for more details). The genetic approach could potentially develop into an ideal method to selectively label and monitor different classes of neurons. Thus, protein sensors (voltage-sensitive fluorescent proteins, VSFPs) are being extensively investigated and some of the prototype molecules have shown considerable promise in their ability to detect V_m transients (see Chap. 20). At present, however, this approach to monitoring V_m signaling in individual neurons is limited in several ways. First, the sensitivity and the response time of genetically encoded protein sensors are, as yet, insufficient. Kinetics in particular need to be two orders of magnitude look for the kind of experiments envisioned here. In addition, the problems regarding linearity of response, adequate expression in vertebrate neurons, and possible undesirable effects (e.g. significant capacitive load) on neuronal physiology have not been resolved. Thus, the practical application of this approach is just beginning to be realized (Cao et al. 2013; Jin et al. 2012).

References

Acker CD, Antic SD (2009) Quantitative assessment of the distributions of membrane conductances involved in action potential backpropagation along basal dendrites. J Neurophysiol 101 (3):1524–1541, http://www.pubmedcentral.nih.gov/articlerender.fcgi?artid=2666409&tool=pmcentrez&rendertype=abstract. Accessed 13 Jan 2014

Acker CD, Yan P, Loew LM (2011) Single-voxel recording of voltage transients in dendritic spines. Biophys J 101(2):L11–L13, http://www.pubmedcentral.nih.gov/articlerender.fcgi?artid=3136788&tool=pmcentrez&rendertype=abstract. Accessed 26 Jan 2014

Antic S et al (2000) Functional profile of the giant metacerebral neuron of Helix aspersa: temporal and spatial dynamics of electrical activity in situ. J Physiol 527(Pt 1):55–69, http://www.pubmedcentral.nih.gov/articlerender.fcgi?artid=2270048&tool=pmcentrez&rendertype=abstract. Accessed 23 Nov 2014

Antic S, Major G, Zecevic D (1999) Fast optical recordings of membrane potential changes from dendrites of pyramidal neurons. J Neurophysiol 82(3):1615–1621, http://www.ncbi.nlm.nih.gov/pubmed/10482775. Accessed 23 Nov 2014

Antic S, Zecević D (1995) Optical signals from neurons with internally applied voltage-sensitive dyes. J Neurosci 15(2):1392–1405, http://www.ncbi.nlm.nih.gov/pubmed/7869106. Accessed 29 Jul 2014

Antic SD (2003) Action potentials in basal and oblique dendrites of rat neocortical pyramidal neurons. J Physiol 550(Pt 1):35–50, http://www.pubmedcentral.nih.gov/articlerender.fcgi?artid=2343022&tool=pmcentrez&rendertype=abstract. Accessed 23 Nov 2014

Araya R, Jiang J et al (2006a) The spine neck filters membrane potentials. Proc Natl Acad Sci U S A 103(47):17961–17966, http://www.pubmedcentral.nih.gov/articlerender.fcgi?artid=1693855&tool=pmcentrez&rendertype=abstract. Accessed 22 Jan 2014

Araya R, Eisenthal KB, Yuste R (2006b) Dendritic spines linearize the summation of excitatory potentials. Proc Natl Acad Sci U S A 103(49):18799–18804, http://www.pubmedcentral.nih.gov/articlerender.fcgi?artid=1693742&tool=pmcentrez&rendertype=abstract. Accessed 31 Jan 2014

Bennett CB, Muschol M (2009) Large neurohypophysial varicosities amplify action potentials: results from numerical simulations. Endocrinology 150(6):2829–2836, http://www.ncbi.nlm.nih.gov/pubmed/19213831. Accessed 31 Jan 2014

Bloodgood BL, Sabatini BL (2005) Neuronal activity regulates diffusion across the neck of dendritic spines. Science (New York, N Y) 310(5749):866–869, http://www.ncbi.nlm.nih.gov/pubmed/16272125. Accessed 24 Jan 2014

Bradley J et al (2009) Submillisecond optical reporting of membrane potential in situ using a neuronal tracer dye. J Neurosci 29(29):9197–9209, http://www.pubmedcentral.nih.gov/articlerender.fcgi?artid=2909666&tool=pmcentrez&rendertype=abstract. Accessed 31 Jan 2014

Canepari M et al (2010) Imaging inhibitory synaptic potentials using voltage sensitive dyes. Biophys J 98(9):2032–2040, http://dx.doi.org/10.1016/j.bpj.2010.01.024. Accessed 27 Jan 2014

Canepari M, Djurisic M, Zecevic D (2007) Dendritic signals from rat hippocampal CA1 pyramidal neurons during coincident pre- and post-synaptic activity: a combined voltage- and calcium-imaging study. J Physiol 580(Pt. 2):463–484, http://www.pubmedcentral.nih.gov/articlerender.fcgi?artid=2075540&tool=pmcentrez&rendertype=abstract. Accessed 31 Jan 2014

Canepari M, Vogt K, Zecevic D (2008) Combining voltage and calcium imaging from neuronal dendrites. Cell Mol Neurobiol 28(8):1079–1093, http://www.pubmedcentral.nih.gov/articlerender.fcgi?artid=3143714&tool=pmcentrez&rendertype=abstract. Accessed 31 Jan 2014

Canepari M, Vogt KE (2008) Dendritic spike saturation of endogenous calcium buffer and induction of postsynaptic cerebellar LTP. PLoS One 3(12), e4011, http://www.pubmedcentral.nih.gov/articlerender.fcgi?artid=2603473&tool=pmcentrez&rendertype=abstract. Accessed 31 Jan 2014

Cao G et al (2013) Genetically targeted optical electrophysiology in intact neural circuits. Cell 154(4):904–913, http://www.ncbi.nlm.nih.gov/pubmed/23932121. Accessed 21 Jan 2014

Caporale N, Dan Y (2008) Spike timing-dependent plasticity: a Hebbian learning rule. Annu Rev Neurosci 31:25–46, http://www.ncbi.nlm.nih.gov/pubmed/18275283. Accessed 25 Jan 2014

Chanda B et al (2005) A hybrid approach to measuring electrical activity in genetically specified neurons. Nat Neurosci 8(11):1619–1626, http://www.ncbi.nlm.nih.gov/pubmed/16205716. Accessed 24 Jan 2014

Cherubini E, Gaiarsa JL, Ben-Ari Y (1991) GABA: an excitatory transmitter in early postnatal life. Trends Neurosci 14(12):515–519, http://www.ncbi.nlm.nih.gov/pubmed/1726341. Accessed 25 Nov 2014

Cohen LB et al (1974) Changes in axon fluorescence during activity: molecular probes of membrane potential. J Membr Biol 19(1):1–36, http://www.ncbi.nlm.nih.gov/pubmed/4431037

Cohen LB, Salzberg BM (1978) Optical measurement of membrane potential. Rev Physiol. Biochem Pharmacol 83:35–88

Cohen LB, Lesher S (1986) Optical monitoring of membrane potential: methods of multisite optical measurement. Soc Gen Physiol Ser 40:71–99, http://www.ncbi.nlm.nih.gov/pubmed/3520842. Accessed 13 Nov 2014

Daintry J (ed) (1984) Laser speckle and related phenomena, 2nd edn. Springer-Verlag, Berlin, http://www.ncbi.nlm.nih.gov/pubmed/18213052. Accessed 13 Nov 2014

Davila HV et al (1974) Changes in ANS and TNS fluorescence in giant axons from Loligo. J Membr Biol 15(1):29–46, http://www.ncbi.nlm.nih.gov/pubmed/4837989. Accessed 13 Nov 2014

Djurisic M et al (2008) Functional structure of the mitral cell dendritic tuft in the rat olfactory bulb. J Neurosci 28(15):4057–4068, http://www.ncbi.nlm.nih.gov/pubmed/18400905. Accessed 31 Jan 2014

Djurisic M et al (2004) Voltage imaging from dendrites of mitral cells: EPSP attenuation and spike trigger zones. J Neurosci 24(30):6703–14, http://www.ncbi.nlm.nih.gov/pubmed/15282273. Accessed 31 Jan 2014

Dombeck DA et al (2006) Optical recording of fast neuronal membrane potential transients in acute mammalian brain slices by second-harmonic generation microscopy. J Neurophysiol 94 (5):3628–3636, http://www.ncbi.nlm.nih.gov/pubmed/16093337. Accessed 23 Nov 2014

Fink AE et al (2012) Two-photon compatibility and single-voxel, single-trial detection of subthreshold neuronal activity by a two-component optical voltage sensor. PLoS One 7(8):e41434, http://www.pubmedcentral.nih.gov/articlerender.fcgi?artid=3411718&tool=pmcentrez&rendertype=abstract. Accessed 20 Jan 2014

Foust A et al (2010) Action potentials initiate in the axon initial segment and propagate through axon collaterals reliably in cerebellar Purkinje neurons. J Neurosci 30(20):6891–6902, http://www.pubmedcentral.nih.gov/articlerender.fcgi?artid=2990270&tool=pmcentrez&rendertype=abstract. Accessed 31 Jan 2014

Geiger RP et al (2006) Patch-clamp recording from mossy fiber terminals in hippocampal slices. Nat Protoc 1(4):2075–81, http://www.ncbi.nlm.nih.gov/pubmed/17487197. Accessed 24 Jan 2014

González JE, Tsien RY (1997) Improved indicators of cell membrane potential that use fluorescence resonance energy transfer. Chem Biol 4(4):269–277, http://www.ncbi.nlm.nih.gov/pubmed/9195864. Accessed 13 Nov 2014

González JE, Tsien RY (1995) Voltage sensing by fluorescence resonance energy transfer in single cells. Biophys J 69(4):1272–1280, http://www.pubmedcentral.nih.gov/articlerender.fcgi?artid=1236357&tool=pmcentrez&rendertype=abstract

Grinvald A et al (1982) Improved fluorescent probes for the measurement of rapid changes in membrane potential. Biophys J 39(3):301–308, http://www.pubmedcentral.nih.gov/articlerender.fcgi?artid=1328947&tool=pmcentrez&rendertype=abstract. Accessed 23 Nov 2014

Grinvald A et al (1987) Optical recording of synaptic potentials from processes of single neurons using intracellular potentiometric dyes. Biophys J 51(4):643–651, http://www.pubmedcentral.nih.gov/articlerender.fcgi?artid=1329936&tool=pmcentrez&rendertype=abstract. Accessed 13 Nov 2014

Grinvald A, Farber IC (1981) Optical recording of calcium action potentials from growth cones of cultured neurons with a laser microbeam. Science (New York, N Y) 212(4499):1164–1167, http://www.ncbi.nlm.nih.gov/pubmed/7233210. Accessed 13 Nov 2014

Grinvald A, Ross WN, Farber I (1981) Simultaneous optical measurements of electrical activity from multiple sites on processes of cultured neurons. Proc Natl Acad Sci U S A 78 (5):3245–3249, http://www.pubmedcentral.nih.gov/articlerender.fcgi?artid=319538&tool=pmcentrez&rendertype=abstract

Gulledge AT, Carnevale NT, Stuart GJ (2012) Electrical advantages of dendritic spines. PLoS One 7(4), e36007, http://www.pubmedcentral.nih.gov/articlerender.fcgi?artid=3332048&tool=pmcentrez&rendertype=abstract. Accessed 31 Jan 2014

Gupta RK et al (1981) Improvements in optical methods for measuring rapid changes in membrane potential. J Membr Biol 58(2):123–37, http://www.ncbi.nlm.nih.gov/pubmed/7218335. Accessed 13 Nov 2014

Harnett MT et al (2012) Synaptic amplification by dendritic spines enhances input cooperativity. Nature 491(7425):599–602, http://dx.doi.org/10.1038/nature11554. Accessed 23 Jan 2014

Holthoff K, Zecevic D, Konnerth A (2010) Rapid time course of action potentials in spines and remote dendrites of mouse visual cortex neurons. J Physiol 588(Pt 7):1085–1096, http://www.pubmedcentral.nih.gov/articlerender.fcgi?artid=2852997&tool=pmcentrez&rendertype=abstract. Accessed 31 Jan 2014

Hu W et al (2009) Distinct contributions of Na(v)1.6 and Na(v)1.2 in action potential initiation and backpropagation. Nat Neurosci 12(8):996–1002, http://www.ncbi.nlm.nih.gov/pubmed/19633666. Accessed 31 Jan 2014

Iwasato T et al (2000) Cortex-restricted disruption of NMDAR1 impairs neuronal patterns in the barrel cortex. Nature 406(6797):726–731, http://www.pubmedcentral.nih.gov/articlerender.fcgi?artid=3558691&tool=pmcentrez&rendertype=abstract

Jack J, Noble D, Tsien R (1975) Electric current flow in excitable cells. Oxford Univ Press, London

Jin L et al (2012) Single action potentials and subthreshold electrical events imaged in neurons with a fluorescent protein voltage probe. Neuron 75(5):779–785, http://www.pubmedcentral.nih.gov/articlerender.fcgi?artid=3439164&tool=pmcentrez&rendertype=abstract. Accessed 20 Jan 2014

Kampa BM, Stuart GJ, Kampa M (2006) Calcium spikes in basal dendrites of layer 5 pyramidal neurons during action potential bursts. J Neurosci 26(28):7424–32, http://www.ncbi.nlm.nih.gov/pubmed/16837590. Accessed 31 Jan 2014

Kerr JND, Denk W (2008) Imaging in vivo: watching the brain in action. *Nature reviews*. Neuroscience 9(3):195–205, http://www.ncbi.nlm.nih.gov/pubmed/18270513. Accessed 9 July 2014

Koch C, Zador A (1993) The function of dendritic spines: devices subserving biochemical rather than electrical compartmentalization. J Neurosci 13(2):413–22, http://www.ncbi.nlm.nih.gov/pubmed/8426220. Accessed 23 Aug 2014

Konnerth A, Obaid AL, Salzberg BM (1987) Optical recording of electrical activity from parallel fibres and other cell types in skate cerebellar slices in vitro. J Physiol 393:681–702, http://www.pubmedcentral.nih.gov/articlerender.fcgi?artid=1192418&tool=pmcentrez&rendertype=abstract. Accessed 13 Nov 2014

Kuhn B, Fromherz P, Denk W (2004) High sensitivity of Stark-shift voltage-sensing dyes by one- or two-photon excitation near the red spectral edge. Biophys J 87(1):631–639, http://www.pubmedcentral.nih.gov/articlerender.fcgi?artid=1304385&tool=pmcentrez&rendertype=abstract. Accessed 31 Jan 2014

Larkum ME et al (2009) Synaptic integration in tuft dendrites of layer 5 pyramidal neurons: a new unifying principle. Science (New York, N Y) 325(5941):756–60, http://www.ncbi.nlm.nih.gov/pubmed/19661433. Accessed 28 Jan 2014

Loew LM (1982) Design and characterization of electrochromic membrane probes. J Biochem Biophys Methods 6(3):243–260, http://www.ncbi.nlm.nih.gov/pubmed/7130621

Matsukawa H et al (2003) Motor dysfunction and altered synaptic transmission at the parallel fiber-Purkinje cell synapse in mice lacking potassium channels Kv3.1 and Kv3.3. J Neurosci 23 (20):7677–7684, http://www.ncbi.nlm.nih.gov/pubmed/12930807

Milojkovic BA, Zhou W-L, Antic SD (2007) Voltage and calcium transients in basal dendrites of the rat prefrontal cortex. J Physiol 585(Pt 2):447–468, http://www.pubmedcentral.nih.gov/articlerender.fcgi?artid=2375496&tool=pmcentrez&rendertype=abstract. Accessed 31 Jan 2014

Moreaux L et al (2000) Membrane imaging by simultaneous second-harmonic generation and two-photon microscopy. Optics Lett 25(5):320–2, http://www.ncbi.nlm.nih.gov/pubmed/18059867. Accessed 13 Nov 2014

Nevian T et al (2007) Properties of basal dendrites of layer 5 pyramidal neurons: a direct patch-clamp recording study. Nat Neurosci 10(2):206–214, http://www.ncbi.nlm.nih.gov/pubmed/17206140. Accessed 31 Jan 2014

Noguchi J et al (2005) Spine-neck geometry determines NMDA receptor-dependent Ca2+ signaling in dendrites. Neuron 46(4):609–622, http://www.ncbi.nlm.nih.gov/pubmed/15944129. Accessed 24 Jan 2014

Nuriya M et al (2006) Imaging membrane potential in dendritic spines. Proc Natl Acad Sci U S A 103(3):786–790, http://www.pubmedcentral.nih.gov/articlerender.fcgi?artid=1334676&tool=pmcentrez&rendertype=abstract

Obaid A, Shimizu H, Salzberg B (1982) Intracellular staining with potentiometric dyes: optical signals from identified leech neurons and their processes. Biol Bull 163:388

Palmer LM, Stuart GJ (2009) Membrane potential changes in dendritic spines during action potentials and synaptic input. J Neurosci 29(21):6897–6903, http://www.ncbi.nlm.nih.gov/pubmed/19474316. Accessed 31 Jan 2014

Palmer LM, Stuart GJ (2006) Site of action potential initiation in layer 5 pyramidal neurons. J Neurosci 26(6):1854–1863, http://www.ncbi.nlm.nih.gov/pubmed/16467534. Accessed 31 Jan 2014

Parsons TD et al (1991) Long-term optical recording of patterns of electrical activity in ensembles of cultured Aplysia neurons. J Neurophysiol 66(1):316–333, http://www.ncbi.nlm.nih.gov/pubmed/1919674

Parsons TD et al (1989) Optical recording of the electrical activity of synaptically interacting Aplysia neurons in culture using potentiometric probes. Biophys J 56(1):213–221, http://www.pubmedcentral.nih.gov/articlerender.fcgi?artid=1280466&tool=pmcentrez&rendertype=abstract. Accessed 13 Nov 2014

Popovic MA et al (2011) The spatio-temporal characteristics of action potential initiation in layer 5 pyramidal neurons: a voltage imaging study. J Physiol 589(Pt 17):4167–4187, http://www.pubmedcentral.nih.gov/articlerender.fcgi?artid=3180577&tool=pmcentrez&rendertype=abstract. Accessed 31 Jan 2014

Rall W (1974) Cellular mechanisms subserving changes in neuronal activity. In: Woody E (ed) Brain info. University of California, Los Angeles

Roberts M (2004) Signals and systems: analysis using transform methods and MATLAB. McGraw-Hill Professional, New York

Rohr S, Salzberg BM (1994) Characterization of impulse propagation at the microscopic level across geometrically defined expansions of excitable tissue: multiple site optical recording of transmembrane voltage (MSORTV) in patterned growth heart cell cultures. J Gen Physiol 104(2):287–309, http://www.pubmedcentral.nih.gov/articlerender.fcgi?artid=2229204&tool=pmcentrez&rendertype=abstract

Ross WN, Krauthamer V (1984) Optical measurements of potential changes in axons and processes of neurons of a barnacle ganglion. J Neurosci 4(3):659–72, http://www.ncbi.nlm.nih.gov/pubmed/6707730 [Accessed November 13, 2014]

Rowan MJM, Tranquil E, Christie JM (2014) Distinct Kv channel subtypes contribute to differences in spike signaling properties in the axon initial segment and presynaptic boutons of cerebellar interneurons. J Neurosci 34(19):6611–6623, http://www.ncbi.nlm.nih.gov/pubmed/24806686. Accessed 12 Sept 2014

Sabatini BL, Oertner TG, Svoboda K (2002) The life cycle of Ca(2+) ions in dendritic spines. Neuron 33(3):439–452, http://www.ncbi.nlm.nih.gov/pubmed/11832230

Sala C, Segal M (2014) Dendritic spines: the locus of structural and functional plasticity. Physiol Rev 94(1):141–88, http://www.ncbi.nlm.nih.gov/pubmed/24382885. Accessed 13 July 2014

Salzberg BM (1977) Optical recording of neuronal activity in an invertebrate central nervous system: simultaneous monitoring of several neurons. J Neurophysiol 40:1281–1291

Salzberg B (1978) Optical signals from giant axon following perfusion or superfusion with potentiometric probes. Biol Bull 155:463–464

Segev I, Rall W (1998) Excitable dendrites and spines: earlier theoretical insights elucidate recent direct observations. Trends Neurosci 21(11):453–60, http://www.ncbi.nlm.nih.gov/pubmed/9829684

Shoham D et al (1999) Imaging cortical dynamics at high spatial and temporal resolution with novel blue voltage-sensitive dyes. Neuron 24(4):791–802, http://www.ncbi.nlm.nih.gov/pubmed/10624943

Shu Y et al (2006) Modulation of intracortical synaptic potentials by presynaptic somatic membrane potential. Nature 441(7094):761–5, http://www.ncbi.nlm.nih.gov/pubmed/16625207. Accessed 21 Jan 2014

Shu Y et al (2007) Properties of action-potential initiation in neocortical pyramidal cells: evidence from whole cell axon recordings. J Neurophysiol 97(1):746–60, http://www.ncbi.nlm.nih.gov/pubmed/17093120. Accessed 31 Jan 2014

Stuart GJ, Dodt HU, Sakmann B (1993) Patch-clamp recordings from the soma and dendrites of neurons in brain slices using infrared video microscopy. Pflügers Arch 423(5-6):511–8, http://www.ncbi.nlm.nih.gov/pubmed/8351200. Accessed 12 Nov 2014

Stuart GJ, Sakmann B (1994) Active propagation of somatic action potentials into neocortical pyramidal cell dendrites. Nature 367(6458):69–72, http://www.ncbi.nlm.nih.gov/pubmed/8107777. Accessed 12 Nov 2014

Svoboda K, Tank DW, Denk W (1996) Direct measurement of coupling between dendritic spines and shafts. Science (New York, N Y) 272(5262):716–719, http://www.ncbi.nlm.nih.gov/pubmed/8614831. Accessed 24 Aug 2014

Tønnesen J et al (2014) Spine neck plasticity regulates compartmentalization of synapses. Nat Neurosci 17(5):678–685, http://www.ncbi.nlm.nih.gov/pubmed/24657968. Accessed 10 July 2014

Vetter P, Roth A, Häusser M (2001) Propagation of action potentials in dendrites depends on dendritic morphology. J Neurophysiol 85(2):926–37, http://www.ncbi.nlm.nih.gov/pubmed/11160523

Waggoner A, Grinvald A (1977) Mechanisms of rapid optical changes of potential sensitive dyes. Ann N Y Acad Sci 303:217–242

Wang D et al (2010) Improved probes for hybrid voltage sensor imaging. Biophys J 99(7):2355–2365, http://www.pubmedcentral.nih.gov/articlerender.fcgi?artid=3042572&tool=pmcentrez&rendertype=abstract. Accessed 31 Jan 2014

Willadt S et al (2014) Combined optogenetics and voltage sensitive dye imaging at single cell resolution. Front Cell Neurosci 8:311, http://www.pubmedcentral.nih.gov/articlerender.fcgi?artid=4189389&tool=pmcentrez&rendertype=abstract. Accessed 20 Nov 2014

Wu J, Cohen L, Falk C (1999) Fast multisite optical measurement of membrane potential with two examples. In: Mason WT (ed) Fluorescence and luminescence probes for biological activity. Academic, London

Wuskell JP et al (2006) Synthesis, spectra, delivery and potentiometric responses of new styryl dyes with extended spectral ranges. J Neurosci Methods 151(2):200–215, http://www.ncbi.nlm.nih.gov/pubmed/16253342. Accessed 31 Jan 2014

Yu Y, Shu Y, McCormick DA (2008) Cortical action potential backpropagation explains spike threshold variability and rapid-onset kinetics. J Neurosci 28(29):7260–7272, http://www.pubmedcentral.nih.gov/articlerender.fcgi?artid=2664555&tool=pmcentrez&rendertype=abstract. Accessed 31 Jan 2014

Yuste R, Denk W (1995) Dendritic spines as basic functional units of neuronal integration. Nature 375(6533):682–684, http://www.ncbi.nlm.nih.gov/pubmed/7791901

Zecević D (1996) Multiple spike-initiation zones in single neurons revealed by voltage-sensitive dyes. Nature 381(6580):322–325, http://www.ncbi.nlm.nih.gov/pubmed/8692270. Accessed 29 July 2014

Zhou W-L et al (2008) Dynamics of action potential backpropagation in basal dendrites of prefrontal cortical pyramidal neurons. Eur J Neurosci 27(4):923–936, http://www.pubmedcentral.nih.gov/articlerender.fcgi?artid=2715167&tool=pmcentrez&rendertype=abstract. Accessed 31 Jan 2014

Zhou W-L et al (2007) Intracellular long-wavelength voltage-sensitive dyes for studying the dynamics of action potentials in axons and thin dendrites. J Neurosci Methods 164(2):225–239, http://www.pubmedcentral.nih.gov/articlerender.fcgi?artid=2001318&tool=pmcentrez&rendertype=abstract. Accessed 31 Jan 2014

Chapter 4
Combining Membrane Potential Imaging with Other Optical Techniques

Nadia Jaafari, Kaspar E. Vogt, Peter Saggau, Loew M. Leslie, Dejan Zecevic, and Marco Canepari

Abstract Membrane potential imaging using voltage-sensitive dyes can be combined with other optical techniques for a variety of applications. Combining voltage imaging with Ca^{2+} imaging allows correlating membrane potential changes with intracellular Ca^{2+} signals or with Ca^{2+} currents. Combining voltage imaging with uncaging techniques allows analyzing electrical signals elicited by photorelease of a particular molecule. This approach is also a useful tool to calibrate the change in fluorescence intensity in terms of membrane potential changes from different sites permitting spatial mapping of electrical activity. Finally, combining voltage imaging with optogenetics, in particular with channelrhodopsin stimulation, opens the gate to novel investigations of brain circuitries by allowing measurements of synaptic signals mediated by specific sets of neurons. Here we describe in detail the methods of membrane potential imaging in combination with other optical techniques and discus some important applications.

N. Jaafari • M. Canepari (✉)
Inserm U836, Grenoble Institute of Neuroscience, Team 3, Grenoble Cedex 09, France

Laboratoire Interdisciplinare de Physique (CNRS UMR 5588), Université Joseph Fourier, Bat. E45, 140 avenue de la physique, Domaine univ., 38400 St Martin d'Hères, B.P. 87, 38402 St Martin d'Hères cedex, France

Laboratories of Excellence, Ion Channel Science and Therapeutics, France
e-mail: marco.canepari@ujf-grenoble.fr

K.E. Vogt
International Institute for Integrative Sleep Medicine, University of Tsukuba, Tsukuba, Ibaraki, Japan

P. Saggau
Allen Institute for Brain Science, 551N 34th Street, Seattle, WA 98103, USA

L.M. Leslie
Department of Cell Biology, R. D. Berlin Center for Cell Analysis and Modeling, University of Connecticut Health Center, Farmington, CT 06030-1507, USA

D. Zecevic
Department of Cellular and Molecular Physiology, Yale University School of Medicine, New Haven, CT 06520, USA

M. Canepari et al. (eds.), *Membrane Potential Imaging in the Nervous System and Heart*, Advances in Experimental Medicine and Biology 859, DOI 10.1007/978-3-319-17641-3_4

Keywords Calcium imaging • Calcium channels • Photolysis • Signal calibration • Channelrhodopsin

1 Introduction

The advantages of the optical measurements of membrane potential (V_m), over electrode recordings, are the possibility to record voltage changes at high spatial resolution and from a very large number of sites and the ability to probe structures too small to tolerate electrodes. Combining this measurement with another optical measurement or with an optical stimulation at the same spatial resolution and scale allows investigating and understanding the voltage-dependence of a variety of physiological phenomena or the changes of V_m produced by activation of a particular signaling pathway as well as by a well-defined stimulation protocol. In this chapter we describe in detail the main optical techniques that have been so far successfully combined with V_m imaging: calcium (Ca^{2+}) imaging, photorelease of caged compounds and optogenetic photostimulation. At the end of this chapter we discuss the case of V_m imaging combined with channelrhodopsin (ChR) stimulation, which is becoming a fundamental experimental approach in neurophysiology.

Combined optical measurements have been widely used in biology for a long time. In molecular biology, for instance, the concomitant localization of two or more proteins is achieved by using fluorescent antibodies with different absorption and emission spectra. Combined imaging of two different physiological parameters requiring sequences of signal acquisitions was more difficult to achieve until 20 years ago. Thus, before the introduction of more sophisticated imaging systems, combined optical recordings at relatively slow rate and using two indicators have been utilized to measure intracellular pH and Ca^{2+} (Martinez-Zaguilan et al. 1991) as well as slow V_m changes and Ca^{2+} signals (Kremer et al. 1992). Combined functional imaging is therefore an experimental approach that evolved in the last two decades in parallel with advancements in imaging technology. The correlation of a V_m recording with a measurement of a Ca^{2+} signal is perhaps the most important case of dual optical measurement of a functional signal (Canepari et al. 2008). Indeed, crucial information on Ca^{2+} influx from the extracellular space comes from the fact that Ca^{2+} signals are directly correlated with V_m when they originate from Ca^{2+} channels in the plasma membrane. This category includes voltage-gated Ca^{2+} channels (VGCCs), glutamate receptors (in particular NMDA receptors) and other Ca^{2+} permeable pores. In all these cases, the charge influx associated with Ca^{2+} contributes, to a different extent, to the current underlying depolarization. In addition, the biophysical properties of several Ca^{2+} signals in the plasma membrane also depend on membrane depolarization. For instance, Ca^{2+} influx via VGCCs is voltage-dependent since depolarization is necessary for channel opening. Similarly, Ca^{2+} influx via NMDA receptors is voltage-dependent since depolarization is necessary to unblock the channel from Mg^{2+}. In contrast, Ca^{2+} signals are not directly correlated with V_m when they are due to Ca^{2+} flux through internal membranes. For example Ca^{2+} release from internal store is not directly correlated with V_m. In this case, Ca^{2+} flux does not contribute to the current underlying the V_m change.

Among photostimulations combined to V_m imaging, chemical stimulation delivered by photorelease of a molecule from a caged compound is a powerful tool to explore many physiological phenomena. Notably, uncaging of principal neurotransmitter such as glutamate or GABA, allows investigating how local activation of synaptic receptors produces or modulates V_m in different sub-cellular sites (Vogt et al. 2011b). V_m imaging can be also combined with ChR stimulation expressed in sub-sets of well identified neurons. This application may provide new information on the function of local neuronal circuits (Tsuda et al. 2013).

The techniques of V_m imaging described in this chapter are achieved by recording signals from organic voltage-sensitive dyes. Unless specified otherwise, this type of signal is expressed as a fractional change in fluorescence intensity $\Delta F/F_0 = (F(t) - F(t_0))/F(t_0)$, where F(t) is the recorded fluorescence intensity at a given time and $F(t_0)$ is the fluorescence intensity at the resting membrane potential (resting fluorescence). Finally, most of the examples in this chapter are recordings from individually-stained single-neurons described in the previous chapter.

2 Principles of Combining Two Optical Techniques

The combination of two optical techniques is typically achieved by using two photosensitive molecules. These molecules can be indicators, for instance fluorescence dyes, or molecules that produce a stimulation of the cell upon light activation. If one of the two molecules is used for photostimulation, for instance a caged compound or a photo-sensitive protein, achieving simultaneous imaging requires that this molecule is not excited by the same wavelength used for excitation of the imaging indicator. This requirement is straightforward to achieve for many caged compounds that are activated exclusively by UV light and that can be combined with indicators excited in the visible range. In contrast, combining imaging techniques with ChR stimulation is more difficult since ChR is activated by light over a broad visible spectrum. If the two molecules are both fluorescent indicators, for instance a voltage-sensitive dye and a Ca^{2+} indicator, combined imaging can be achieved in three different ways depending on the optical properties of the two indicators. First, the two indicators may have separate absorption spectra but a largely overlapping region in the emission spectra. Second, the two indicators may have a large overlapping region in their absorption spectra but well separated emission spectra. Third, the two indicators may have a separation both in the absorption and in the emission spectra.

Figure 4.1 shows three examples of spectral representations for the different cases and the optical arrangements to combine the recording of the two signals. In the first case (Fig. 4.1a), the fluorescence from the two dyes can be excited using two separate wavelengths and the emitted light from the two indicators is recorded

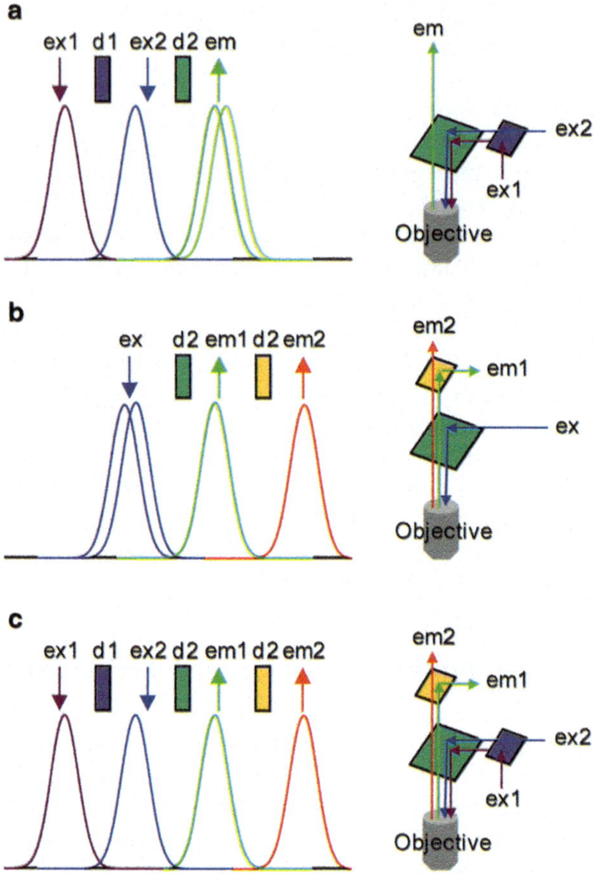

Fig. 4.1 Alternatives for combined imaging using two-indicators. (**a**) (*Left*) Ideal absorption and emission spectra of two indicators with separate absorption spectra and overlapping emission spectra. *Arrows* indicated by "ex" and "em" are positioned over ideal excitation and emission. *Rectangles* indicated by "d" are in position of ideal dichroic mirrors. (*Right*) Optical arrangements for excitation and recording of fluorescence using either one or two light sources. (**b**) (*Left*) Same as (**a**) but for two indicators with overlapping absorption spectra and separate emission spectra. (*Right*) Optical arrangements for excitation and recording of fluorescence using either one or two light detection systems. (**c**) (*Left*) Same as (**a**) and (**b**) but for two indicators with separate absorption and emission spectra. (*Right*) Optical arrangements for excitation and recording of fluorescence using either one or two light sources and light detection systems

using a single detector. In this case, it is never possible to achieve a simultaneous measurement since the imaging experiment requires alternating between the two different excitation wavelengths. In the second case (Fig. 4.1b), because the excitation spectra are overlapping, fluorescence from both indicators can be excited using the same excitation wavelength. Emitted light can be recorded by two different detectors after splitting the emitted light with a secondary dichroic mirror. Thus, by synchronizing data acquisition of the two detection systems, it is possible

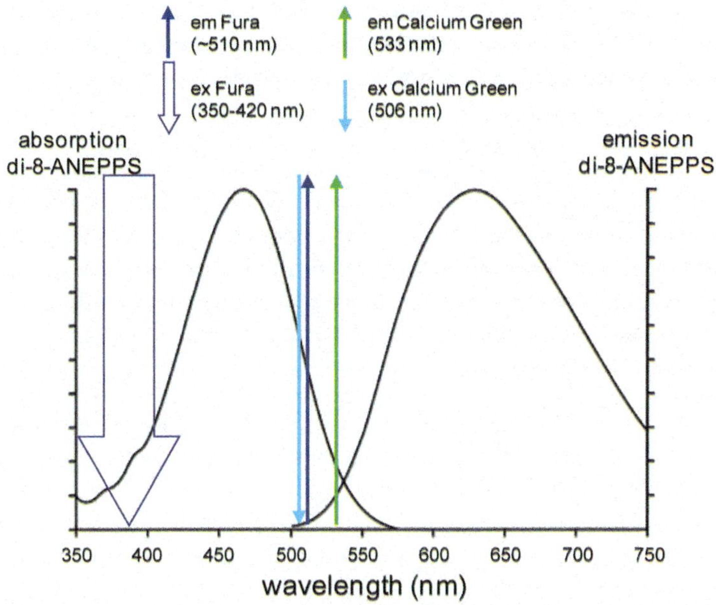

Fig. 4.2 Absorption and emission spectra of di-8 ANEPPS in combination with the ideal absorption and emission of Calcium Green and Fura calcium indicators. *Left* and *right* curves are the absorption and emission spectra as reported by Invitrogen—Molecular Probes. The *thick downward arrow* represents the excitation of the ratiometric indicator Fura. The *thin downward arrows* represent the excitation peak of Calcium Green. The *upward arrows* represent the emission peaks of the two indicators

to achieve simultaneous imaging of the fluorescence signals from the two indicators. Finally, in the third case (Fig. 4.1c), the combined excitation and detection of the two fluorescence systems can be done using two excitation wavelengths and two detection systems simultaneously.

The optimization of recording signals from two indicators depends on the availability of probes with narrow excitation and emission bands. This is not the case for the available fast voltage sensitive dyes. Most of the styryl voltage dyes are characterized by a broad excitation spectrum in the visible range and a broad emission spectrum in the red-IR region (Fluhler et al. 1985). Figure 4.2 shows the absorption and emission spectra of the hydrophobic voltage-sensitive dye di-8-ANEPPS which are similar to those of di-4-ANEPPS and of the water soluble indicators JPW-1114 (di-2-ANEPEQ) and JPW-3028 (di-1-ANEPEQ) all based on the same chromophore. Similar spectral properties are also shared by many of the commonly used "RH" dyes such as RH-795, RH-237, RH-421 and RH-414. With these dyes, well separated spectra are difficult to obtain using commercially available Ca^{2+} indicators excited in the visible spectral range. Nevertheless, the combined recordings are still possible due to a large Stokes shift of charge-shift voltage probes (~150 nm). In addition, more efficient combined recordings can be achieved using Ca^{2+} indicators excited in the UV spectral range.

Both voltage dyes and Ca^{2+} indicators can be loaded into cells either by extracellular bath application of the indicator or by intracellular injection using either a patch-electrode or a sharp microelectrode. For extracellular loading, Ca^{2+} dyes are commercially available in membrane-permeant AM-ester form. Once in the cell, the acetyl-ester is hydrolyzed by the endogenous esterases and the released free indicator remains in the cytoplasm (Yuste 2000). For intracellular loading, Ca^{2+} indicators are available as water-soluble potassium salts. When injected into cells from a patch electrode, the dye can quickly equilibrate in the cytosol at a given concentration permitting quantitative estimate of Ca^{2+} signals (Eilers and Konnerth 2000). In general, intracellular loading of the Ca^{2+} indicator is done in conjunction with the intracellular loading of the voltage sensitive dye. It is also possible to load the cell with the voltage sensitive dye first and include the Ca^{2+} indicator in the pipette used for re-patching the labeled cell. In this protocol, V_m imaging is carried out first, immediately after re-patching, during which time the cell is loaded with the Ca^{2+} indicator from the recording pipette.

3 Combining V_m and Ca^{2+} Imaging

3.1 Combining V_m and Ca^{2+} Imaging in Sequential Way

Commonly used voltage sensitive dyes such as di-4-ANEPPS, di-8-ANEPPS and Di-2-ANEPEQ (JPW-1114) have wide excitation spectrum in the blue/green region and wide emission spectrum in the red/IR region (Fig. 4.2). If combined with Fluo, Calcium Green or Oregon Green Ca^{2+} indicators, both dyes can be excited by a blue wavelength (which, however, is not optimal for the voltage-sensitive dye) and green (Ca^{2+}) and red (V_m) fluorescence can be recorded separately. Voltage sensitive dye excitation is optimal at 532 nm (Canepari et al. 2010). Thus, if combined with UV-excitable Fura indicators, the voltage-sensitive dye can be excited at its optimal visible wavelength and, as with the previous combination, the two emission signals can be recorded separately.

Although both types of combinations allow simultaneous measurement of V_m and Ca^{2+} fluorescence, many studies were initially performed through sequential recordings of the two optical signals. Combined V_m and Ca^{2+} imaging from individual neurons using blue-excitable/green-emitting Ca^{2+} indicators has been also used in several other studies. In the barrel cortex, the high-affinity indicator Oregon Green BAPTA-1 was combined with the voltage sensitive dye JPW-1114 in recordings from individual neurons and with the voltage indicator RH-1691 in in vitro and in vivo network recordings (Berger et al. 2007). In prefrontal cortical neurons, dendritic recordings were obtained combining JPW-1114 either with the high-affinity Ca^{2+} indicator Calcium Green-1 or with the low-affinity Ca^{2+} indicator Fluo-5F (Milojkovic et al. 2007).

Table 4.1 Dissociation constants and buffering capacities of Fura dyes

Dye	K_d (μM)	K_{dye}(300 μM)	K_{dye}(1 mM)
Fura-2	0.224[a]	~1,300	~4,500
Fura-5F	0.4[a]	~750	~2,500
Bis-fura-2	0.525[a]	~570	~1,900
Fura-4F	0.77[a]	~390	~1,300
Fura-6F	5.3[a]	~57	~190
Fura-FF	10[b]	~30	~100
Mag-fura-2	25[c]–40[d]	~7.5–12	~25–40

A list of commercially available Fura dyes with dissociation constant (K_d) and buffering capacities (K_{dye}) at the two concentrations of 300 μM and 1 mM
[a]Invitrogen—Molecular Probes handbook (for fura-2 and bis-fura-2 at 1 mM Mg^{2+})
[b]Schneggenburger et al. 1999
[c]Hyrc et al. 2000
[d]Naraghi 1997

The use of UV-excitable Fura indicators minimizes the spectral overlap with the voltage-sensitive dye. Additionally, the voltage-sensitive dye is not excited during Ca^{2+}-imaging preventing a substantial photodynamic damage that can occur during relatively long recording periods (Canepari et al. 2008). UV-excitable Fura indicators with different equilibrium constants (K_d) are commercially available (see Table 4.1). This permits the choice of the most suitable indicator for the measurement of either the Ca^{2+} influx (high-affinity indicator) or the change in intracellular free Ca^{2+} concentration (low-affinity indicator).

In contrast to the indicators excited in the visible range, Fura indicators allow ratiometric measurements. When excited above their isosbestic wavelength (~360 nm), the changes of fluorescence associated with an increase in Ca^{2+} concentration are negative. Imaging above the isosbestic wavelength has several advantages. First, the resting fluorescence, which corresponds to nominally 0 Ca^{2+}, is significantly higher than with dyes that increase their fluorescence with Ca^{2+}. Second, the achievable imaging contrast is high because healthy cells are characterized by very low resting Ca^{2+}, and the dye in adjacent structures, exposed to millimolar Ca^{2+} concentrations, is essentially not fluorescent. Third, since the dynamic range of the fluorescence is ~1 after subtraction of fluorescence at saturating Ca^{2+}, these measurements allow a better quantitative estimation of the Ca^{2+} signals. An example of V_m and Ca^{2+} optical signals recorded sequentially from the two indicators in three different regions on the dendritic tree of a Purkinje neuron stained by JPW-1114 and Fura-FF is illustrated in Fig. 4.3. Using an excitation band of 387 ± 5 nm for Fura-FF (Fig. 4.3a), the Ca^{2+} increase associated with a climbing fibre excitatory postsynaptic potential (EPSP) corresponds to a negative $\Delta F/F_0$ (Fig. 4.3b). Sequential V_m and Ca^{2+} recording using Fura Ca^{2+} indicators has been done in CA1 hippocampal pyramidal neurons (Canepari et al. 2007), in prefrontal cortex layer-5 pyramidal neurons (Milojkovic et al. 2007) and in cerebellar Purkinje neurons (Canepari and Vogt 2008). Sequential recording (as well as signal averaging) is meaningful only if repeated application of the same stimulation protocol

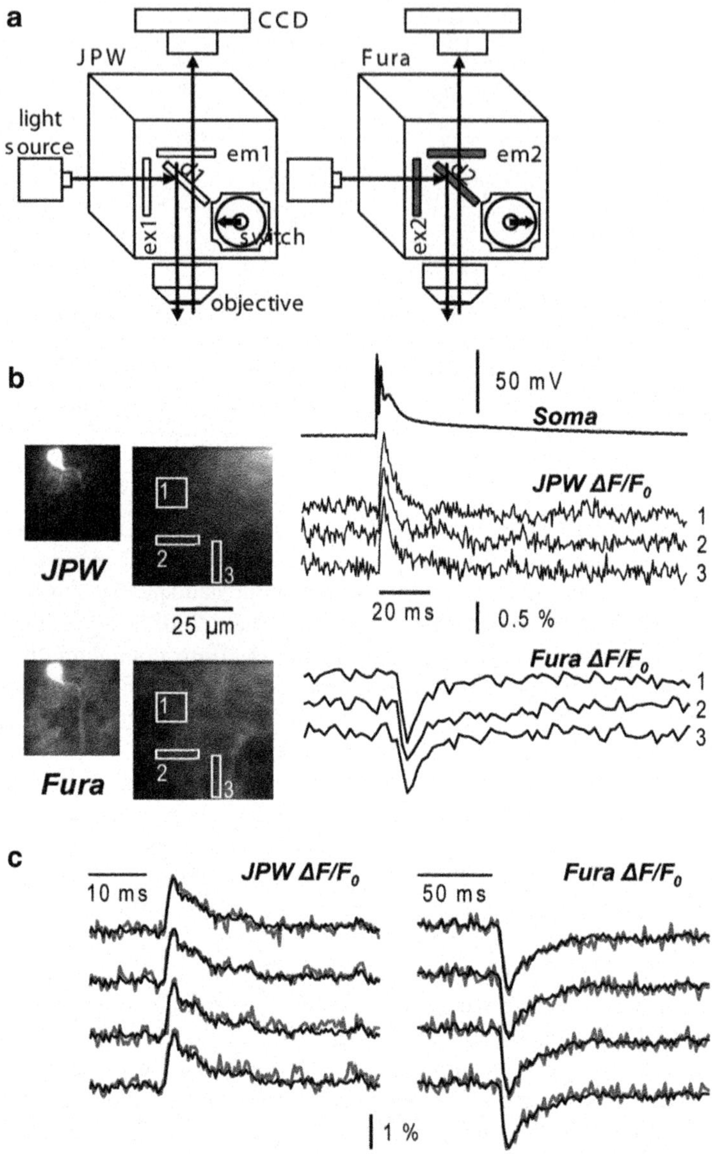

Fig. 4.3 Combining V_m and Ca^{2+} imaging using JPW-1114 and a UV excitable Ca^{2+} indicator. (**a**) Schematic of the imaging apparatus for sequential V_m and Ca^{2+} imaging using a Fura indicator. Filters for voltage-sensitive dye (JPW): ex1 = 525 ± 25 nm, d1 > 570 nm and em1 > 610 nm. Filters for Ca^{2+} indicator (Fura): ex2 = 387 ± 6 nm, d1 > 470 nm and em1 = 510 ± 42 nm. (**b**) V_m and Ca^{2+} fractional changes of fluorescence from cerebellar Purkinje neuron dendrites related to a climbing fibre EPSP recorded from the locations 1–3 reported in the fluorescence images on the *left*. Somatic recording of the climbing fibre EPSP is the *upper trace*. (**c**) Individual V_m (*left*) and Ca^{2+} (*right*) fractional changes of fluorescence (*gray traces*) related to a climbing fibre EPSP. Superimposed *black traces* are the averages of the four trials

results in the same response. This requirement must be confirmed experimentally by comparing individual recordings as shown in Fig. 4.3c. In this example, the V_m and the Ca^{2+} optical signals were recorded from a dendritic location on a cerebellar Purkinje neuron in response to four repetitions of climbing fibre activation separated by 1 min. The results showed that signals were practically identical in four individual trials (gray traces) permitting the correlation of V_m and Ca^{2+} optical signals recorded sequentially in response to the same stimulus.

An important aspect of combined V_m and Ca^{2+} imaging is the interpretation of Ca^{2+} optical signals, which depends on how much the Ca^{2+} indicator perturbs the physiological Ca^{2+} homeostasis. The buffering capacity of a Ca^{2+} indicator, K_{dye}, defined as the ratio between the dye-bound Ca^{2+} and the free Ca^{2+} in the presence of the indicator depends on the dissociation constant and on the concentration of the indicator (Table 4.1). The perturbation of the physiological Ca^{2+} introduced by the Ca^{2+} indicator can be evaluated by comparing the parameter K_{dye} with the endogenous buffering capacity of the cell (K_{cell}). The interpretation of Ca^{2+} optical signals is simplified when K_{dye} is either much larger or much smaller than K_{cell}, i.e. when most of Ca^{2+} binds to the indicator or when the fraction of Ca^{2+} bound to the indicator is negligible compared to the total Ca^{2+} (Canepari et al. 2008). Without addressing in detail the issue of calibration of Ca^{2+} optical signals (Neher 2000) in this volume focused on V_m imaging, we will note that, in the first case, the $\Delta F/F_0$ signal associated with Ca^{2+} increase is approximately linear with the total intracellular Ca^{2+} signal. In the second case, the time course of the ratio $(F - F_{min})/(F_{max} - F)$, where F_{min} and F_{max} are the fluorescence intensities at 0 and saturating Ca^{2+} respectively, is linear with the physiological intracellular free Ca^{2+} concentration change ($\Delta[Ca^{2+}]_i$). In many instances, Ca^{2+} signals are due to Ca^{2+} influx through a Ca^{2+} channel in the plasma membrane. The contribution to the change in membrane potential due to the Ca^{2+} influx depends on the Ca^{2+} permeability of the channel relative to its permeability to other ions, in particular to Na^+ and K^+. This is different for VGCCs, AMPA receptors, NMDA receptors and other Ca^{2+} permeable pores such as transient receptor potential (TRP) channels. *Vice-versa*, because opening or unblocking of Ca^{2+} channels often depends, in a non-linear manner, on the local membrane potential, the Ca^{2+} influx is often, but not always, larger when the voltage related depolarizing optical signal is larger (Canepari et al. 2007). Because this bi-directional relationship is complex, the analysis of V_m and Ca^{2+} signals from multiple sites on individual neurons is not straightforward and requires careful interpretation.

A notable application of combined V_m and Ca^{2+} imaging is the study of synaptic plasticity where the supra-linear Ca^{2+} signal plays an important role. The supra-linear Ca^{2+} signal occurs when two stimulation protocols are combined to evoke the coincident activity (pairing protocol) which results in a Ca^{2+} signal that is larger than the sum of the two Ca^{2+} signals associated with the application of individual stimulation protocols. A supra-linear Ca^{2+} signal is always caused by a supra-linear $\Delta[Ca^{2+}]_i$, but not necessarily by a supra-linear Ca^{2+} influx through the plasma membrane. Indeed, the supra-linear Ca^{2+} signal might be caused by Ca^{2+} release from internal stores or by the saturation of the endogenous Ca^{2+} buffer. However, in the latter case, if the buffering capacity of the dye dominates over the endogenous buffering capacity of the cell and the dye is not saturated, the presence of the indicator will cancel the supra-linear $\Delta[Ca^{2+}]_i$ and no Ca^{2+} dependent $\Delta F/F_0$ will be observed. Thus, in these conditions, detection of supra-linear Ca^{2+} dependent $\Delta F/F_0$ signals which are not due to Ca^{2+} release from internal stores always corresponds to a supra-linear Ca^{2+} influx and it must always correlate with an increase in the depolarizing V_m signal. In contrast, if the buffering of the dye is negligible compared to the endogenous buffering of the cell, a supra-linear $\Delta[Ca^{2+}]_i$ signal that is due to the saturation of the endogenous Ca^{2+} buffer can be measured reliably. The two different types of supra-linear Ca^{2+} signals not involving Ca^{2+} release from stores are illustrated by the two following examples.

In measurements from CA1 hippocampal pyramidal neurons (Canepari et al. 2007), cells were loaded with the voltage-sensitive dye JPW-3028 and 300 μM Bis-Fura-2. The purpose of these experiments was to analyze, over large regions of the dendritic tree, the V_m and Ca^{2+} optical signals associated with back-propagating action potentials, with excitatory postsynaptic potentials (EPSPs) and with pairing of these two membrane potential transients using an LTP induction protocol. Pyramidal neurons in the CA1 region are characterized by relatively low endogenous buffering capacity in the dendrites (~100) and even lower buffering capacity (~20) in the spines (Sabatini et al. 2002). Thus, the buffering capacity of 300 μM Bis-Fura-2 (K_d ~500 in the presence of Mg^{2+}, see Table 4.1) is ~6 times larger than the buffering capacity of the dendrite and ~30 times larger than the buffering capacity of the spine. Ca^{2+} signals associated either with back-propagating action potentials or with EPSPs were mediated by Ca^{2+} influx through voltage-gated Ca^{2+} channels and/or through NMDA receptors. As shown in Fig. 4.4a, the pairing of the two stimulating protocols elicited a supra-linear Ca^{2+} signal. In the presence of the NMDA receptor blocker AP-5, since the measurements were done using a non-saturating concentration of a high-affinity indicator, the supra-linear Ca^{2+} signal must have been caused by supra-linear Ca^{2+} influx mediated by recruitment of additional VGCCs. In agreement with this expectation, at each site where a supra-linear Ca^{2+} signal was observed, the V_m related optical signal during the pairing protocol had a larger peak depolarization compared to the signals associated with unpaired stimulations (Fig. 4.4a).

In measurements from cerebellar Purkinje neurons (Canepari and Vogt 2008), cells were loaded with 1 mM of the Ca^{2+} indicator Fura-FF and with the voltage sensitive dye JPW-1114. Purkinje neurons have an exceptionally high dendritic K_{cell} estimated at ~2,000 (Fierro and Llano 1996). Therefore, the addition of a low affinity Ca^{2+} indicator such as Fura-FF (K_d ~10), even at millimolar concentrations, does not significantly alter the physiological homeostasis and it is possible to record changes in free intracellular Ca^{2+} concentration ($\Delta[Ca^{2+}]_i$ signals Canepari et al. 2008). In this particular preparation, it was possible to calibrate V_m signals in terms of membrane potential. As shown in Fig. 4.4b, pairing the stimulation of the large climbing fibre synaptic potential with local parallel fibres stimulation generated a local dendritic supra-linear $\Delta[Ca^{2+}]_i$ signal which was independent of Ca^{2+} release from stores (Brenowitz and Regehr 2005). Although the $\Delta[Ca^{2+}]_i$ signal originated from Ca^{2+} influx through VGCCs, in Purkinje neurons, in contrast

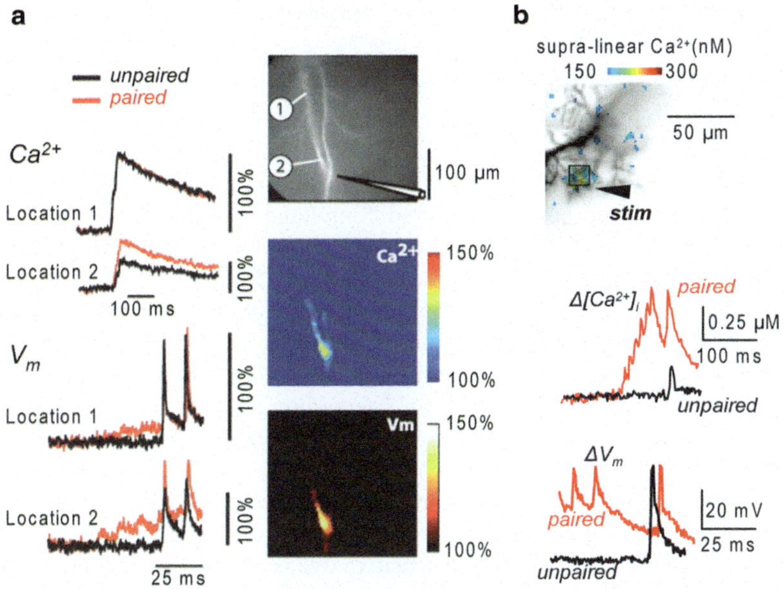

Fig. 4.4 Combining V_m and Ca^{2+} imaging using high- and low-affinity Fura indicators. (**a**) (*Left*) Ca^{2+} and V_m fractional changes of fluorescence corresponding to back-propagating action potentials (bAPs) and paired EPSP–bAP signals from two locations on the dendritic tree of a CA1 hippocampal pyramidal neuron. Signals are superimposed to reveal the region-specific increase in peak depolarization during paired activity. (*Right*) Fluorescence image of the dendritic arbor of three hippocampal pyramidal neurons. The stimulating electrode is indicated schematically. (*Middle and bottom panels*) Color-coded spatial maps of the Ca^{2+} signal and of the V_m signal during paired activity respectively; signals were scaled using the signals in unpaired conditions. Experiments are in the presence of the NMDA receptor blocker AP-5. Supra-linear Ca^{2+} correlated with larger depolarization. Reproduced from Canepari et al. 2007 with the permission of Wiley-Blackwell. (**b**) (*Top*) Recorded dendrites of a cerebellar Purkinje neuron; supra-linear $\Delta[Ca^{2+}]_i$ from the difference between $\Delta[Ca^{2+}]_i$ associated with the pairing protocol and $\Delta[Ca^{2+}]_i$ associated with the unpaired PF-EPSPs and CF-EPSP in color-coded scale. (*Bottom*) $\Delta[Ca^{2+}]_i$ and ΔV_m from the region square region depicted above associated with one unpaired (*black traces*) and one paired CF-EPSP. Supra-linear Ca^{2+} is not associated with larger depolarization. Modified from Canepari and Vogt 2008

to CA1 pyramidal neurons, the V_m signal during the pairing protocol did not have larger peak depolarization compared to the V_m signals associated with unpaired stimulations (Fig. 4.4b). In this case the Ca^{2+} influx associated with the climbing fibre synaptic potential was the same in paired and unpaired conditions. The corresponding $\Delta[Ca^{2+}]_i$ signal, however, was larger following paired stimulation because the Ca^{2+} influx associated with the parallel fibre stimulation locally and transiently saturated the endogenous Ca^{2+} buffer (Canepari and Vogt 2008). This supra-linear Ca^{2+} signal could be correctly interpreted only when recorded using a low-affinity indicator. When the cell was injected with 10 mM Bis-Fura-2, K_{dye} became ~10 times larger than K_{cell} and supra-linear Ca^{2+} signals were abolished (Canepari and Vogt 2008).

3.2 Simultaneous V_m and Ca^{2+} Imaging

Sequential V_m and Ca^{2+} imaging produce valid results if identical responses are evoked in repetitive measurements. Many physiological processes, however, are stochastic, showing large trial-to-trial variability. In these cases, the study of the time variability contains important biological information and simultaneous V_m and Ca^{2+} optical measurement is required. Simultaneous V_m and Ca^{2+} imaging was initially achieved using the Ca^{2+} indicator Calcium Orange, which has excitation peak at ~550 nm and emission peak at ~580 nm (Sinha et al. 1995; Sinha and Saggau 1999), or with blue-excitable indicators (Bullen et al. 1997; Bullen and Saggau 1998). These possibilities may have, however, some inherent problems. First, the small overlap between the emission spectra of the two dyes may produce a "bleeding" of the V_m signal into the Ca^{2+} recording channel that contaminates the signal. In some situations, such emission cross talk can be mathematically corrected (Sinha et al. 1995). Second, while V_m recordings of synaptic or action potentials can be often achieved within short time, the complete time-course of the corresponding Ca^{2+} signal may require longer acquisition intervals (Canepari and Vogt 2008). In this case, the voltage sensitive dye is exposed to high intensity light for a longer time than what would be required for the V_m measurement alone, increasing both bleaching of the dye and phototoxicity. Simultaneous V_m and Ca^{2+} imaging was also performed using the combination of an absorption voltage sensitive dye with a fluorescent Ca^{2+} indicator. Using bath application of the absorption dye RH-482 and a low-affinity Ca^{2+} indicator Magnesium Green, Sabatini and Regehr could correlate the Ca^{2+} signal underlying synaptic release with the excitation of pre-synaptic terminals in cerebellar synapses (Sabatini and Regehr 1996; Sabatini and Regehr 1997). In these experiments, simultaneous Ca^{2+} and V_m measurements were done by using excitation light wavelength of ~500 nm and recording simultaneously fluorescence from the Ca^{2+} indicator Magnesium Green and the transmitted light from voltage-sensitive dye RH-482 at 710 nm from a single location with two photodiodes. The important limitation of this approach is that the absorption measurements from multiple sites on individual neurons are substantially less sensitive compared to fluorescence measurements (Antic and Zecevic 1995). In the above example, extensive temporal and spatial signal averaging over a large region of the cerebellar slice, which could be obtained with absorption signals, permitted signal detection from presynaptic terminals (Sabatini and Regehr 1997).

The above mentioned problems of simultaneous V_m and Ca^{2+} imaging using the same excitation wavelength can be overcome by using a Fura indicator. The realization of this measurement, however, required the design of a dedicated apparatus (Vogt et al. 2011a) to optimize the signal-to-noise ratio (S/N). In principle, di-8-ANEPPS and JPW-1114 could be excited by blue light in the configuration shown in Fig. 4.1c. However, narrow-band laser excitation at 532 nm or at 543 nm is associated with much larger sensitivity (Canepari et al. 2010) permitting

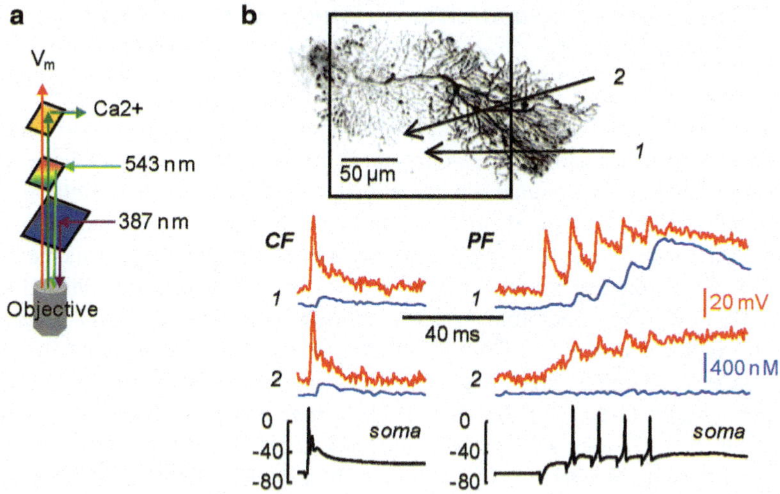

Fig. 4.5 Simultaneous V_m and Ca^{2+} imaging. (**a**) Schematic of the optical configuration to achieve simultaneous V_m and Ca^{2+} imaging; 543 and 387 nm are used for voltage sensitive dye and Fura indicator excitations respectively; low *green* and *red* emissions of the two indicators are separated and detected with two cameras. (**b**) (*Top*) Reconstruction of a cerebellar Purkinje neuron with the dendritic area in recording position outlined; two sites of interest (*1* and *2*) are indicated by the *arrows*. (*Bottom*) ΔV_m (*red traces*) and $\Delta[Ca^{2+}]_i$ (*blue traces*) signals associated with a climbing EPSP (*left*) or with a train of five parallel fibre EPSPs at 100 Hz (*right*) from the two sites of interest; somatic electrical recordings shown below. Modified from Vogt et al. 2011a

single-trial recordings. These excitation wavelengths are longer than the emission peak of Fura indicators. To allow transmission of V_m and Ca^{2+} emission wavelengths, while reflecting intermediate excitation wavelength (543 nm), the apparatus included a dual-band pass dichroic mirror as shown in the schematic in Fig. 4.5a. Multiple wavelengths dichroic mirrors are commercially available from several companies (e. g. Semrock). The mirror that was used in this case reflected 543 nm and transmitted lower green and red wavelengths corresponding to the emissions of Fura and JPW1114 indicators respectively. This optimal configuration allowed resolving dendritic excitation and associated Ca^{2+} signals from relatively small regions in single trial measurements. In the example shown in Fig. 4.5b, simultaneous ΔV_m and $\Delta[Ca^{2+}]_i$ signals, calibrated as described in Canepari and Vogt (2008), associated with one climbing fibre-EPSP and a train of five parallel fibre EPSPs at 100 Hz were compared in two small dendritic regions. ΔV_m and $\Delta[Ca^{2+}]_i$ signals associated with large depolarization could be observed in both regions for the climbing fibre EPSP but only near the parallel fibre stimulating electrode, for the parallel fibre EPSPs train. These measurements, where signals of ~1 % change of fluorescence could be resolved from regions of few microns, represent the current state-of-the-art in sensitivity of simultaneous V_m and Ca^{2+} imaging from single cells.

3.3 Simultaneous Optical Measurement of V_m and Ca^{2+} Currents

Simultaneous V_m and Ca^{2+} imaging can be also used for the study of native fast Ca^{2+} currents based on derivation of Ca-currents from optical signals., We have already stated that the $\Delta F/F_0$ Ca^{2+} signal is proportional to the change in concentration of Ca^{2+} ions bound to the indicator. If, in addition, the Ca^{2+} bound to the indicator is also proportional to the total free Ca^{2+} flux entering the cell through the plasma membrane, then the Ca^{2+} current can be estimated by the derivative of $\Delta F/F_0$ (Sabatini and Regehr 1998). This estimate is possible with low-affinity indicators (Kao and Tsien 1988) where the equilibration of the dye-Ca^{2+} binding reaction is faster than the Ca^{2+} influx. It was recently demonstrated that the reaction between Ca^{2+} and the low-affinity indicator Oregon Green 488 BAPTA-5N is faster than a typical VGCC-mediated current, i.e. the $\Delta F/F_0$ signal from this indicator matches the kinetics of the Ca^{2+} current integral (Jaafari et al. 2014). It follows that the kinetics of the $\Delta F/F_0$ derivative is proportional to the kinetics of the Ca^{2+} current and can be calibrated in terms of "Ca^{2+} current volume density". The simultaneous V_m optical measurement allows studying the voltage dependence of the current, i.e. the membrane potential of activation and de-activation of the channel. Technically, simultaneous V_m and Ca^{2+} imaging using Oregon Green 488 BAPTA-5N requires blue illumination for both indicators (see previous paragraph) and high-speed acquisitions (20 kHz frame rate) to investigate the time-course of the Ca^{2+} current at high temporal resolution. Figure 4.6a shows an example of a CA1 hippocampal pyramidal neuron filled with 1 mM of Oregon

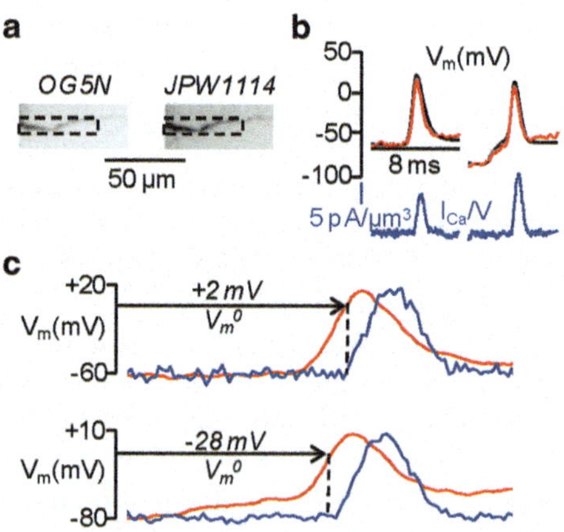

Fig. 4.6 Dendritic Ca^{2+} currents associated with an action potential at different starting V_m. (**a**) Fluorescence images of a CA1 hippocampal pyramidal neuron filled with Oregon Green 488 BAPTA-5 N (*left image*) and the voltage sensitive dye JPW1114 (*right image*); the region of interest from where fluorescence was averaged is outlined. (**b**) Ca^{2+} currents (*blue*) associated with one action potential starting either at $V_m = -60$ or -80 mV; V_m recorded optically (*red*) and with the electrode (*black*) superimposed. (**c**) Ca^{2+} currents normalized and superimposed to the action potential in the two cases; V_m at the current onset (V_m^0) indicated. Modified from Jaafari et al. 2014

Green 488 BAPTA-5N and the voltage sensitive dye JPW1114. Figure 4.6b shows the Ca^{2+} current volume density (I_{Ca}/V) associated with an action potential monitored optically in the proximal apical dendritic segment of a CA1 hippocampal pyramidal neuron. The I_{Ca}/V and V_m signals were measured under two conditions: for an action potential starting at $V_m = -60$ mV and for an action potential starting at $V_m = -80$ mV (Fig. 4.6b). When the action potential started at $V_m = -80$ mV, I_{Ca}/V was larger and started at a more hyperpolarized V_m (Fig. 4.6c), indicating that at $V_m = -80$ mV the Ca^{2+} current contained a component mediated by low-voltage activated (T-type) Ca^{2+} channels (Cueni et al. 2009). The example shown in Fig. 4.6 illustrates how simultaneous optical measurement of V_m and Ca^{2+} currents can drastically improve our understanding of the physiological function of Ca^{2+} channels by permitting the exploration of local biophysics of native channels during physiological activity.

4 Combining V_m Imaging with Photostimulation

4.1 Combining V_m Imaging with Glutamate or GABA Photorelease

Membrane potential imaging can be combined with uncaging techniques to investigate how the photorelease of biologically active molecules changes the membrane potential in different regions of a cell. In particular, the caged compounds 4-Methoxy-7-nitroindolinyl-caged-L-glutamate (MNI-glutamate, Canepari et al. 2001) and 1-(4-Aminobutanoyl)-4-[1,3-*bis*(dihydroxyphosphoryloxy)propan-2-yloxy]-7-nitroindoline (DPNI-GABA, Trigo et al. 2009) can be used to obtain fast photorelease of either L-glutamate or GABA in sub-cellular regions of variable size and to investigate the regulation of membrane potential. An example of this methodological approach is the study of shunting inhibition (Mann and Paulsen 2007). Propagation of action potentials in the axon or back-propagation of action potentials in the dendrites can be strongly regulated by local changes of membrane conductance due to opening of $GABA_A$ receptors. Simultaneous V_m imaging and GABA photorelease is an ideal approach to investigate shunting inhibition. In the CA1 hippocampal pyramidal neuron shown in Fig. 4.7a, the effect of local GABA photorelease on the back-propagating action potential in the apical dendrite was explored. DPNI-GABA was applied using a patch pipette positioned close to the distal part of the dendrite at ~200 μm from the soma. The fluorescence change associated with a back-propagating action potential under control conditions (Fig. 4.7b, left traces) was compared with a back-propagating action potential occurring 15 ms after an episode of GABA photorelease (Fig. 4.7b, right traces) from four regions of interest. The amplitude of the somatic action potential was the same in both cases, but local GABA uncaging decreased the amplitude of the fluorescence change associated with the back-propagating action potential progressively with the distance from the soma. Although shunting inhibition induced by GABA uncaging is produced by

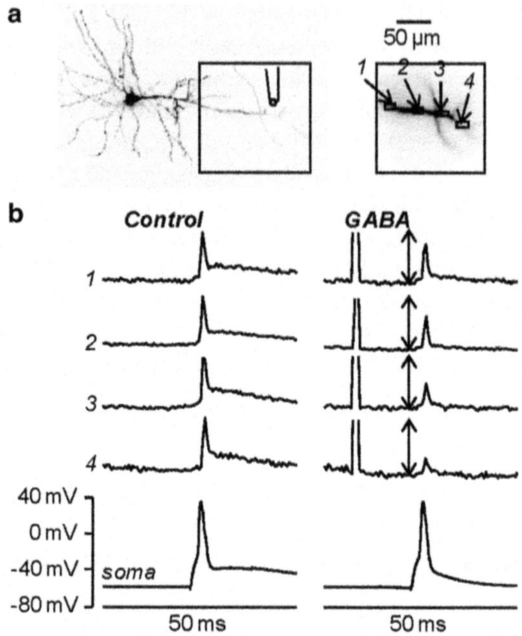

Fig. 4.7 Investigating shunting inhibition using combined V_m imaging and local GABA photorelease. (**a**) Image of a CA1 hippocampal pyramidal neuron; the dendritic area in recording position is outlined and four regions of interest are shown on the *right*; the position of the pipette used for caged-glutamate application is illustrated. (**b**) V_m optical signals from the regions *1–4* and somatic recordings associated with one back-propagating action potential elicited by somatic current injection in control condition (*left*) and 15 ms after an episode of 1 ms GABA photorelease (*right*). The optical signals are normalized to the peaks of the spike in control condition; the length of the *double-arrow* corresponds to the peak. Modified from Vogt et al. 2011b

non-physiological co-activation of synaptic and extra-synaptic GABA receptors, this experiment represents a useful model of this phenomenon since it permits a clear separation of presynaptic and postsynaptic effects, a clear localization of the activated GABA receptors and the possibility to perform pharmacological studies that cannot be done using presynaptic stimulation. Thus, combining membrane potential imaging with GABA photorelease shows the spatial restriction of the shunting effect and the resulting rapid loss in action potential backpropagation.

4.2 *Using Glutamate Uncaging to Calibrate Voltage-Sensitive Dyes Signals*

The calibration of V_m optical signals on an absolute scale (in mV) provides direct measurements of V_m signal amplitudes from multiple locations. This information can be used to construct and analyze the dynamic spatial maps of signal amplitudes throughout neuronal processes. The temporally and spatially well resolved maps of activity are one of the key elements in understanding the rules governing dendritic signal processing and plasticity. The absolute calibration of optical V_m signals, however, is not straightforward and often the exact calibration is not possible. Voltage-sensitive dye recording of membrane potential transients belongs to a class of indirect measurements. The quantity that is being measured directly, by photodetectors, is light intensity, and the quantity that needs to be monitored is derived from a known relationship between the light intensity and V_m. It is convenient that the relationship between light intensity and V_m is strictly linear over the entire physiological range for many voltage sensitive dyes (e.g. Gupta et al. 1981; Loew and Simpson 1981; Wu and Cohen 1993). In a multisite recording, the fractional change in light intensity is proportional to V_m, but also, to a different extent at different sites, to an additional factor: the ratio of inactive dye to active dye. The inactive dye is bound to connective tissue or any other membranes that do not change potential while the active dye is bound to the excitable membrane being monitored. The inactive dye contributes to the resting fluorescence only, and the light from active dye contributes to the resting fluorescence and also carries the signal. In experiments utilizing intracellular application of the dye, inactive dye would be bound to intracellular membranes and organelles. It is a general rule that the ratio of active dye to inactive dye is different for different regions of the object and is also unknown. It follows that the sensitivity of recording from different regions of the object will be different and the calibration of all detectors cannot be done by calibrating the optical signal from any single site. In this situation, the calibration is absolute and straightforward if a calibrating electrical signal that has known amplitude at all locations is available. An all-or-none action potential signal is ideal for this purpose and can be used to create a sensitivity profile of the measuring system. This type of calibration was used to scale the amplitudes of sub-threshold signals in mitral cells of the olfactory bulb (Djurisic et al. 2004). If action potentials are not available, another type of calibrating electrical signal can be a slow electrical change spreading with minimal attenuation over relatively long distances. Such condition occurs, for instance, in cerebellar Purkinje neurons (Roth and Häusser 2001; Canepari et al. 2008).

The possibility to achieve uniform depolarization over the entire dendritic tree using L-glutamate photolysis may extend the possibility to calibrate fluorescence signals to other cell types. This calibration procedure is based on the principle that if the ionotropic glutamate receptor becomes the dominant conductance in a particular neuronal compartment, its reversal potential will determine the membrane potential of the compartment. Thus, if dominance of glutamate receptor conductance is obtained by L-glutamate uncaging, the resulting V_m change in different compartments of the neuron will be the same and can be used to calibrate voltage-sensitive dye imaging data. The idea of a calibration protocol based on L-glutamate uncaging relies on saturating depolarization in the visualized area (Vogt et al. 2011b). The fractional change of fluorescence will then correspond to the resting membrane potential in each visualized site where photolysis occurs.

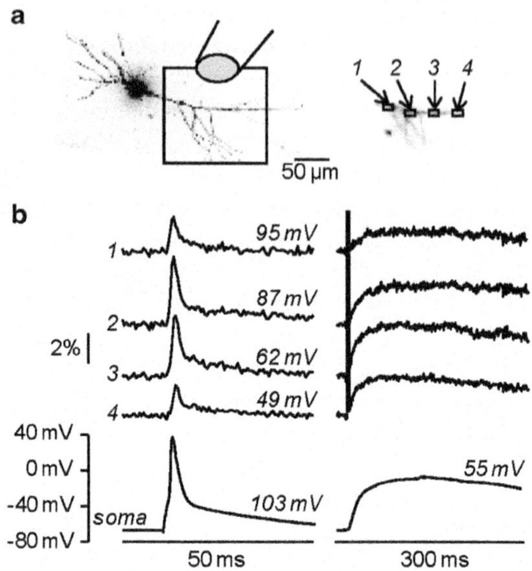

Fig. 4.8 Calibration of fractional changes of fluorescence in terms of membrane potential using L-glutamate (**a**) Image of a CA1 hippocampal pyramidal neuron; the dendritic area in recording position is outlined and four regions of interest are shown on the *right*; caged-glutamate is applied over the visualized area of the dendrite and wide-field illumination was used for photolysis. (**b**) (*Left*) V_m optical signals from the regions of interest *1–4* and somatic recordings associated with one back-propagating action potential. (*Right*) After addition of 1 μM TTX and 50 μM CTZ, V_m optical signals from the same regions and somatic recordings associated with 1 ms L-glutamate photo-release. The peaks of the somatic signal are indicated over their respective traces; the sizes of the estimated peak spikes in the regions of interest are indicated over their respective traces. Modified from Vogt et al. 2011b

In the CA1 hippocampal pyramidal neuron of Fig. 4.8a, one action potential was evoked by somatic current injection. The fractional changes of fluorescence associated with the back-propagating action potential from four ROIs on the apical dendrites are shown in Fig. 4.8b (left traces). The amplitude of the signal is variable and could not be directly correlated with the absolute change of membrane potential. After addition of 1 μM TTX, to block action potentials, and of 50 μM CTZ, to prolong depolarization by blocking AMPA receptor desensitization, L-glutamate was photoreleased over the whole field of view. The recording was performed starting from the resting V_m of the cell. The size of the fluorescence change at each site and the absolute somatic depolarization elicited by uncaging increased with the amount of released L-glutamate, i.e. with the exposure time of the UV light, eventually reaching saturation. The saturating V_m corresponded to the reversal potential of glutamate receptors. Using this information, the amplitude of the back-propagating action potential signals was converted into the change in membrane potential. This method of calibration of voltage-sensitive dyes signals, based on L-glutamate photorelease, can only be applied to cell types characterized by dominant glutamate conductance.

4.3 Combining V_m Imaging with ChR Stimulation

For control of neuronal activity, optogenetic approaches based on light-sensitive channels and other photo-reactive molecules enable highly selective control of genetically-defined populations of neurons (Mancuso et al. 2011; Yizhar et al. 2011). Such techniques offer high time resolution and high spatial precision, allowing stimulation of either individual neurons or large neuronal populations. A commonly used protein for this application is Channelrhodopsin2 (ChR2, Zhang et al. 2006). Combination of optical stimulation using ChR2, with membrane potential imaging is in principle a potent approach to investigate electrical activity both at cellular and at network level. The accomplishment of this type of experiment, however, requires voltage-sensitive dyes that are excitable by red wavelengths since activation of ChR2 occurs at visible wavelengths <600 nm. Indicators with these characteristics have been developed by Rina Hildensheim and Amiram Grinvald to investigate electrical activity from the entire cortex *in vivo* without optical interference of hemoglobin absorption (Grinvald and Hildensheim 2004). More recently, red or near-IR excitable or fluorescent indicators have been developed by Joseph Wuskell, Ping Yan and Leslie Loew to monitor electrical activity in the heart (Matiukas et al. 2005) and in the brain (Wuskell et al. 2006) or to be injected in individual cells (Zhou et al. 2007; Yan et al. 2012). Among these new voltage sensitive dyes, five have been specifically tested to be used in combination with ChR2 stimulation (Kee et al. 2008). In the example shown in Fig. 4.9a (modified from Tsuda et al. 2013), ChR2 was selectively expressed in interneurons located in the molecular layer, providing inhibitory post-synaptic potentials (IPSPs) to cerebellar Purkinje neurons when the light-sensitive channel was excited. In this experiment, the cerebellar slice was stained with the voltage sensitive dye Di-2-ANBDQPQ (Zhou et al. 2007) and excited at 658 nm, beyond the excitation spectrum of ChR2 (Kee et al. 2008). Transient photo-stimulation of the molecular layer with blue light produced an IPSP in the nearby Purkinje neurons. As shown in Fig. 4.9a, the IPSP elicited by photo-stimulation of the molecular layer could be measured optically. Notably, the kinetics of the optical signal was identical to that of the IPSP measured with a patch electrode. In another study, ChR2 was selectively expressed in GABAergic interneurons of the hippocampal *stratum lacunosum moleculare* and the regulation of electrical activity in different areas was characterized by staining the slice with the indicator DI-4-ANBDQPQ (Kee et al. 2008) and by performing V_m imaging (Leão et al. 2012). Thus, using this strategy, it was possible to determine that interneurons from the stratum lacunosum molecular differentially modulate CA3 and entorhinal inputs to hippocampal CA1 neurons.

Using water soluble indicators excitable with red light (Zhou et al. 2007; Yan et al. 2012), electrical activity can be imaged in combination with ChR2 photostimulation from individually stained cells (Willadt et al. 2014). The clear advantage of this approach is that the higher sensitivity of V_m imaging in individually stained neurons permits optical measurements of smaller membrane potential

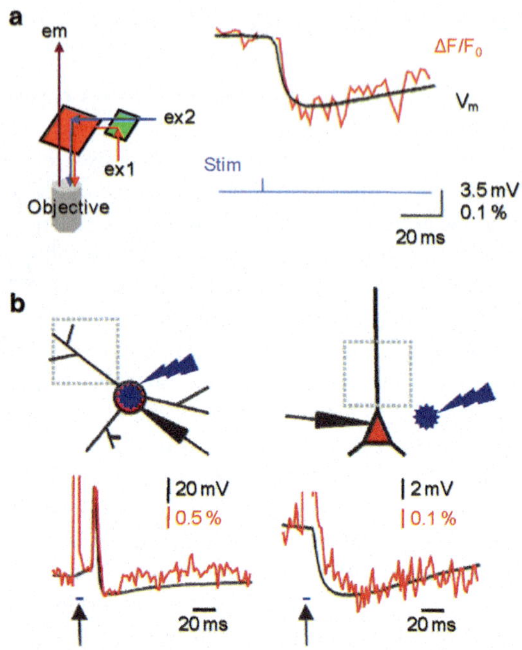

Fig. 4.9 Combining V_m imaging with ChR2 stimulation. (**a**) (*Left*) Schematic of the optical configuration to achieve this combination; *red* and *blue* illumination are used for voltage sensitive dye and ChR2 excitations respectively; IR fluorescence emission is detected. (*Right*) IPSP recorded electrically (with a patch electrode) and optically from a the soma of a Purkinje neuron following ChR2 excitation at interneurons located in the molecular layer; modified from Tsuda et al. 2013 (**b**) Imaging of light evoked action potential in a ChR2 expressing interneuron (*left*) and of IPSP in CA1 pyramidal neuron induced by ChR stimulation at interneurons (*right*). (*Top*) Schematic illustrations of the experiment on the top with site of blue flash and imaging region. (*Bottom*) Simultaneous optical dendritic and electrical somatic recordings; *arrows* indicate blue light application (5 ms); modified from Willadt et al. 2014

changes from well identified sub-cellular compartments. In the examples shown in Fig. 4.9b, ChR2 was expressed by all GABAergic interneurons and individual hippocampal cells were filled with the indicator Di-2-ANBDQPTEA. When electrical and optical recordings were performed from a GABAergic interneuron, ChR2 photo-stimulation of the recorded cell generated an action potential that could be monitored both electrically and optically from proximal dendritic segment (Fig. 4.9b, left). In contrast, when electrical and optical recordings were performed from a CA1 pyramidal neuron, ChR2 photo-stimulation of pre-synaptic interneurons generated an IPSP that was recorded both electrically and optically from the proximal part of the apical dendrite (Fig. 4.9b, right). In conclusion, these initial studies indicate that combining membrane potential imaging with optogenetic photostimulation is likely to become a standard approach in the near future.

References

Antic S, Zecevic D (1995) Optical signals from neurons with internally applied voltage-sensitive dyes. J Neurosci 15:1392–1405

Berger T, Borgdorff A, Crochet S, Neubauer FB, Lefort S, Fauvet B, Ferezou I, Carleton A, Lüscher HR, Petersen CC (2007) Combined voltage and calcium epifluorescence imaging in vitro and in vivo reveals subthreshold and suprathreshold dynamics of mouse barrel cortex. J Neurophysiol 97:3751–3762

Brenowitz SD, Regehr WG (2005) Associative short-term synaptic plasticity mediated by endocannabinoids. Neuron 45:419–431

Bullen A, Patel SS, Saggau P (1997) High-speed, random-access fluorescence microscopy: I. High-resolution optical recording with voltage-sensitive dyes and ion indicators. Biophys J 73:477–491

Bullen A, Saggau P (1998) Indicators and optical configuration for simultaneous high-resolution recording of membrane potential and intracellular calcium using laser scanning microscopy. Pflugers Arch 436:788–796

Canepari M, Djurisic M, Zecevic D (2007) Dendritic signals from rat hippocampal CA1 pyramidal neurons during coincident pre- and post-synaptic activity: a combined voltage- and calcium imaging study. J Physiol 580:463–484

Canepari M, Nelson L, Papageorgiou G, Corrie JE, Ogden D (2001) Photochemical and pharmacological evaluation of 7-nitroindolinyl-and 4-methoxy-7-nitroindolinyl-amino acids as novel, fast caged neurotransmitters. J Neurosci Meth 112:29–42

Canepari M, Vogt KE (2008) Dendritic spike saturation of endogenous calcium buffer and induction of postsynaptic cerebellar LTP. PLoS One 3, e4011

Canepari M, Vogt K, Zecevic D (2008) Combining voltage and calcium imaging from neuronal dendrites. Cell Mol Neurobiol 58:1079–1093

Canepari M, Willadt S, Zecevic D, Vogt KE (2010) Imaging inhibitory synaptic potentials using voltage sensitive dyes. Biophys J 98:2032–2040

Cueni L, Canepari M, Adelman JP, Lüthi A (2009) Ca2+ signaling by T-type Ca2+ channels in neurons. Pflugers Arch 457:1161–1172

Djurisic M, Antic S, Chen WR, Zecevic D (2004) Voltage imaging from dendrites of mitral cells: EPSP attenuation and spike trigger zones. J Neurosci 24:6703–6714

Eilers J, Konnerth A (2000) Dye loading with patch pipettes. In: Yuste R, Lanni F, Konnerth A (eds) Imaging neurons a laboratory manual. Cold Spring Harbour Laboratory Press, New York, NY

Fierro L, Llano I (1996) High endogenous calcium buffering in Purkinje cells from rat cerebellar slices. J Physiol 496:617–625

Fluhler E, Burnham VG, Loew LM (1985) Spectra, membrane binding, and potentiometric responses of new charge shift probes. Biochemistry 24:5749–5755

Grinvald A, Hildesheim R (2004) VSDI: a new era in functional imaging of cortical dynamics. Nat Rev Neurosci 5: 874–885

Gupta RK, Salzberg BM, Grinvald A, Cohen LB, Kamino K, Lesher S, Boyle MB, Waggoner AS, Wang CH (1981) Improvements in optical methods for measuring rapid changes in membrane potential. J Mem Biol 58:123–137

Hyrc KL, Bownik JM, Goldberg MP (2000) Ionic selectivity of low-affinity ratiometric calcium indicators: mag-Fura-2, Fura-2FF and BTC. Cell Calcium 27:75–86

Jaafari N, De Waard M, Canepari M (2014) Imaging fast calcium currents beyond the limitations of electrode techniques. Biophys J 107:1280–1288

Kao JP, Tsien RY (1988) Ca2+ binding kinetics of fura-2 and azo-1 from temperature-jump relaxation measurements. Biophys J 53:635–639

Kee MZ, Wuskell JP, Loew LM, Augustine GJ, Sekino Y (2008) Imaging activity of neuronal populations with new long-wavelength voltage-sensitive dyes. Brain Cell Biol 36:157–172

Kremer SG, Zeng W, Skorecki KL (1992) Simultaneous fluorescence measurement of calcium and membrane potential responses to endothelin. Am J Physiol 263:1302–1309

Leão RN, Mikulovic S, Leão KE, Munguba H, Gezelius H, Enjin A, Patra K, Eriksson A, Loew LM, Tort (2012) OLM interneurons differentially modulate CA3 and entorhinal inputs to hippocampal CA1 neurons. Nat Neurosci 15:1524–1530

Loew LM, Simpson LL (1981) Charge-shift probes of membrane potential: a probable electrochromic mechanism for p-aminostyrylpyridinium probes on a hemispherical lipid bilayer. Biophys J 34:353–365

Mancuso JJ, Kim J, Lee S, Tsuda S, Chow NB, Augustine GJ (2011) Optogenetic probing of functional brain circuitry. Exp Physiol 96:26–33

Mann EO, Paulsen O (2007) Role of GABAergic inhibition in hippocampal network oscillations. Trends Neurosci 30:343–349

Martinez-Zaguilan R, Martinez GM, Lattanzio F, Gillies RJ (1991) Simultaneous measurement of intracellular pH and Ca2+ using the fluorescence of SNARF-1 and fura-2. Am J Physiol 260:297–307

Matiukas A, Mitrea BG, Pertsov AM, Wuskell JP, Wei MD, Watras J, Millard AC, Loew LM (2005) New near-infrared optical probes of cardiac electrical activity. Am J Physiol Heart Circ Physiol 290:H2633–H2643

Milojkovic BA, Zhou WL, Antic SD (2007) Voltage and calcium transients in basal dendrites of the rat prefrontal cortex. J Physiol 585:447–468

Naraghi M (1997) T-jump study of calcium binding kinetics of calcium chelators. Cell Calcium 22:255–268

Neher E (2000) Some quantitative aspects of calcium fluorimetry. In: Yuste R, Lanni F, Konnerth A (eds) Imaging neurons a laboratory manual. Cold Spring Harbour Laboratory Press, New York, NY

Roth A, Häusser M (2001) Compartmental models of rat cerebellar Purkinje cells based on simultaneous somatic and dendritic patch-clamp recordings. J Physiol 535:445–472

Sabatini BS, Oertner TG, Svoboda K (2002) The life cycle of Ca^{2+} ions in dendritic spines. Neuron 33:439–452

Sabatini BL, Regehr WG (1996) Timing of neurotransmission at fast synapses in the mammalian brain. Nature 384:170–172

Sabatini BL, Regehr WG (1997) Control of neurotransmitter release by presynaptic waveform at the granule cell to purkinje cell synapse. J Neurosci 17:3425–3435

Sabatini BL, Regehr WG (1998) Optical measurement of presynaptic calcium currents. Biophys J 74:1549–1563

Schneggenburger R, Meyer AC, Neher E (1999) Released fraction and total size of a pool of immediately available transmitter quanta at a calyx synapse. Neuron 23:399–409

Sinha SR, Patel SS, Saggau P (1995) Simultaneous optical recording of evoked and spontaneous transients of membrane potential and intracellular calcium concentration with high spatio-temporal resolution. J Neurosci Meth 60:46–49

Sinha SR, Saggau P (1999) Simultaneous optical recording of membrane potential and intracellular calcium from brain slices. Methods 18:204–214

Trigo FF, Papageorgiou G, Corrie JE, Ogden D (2009) Laser photolysis of DPNI-GABA, a tool for investigating the properties and distribution of GABA receptors and for silencing neurons in situ. J Neurosci Meth 181:159–169

Tsuda S, Kee MZ, Cunha C, Kim J, Yan P, Loew LM, Augustine GJ (2013) Probing the function of neuronal populations: combining micromirror-based optogenetic photostimulation with voltage-sensitive dye imaging. Neurosci Res 75:76–81

Vogt KE, Gerharz S, Graham J, Canepari M (2011a) High-resolution simultaneous voltage and Ca^{2+} imaging. J Physiol 589:489–494

Vogt KE, Gerharz S, Graham J, Canepari M (2011b) Combining membrane potential imaging with L-glutamate or GABA photorelease. PLoS One 6, e24911

Willadt S, Canepari M, Yan P, Loew LM, Vogt KE (2014) Combined optogenetics and voltage sensitive dye imaging at single cell resolution. Front Cell Neurosci 8:311

Wu JY, Cohen LB (1993) Fast multisite optical measurement of membrane potential. In: Mason WT (ed) Biological techniques: fluorescent and luminescent probes for biological activity. Academic, New York, NY

Wuskell JP, Boudreau D, Wei MD, Jin L, Engl R, Chebolu R, Bullen A, Hoffacker KD, Kerimo J, Cohen LB, Zochowski MR, Loew LM (2006) Synthesis, spectra, delivery and potentiometric responses of new styryl dyes with extended spectral ranges. J Neurosci Methods 151:200–215

Yan P, Acker CD, Zhou WL, Lee P, Bollensdorff C, Negrean A, Lotti J, Sacconi L, Antic SD, Kohl P, Mansvelder HD, Pavone FS, Loew LM (2012) Palette of fluorinated voltage-sensitive hemicyanine dyes. Proc Natl Acad Sci U S A 109:20443–20448

Yizhar O, Fenno LE, Davidson TJ, Mogri M, Deisseroth K (2011) Optogenetics in neural systems. Neuron 71:9–34

Yuste R (2000) Loading brain slices with AM Esters of calcium indicators. In: Yuste R, Lanni F, Konnerth A (eds) Imaging neurons a laboratory manual. Cold Spring Harbour Laboratory Press, New York, NY

Zhang F, Wang LP, Boyden ES, Deisseroth K (2006) Channelrhodopsin-2 and optical control of excitable cells. Nat Methods 3:785–792

Zhou W-L, Yan P, Wuskell JP, Loew LM, Antic SD (2007) Intracellular long-wavelength voltage-sensitive dyes for studying the dynamics of action potentials in axons and thin dendrites. J Neurosci Meth 164:225–239

Chapter 5
Monitoring Spiking Activity of Many Individual Neurons in Invertebrate Ganglia

W.N. Frost, C.J. Brandon, A.M. Bruno, M.D. Humphries, C. Moore-Kochlacs, T.J. Sejnowski, J. Wang, and E.S. Hill

Abstract Optical recording with fast voltage sensitive dyes makes it possible, in suitable preparations, to simultaneously monitor the action potentials of large numbers of individual neurons. Here we describe methods for doing this, including considerations of different dyes and imaging systems, methods for correlating the optical signals with their source neurons, procedures for getting good signals, and the use of Independent Component Analysis for spike-sorting raw optical data into single neuron traces. These combined tools represent a powerful approach for large-scale recording of neural networks with high temporal and spatial resolution.

Keywords Imaging • Voltage sensitive dye • Invertebrate • Optical recording • Ganglia • Absorbance dye • Spike sorting • Independent component analysis • Clustering

---

W.N. Frost (✉) • C.J. Brandon • J. Wang • E.S. Hill
Department of Cell Biology and Anatomy, The Chicago Medical School, Rosalind Franklin University of Medicine and Science, 3333 Green Bay Road, North Chicago, IL 60064, USA
e-mail: william.frost@rosalindfranklin.edu

A.M. Bruno
Department of Neuroscience, The Chicago Medical School, Rosalind Franklin University of Medicine and Science, North Chicago, IL 60064, USA

M.D. Humphries
Faculty of Life Sciences, University of Manchester, Manchester, UK

C. Moore-Kochlacs
Department of Mathematics and Statistics, Boston University, Boston, MA 02215, USA

McGovern Institute, Massachusetts Institute of Technology, Cambridge, MA 02139, USA

T.J. Sejnowski
Howard Hughes Medical Institute, The Salk Institute for Biological Studies, La Jolla, CA 92037, USA

Division of Biological Sciences, University of California San Diego, La Jolla, CA 92093, USA

M. Canepari et al. (eds.), *Membrane Potential Imaging in the Nervous System and Heart*, Advances in Experimental Medicine and Biology 859,
DOI 10.1007/978-3-319-17641-3_5

1 Introduction

Efforts to understand how the brain works require tools for observing circuits in action. Conventional electrophysiological methods can monitor no more than a few neurons at a time, providing a severely restricted perspective on network function. Optical recording with fast voltage-sensitive dyes (fVSDs) offers a way past this limitation, by revealing the action potentials of dozens to hundreds of individual neurons during behaviorally-relevant motor programs (Fig. 5.1). The availability of methods for monitoring large-scale brain activity with single-cell and sub-millisecond resolution is likely to advance our understanding of network function far beyond that attained using traditional methods (Yuste 2008).

It has been three decades since fVSDs were first used to monitor the simultaneous firing of multiple individual neurons (Grinvald et al. 1977; Salzberg et al. 1977). While most such studies have used invertebrate preparations, particularly those with large neurons whose action potentials fully invade the soma (Boyle et al. 1983; London et al. 1987; Cohen et al. 1989; Zecevic et al. 1989; Nakashima et al. 1992; Wu et al. 1994; Nikitin and Balaban 2000; Zochowski et al. 2000a; Brown et al. 2001; Kojima et al. 2001; Frost et al. 2007), a few studies have accomplished this in vertebrate preparations, such as the enteric nervous system ganglia (Neunlist et al. 1999; Obaid et al. 1999; Vanden Berghe et al. 2001; Schemann et al. 2002). See (Hill et al. 2014) for a recent review of VSD imaging in small neuronal networks.

Given the enormous potential of optical recording with fVSDs for studying neural networks, why haven't more laboratories adopted this approach? A key reason is that the optical signals corresponding to action potentials in individual neurons are miniscule—often ranging from .001 to less than .0001 of the resting light level—making their detection quite challenging. Another reason is the difficulty of combining high resolution imaging with intracellular recording from multiple neurons. Although such combined methodology would be highly useful for circuit mapping, the light-efficient compound microscopes used for imaging lack stereopsis, making integrating multiple intracellular electrodes so difficult that few bother. Ideally, one could penetrate and drive a known neuron, image its followers, and then penetrate those followers with a second electrode to test for a direct synaptic connection (Taylor et al. 2003). A third difficulty with using optical recording for network studies is that individual detectors often record multiple neurons, yielding mixed signals. Conversely, multiple detectors often record the same neuron, yielding redundant traces. As a result, it can be very difficult to know the number of neurons included in the optical data set.

Fortunately, it is possible to overcome these difficulties. Here we describe procedures for obtaining a satisfactory signal-to-noise ratio using fVSDs, for easier integration of sharp electrodes into imaging experiments, and for the transformation of raw data sets of mixed and redundant traces into new sets

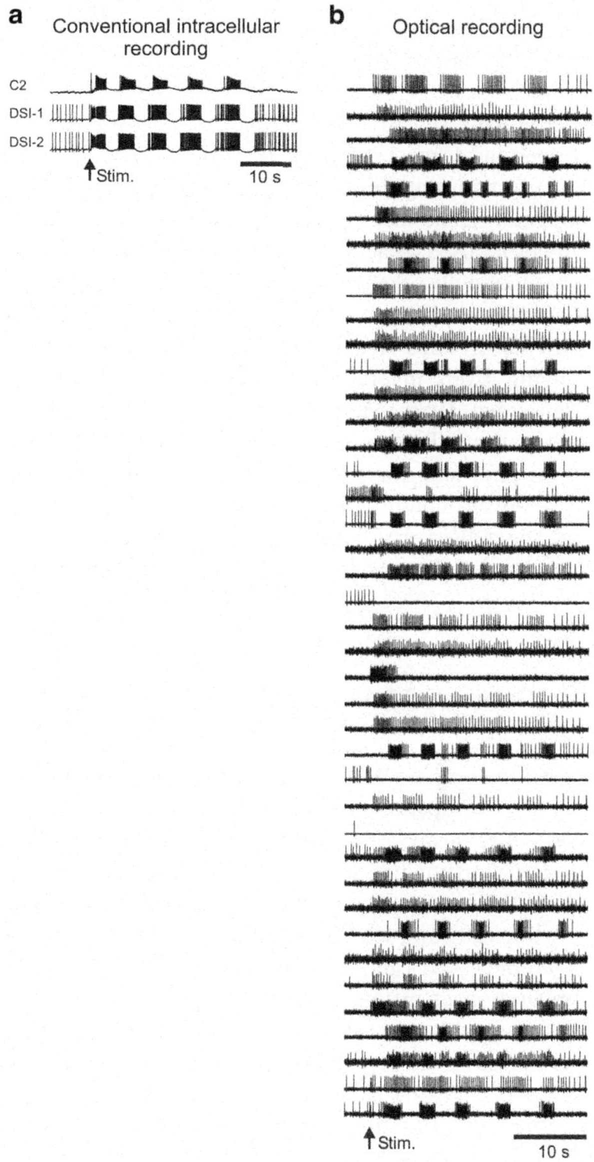

Fig. 5.1 Imaging with fast voltage sensitive dyes allows simultaneous recording of the individual action potentials generated by large numbers of neurons. (**a**) Conventional sharp electrode intracellular recording of the *Tritonia* swim motor program. A 1 s 10 Hz stimulus to Pedal Nerve 3 triggered a five cycle swim motor program, recorded in the C2 and DSI interneurons of the swim central pattern generator. Experiments using sharp electrodes are typically limited to recording 3–4 neurons simultaneously. (**b**) Optical recording from the dorsal pedal ganglion of the *Tritonia* swim motor program, after processing with Independent Component Analysis, showing 41 of the 63 neurons detected. A 2 s 10 Hz stimulus to Pedal Nerve 3 was used to trigger this five cycle motor program, which was recorded with a RedShirtImaging NeuroPDA-III 464 photodiode array and the fast absorbance voltage sensitive dye RH155

containing a single neuron per trace. This discussion is particularly relevant for investigators considering large scale optical recording with fVSDs with single cell resolution, such as is readily achievable in invertebrate ganglia and in certain vertebrate peripheral nervous system preparations. Here we focus primarily on the use of photodiode arrays.

2 Choice of Fluorescence vs. Absorbance Fast Voltage Sensitive Dyes

Fast voltage-sensitive dyes come in two main types: absorbance and fluorescence (Ebner and Chen 1995; Zochowski et al. 2000b). Both change their light response linearly with membrane potential, and do so fast enough to trace out each action potential. Generally speaking, absorbance dyes have been preferred for network studies focused on action potentials. Direct comparisons in the vertebrate brain slice have found absorbance dyes to yield larger signal-to-noise with less phototoxicity than the tested fluorescence dyes (Jin et al. 2002; Chang and Jackson 2003). Two such absorbance dyes, RH155 and RH482, have been applied to preparations ranging from invertebrate ganglia (Yagodin et al. 1999), to vertebrate slice (Senseman 1996; Momose-Sato et al. 1999; Yang et al. 2000), to cell culture preparations (Parsons et al. 1991). We have similarly used both dyes successfully in *Tritonia* and *Aplysia*. Recently, fast fluorescent dyes have been developed that have lower phototoxicity and other desired qualities than previous dyes (for example, see (Obaid et al. 2004; Carlson and Coulter 2008; Miller et al. 2012)). We have not yet tried these on our invertebrate preparations.

3 Choice of Imaging System

Most imaging systems employ either a camera or photodiode array. Both designs operate as a spatial array of independent sensors that convert light to voltage, with the entire array sampled at a fixed frequency. Two systems we have tried by RedShirtImaging are the NeuroCMOS-SM128 camera and the NeuroPDA-III photodiode array. The CMOS-SM128 has 16,384 independent sensors arranged on a 128 × 128 chip, with the full set sampled at 2,500 Hz. Using the RedShirtImaging Neuroplex software, the data can be viewed either as a camera image of the preparation at different points in time, with pixel color or grayscale brightness indicating the membrane potential of that location, or as a set of up to 16,384 V traces (one trace per pixel). The latter can be spatially averaged using either grid-based binning or user-outlined regions of interest, such as single neurons. By contrast, the RedShirtImaging NeuroPDA-III is organized as a hexagonal array of 464 sensors (Fig. 5.2), with the full set sampled at 1,600 Hz. The PDA data are shown as miniature traces organized on screen in the positions of the diodes that recorded them. Because the PDA does not have the spatial resolution to form a useful image, the 464 miniature traces are superimposed on an image of the preparation taken by a digital camera that is also used to focus the PDA on the neurons of interest. As with the camera system, diodes can be selected individually or in groups for display of their recorded voltage traces, raw or filtered, individual or spatially averaged, in a separate window in Neuroplex (Fig. 5.3).

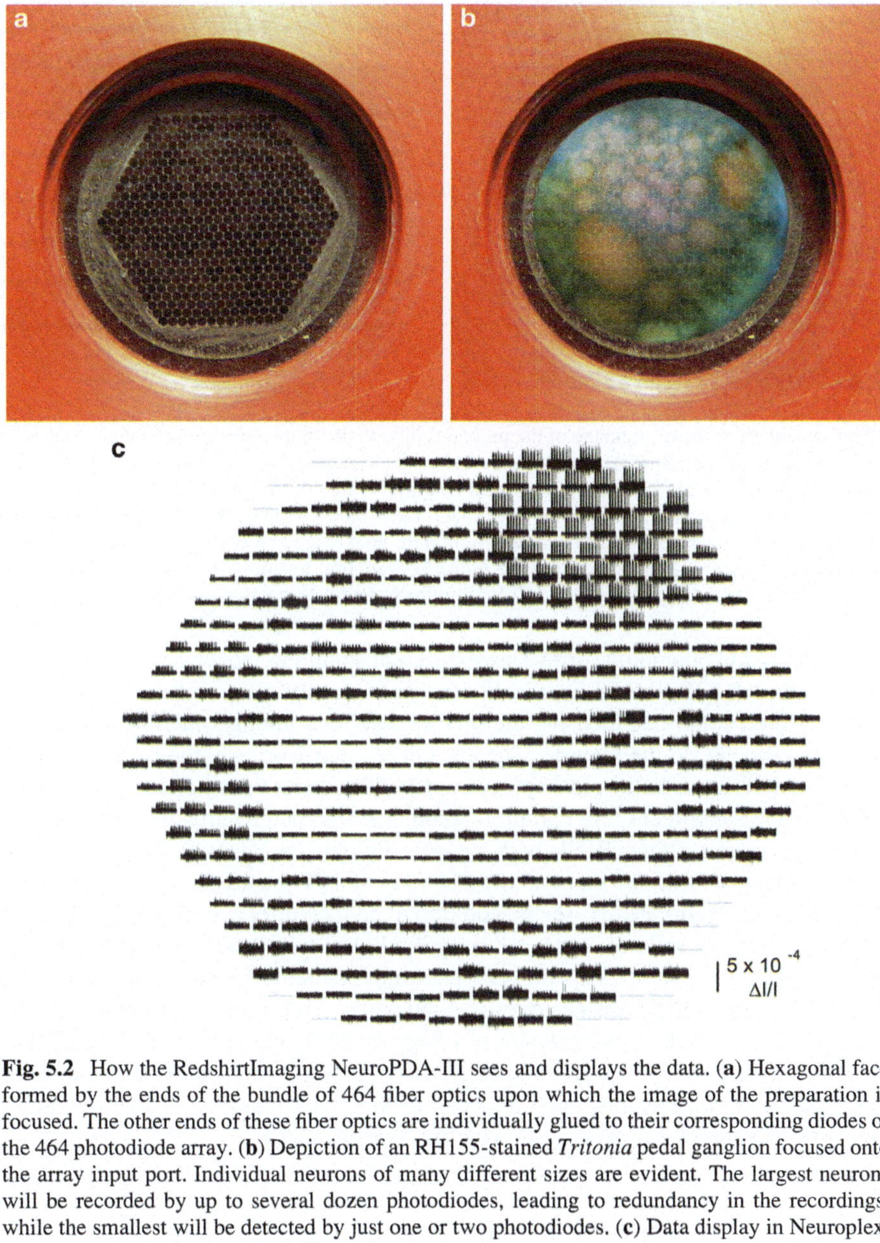

Fig. 5.2 How the RedshirtImaging NeuroPDA-III sees and displays the data. (**a**) Hexagonal face formed by the ends of the bundle of 464 fiber optics upon which the image of the preparation is focused. The other ends of these fiber optics are individually glued to their corresponding diodes of the 464 photodiode array. (**b**) Depiction of an RH155-stained *Tritonia* pedal ganglion focused onto the array input port. Individual neurons of many different sizes are evident. The largest neurons will be recorded by up to several dozen photodiodes, leading to redundancy in the recordings, while the smallest will be detected by just one or two photodiodes. (**c**) Data display in Neuroplex. As soon as the data are acquired, Neuroplex displays them as miniature traces organized according to the position of each acquiring diode in the array. Shown is a 35 s optical recording of a swim motor program, obtained from a different preparation than the one photographed in panel (**b**). Nearly 50 diodes detected a single large neuron located in the upper right portion of the array. Clicking on specific traces expands them in a separate window in Neuroplex (see Fig. 5.3)

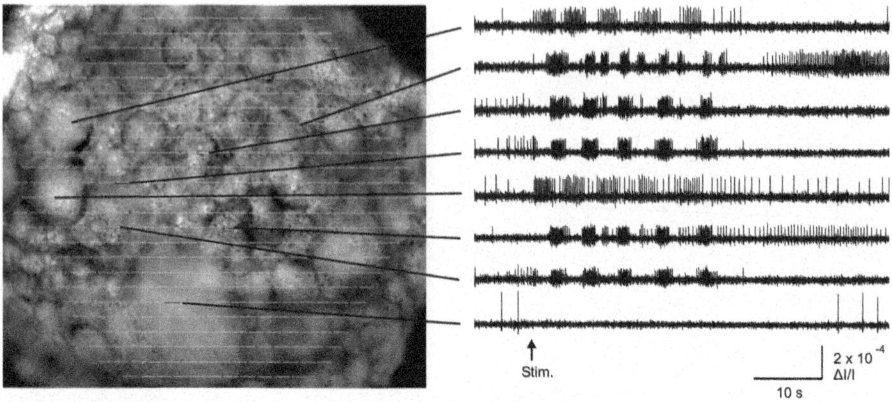

Fig. 5.3 Superimposing an image of the preparation with the diode array makes it easy to inspect the firing of different neurons. The *left panel* shows an image of the dorsal surface of the pedal ganglion, superimposed in Neuroplex with a display of the 464 acquired optical traces in their corresponding diode positions (*green lines*). The *right panel* shows how clicking on diodes of interest displays their optical data. These are filtered, unaveraged recordings from single diodes, each showing the firing recorded from the indicated ganglion location. A five cycle swim motor program was elicited by a 10 Hz, 1 s, 10 V stimulus to Pedal Nerve 3, delivered at the *arrow*

Cameras and photodiode arrays have different strengths and weaknesses. Cameras such as the CMOS-SM128 have sufficient pixel density to provide a recognizable image, which simplifies focusing the imager and determining which regions and/or neurons gave rise to the recorded signals. On the negative side, the camera's larger sensor number leads to file sizes approximately 50× larger than those acquired by the PDA for the same recording duration. Another difference regards the sensitivity of the two types of systems. While both systems digitize the data with 14 bit resolution (the CMOS system in the camera, the PDA via an A/D board in the computer), the NeuroPDA-III gains nearly seven additional bits of effective resolution by first AC-coupling each trace to subtract the resting light level, and then amplifying it by 100× before digitization. This makes PDAs more suitable for detecting the smaller signals provided by absorbance dyes if digitization resolution is a limiting factor, a point made by several others (Sinha and Saggau 1999; Vanden Berghe et al. 2001; Kosmidis et al. 2005). It is worth noting, however, that AC-coupled PDAs are unsuited for studies of slow events, such as slow EPSPs. The majority of our remaining discussion concerns our experience with the NeuroPDA-III.

4 Two Microscope Designs for Optical Recording

We use two different microscope designs, depending on the degree to which intracellular electrodes will be used in the experiment. In both systems (Fig. 5.4), the microscope is mounted on an EXFO-Burleigh motorized translator attached to a

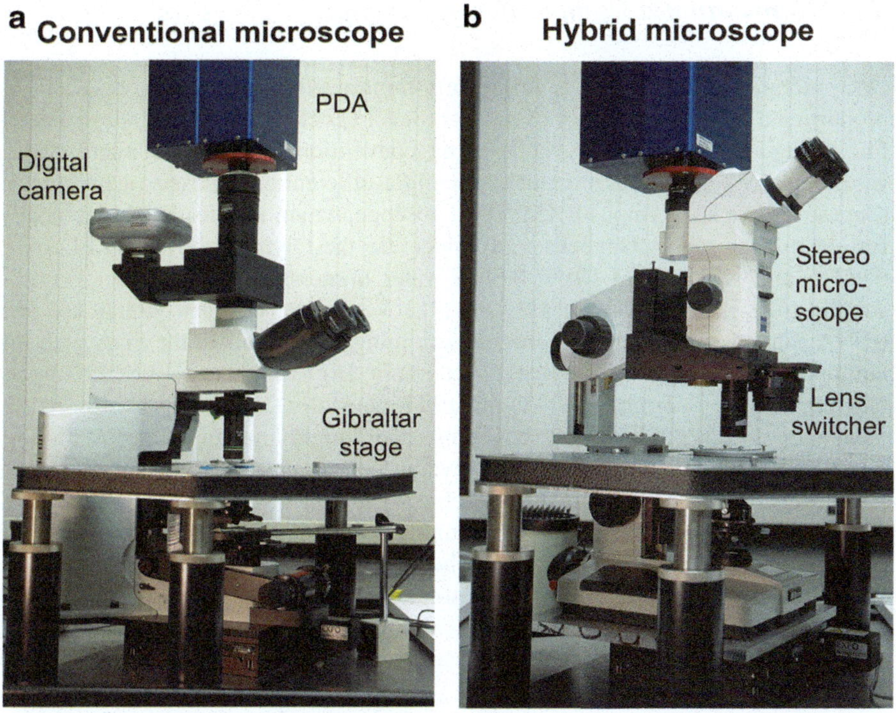

Fig. 5.4 Two microscope designs used for imaging. (**a**) For experiments involving minimal-to-no conventional intracellular recording we use a standard compound microscope. A parfocal digital camera is used to focus the PDA, and to provide a picture of the preparation to superimpose with the array data. (**b**) For experiments involving significant intracellular recording with the imaging we use a custom microscope consisting of the lower half of a compound microscope, to provide the bright transillumination needed for absorbance fVSDs, mated to and optically aligned with a stereomicroscope. A linear sliding lens switcher is used to alternate between a 0.6 NA 20× objective for the imaging, and a parfocal stereomicroscope objective for penetrating neurons with sharp electrodes. During experiments the imaging lens is fitted with a custom water immersion cap to eliminate image movement related to surface ripples. Both microscopes are mounted on motorized translators to allow the viewing/imaging area to be changed without disturbing impaled neurons. During experiments, the upper superstructure is further stabilized by inserting two foam-topped posts between the Gibraltar platform and the front corners of the lens switcher. A digital eyepiece camera is used with this microscope to obtain the images of the preparation to be superimposed with the array data

Burleigh Gibraltar stage that provides a large stable platform for mounting the manipulators, recording chamber, suction electrodes, bath ground, bath temperature probe, optional epi-illumination, perfusion tubes, and cooling system used in our *Tritonia* or *Aplysia* experiments. The translator allows penetration of a neuron in one ganglion, and then subsequent movement of the microscope and PDA without disturbing the impalement, in order to optically record from neurons in other ganglia that are affected by spike trains driven in the impaled neuron.

4.1 Conventional Compound Microscope

Once vibration-related noise is sufficiently minimized, the signal-to-noise ratio is proportional to the square root of the light level (Zochowski et al. 2000b). Therefore, imaging is traditionally done using conventional compound microscopes, whose high NA objectives are much more light-efficient than those of stereomicroscopes. We use an Olympus BX51WI microscope, which was designed specifically for electrophysiology. Our lenses of choice are the Olympus XLUMPLFL20xW 0.95NA and XLUMPLFL10xW 0.6NA water immersion objectives. In *Tritonia* ganglia transilluminated with a 100 W, 12 V tungsten-halogen lamphouse and 0.9NA condenser, these lenses provide saturating light levels to the PDA with the lamp power set to slightly above half-maximum. High efficiency optics are important because they permit lower illumination intensities, translating to slower dye bleaching, lower phototoxicity concerns, and longer imaging time. With this microscope and optics we can obtain several 120 s acquisitions at near-saturating light levels for the PDA before dye bleaching begins to degrade the action potential signals. We have recently begun using a high power 735 nm LED light source (Thor Labs M735L2-C1) to transilluminate the preparation, driven with a DC power supply (Thor Labs DC2100). This stimulator-controlled illumination method avoids the need for both a bandpass optical filter and an electromechanical shutter, the latter of which can introduce unwanted vibrations that can potentially add artifact signals to the imaging data.

4.2 Custom Hybrid Microscope

Despite their advantages, conventional compound microscopes lack stereopsis, making it difficult to incorporate multiple intracellular electrodes into imaging experiments. This is a major hindrance for many potential uses of optical recording, such as circuit mapping, where one would like to locate and immediately penetrate specific neurons identified in the optical recordings. Another aim is to drive a known neuron with one intracellular electrode while imaging its follower neurons, and then penetrate those with a second intracellular electrode. To carry out such experiments, we constructed a hybrid microscope that combines desirable features of both stereo and conventional compound microscopes.

We began by cutting off the top of a Nikon Optiphot microscope and then mounted a Zeiss SV-11 stereomicroscope onto a vertical post bolted to its stump. This allows the preparation to be viewed through the stereo objective, with the full depth perception necessary to penetrate specific neurons with intracellular electrodes. At the same time we retained the compound microscope's lamphouse and condenser, to deliver maximum light to the preparation for imaging. A key part of this design is the linear sliding lens switcher (Combizoom 400, Kramer Scientific), which allows the user to alternate between a high numerical aperture (NA) objective for imaging and a conventional parfocal stereomicroscope objective for impaling neurons. This hybrid microscope design has been described in detail in a separate publication (Frost et al. 2007).

While the ability to integrate multiple intracellular electrodes into optical recording experiments is a big plus, the hybrid microscope has certain disadvantages compared with the traditional compound microscope. First, light transmission is lower through the zoom optics of the stereomicroscope body, requiring delivery of brighter illumination to the preparation to achieve adequate signal-to-noise ratios. To get sufficient illumination, we power a traditional 12 V 100 W tungsten halogen bulb with a 16 V racing car battery stepped down to 14.5 V. This has the advantage of true DC power, and hence no ripple in the light to the imager. However, the brighter light comes at the cost of faster photobleaching and thus less imaging time per experiment (four to five 35 s acquisitions vs. several 120 s acquisitions with the conventional compound microscope setup). Second, the method of hanging the stereomicroscope and PDA from the rear post leads to increased vibration-based noise in the optical recordings. We minimize the vibration by inserting support posts beneath the front of the lens switcher, effectively supporting the microscope in three places. This design has greatly facilitated our use of multiple intracellular electrodes during imaging experiments. With intracellular recording comes the need to expose preparations to light while positioning electrodes and penetrating neurons. To avoid bleaching the fVSD while working with intracellular electrodes, we illuminate the preparation through a green filter that blocks the red wavelength used for imaging with RH155.

5 Correlating Optical Signals with Specific Neurons in the Preparation

Because of its poor spatial resolution, a PDA doesn't provide a useful image—its frame view mode looks like a hexagonal checkerboard of grayscale squares, corresponding to the brightness recorded by the different diodes at one point in time. Therefore, identifying the neurons the data are coming from is more difficult with a PDA than with camera-based imagers, which have much higher spatial resolution. To do this, we take a picture of the preparation and superimpose it in Neuroplex with a diode map linked to the optical data. Clicking on any diode/neuron then displays the optical data recorded at that location (Fig. 5.3). The image superimposition involves a three step procedure:

During experiments with the PDA, the focus point for the imaging is set using the live image provided by the microscope camera used to take the superimposed photograph. To ensure that the PDA and camera are parfocal, we focus the camera on a calibration slide containing a black circle, and then adjust the focusing collar under the PDA so that the circle's edge shows the most abrupt black-to-white transition on the checkerboard pattern provided by the PDA. An alternative focusing method is to remove the PDA from its photoport and place a ground glass surface at the exact position the array face normally sits. Adjustments are then made to bring this image and that seen by the camera into parfocality. A third method is suggested by (Obaid et al. 2004), who used a custom microscope photoport component which, when slid into place, provides a direct view of the preparation image focused onto the photodiode array face.

When using sharp electrodes in combination with imaging, ganglia are surgically desheathed to allow easy penetration of the neurons. However, such preparations tend to be less planar than when sheathed, with the result that photographs taken through the objective lens will have just a portion of the surface neurons in focus. To obtain a clear picture of the entire visible ganglion surface to superimpose with the PDA data one can take a stack of several images at different focus depths and then combine them into a single in-focus image using additional software, such as the fast focus enhancement feature of Rincon (Fig. 5.5; ImagingPlanet, Goleta, CA). When working with sheathed preparations, we typically image through a coverslip fragment pressed down on the ganglion, which flattens it further. This has the dual advantages of yielding more in-focus neurons, and also a higher yield of optically recorded neurons (see Sect. 6.5).

In order to correlate the imaging data with specific neurons, the above photograph must be precisely superimposed with the diode map. To do this, we collect both a camera picture and PDA readings of the light shining through three closely

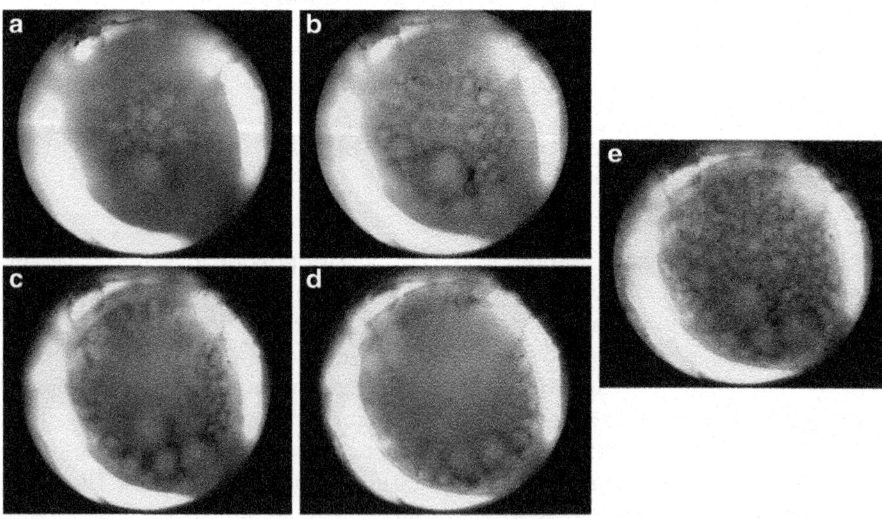

Fig. 5.5 Use of software to obtain a single in-focus image to superimpose with the photodiode array data. In order to obtain a suitable picture of the highly convex *Tritonia* ganglion surface to superimpose with the array data in Neuroplex during experiments, we use software that combines the in-focus portions of several images taken at different focus positions, while discarding the *blurred areas*. (**a**) Uppermost surface of the sheathed *Tritonia* dorsal pedal ganglion. (**b**) Focused 100 μm lower. (**c**) 200 μm lower. (**d**) 300 μm lower. (**e**) Combination of 13 images taken at 25 μm intervals, showing all neurons visible on the ganglion surface

spaced pinholes in a foil sheet. The magnification and position of the camera image is then adjusted in Neuroplex until it exactly overlies the PDA image of the pinholes, rotating the PDA as needed, after which it is locked into position. In subsequent experiments these same coordinates can then be used to accurately superimpose the camera image of the preparation on the array data.

6 Getting Good Signals

When recording with fVSDs, neuronal action potentials may produce signals in the range of .001 to less than .0001 of the resting light level. Discriminating such small signals is a challenge. Here we describe some key steps that make this possible using the absorbance dye RH155 in the marine mollusks *Tritonia diomedea* and *Aplysia californica*.

6.1 Choose an Optimal Transmission Filter

The ability to see the small signals provided by voltage sensitive absorption dyes requires the use of illumination wavelengths optimized for each dye. RH155 and RH482, for example, have biphasic absorption curves – below a certain wavelength they decrease their absorption with depolarization, above it they increase absorption (Parsons et al. 1991; Jin et al. 2002). To avoid signal cancellation, when illuminating with white light the bandpass filter selected for use with such dyes must thus be centered on one side or the other of the crossover wavelength, with an appropriately limited bandwith. After trying several, we chose a 725/25 bandpass filter (Chroma Technology) for our work with *Tritonia* and *Aplysia* using RH155 when using tungsten halogen illumination. Because the spectral characteristics of voltage sensitive dyes can shift in different solutions and tissues, the filter properties needed to provide optimal neuronal signals should be determined in the preparation of interest (Wang et al. 2009). As mentioned in Sect. 4.1, the use of an appropriate high-power LED illumination system avoids the need for a bandpass filter.

6.2 Get the Light as Close to Saturation as Possible

Once vibration-related noise has been sufficiently minimized (see below), signal-to-noise ratio increases in proportion to the square root of the illumination intensity (Zochowski et al. 2000b). This means that optimal optical recordings will be obtained when the resting light level reaching the imager is as close to the device's saturation level as possible. There are two steps for achieving this: (1) maximize the light to the preparation, and (2) maximize the light transmitted from the preparation to the imager.

6.2.1 Maximize the Light to the Preparation

The lamphouse should be adjusted to position the bulb filament directly in the center of the field of view. If it has an adjustable back mirror, this should be positioned to make the image as bright as possible. Built-in diffusion and heat filters, the latter which attenuate the wavelength we use with RH155 (725 nm), can be removed from the light path between the lamphouse and the condenser to maximize light transmission. Alternatively, the built-in heat filter could be replaced with one chosen to block IR wavelengths while passing those needed for the particular dye being used. Different condensers concentrate the light into smaller or larger areas, reflecting the magnification range of the objectives they were designed to work with. The typical condenser illuminates an area of the recording chamber far larger than the area seen by the 20× objective we use in most experiments. After trying several condensers, the brightest light delivery with tungsten halogen illumination was obtained with a simple 0.9 NA Abbe achromat condenser having a swing-in top lens. This style is designed for use with low magnification objectives with the top lens removed, and higher magnification lenses with the top lens in place, which concentrates the light into a smaller area, closer to the field of view of our 20× imaging objective. The brighter light provided by our LED illumination system allows the use of conventional condensers. Condenser height is empirically adjusted to produce the brightest illumination of the preparation, which occurs near the point of Kohler illumination, when the base diaphragm of the microscope is in sharp focus. With the swing-in top lens condenser, this occurs close enough to the preparation to require a recording chamber having a cover slip bottom topped by a layer of Sylgard just thick enough to hold the minuten pins used to position the preparation. Under Kohler illumination, the microscope base diaphragm can be adjusted to constrict the illuminated area to the exact area being imaged. This prevents bleaching of other areas in the preparation, such as nearby ganglia, allowing them to be imaged later in the experiment.

6.2.2 Maximize the Light from the Preparation to the Imager

Increased efficiency in this part of the light path allows the desired near-saturating light levels to be obtained with lower lamp brightness, and thus slower dye bleaching and reduced potential phototoxicity. It also translates to longer imaging time, beneficial for monitoring long duration events such as the *Tritonia* swim motor program. An important way to maximize light transmission in this part of the pathway is to use the highest NA objective lens available for the desired magnification, matched with a condenser with a comparable NA. A photoport should also be used that allows 100 % of the light to be directed to the imager.

6.3 Reduce Vibration-Related Noise

The major source of noise in PDA experiments arises from vibration, which causes small oscillatory movements of contrast edges in the preparation on the photodiodes, producing oscillations in the recordings that can be many times the size of the optically recorded action potentials. Common sources of vibration include the opening of the shutter controlling the light to the preparation, pulsations related to perfusion pumps, and movements of the floor and walls caused by the building air handling systems, as well as by the movements of people in and outside the laboratory. A vibration isolation table is essential for minimizing externally-originating vibrations. We currently use a spring-based isolation table (Kinetics Systems MK26), which has been suggested by others to provide vibration isolation superior to that provided by conventional air tables for VSD imaging (Jin et al. 2002). In addition, we (1) work on the ground floor, (2) use a compound microscope designed with low vibration and electrophysiology in mind (Olympus BX51WI), (3) use commercial water immersion objectives, or affix custom made water immersion caps to our non-water immersion objectives to eliminate vibrations at the air/solution interface, (4) mount the imaging shutter on the frame of the air table rather than on its surface, or use an LED illumination system that makes a shutter unnecessary, (5) ensure that cables connected to outside devices contact the table before reaching the microscope and imager, and (6) for the hybrid microscope, which otherwise would be supported only by the back mounting post, we insert two metal posts topped with foam pads between the frame of the lens switcher and the Gibraltar stage. The Neuroplex software can readily reveal the effects of these vibration-reducing efforts by displaying a Fourier analysis of the frequency components in the optical recordings.

6.4 Issues Related to Preparations Requiring Cold Saline

For preparations requiring cold saline, such as *Tritonia* (11 °C), condensation can form on the bottom side of the recording chamber, interfering with imaging. To prevent this, we place a drop of fluid on the condenser lens, which then fuses to the bottom of the recording chamber when the condenser is raised to its working position. In times of high humidity, significant condensation can also form on the inside of water immersion objective lenses, or on the inside of custom made water immersion lens caps. During such conditions, we run a dehumidifier in the laboratory, and also place the objective and any lens cap in a dessicant-containing sealed plastic bag the night before the experiment. Just before recording, we attach the cap within the bag before placing the lens on the microscope. Our custom lens cap has a lightly greased O-ring within it to both hold it securely on the lens and to establish an airtight seal.

6.5 Other Methods for Increasing the Number of Neurons Detected in the Imaging

While high-NA objective lenses are highly desirable for their light efficiency, this comes at the cost of a narrower depth of focus. This can be advantageous—the focus point of the microscope provides information on the depth location of the optically recorded neurons. A disadvantage is that signal loss from out-of-focus neurons means that many are not detected. A simple method to increase the number of neurons detected optically is to gently flatten the ganglion using a coverslip fragment held in place by Vaseline placed beside the ganglion on the chamber floor. In addition, we have recently begun selectively applying dye to just the ganglion to be imaged, by pushing it through a polyethelene tube pressed down over the ganglion. These procedural refinements are currently providing yields of 150–200 recorded neurons in our *Aplysia* preparations.

7 Spike-Sorting the Raw Optical Data with Independent Component Analysis

Many scientific issues can be addressed by simply inspecting the filtered, but otherwise raw diode recordings. For circuit mapping, previously unknown neurons that fire during a given motor program are easily identified. Neurons excited or inhibited by neurons driven with intracellular electrodes are also easily spotted. Other issues, however, cannot readily be addressed through simple visual inspection of the data traces. For example, it is nearly impossible to determine the number of neurons present in a typical optical recording. The larger neurons appear redundantly in many traces, while many traces contain action potentials from multiple neurons (Fig. 5.6a). To take full advantage of the power of optical recording, we need a method for spike-sorting these mixed and redundant data sets.

Independent Component Analysis (ICA) has been introduced as a rapid and automated method for spike-sorting optical recording data sets obtained with fVSDs (Brown et al. 2001; Brown et al. 2008). ICA is a blind source separation methodology for separating signal mixtures into their original sources. ICA finds a linear transformation of the raw traces that minimizes the mutual information shared between the components, separating the redundant and mixed neuronal source signals into separate, single-neuron traces (Fig. 5.6b1). For an imager with N pixels or diodes, ICA will return up to N independent components, which can include artifacts and other noise sources, effectively removing them from the neuronal recordings (Brown et al. 2001).

We use infomax ICA (Bell and Sejnowski 1995; Delorme and Makeig 2004). With a current laptop computer running Windows, ICA can process a 120 s file of 464 optical traces digitized at 1,600 Hz into the set of independent components in a

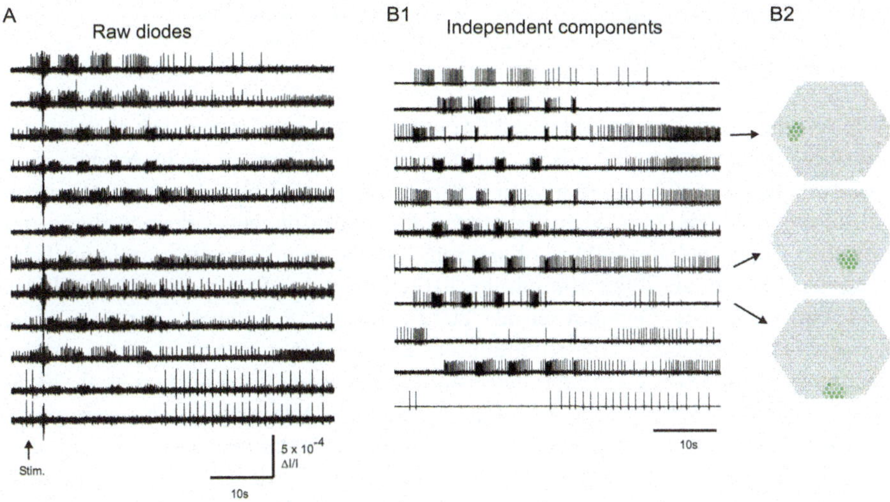

Fig. 5.6 Use of Independent Component Analysis to sort the raw data into a set of single-neuron traces. (**a**) Selected traces from the raw diode data, after processing with 5 Hz high-pass and 100 Hz low-pass Butterworth software filters in Neuroplex. A 10 Hz, 2 s stimulus was applied to Pedal Nerve 3 at the *arrow*, eliciting a four cycle swim motor program. The raw traces contain redundantly recorded neurons (see the large action potentials in traces 1 and 2, and 11 and 12), as well as mixtures of more than one neuron per trace (most apparent in traces 2, 9, 10 and 11). While neurons of interest are readily identified in the raw data, it is impossible to determine how many neurons were recorded by visual inspection alone. (**b1**) Selected traces of the 61 components containing neurons that were extracted by ICA from the 464 diode raw data set. ICA transforms the recorded set of signal mixtures into their original single-neuron sources. The *top trace* is the independent component corresponding to the neuron with large action potentials recorded redundantly by the top two diodes in panel (**a**). The *bottom trace* is the independent component corresponding to the neuron recorded redundantly by the bottom two diodes in panel (**a**). ICA is an automated process that can be run on any MATLAB-equipped personal computer. Note that the prominent artifact present shortly after the stimulus in most of the raw data traces is absent from the component traces, having been extracted as a separate independent component (not shown.) (**b2**) Array maps of the diodes contributing to three components, indicating the XY position and approximate size of the source neurons

few minutes (Hill et al. 2010). We have empirically found that ICA resolves more neurons when provided with data sets of several tens of seconds duration that also include spontaneous activity (Hill et al. 2012). MATLAB programs have been written (C.M.-K., M.H.) that return maps showing the position of each resolved component on the diode array (Fig. 5.6b2).

Using ICA we are able to (1) eliminate redundancy by combining neurons recorded by multiple diodes into single traces, (2) sort multiple neurons recorded by single diodes into separate single-neuron traces, (3) provide the array location of each resolved neuron, and (4) remove artifacts from the traces. ICA thus represents a powerful, fast and automated tool for spike-sorting data sets obtained with photodiode arrays and fast voltage sensitive dyes.

7.1 Extracting and Mapping Neuronal Ensembles from Spike-Sorted Data

While some topics can be addressed through simple visual inspection of the action potential traces returned by ICA, further analysis is needed to investigate most issues related to network organization (Briggman et al. 2006). For example, many clustering methods have been developed that partition network recordings into separate functional ensembles of neurons having significantly correlated firing patterns (Feldt et al. 2009; Humphries 2011; Lopes-dos-Santos et al. 2011; Lyttle and Fellous 2011; Gerstein et al. 2012). Figure 5.7 shows our use of an unsupervised consensus clustering approach applied to a recording of the *Aplysia* locomotion motor program recorded from the pedal ganglion (Bruno et al. 2015). The resolved ensembles can be plotted as action potential rasters, a graphical network layout, and with respect to the position of each neuron in the ganglion. Ensemble organization can then be the starting point for deeper studies of network structure across time for individual motor programs and also across multiple motor programs for studies of how they change with experience. Major strides in understanding the brain will require the continued development of specialized analytical tools for analyzing large-scale recordings of neural activity (Koch 2012; Alivisatos et al. 2013).

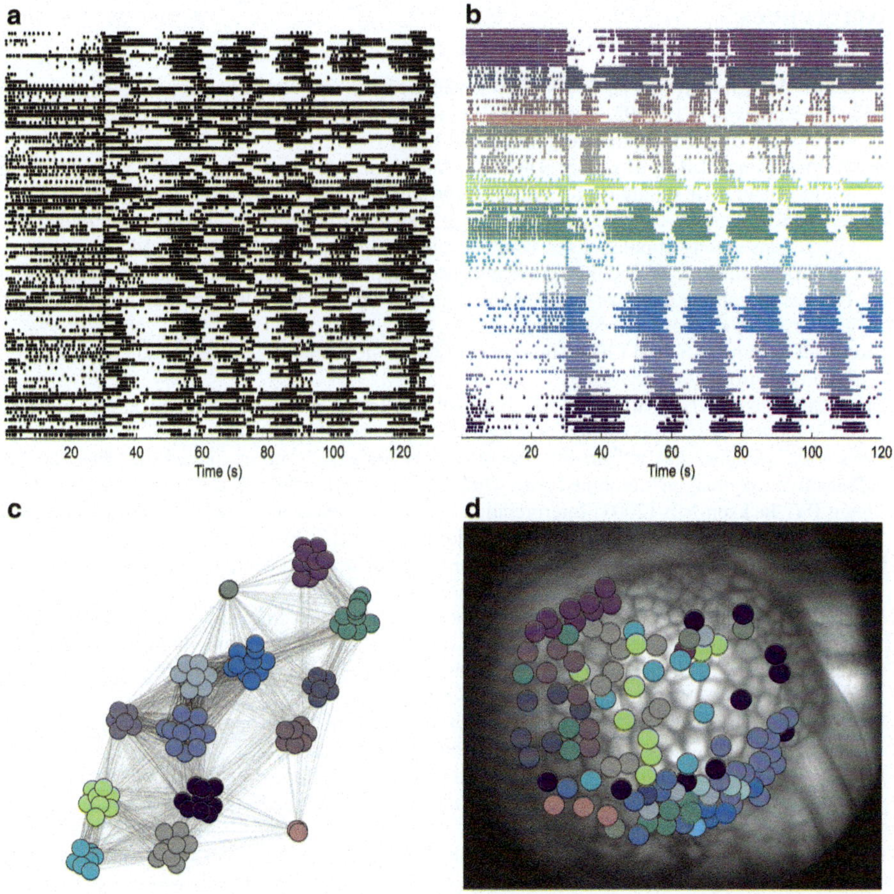

Fig. 5.7 Use of clustering to identify and map functional neuronal ensembles participating in the *Aplysia* locomotion motor program. (**a**) Action potential recording of 109 neurons in the dorsal right pedal ganglion. 30 s into the 2 min recording, a nerve stimulus was applied to trigger the rhythmic motor program. (**b**) Unsupervised clustering yielded 14 ensembles based on their degree of significantly correlated firing. Different colors represent each ensemble. The graph theoretic based consensus clustering method is described in detail elsewhere (Bruno et al. 2015). (**c**) Network visualization displaying the degree of firing correlation among the recorded neurons. *Gray-scale* of the *lines* indicates the degree of correlation, with *darker lines* indicating more highly correlated firing (Traud et al. 2009). (**d**) All recorded neurons are shown at their locations on the ganglion (*medial side at left*)

Acknowledgments Supported by NIH NS060921, NSF 1257923, Dart Foundation and Grass Foundation Marine Biological Laboratory summer fellowships, the Fred B. Snite Foundation, Rosalind Franklin University of Medicine and Science (WF), NIH F31NS079036 (AB), MRC Senior non-Clinical Fellowship (MH) and the Howard Hughes Medical Institute (TS).

References

Alivisatos AP, Chun M et al (2013) Neuroscience: the brain activity map. Science 339:1284–1285

Bell AJ, Sejnowski TJ (1995) An information-maximization approach to blind separation and blind deconvolution. Neural Comput 7(6):1129–1159

Boyle MB, Cohen LB et al (1983) The number and size of neurons in the CNS of gastropod molluscs and their suitability for optical recording of activity. Brain Res 266(2):305–317

Briggman KL, Abarbanel HD et al (2006) From crawling to cognition: analyzing the dynamical interactions among populations of neurons. Curr Opin Neurobiol 16(2):135–144

Brown GD, Yamada S et al (2008) Independent component analysis of optical recordings from Tritonia swimming neurons, Technical Report INC-08-001, Institute for Neural Computation, University of California at San Diego

Brown GD, Yamada S et al (2001) Independent component analysis at the neural cocktail party. Trends Neurosci 24(1):54–63

Bruno AM, Frost WN et al (2015) Modular deconstruction reveals the dynamical and physical building blocks of a locomotion motor program. Neuron 86:304–18

Carlson GC, Coulter DA (2008) In vitro functional imaging in brain slices using fast voltage-sensitive dye imaging combined with whole-cell patch recording. Nat Protoc 3(2):249–255

Chang PY, Jackson MB (2003) Interpretation and optimization of absorbance and fluorescence signals from voltage-sensitive dyes. J Membr Biol 196(2):105–116

Cohen L, Hopp HP et al (1989) Optical measurement of action potential activity in invertebrate ganglia. Annu Rev Physiol 51:527–541

Delorme A, Makeig S (2004) EEGLAB: an open source toolbox for analysis of single-trial EEG dynamics including independent component analysis. J Neurosci Methods 134(1):9–21

Ebner TJ, Chen G (1995) Use of voltage-sensitive dyes and optical recordings in the central nervous system. Prog Neurobiol 46(5):463–506

Feldt S, Waddell J et al (2009) Functional clustering algorithm for the analysis of dynamic network data. Phys Rev E Stat Nonlin Soft Matter Phys 79(5 Pt 2):056104

Frost WN, Wang J et al (2007) A stereo-compound hybrid microscope for combined intracellular and optical recording of invertebrate neural network activity. J Neurosci Methods 162(1–2):148–154

Gerstein GL, Williams ER et al (2012) Detecting synfire chains in parallel spike data. J Neurosci Methods 206(1):54–64

Grinvald A, Salzberg BM et al (1977) Simultaneous recording from several neurones in an invertebrate central nervous system. Nature 268(5616):140–142

Hill ES, Bruno AM et al (2014) Recent developments in VSD imaging of small neuronal networks. Learn Mem 21(10):499–505

Hill ES, Bruno AM et al (2012) ICA applied to VSD imaging of invertbrate neuronal networks. In: Ganesh N (ed) Independent component analysis for audio and biosignal applications. InTech: Chapter 12, p. 235–246

Hill ES, Moore-Kochlacs C et al (2010) Validation of independent component analysis for rapid spike sorting of optical recording data. J Neurophysiol 104(6):3721–3731

Humphries MD (2011) Spike-train communities: finding groups of similar spike trains. J Neurosci 31(6):2321–2336

Jin W, Zhang RJ et al (2002) Voltage-sensitive dye imaging of population neuronal activity in cortical tissue. J Neurosci Methods 115(1):13–27

Koch C (2012) Modular biological complexity. Science 337(6094):531–532

Kojima S, Hosono T et al (2001) Optical detection of neuromodulatory effects of conditioned taste aversion in the pond snail Lymnaea stagnalis. J Neurobiol 49(2):118–128

Kosmidis EK, Cohen LB et al (2005) Imaging with voltage-sensitive dyes: spike signals, population signals, and retrograde transport. In: Yuste R, Konnerth A (eds) Imaging in neuroscience and development: a laboratory manual. Cold Spring Harbor Laboratory Press, Cold Spring Harbor, NY, pp 445–455

London JA, Zecevic D et al (1987) Simultaneous optical recording of activity from many neurons during feeding in Navanax. J Neurosci 7(3):649–661

Lopes-dos-Santos V, Conde-Ocazionez S et al (2011) Neuronal assembly detection and cell membership specification by principal component analysis. PLoS One 6(6), e20996

Lyttle D, Fellous JM (2011) A new similarity measure for spike trains: sensitivity to bursts and periods of inhibition. J Neurosci Methods 199(2):296–309

Miller EW, Lin JY et al (2012) Optically monitoring voltage in neurons by photo-induced electron transfer through molecular wires. Proc Natl Acad Sci U S A 109(6):2114–2119

Momose-Sato Y, Sato K et al (1999) Evaluation of voltage-sensitive dyes for long-term recording of neural activity in the hippocampus. J Membr Biol 172(2):145–157

Nakashima M, Yamada S et al (1992) 448-Detector optical recording system: development and application to *Aplysia* gill-withdrawal reflex. IEEE Trans Biomed Eng 39:26–36

Neunlist M, Peters S et al (1999) Multisite optical recording of excitability in the enteric nervous system. Neurogastroenterol Motil 11(5):393–402

Nikitin ES, Balaban PM (2000) Optical recording of odor-evoked responses in the olfactory brain of the naive and aversively trained terrestrial snails. Learn Mem 7(6):422–432

Obaid AL, Koyano T et al (1999) Spatiotemporal patterns of activity in an intact mammalian network with single-cell resolution: optical studies of nicotinic activity in an enteric plexus. J Neurosci 19(8):3073–3093

Obaid AL, Loew LM et al (2004) Novel naphthylstyryl-pyridium potentiometric dyes offer advantages for neural network analysis. J Neurosci Methods 134(2):179–190

Parsons TD, Salzberg BM et al (1991) Long-term optical recording of patterns of electrical activity in ensembles of cultured *Aplysia* neurons. J Neurophysiol 66:316–333

Salzberg BM, Grinvald A et al (1977) Optical recording of neuronal activity in an invertebrate central nervous system: simultaneous monitoring of several neurons. J Neurophysiol 40(6):1281–1291

Schemann M, Michel K et al (2002) Cutting-edge technology. III. Imaging and the gastrointestinal tract: mapping the human enteric nervous system. Am J Physiol Gastrointest Liver Physiol 282(6):G919–G925

Senseman DM (1996) High-speed optical imaging of afferent flow through rat olfactory bulb slices: voltage-sensitive dye signals reveal periglomerular cell activity. J Neurosci 16(1):313–324

Sinha SR, Saggau P (1999) Optical recording from populations of neurons in brain slices. In: Johanson H, Windhorst U (eds) Modern techniques in neuroscience research. p. 459–486

Taylor AL, Cottrell GW et al (2003) Imaging reveals synaptic targets of a swim-terminating neuron in the leech CNS. J Neurosci 23(36):11402–11410

Traud AL, Frost C et al (2009) Visualization of communities in networks. Chaos 19(4):041104

Vanden Berghe P, Bisschops R et al (2001) Imaging of neuronal activity in the gut. Curr Opin Pharmacol 1(6):563–567

Wang Y, Jing G et al (2009) Spectral characterization of the voltage-sensitive dye di-4-ANEPPDHQ applied to probing live primary and immortalized neurons. Opt Express 17(2):984–990

Wu J, Cohen LB et al (1994) Neuronal activity during different behaviors in *Aplysia*: a distributed organization. Science 263:820–823

Yagodin S, Collin C et al (1999) Mapping membrane potential transients in crayfish (Procambarus clarkii) optic lobe neuropils with voltage-sensitive dyes. J Neurophysiol 81(1):334–344

Yang S, Doi T et al (2000) Multiple-site optical recording of mouse brainstem evoked by vestibulocochlear nerve stimulation. Brain Res 877(1):95–100

Yuste R (2008) Circuit neuroscience: the road ahead. Front Neurosci 2(1):6–9

Zecevic D, Wu J et al (1989) Hundreds of neurons in the *Aplysia* abdominal ganglion are active during the gill-withdrawal reflex. J Neurosci 9:3681–3689

Zochowski M, Cohen LB et al (2000a) Distributed and partially separate pools of neurons are correlated with two different components of the gill-withdrawal reflex in Aplysia. J Neurosci 20(22):8485–8492

Zochowski M, Wachowiak M et al (2000b) Imaging membrane potential with voltage-sensitive dyes. Biol Bull 198(1):1–21

Part III
Monitoring Activity of Networks and Large Neuronal Populations

Chapter 6
Monitoring Integrated Activity of Individual Neurons Using FRET-Based Voltage-Sensitive Dyes

Kevin L. Briggman, William B. Kristan, Jesús E. González, David Kleinfeld, and Roger Y. Tsien

Abstract Pairs of membrane-associated molecules exhibiting fluorescence resonance energy transfer (FRET) provide a sensitive technique to measure changes in a cell's membrane potential. One of the FRET pair binds to one surface of the membrane and the other is a mobile ion that dissolves in the lipid bilayer. The voltage-related signal can be measured as a change in the fluorescence of either the donor or acceptor molecules, but measuring their ratio provides the largest and most noise-free signal. This technology has been used in a variety of ways; three are documented in this chapter: (1) high throughput drug screening, (2) monitoring the activity of many neurons simultaneously during a behavior, and (3) finding synaptic targets of a stimulated neuron. In addition, we provide protocols for using the dyes on both cultured neurons and leech ganglia. We also give an updated description of the mathematical basis for measuring the coherence between electrical and optical

K.L. Briggman
Circuit Dynamics and Connectivity Unit, National Institute of Neurological Disorders and Stroke, National Institutes of Health, Bethesda, MD 20892, USA
e-mail: kevin.briggman@nih.gov

W.B. Kristan (✉)
Neurobiology Section, Division of Biological Sciences, University of California, San Diego 9500 Gilman Drive, La Jolla, CA 92093-0357, USA
e-mail: wkristan@ucsd.edu

J.E. González
Avelas Biosciences, 11099 North Torrey Pines Road, Suite 290, La Jolla, CA 92037, USA
e-mail: txsandpiper@sbcglobal.net

D. Kleinfeld
Department of Physics, University of California, San Diego, 9500 Gilman Drive, La Jolla, CA 92093-0374, USA

R.Y. Tsien
Department of Pharmacology, Howard Hughes Medical Institute, University of California, San Diego 310 George Palade Labs, 9500 Gilman Drive, La Jolla, CA 92093-0647, USA

M. Canepari et al. (eds.), *Membrane Potential Imaging in the Nervous System and Heart*, Advances in Experimental Medicine and Biology 859,
DOI 10.1007/978-3-319-17641-3_6

signals. Future improvements of this technique include faster and more sensitive dyes that bleach more slowly, and the expression of one of the FRET pair genetically.

Keywords Fluorescence resonance energy transfer (FRET) • Voltage-sensitive dyes • Invertebrate ganglia • Invertebrate neurons • Genetically encoded probes • Functional connectivity

1 Introduction

Fluorescence resonance energy transfer (FRET) is a physical dipole–dipole coupling between the excited state of a donor fluorophore and an acceptor chromophore that causes relaxation of the donor to a non-fluorescent ground state which excites fluorescence in the acceptor. Initially described by Förster (1948), FRET has been extensively reviewed (Stryer 1978; Clegg 1995; Selvin 2000; Lakowicz 2006). In practical terms, the efficiency of FRET depends on the properties of the chromophores and the distance between them, measured as the Förster radius (R_o): the distance between the donor and the acceptor at which half the energy is transferred. The magnitude of R_o depends on the donor quantum yield and the spectral overlap between the donor emission and the acceptor absorbance spectra. For commonly used synthetic FRET pairs, R_o values range from 2 to 8 nm (Wu and Brand 1994). FRET efficiency is inversely proportional to the sixth power of the donor and acceptor distance, providing a sensitive readout of intermolecular distances near R_o. Experimentally, FRET is measured either as the decrease in the lifetime or intensity of donor fluorescence after the addition of acceptor, or as the increase in acceptor fluorescence after the addition of donor. Because typical R_o values are similar to protein and membrane dimensions, FRET has proven useful in a wide variety of biochemical and cellular applications to investigate protein–protein interactions, protein and DNA conformational analysis, and membrane topography (Stryer 1978; Clegg 1995; Selvin 2000; Lakowicz 2006). Furthermore, with the advent of a large variety of fluorescent protein color variants, FRET has been natural method to probe cellular biochemistry (Piston and Kremers 2007). Here we review how FRET probes have been used to measure cellular membrane potentials.

Voltage sensitive dyes (VSD) based upon FRET are composed of two molecules, with either the donor or acceptor being a hydrophobic anion introduced into the plasma membrane acting as the voltage sensor by translocating between the energy minima at the intracellular and extracellular membrane-water interfaces (Fig. 6.1a) (Gonzalez and Tsien 1995). When the transmembrane potential changes, the hydrophobic anion redistributes as an exponential function of the potential according to the Nernst equation. A second impermeant fluorophore is attached to one face of the membrane where it can undergo FRET with the mobile molecule, in proportion to the distance between the two molecules. When the impermeant fluorophore is bound to the extracellular membrane surface and the cell is at its normal (negative) resting potential, the anions are predominately near the extracellular face of the membrane, so that the two molecules produce efficient FRET.

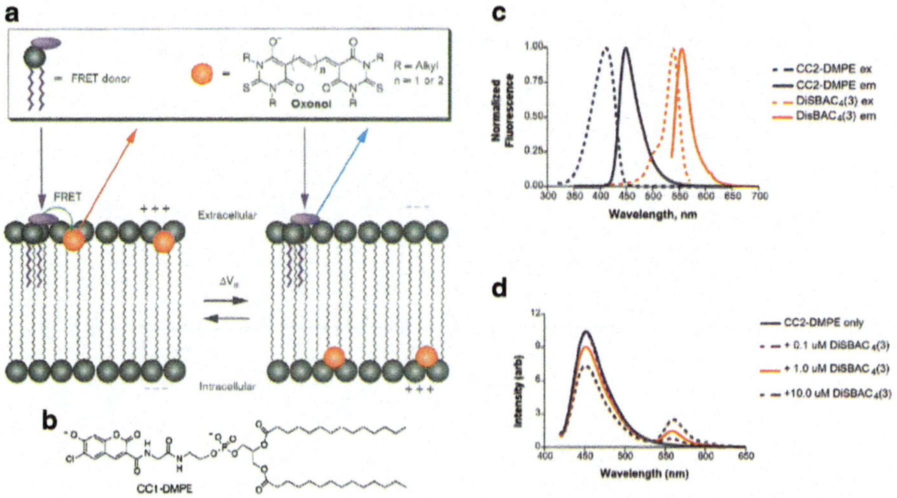

Fig. 6.1 Properties of a FRET VSD dye pair. (**a**) Most FRET VSDs incorporate an immobile donor (*purple*) and a mobile anion (*orange*). A variety of molecules can be used for this purpose. Oxonol derivatives are a versatile choice for the mobile anion. When the donor fluorophore is excited (*purple arrow*), it will either emit blue photons when the mobile anion is near the intracellular surface of the membrane (*blue arrow in the right diagram*) or produce FRET when the mobile ion is near the extracellular surface (*orange arrow in the left diagram*). (**b**) The structure of a FRET donor, CC1-DMPE, a coumarin-labelled phospholipid. (**c**) The excitation and emission spectra for the CC2-DMPE and $DiSBAC_4(3)$ dye pair. The overlap between donor emission and acceptor excitation determines the FRET efficiency; the more complete the overlap, the more efficient is FRET. The emission spectra for these two dyes are well separated, making them a useful dye pair. (**d**) Emission spectra of the two dyes as $DiSBAC_4(3)$ was loaded into cell membranes. Increased concentrations of $DiSBAC_4(3)$ progressively quenched the CC2-DMPE emission. Panels (**a**) and (**b**) are reproduced with permission from Gonzalez and Tsien 1997, and panels (**c**) and (**d**) are reproduced with permission from Gonzalez and Maher 2002

When the membrane depolarizes, for whatever reason, the anions equilibrate at a higher density at the intracellular membrane surface, thereby decreasing FRET. When both molecules are fluorescent, all increases in the acceptor emission are at the expense of the donor emission, and vice versa (Fig. 6.2a, b), thus providing a ratiometric signal of the membrane potential change.

FRET-based dyes are sometimes called 'slow' dyes, which is true relative to electrochromic dyes that have typical response times of a few μs (Ebner and Chen 1995; Baker et al. 2005), but FRET dye time constants actually span a large range, from 400 μs to 500 ms, depending on the properties of the mobile anion. While not the fastest, FRET dyes have yielded some of the largest observed fractional fluorescence changes, ranging from 10 to 20 % per 100 mV in intact tissue (Cacciatore et al. 1999) to 100–300 % per 100 mV in isolated cells (Gonzalez and Maher 2002).

Following a brief account of the development of FRET VSDs, we discuss their temporal resolution, sensitivity, and phototoxicity. We then present examples of how these dyes have been used to image neurons. Finally, we provide detailed staining protocols as a reference and starting points for use in other systems.

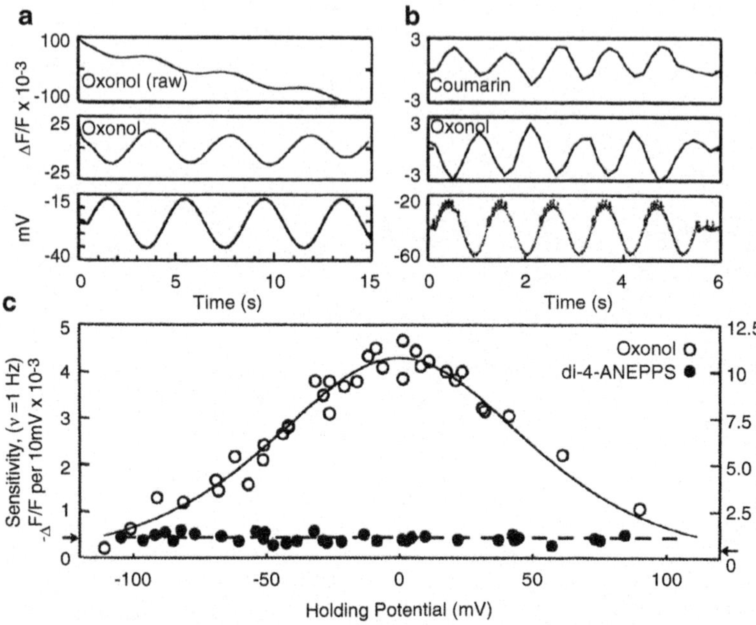

Fig. 6.2 Voltage dependent signals of FRET VSDs. (**a**) Leech neurons were voltage clamped and the membrane potentials was sinusoidally varied about a holding potential (*bottom trace*). The raw oxonol trace (*top trace*) was filtered (*middle trace*) to remove the slow bleaching artifact visible in the *top trace*. (**b**) Sequential recordings of the coumarin and oxonol emission wavelengths to the same sinusoidally varying intracellular stimulus demonstrating that, because the two dyes produce out-of-phase signals, the sensitivity can be enhanced by measuring the ratio of the two signals. (**c**) The sensitivity of the FRET signal is a function of the membrane potential measured at various holding potentials. The data are well fit by the expected redistribution of oxonol across the membrane during changes in membrane potential (*open circles*). FRET VSDs in leech ganglia are significantly more sensitive than other VSDs such as di-4-ANEPPS (*filled circles*). Panels reproduced with permission from Cacciatore et al. 1999

2 Development of FRET Dye Pairs

Gonzalez and Tsien were first to image voltage-dependent responses from a FRET-based VSD (Gonzalez and Tsien 1995). They used negatively charged hydrophobic oxonol derivatives, bis-(1,3-dialkyl-2-thiobarbiturate)-trimethineoxonol, ($DiSBAC_x(3)$, where x refers to the number of carbons on the alkyl group and three refers to the number of polymethine carbons), as the mobile voltage sensor. Negatively charged oxonol molecules such as $DiSBAC_2(3)$ and bis-(1,3-dibutylbarbituric acid) trimethine oxonol [$DiBAC_4(3)$] have previously been used as redistribution VSDs (Rink et al. 1980; Ebner and Chen 1995). They move between intracellular and extracellular compartments in response to membrane potential changes with very slow kinetics (seconds to minutes). An impermeant oxonol has also been used as a fast absorption dye (Grinvald et al. 1981; Wu et al. 1994a; Momose-Sato et al. 1999).

Fluorescence changes result from the fact that oxonols are non-fluorescent in water but are highly fluorescent in hydrophobic environments such as membranes. An anion was chosen, rather than a cation, to take advantage of the dipole potential generated from the polar lipid head-groups, presumably from the ester carbonyl groups, which greatly speeds anion translocation. The rate differences for isostructural borates and phosphonium ions are orders of magnitude faster for the negatively charged borates (Flewelling and Hubbell 1986). Although the mobile anion can in principle serve as either the FRET donor or acceptor, it is most commonly used as the FRET acceptor. As discussed below, there are theoretical and practical advantages for the reverse configuration. The FRET donor was initially chosen to be fluorescein-labeled wheat germ agglutinin (FL-WGA) that bound to N-acetylglucosamine groups on the extracellular face of plasma membranes (Gonzalez and Tsien 1995). The same study showed that using longer chain alkyl groups not only increased the oxonol hydrophobicity but also resulted in faster membrane translocation speed, which improved the FRET temporal resolution, presumably by burying the oxonol deeper into the low dielectric region of the membrane. The optimal chain length was six and $DiSBAC_6(3)$ had a translocation time constant ~2 ms.

The original dye pair was improved by replacing FL-WGA with a fluorescent phospholipid, N-(6-chloro-7-hydroxy-2-oxo-2H-l-benzopyran-3-carboxamidoacetyl)-dimyristoylphosphatidylethanolamine [CC1-DMPE] (Gonzalez and Tsien 1997) (Fig. 6.1b). A second chlorocoumarin phospholipid CC2-DMPE is commercially available from Life Technologies (formerly Invitrogen); it has similar properties to CC1-DMPE. The coumarin has a high quantum yield (indistinguishable from 1) and makes an excellent FRET partner with trimethine thiobarbiturate oxonols. The fluorescence emission maxima are separated by 100 nm (Fig. 6.1c), thus enabling efficient collection of both donor and acceptor photons. Another key property of chlorocoumarin phospholipid donors is that they have two negative charges at physiological pH. This is important because it anchors the fluorescent head-group at the extracellular surface and prevents probe translocation across the plasma membrane which would greatly degrade the voltage-sensitive FRET signal. In addition to the phosphate charge, the chloro-group reduces the pKa of the hydroxy coumarin (Fig. 6.1b) to ~5, which results in a second negative charge at physiological pH. A pyrene phospholipid has also been developed as a pH insensitive FRET donor that is compatible with low pH conditions used to activate acid sensitive channels (Maher et al. 2007). The use of a phospholipid placed the donor fluorophore closer to the membrane-water interface, thereby decreasing the minimal distance between donor and acceptor, which may account for the increased voltage sensitivity with the fluorescent impermeant phospholipid (Gonzalez and Tsien 1997). In addition to the original oxonol dye $DiSBAC_6(3)$, longer wavelength pentamethine oxonols were developed (the oxonol shown in Fig. 6.1a has $n = 2$) (Gonzalez and Tsien 1997). Increasing the polymethine chain length from trimethine oxonol [$DiSBAC_6(3)$] to pentamethine oxonol [$DiSBAC_6(5)$] increases charge delocalization, thereby further lowering the activation energy for translocation. This change yielded a time constant for $DiSBAC_6(5)$ of ~0.4 ms.

The FRET-based VSD strategy has more recently been used in hybrid voltage sensor systems, incorporating fast anions, such as dipicrylamine (DPA) or $DiBAC_4(5)$, with a genetically expressed donor fluorophore such as farnesylated enhanced GFP (eGFP-F) (Chanda et al. 2005). DPA is an absorption dye that does not fluoresce; it serves only to quench the fluorescence of the donor molecule when FRET occurs. The structure of the donor fluorophore has been additionally modified to be able to record action potentials (DiFranco et al. 2007; Sjulson and Miesenbock 2008).

Together these developments and optimizations of FRET-based VSDs demonstrate the versatility in choosing the donor and acceptor for a variety of applications. Most importantly, the source of the voltage sensitivity is well understood making the rational design of future improvements possible.

3 Response Time of FRET Indicators

Because the lifetime of FRET is short, on the order of nanoseconds, the time constant of FRET VSDs is determined by the rate that the hydrophobic anion equilibrates across the membrane when the membrane potential changes (Gonzalez and Tsien 1995). As mentioned above, increasing the length of the alkyl side chains of oxonol increases hydrophobicity and therefore increases its response speed. For example, $DiSBAC_2(3)$ and $DiSBAC_6(3)$ have time constants of 500 and 2 ms, respectively (Gonzalez and Maher 2002). However, increased hydrophobicity comes at the cost of reduced aqueous solubility, limiting loading concentrations (Table 6.1). Using pluronic F-127 and ß-cyclodextrin brings the more hydrophobic oxonols into contact with cells, which helps to load these dyes. The optimal loading concentration for any FRET acceptor should be high enough to quench a large fraction of donor fluorescence (Fig. 6.1d), to be sure that the acceptor is in the same membrane as the donor, but not so high that it adds a large capacitance to the cell membranes. In practice, loading concentrations for $DiSBAC_2(3)$ and $DiSBAC_4(3)$ oxonol should not exceed 20 μM (Cacciatore et al. 1999), although

Table 6.1 Oxonol derivatives that can be used as the mobile anion and their associated time constants and measured sensitivities

	Commerc. available	Pluronic loading	λ_{ex} (nm)	λ_{em} (nm)	T_c (ms)	V_m sensitivity % ΔR per mV
$DiSBAC_2(3)$	Yes	Optional	540	560	500	1–3
$DiSBAC_4(3)$	Yes	Suggested	540	560	20	0.6–1
$DiSBAC_6(3)$	No	Required	540	560	2	0.4–0.8
$DiSBAC_2(5)$	No	Optional	640	660	50	0.5–2[a]
$DiSBAC_4(5)$	No	Suggested	640	660	2	<0.4[a]
$DiSBAC_6(5)$	No	Required	640	660	0.40	<0.2[a]

[a]Not optimized

Table reproduced with permission from Gonzalez and Maher 2002 the effect of voltage sensitive anions on membrane capacitance should always be assayed with intracellular electrical recordings when applied to a new preparation. For example, concentrations of DPA above 2 μM inhibit action potentials in *Drosophila* neurons (Sjulson and Miesenbock 2008).

Increasing charge delocalization by changing the polymethine chain length of oxonol from trimethine to pentamethine increases the translocation speed 2- to 20-fold (Table 6.1). $DiSBAC_6(5)$ has a time constant of ~0.4 ms compared to 2 ms for $DiSBAC_6(3)$ (Gonzalez and Tsien 1997). The time constant of DPA used in hybrid voltage sensor systems is ~0.5 ms (Chanda et al. 2005). All the quoted oxonol time constants were measured at 20 °C; increasing temperature to 29 °C speeds up the translocation rate two to threefold (Gonzalez and Tsien 1995).

4 Sensitivity of FRET Indicators

The efficiency of FRET depends on the Förster distance, R_o, the distance of at which FRET is 50 % efficient. To ensure that FRET occurs selectively between the donor and acceptor on the same side of the membrane, ideally R_o needs to be less than the width of the intra- and extracellular binding sites of the membrane (3–5 nm). One way to decrease R_o while maintaining strongly absorbing fluorophores with high quantum yield is to decrease spectral overlap between the donor emission and acceptor excitation spectra (Fig. 6.1c). For the original fluorescein-labeled wheat germ agglutinin (FL-WGA) and oxonol pairs, the measured fluorescence changes where more than 20-fold less than expected from a Nernst equation-based mechanism, meaning the voltage sensitivity of oxonol was not efficiently transduced into a usable FRET signal (Gonzalez and Tsien 1995). Bringing the donor closer to the membrane and decreasing the donor and acceptor spectral overlap, by replacing FL-WGA with CC1-DMPE reduced the minimal distance between the two fluorophores, increasing FRET efficiency and, thereby, the sensitivity of the measurement.

The modification of $DiSBAC_6(3)$ to $DiSBAC_6(5)$, in addition to increasing translocation speed, shifted the absorption spectrum of oxonol to longer wavelengths by 100 nm (Gonzalez and Tsien 1997). The decreased overlap between the CC1-DMPE emission and $DiSBAC_6(5)$ absorption spectra further reduced R_o to 3.7 nm. The reduction in FRET due to the decreased spectral overlap is compensated by an increase in FRET selectivity for oxonol molecules near the extracellular face of the membrane. Separating the spectral overlap between donor and acceptor has the added benefit of limiting inadvertent illumination of the acceptor. Unfortunately, the pentamethine oxonols bleach much faster than trimethine oxonols and this imposes a significant limitation on using this faster oxonol.

The voltage sensitivity of FRET VSDs is maximal when the acceptor can quench most of the donor emission when the two molecules are in close proximity. Thus, the stoichiometry of the loading concentrations of the donor and acceptor needs to be optimized depending on cell type. In theory, it is preferable to have the mobile anion serve as FRET donor because efficient FRET relies on high acceptor concentrations. Lowering oxonol concentration limits photodamage, electrostatic repulsion between oxonol molecules, and the capacitive load on cells. A proof of this principle used $DiSBAC_6(3)$ as the mobile FRET donor and Cy5-labelled DMPE as the immobile acceptor for which ratiometric sensitivities of 5–15 % per 100 mV were recorded (Gonzalez and Tsien 1997). Using the mobile ion as the donor and lowering its concentration, however, produced rapid bleaching. Nevertheless, because there are multiple benefits to using the mobile voltage sensor as FRET donor, further development of this approach is warranted (Dumas and Stoltz 2005), including the development of a more photostable mobile fluorescent anion.

The loading concentrations of $DiSBAC_2(3)$ and $DiSBAC_4(3)$ oxonols are limited to <20 μM for reasons discussed above. For more hydrophobic oxonols, such as $DiSBAC_6(3)$ which have been used primarily in cell culture, the loading concentrations are much lower (<2 μM) due to their lower aqueous solubility and higher membrane partitioning. Loading concentrations of CC1-DMPE are limited to <20 μM in leech neurons because higher concentrations lowered the input resistance of the neurons (Cacciatore et al. 1999). The use of the modified compounds, CC2-DMPE or CC3-DMPE, allows concentrations of up to 160 μM to be used without significantly reducing membrane resistance.

The voltage dependence of the sensitivity can be fit by the Boltzman statistics for a two-state system under the influence of a potential (Fig. 6.2c) (Gonzalez and Tsien 1995; Cacciatore et al. 1999). The most linear and sensitive region of this curve lies within the physiological range. As with all VSDs, measured sensitivities vary between cell type and experimental preparation. The largest ratiometric sensitivities of 100–300 % per 100 mV have been measured in isolated cell lines using CC2-DMPE and $DiSBAC_2(3)$ (Gonzalez and Maher 2002). The use of faster, more hydrophobic oxonols, $DiSBAC_4(3)$ and $DiSBAC_6(3)$, yield sensitivities of 60–100 % per 100 mV and 40–80 % per 100 mV, respectively (Table 6.1). This reflects the tradeoff between speed and sensitivity as the hydrophobicity of the voltage sensor increases. Sensitivities are reduced for intact tissues such as a leech ganglion due to background staining. Optimal ratioed sensitivities of leech neurons using CC2-DMPE and $DiSBAC_2(3)$ are 10–20 % per 100 mV (Fig. 6.2c) (Cacciatore et al. 1999; Briggman et al. 2005). By comparison, the styryl VSD di-4-ANEPPS had sensitivities in leech neurons of around 0.5 % per 100 mV.

5 Genetically Encoded FRET Sensors

Two major limitations of synthetic membrane potential dyes are poor tissue penetration and indiscriminate cellular staining, both of which produce reduced signal above background. A specific challenge for neuroscience applications is labeling target neurons in among non-neuronal cells such as glia. To address these issues, several approaches have been explored and developed that utilize genetically encoded fluorescent proteins (FP). Initial efforts involved creating fusion proteins between FP and voltage-gated ion channels and then expressing the construct in the cell or tissue of interest. One general strategy is to turn membrane potential changes into conformational changes in voltage-dependent ion channels, which is then transduced into optical changes. Examples of such voltage-sensitive proteins include: Shaker potassium channels (Siegel and Isacoff 1997; Guerrero et al. 2002); a sodium channel (Ataka and Pieribone 2002); and the voltage sensor domain (VSD) of Kv2.1 potassium channel (Sakai et al. 2001; Knopfel et al. 2003). Although detectable, the voltage-sensitive fluorescence changes are small and often have complex kinetics resulting from multiple voltage-dependent conformational changes of the VSD, channel, and FP. With the goals of creating ratiometric probes and increasing the voltage-sensitive fluorescence change, constructs were made to express FP FRET pairs attached to voltage-sensitive proteins. These efforts in combination with using the VSD from a voltage sensor containing phosphase from *Ciona intestinalis* (Ci-VSD) have led to development of new FRET-based genetically encoded probes VSFP2.x (Dimitrov et al. 2007) and Mermaid (Tsutsui et al. 2008). These probes have improved voltage-sensitive fluorescence changes with ratiometric changes of ~10 % per 100 mV (Mutoh et al. 2009). Using a VSD rather than modified channel proteins has the advantages of (1) not having a channel pore, with it's own ionic fluxes, and (2) VSDs are much smaller proteins, which should make them easier to express in cells.

Another FRET or energy transfer approach is a hybrid system that uses a genetically encoded membrane-bound donor FP that undergoes voltage-sensitive energy transfer with a hydrophobic anion. A hybrid approach has the potential to offer both cellular targeting of the photon acceptor along with the high sensitivity inherent in using a mobile voltage sensor anion. A FRET hybrid approach between Lyn-domain targeted GFP and DiSBAC oxonols produces large voltage-sensitive FRET response in mammalian cells and is compatible with high-throughput screening (Tsien and Gonzalez 2002). Recently, a hybrid voltage sensor (hVOS) was constructed that uses a membrane-targeted GFP as the donor and dipicrylamine (DPA) as the acceptor; this combination has a high sensitivity (34 % dF/F) in neurons (Chanda et al. 2005). DPA translocates across membranes with sub-millisecond speed and is not fluorescent, so it functions as a quencher so that all fluorescence changes are from the FP (Blunck et al. 2006). Although high sensitivity is possible, the optimal DPA membrane surface densities occur at such high concentrations that they cause toxicity or introduce a huge capacitive load to the *Drosophila* antennal lobe neurons (Sjulson and Miesenbock 2008). A second hybrid combination used farnesylated FP and the fluorescent oxonol acceptor $DiBAC_4(5)$ to measure membrane potential changes in the transverse tubular system of skeletal muscle fibers of mice (DiFranco et al. 2007). Both combinations increase the capacitance significantly, although action potential signal were observed with DPA. Progress in molecular biology and FPs has catalyzed the search for genetically encoded fluorescent reporters of membrane potential. Still the technical constraints of recording millisecond changes in intact neuronal systems without significantly perturbing the biological system are very demanding constraints that have yet to be met.

6 Phototoxicity of FRET Indicators

Because oxonols readily photobleach, they are the primary source of bleaching during extended voltage-sensitive FRET recordings, even when they serve as FRET acceptors. One contribution to photobleaching is the photochemical reaction between excited states of the dyes and oxygen to generate reactive singlet oxygen. Because the mobile anion is located within the plasma membrane, singlet oxygen causes cellular toxicity by reacting with unsaturated lipids and proteins within the cell membrane. Evidence that oxonol is the major contributor to photodynamic damage comes from experiments in which cardiomyocytes (Gonzalez and Tsien 1997) and leech neurons (unpublished observation) stained only with oxonol. Phototoxic effects can be partially mitigated by incubating cells with astaxanthin, a lipid-soluble carotenoid free radical scavenger (Palozza and Krinsky 1992; Gonzalez and Tsien 1997). When applied to cardiomyocytes, astaxanthin increased the usable imaging time by a factor of 10. Astaxanthin has similar beneficial effects to reduce photodamage in second harmonic imaging (Sacconi et al. 2006). A key challenge is to develop additional anti-oxidant compounds that can more readily be loaded into cells.

7 Applications of Synthetic FRET VSDs

FRET VSDs have been used in a variety of cell types with applications including high-throughput pharmacology screens, imaging of single neurons, population imaging of intact nervous systems, multilayered keratinocyte cultures (Burgstahler et al. 2003) and pancreatic islets (Kuznetsov et al. 2005). The primary limitations to using FRET VSDs in complex tissue is being able to deliver uniform and sufficient penetration of the two dye molecules to the target cells while minimizing the nonspecific background staining of non-target cells. Therefore, the most successful applications of FRET VSDs have been in preparations with relatively unhindered access to the cell bodies.

7.1 High Throughput Drug Screening

The use of FRET VSDs has been particularly useful for the characterization of pharmacological compounds on the activity of ion channels and transporters in isolated cell lines (Adkins et al. 2001; Weinglass et al. 2008). Isolated cells are easily stained with the more hydrophobic, fast oxonol derivatives such as $DiSBAC_6(3)$, in part, because they do not need to penetrate into tissue. Because the fast fluorescent dyes are very hydrophobic with essentially no aqueous solubility, they bind to the first membrane they encounter. The combination of high throughput ratiometric FRET detection in microtiter well plates and parallel electrical stimulation has many potential applications, including the rapid characterization of hundreds of voltage-gated sodium channel antagonists (Bugianesi et al. 2006; Huang et al. 2006). Recently, FRET VSDs have been shown to be compatible with the highly miniaturized 1,536 well plate format with an assay for an inward rectifying potassium channel (Solly et al. 2008).

7.2 Monitoring Subthreshold Population Activity

Simultaneous recordings from many neurons in the *Aplysia* abdominal ganglia using the fast oxonol absorption VSD, RH155, demonstrated that population activity can be broad and complex in response to sensory stimuli (Tsau et al. 1994; Wu et al. 1994a; Wu et al. 1994b). FRET VSDs have been used to monitor ongoing network activity in leech segmental ganglia. Because the intrinsic locomotory rhythms in the leech are roughly 1 Hz for swimming and 0.1 Hz for crawling (Kristan et al. 2005), $DiSBAC_2(3)$ is sufficiently fast to capture oscillatory dynamics (Briggman and Kristan 2006). The estimation of the magnitude and phase of the coherence between a recording of the motor output on a motor nerve and optical signals from individual neurons provided a quantitative measure of the neurons active at a particular frequency (Cacciatore et al. 1999). The use of multitaper spectral estimation techniques allowed multiple independent estimates of the coherence to be made which provide an estimate of the significance (Fig. 6.3 and Appendix). Importantly, the phase shift due to the time constant of $DiSBAC_2(3)$ proved to be constant across cell types and could, therefore, be subtracted from the signals from the whole population to determine accurate phase relationships. This approach led to the identification of novel candidate neurons involved in the swim central pattern generator (CPG).

This work was extended by imaging ongoing leech swimming and crawling in a single preparation (Briggman and Kristan 2006). A direct comparison of the fraction of neurons participating in each of the two behaviors revealed a subpopulation of neurons active during both behaviors. Again, this approach led to the identification of novel, multifunctional neurons that had remained unidentified by electrophysiological methods alone. Non-rhythmic membrane potential changes have also been monitored during the behavioral choice between swimming and crawling (Briggman et al. 2005). The multidimensional analysis of population recordings demonstrated the importance of the ability to resolve signals in single trials and led to the identification of a neuron that can be manipulated to bias the choice between swimming and crawling.

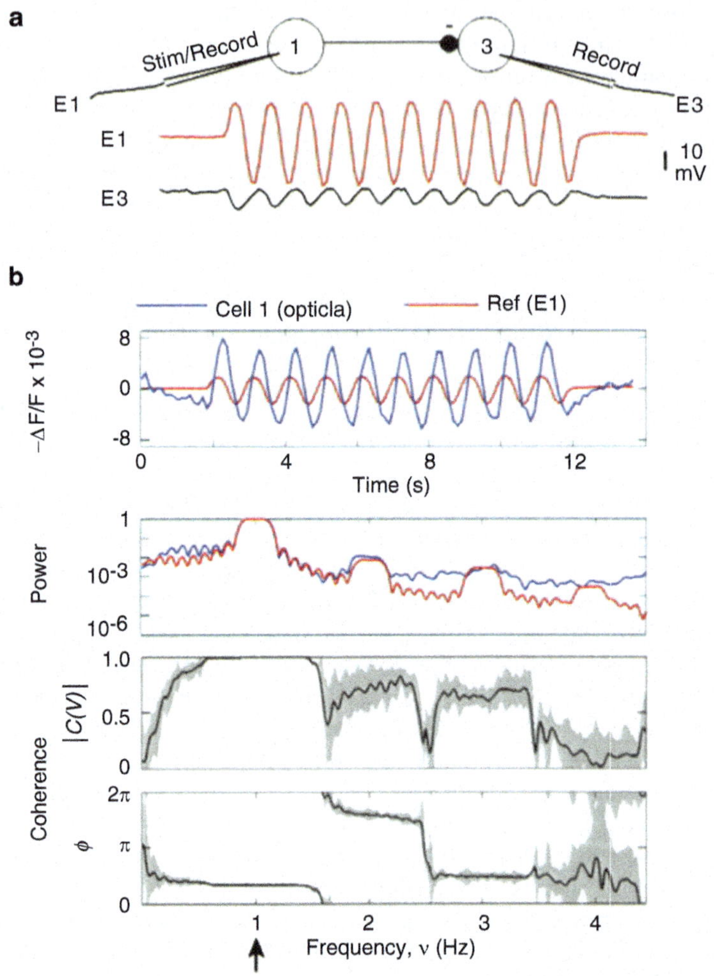

Fig. 6.3 Coherence-based estimates to characterize optical signals. (**a**) The membrane potentials of a synaptically connected pair of neurons recorded intracellularly. While stimulating cell 1 with a sinusoidally varying current. E1 and E2 are the recorded membrane potentials of cells 1 and 3. Cell 1 inhibits cell 3. (**b**) The top trace shows the FRET signal recorded from cell 3, the driven neuron (*black trace*), as its membrane potential was varied sinusoidally (*orange trace*). The significance of the correlation between the optical and underlying electrical signals was quantified by first taking the power spectrum (*middle trace*). As expected, there were strong peaks at 1 Hz, the frequency that the neuron was driven, and at higher harmonics. The amplitude and phase of the coherence (*bottom traces*) were then estimated around the primary peak frequency, 1 Hz. The phase of the coherence at this frequency reflects the phase shift caused by the time constant of the oxonol molecule. Panels reproduced with permission from Cacciatore et al. 1999

7.3 Identifying Functional Connectivity

A particularly useful application of FRET VSDs is the identification of synaptically coupled neurons. The traditional test of functional connectivity, paired electrical recordings from neurons, becomes combinatorially unwieldy as the number of neurons to test increases. An elegant solution to this problem is to drive a presynaptic neuron of interest and use FRET VSDs to image follower neurons (Fig. 6.3a). Because FRET VSDs are sensitive enough to resolve subthreshold fluctuations in membrane potential, it is possible to detect both depolarizing and hyperpolarizing synaptic potentials.

This approach was used to identify postsynaptic targets of a command interneuron, Tr2, located in the head brain of the leech that can terminate swimming (Taylor et al. 2003). A leech segmental ganglion was imaged while Tr2 was driven to produce 1 Hz spike bursts (Fig. 6.4d). The driving frequency of 1 Hz was chosen both to match the time constant of DiSBAC$_2$(3) and to avoid synaptic fatigue. A recording trial typically contained optical traces from 10 to 20 neurons (Fig. 6.4a–c). The coherence was estimated to identify postsynaptic candidate neurons using the driving signal as reference (Fig. 6.4d, e). Following the rapid screening for postsynaptic follower cells, individual candidate neurons were impaled and connectivity was confirmed by recording spike evoked postsynaptic potentials in follower neurons. This technique was used to identify two novel neurons postsynaptic to Tr2 that are involved in the termination of the swim CPG.

8 Advantages/Disadvantages

Because of their promiscuous solubility into all cell membranes, FRET VSDs have been used primarily in situations where the somata of neurons are easily accessible to the staining solutions, such as in isolated cells and in invertebrate ganglia. The limited penetration of hydrophobic dyes into complicated neuropil limits the uniformity of staining and nonspecific binding reduces maximal sensitivities. Another inherent disadvantage is the capacitive load introduced by the voltage sensor that can inhibit fast changes in membrane potential such as action potentials. This ultimately limits the usable concentration range of the mobile anion.

A primary advantage of FRET VSDs is the high sensitivity relative to other classes of VSDs. Subthreshold fluctuations in membrane potential down to a few mV are resolvable in single trials (Taylor et al. 2003). The slow speed of FRET VSDs is a disadvantage, but the speed is, to some extent, tunable by adjusting the hydrophobicity of the voltage sensor. Hence, the time constant can be matched to

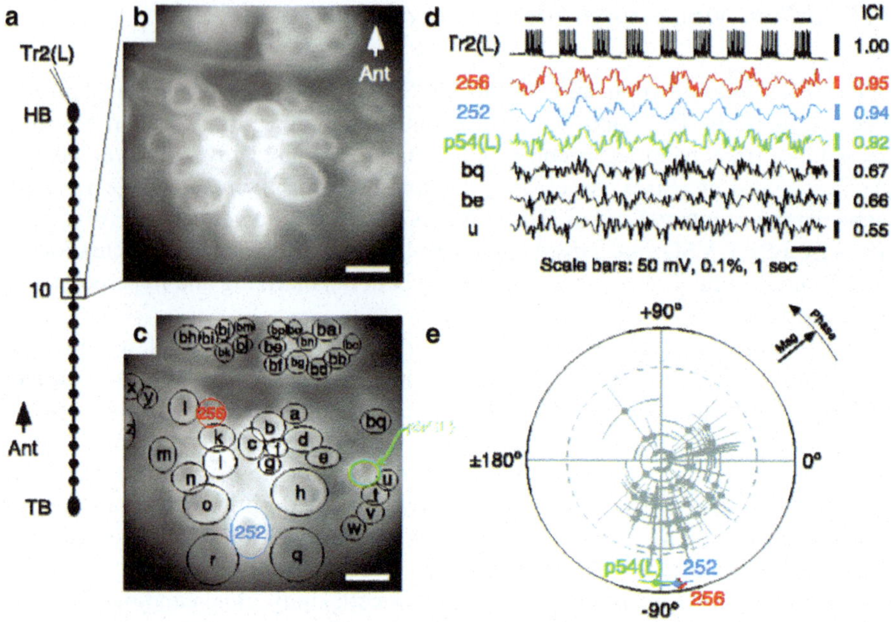

Fig. 6.4 Identification of synaptic connectivity between neurons. (**a**) As indicated by the diagram on the *left*, a neuron (cell Tr2 on the *left side*) in the head brain (HB) of the isolated leech nerve cord was stimulated while imaging a segmental ganglion. (**b**) A *black/white image* of the fluorescence generated by the ventral ganglionic surface stained with the FRET VSDs. (**c**) *Circles* drawn around regions of interest—the neuronal somata of many neurons—on the image. (**d**) Fluorescence changes in six neurons while stimulating cell Tr2. The *top trace* is a voltage recording from cell Tr2 on the *left side* of the head brain, indicating when bursts of action potentials were elicited by passing pulses of current into the soma. The remaining traces are optical recordings from six of the cells *circled* in panel (**c**). The *colored traces* showed significant coherence with the action potential bursts in cell Tr2. (**e**) A polar plot demonstrating the phase and magnitude of the coherence between the stimulated cell and all the optically imaged cells. The *dashed line* denotes the significance threshold. Panels reproduced with permission from Taylor et al. 2003

the desired application: the slower and more sensitive oxonol derivatives are useful for measuring oscillating and steady state membrane potentials, and the faster versions, including hybrid voltage sensor systems, can be used to monitor action potentials. Because FRET VSDs using fluorescent donors and acceptors are ratiometric, they provide several recording advantages: diminished sensitivity to cell motion artifacts, decreased variation in excitation intensity and donor loading concentration, and decreased effects of photobleaching. These dyes have become a valuable adjunct to electrophysiological and behavioral studies, to monitor the activity of many neurons while recording electrically from a select few (Marin-Burgin et al. 2005; Baca et al. 2008).

9 Methodological Details

9.1 *Staining Protocol for Leech Segmental Ganglia*

1. Pin a desheathed leech segmental ganglion onto a Sylgard coated Petri dish and desheath either the ventral or dorsal surface.
2. Prepare a 30 μM solution of CC2-DMPE by dissolving 3 μL of 5 mM CC2-DMPE in DMSO stock solution (aliquoted and stored at −20 °C) and 1 μL of 20 mg/mL pluronic F-127 in DMSO in 500 μL leech saline.
3. Surround the ganglion with a watertight staining chamber consisting of a small plastic cylinder (~1 cm diameter; e.g., the cut end of an Eppendorf tube), sealed to the Sylgard with petroleum jelly. Replace the saline in this chamber with 30 μM CC2-DMPE solution.
4. Stir the staining solution to achieve optimal staining by recirculating the solution using a peristaltic pump. Form a nozzle into the staining chamber by pulling a fine glass microelectrode, breaking off the tip to about a 1 mm diameter, and fire-polishing the tip. Use a dissecting microscope to positioned the nozzle just over the ganglion. Use a glass capillary positioned at the bottom of the chamber for the outflow. Prepare the recirculating system under red light to avoid photobleaching. Stain the ganglion for 30 min. in the dark.
5. Rinse the tissue several times with fresh saline.
6. Prepare a 8 μM solution of the selected oxonol (e.g., $DiSBAC_2(3)$) by dissolving 8 μL of oxonol in DMSO stock solution (aliquoted and stored at −20 °C) in 12 mL leech saline. Bath sonicate this solution for 2–5 min.
7. Prior to adding the oxonol solution, capture an image of the CC2-DMPE stained ganglion to record the baseline (non-FRET) intensity.
8. Replace the leech saline in the chamber with the oxonol solution. Stain with the oxonol solution for 30 min in the dark. Capture images every 5 min to follow the quenching of the CC2-DMPE fluorescence as the oxonol dissolves into membranes. The quenching of the CC2-DMPE fluorescence stabilize by ~50 % within 30 min.
9. Leave the oxonol solution in the bath during recording, replacing it with fresh solution every 30–60 min.

9.2 *Staining Protocol for Mammalian Cells in Culture*

1. Plate cells grown in flasks or harvested from tissue onto glass coverslips or microtiter plates compatible with fluorescence detection equipment. Allow cells to attach and stabilize in culture at 37 °C for 24 h.
2. Wash cells at room temperature with balanced extracellular solution, such as HBSS buffered at pH 7.4.

3. Stain cells for 20–30 min at room temperature with CC2-DMPE (1–20 μM) and oxonol (1–20 μM) solublized in the extracellular solution, using 0.5 % ß-cyclodextrin and 20 μg/mL pluronic F-127 as excipients. Prepare dye stock solutions in DMSO from solid material and aliquot into convenient single-use amounts for storage at −20 °C. Optimized dye concentrations must be empirically determined because they depend on the specific probe and cell type. Optimize the two dyes by varying their concentrations along the axes of a micotiter plate. On coverslips, optimize the dyes by staining with 10 μM CC2-DMPE, then titrating the oxonol dye while monitoring the CC2-DMPE fluorescence; the optimum oxonol concentration reduces CC2-DMPE fluorescence by 50 %. Once optimized, the dyes can be delivered sequentially or simultaneously. Pluronic F-127 assists loading both coumarin DMPE and oxonol dyes; β-cyclodextrin improves the loading of hydrophobic oxonols such as $DiSBAC_6(3)$.
4. Wash away excess dye with a final rinse. For CC2-DMPE and $DiSBAC_6(3)$ the membrane-bound dye remains on plasma membrane for hours at room temperature. The less hydrophobic oxonols like $DiSBAC_2(3)$ will slowly leak out of the cell but this usually does not effect the FRET signal because oxonol is typically used as the acceptor and it is essentially non-fluorescent in water.

9.3 Brightfield Imaging Equipment (for Leech Neurons)

Filter Set: 405 ± 15 nm bandpass excitation filter; 430 nm dichroic mirror; 460 ± 5 nm bandpass emission filter (for CC2-DMPE); 560 nm longpass emission filter (for $DiSBAC_2(3)$).

Illumination: Tungsten halogen lamp powered by a low-ripple power supply.

Objectives: 40×, 0.8 NA or 20×, 0.5 NA water immersion objectives.

Camera: A back-illuminated cooled CCD camera. For ratiometric imaging, a second camera is ideal. However emission-path beam splitters can also been used to record both emission wavelengths with a single camera.

9.4 Fluorescent Plate Reader Equipment (for Cultured Cells)

Filter Set: 405 ± 7.5 nm bandpass excitation filter; 460 ± 22.5 nm bandpass emission filter (for CC2-DMPE); 580 ± 30 nm bandpass emission filter (for $DiSBAC_x(3)$).

Use appropriate dichroic mirrors as needed for equipment.

Illumination: Xenon arc lamp or other violet source.

Simultaneous ratio detection: trifurcated fiber optic bundles with two-wavelength fluorescence emission detected with PMTs.

Appendix

We consider the statistical analysis for determining follower cells (Cacciatore et al. 1999; Taylor et al. 2003; Kleinfeld 2008). The significance of the spectral coherence between the response of any cell, labeled "i", and the driven cell is used to determine if two cells are functionally related and thus are a candidate for a synaptically driven pair. The coherence is a complex function, denoted $C_i(f)$, that it is calculated over the time period of the stimulus, denoted T. We further denote the time series of the optical signals as Vi(t) and the electrical reference drive signal as U(t). The mean value is removed to form:

$$\delta V_i(t) = V_i(t) - \frac{1}{T}\int_0^T dt\ V_i(t)$$

with a similar expression for $\delta U(t)$. The Fourier transform of $\delta Vi(t)$ with respect to the k-th Slepian window (Thomson 1982; Percival and Walden 1993), denoted $w^{(k)}(t)$, is:

$$\delta \tilde{V}_i^{(k)}(f) = \frac{1}{\sqrt{T}}\int_0^T dt\ e^{-i2\pi ft} w^{(k)}(t)\,\delta V_i(t)$$

with a similar expression for $\delta \tilde{U}^{(k)}(f)$. The use of multiple tapers allow averaging over a bandwidth that is set by the number of tapers, K, with the half-bandwidth at half-maximal response given by $\Delta f = (1/T)(K+1)/2$. Our interest lies in the values of $C_i(f)$ for $f = f_{Drive}$ and the confidence limits for these values. We chose the bandwidth so that the estimate of $|C_i(f_{Drive})|$ is kept separate from that of the harmonic $|C_i(2f_{Drive})|$. The choice $\Delta f = 0.4\ f_{Drive}$ works well, so that for $f_{Drive} = 1$ Hz and $T = 9$ s the integer part of 2 • 0.4 • 1 Hz • 9 s—1 yields $K = 6$ tapers (Fig. 6.3). The spectral coherence between the optical signal and the reference is give by:

$$C_i(f) = \frac{\frac{1}{K}\sum_{k=1}^{K} \delta \tilde{V}_i^{(k)}(f)\left[\delta \tilde{U}^{(k)}(f)\right]^*}{\sqrt{\left(\frac{1}{K}\sum_{k=1}^{K}\left|\delta \tilde{V}_i^{(k)}(f)\right|^2\right)\left(\frac{1}{K}\sum_{k=1}^{K}\left|\delta \tilde{U}^{(k)}(f)\right|^2\right)}}.$$

To calculate the standard errors for the coherence estimates, we use the jackknife (Thomson and Chave 1991) and compute delete-one averages of coherence, denoted $C_i^{(n)}(f)$, where *n* is the index of the deleted taper:

$$C_i^{(n)}(f) = \frac{\frac{1}{K-1}\sum_{\substack{k=1\\k\neq n}}^{K} \delta\widetilde{V}_i^{(k)}(f)\left[\delta\widetilde{U}^{(k)}(f)\right]^*}{\sqrt{\left(\frac{1}{K-1}\sum_{\substack{k=1\\k\neq n}}^{k}\left|\delta\widetilde{V}_i^{(k)}(f)\right|^2\right)\left(\frac{1}{K-1}\sum_{\substack{k=1\\k\neq n}}^{K}\left|\delta\widetilde{U}^{(k)}(f)\right|^2\right)}} \quad \forall\, n.$$

Estimating the standard error of the magnitude of $C_i(f)$ requires an extra step since $|C_i(f)|$ is defined on the interval [0, 1] while Gaussian variables exist on $(-\infty, \infty)$. Thus the delete-one estimates, $|C_i^{(n)}(f)|$, were replaced with the transformed values:

$$g\{|C_i|\} = log\left(\frac{|C_i|^2}{1-|C_i|^2}\right).$$

The mean of the transformed variable is:

$$\mu_{i;\,Mag}(f) = \frac{1}{K}\sum_{n=1}^{K} g\left\{C_i^{(n)}(f)\right\}$$

and the standard error of the transformed variable is:

$$\sigma_{i;\,Mag}(f) = \sqrt{\frac{K-1}{K}\sum_{n=1}^{K}\left|g\left\{C_i^{(n)}(f)\right\} - \mu_{i;Mag}(f)\right|^2}.$$

The 95 % confidence interval for the coherence is thus:

$$\left[\sqrt[-1]{1+e^{-(\mu_{i;Mag}-2\sigma_{i;Mag})}},\ \sqrt[-1]{1+e^{-(\mu_{i;Mag}+2\sigma_{i;Mag})}}\right].$$

We now turn to an estimate of the standard deviation of the phase of C(f). Conceptually, the idea is to compute the variation in the relative directions of the delete-one unit vectors $C_i(f)/|C_i(f)|$. The standard error is computed as:

$$\sigma_{i;Phase}(f) = \sqrt{2\frac{K-1}{K}\left(K - \left|\sum_{n=1}^{K}\frac{C_i^{(n)}(f)}{\left|C_i^{(n)}(f)\right|}\right|\right)} \quad \forall\, n.$$

We graph the magnitude and phase of $C_i(f_{Drive})$ for all neurons, along with the confidence interval, on a polar plot (Fig. 6.4e). Finally, we consider whether the coherence of a given cell at f_{Drive} is significantly greater than zero, that is, larger than one would expect to occur by chance from a signal with no coherence. We compared the estimate for each value of $|C_i(f_{Drive})|$ to the null distribution for the magnitude of the coherence, which exceeds

$$|C_i(f_{Drive})| = \sqrt{1 - \alpha^{1/(K-1)}}$$

only in α of the trials (Hannan 1970; Jarvis and Mitra 2001). We use $\alpha = 0.001$ to avoid false-positives. We also calculate the multiple comparisons of α level for each trial, given by $\alpha_{multi} = 1 - (1 - \alpha)N$, where N is the number of cells in the functional image, and verified that it did not exceed $\alpha_{multi} = 0.05$ on any trial.

References

Adkins CE, Pillai GV et al (2001) alpha4beta3delta GABA(A) receptors characterized by fluorescence resonance energy transfer-derived measurements of membrane potential. J Biol Chem 276:38934–38939

Ataka K, Pieribone VA (2002) A genetically targetable fluorescent probe of channel gating with rapid kinetics. Biophys J 82:509–516

Baca SM, Marin-Burgin A, Wagenaar DA, Kristan WB Jr (2008) Widespread inhibition proportional to excitation controls the gain of a leech behavioral circuit. Neuron 57:276–289

Baker BJ, Kosmidis EK et al (2005) Imaging brain activity with voltage- and calcium-sensitive dyes. Cell Mol Neurobiol 25:245–282

Blunck R, Cordero-Morales JF, Cuello LG, Perozo E, Bezanilla F (2006) Detection of the opening of the bundle crossing in KcsA with fluorescence lifetime spectroscopy reveals the existence of two gates for ion conduction. J Gen Physiol 128:569–581

Briggman KL, Kristan WB Jr (2006) Imaging dedicated and multifunctional neural circuits generating distinct behaviors. J Neurosci 26:10925–10933

Briggman KL, Abarbanel HD, Kristan WB Jr (2005) Optical imaging of neuronal populations during decision-making. Science 307:896–901

Bugianesi RM, Augustine PR et al (2006) A cell-sparing electric field stimulation technique for high-throughput screening of voltage-gated ion channels. Assay Drug Dev Technol 4:21–35

Burgstahler R, Koegel H et al (2003) Confocal ratiometric voltage imaging of cultured human keratinocytes reveals layer-specific responses to ATP. Am J Physiol Cell Physiol 284:C944–C952

Cacciatore TW, Brodfuehrer PD et al (1999) Identification of neural circuits by imaging coherent electrical activity with FRET-based dyes. Neuron 23:449–459

Chanda B, Blunck R et al (2005) A hybrid approach to measuring electrical activity in genetically specified neurons. Nat Neurosci 8:1619–1626

Clegg RM (1995) Fluorescence resonance energy transfer. Curr Opin Biotechnol 6:103–110

DiFranco M, Capote J, Quinonez M, Vergara JL (2007) Voltage-dependent dynamic FRET signals from the transverse tubules in mammalian skeletal muscle fibers. J Gen Physiol 130:581–600

Dimitrov D, He Y et al (2007) Engineering and characterization of an enhanced fluorescent protein voltage sensor. PLoS One 2:e440

Dumas D, Stoltz JF (2005) New tool to monitor membrane potential by FRET voltage sensitive dye (FRET-VSD) using spectral and fluorescence lifetime imaging microscopy (FLIM). Interest in cell engineering. Clin Hemorheol Microcirc 33:293–302

Ebner TJ, Chen G (1995) Use of voltage-sensitive dyes and optical recordings in the central nervous system. Prog Neurobiol 46:463–506

Flewelling RF, Hubbell WL (1986) The membrane dipole potential in a total membrane potential model. Applications to hydrophobic ion interactions with membranes. Biophys J 49:541–552
Förster VT (1948) Zwischenmolekulare energiewanderung und fluoreszenz. Ann Phys 6:54–75
Gonzalez JE, Tsien RY (1995) Voltage sensing by fluorescence resonance energy transfer in single cells. Biophys J 69:1272–1280
Gonzalez JE, Tsien RY (1997) Improved indicators of cell membrane potential that use fluorescence resonance energy transfer. Chem Biol 4:269–277
Gonzalez JE, Maher MP (2002) Cellular fluorescent indicators and voltage/ion probe reader (VIPR) tools for ion channel and receptor drug discovery. Receptors Channels 8:283–295
Grinvald A, Ross WN, Farber I (1981) Simultaneous optical measurements of electrical activity from multiple sites on processes of cultured neurons. Proc Natl Acad Sci U S A 78:3245–3249
Guerrero G, Siegel MS, Roska B, Loots E, Isacoff EY (2002) Tuning FlaSh: redesign of the dynamics, voltage range, and color of the genetically encoded optical sensor of membrane potential. Biophys J 83:3607–3618
Hannan EJ (1970) Multiple time series. Wiley, New York, NY
Huang CJ, Harootunian A et al (2006) Characterization of voltage-gated sodium-channel blockers by electrical stimulation and fluorescence detection of membrane potential. Nat Biotechnol 24:439–446
Jarvis MR, Mitra PP (2001) Sampling properties of the spectrum and coherency of sequences of action potentials. Neural Comput 13:717–749
Kleinfeld D (2008) Application of spectral methods to representative data sets in electrophysiology and functional neuroimaging. In: Syllabus for Society for Neuroscience Short Course III on "Neural Signal Processing: Quantitative Analysis of Neural Activity", Society for Neuroscience, vol 3, p. 21–34
Knopfel T, Tomita K, Shimazaki R, Sakai R (2003) Optical recordings of membrane potential using genetically targeted voltage-sensitive fluorescent proteins. Methods 30:42–48
Kristan WB Jr, Calabrese RL, Friesen WO (2005) Neuronal control of leech behavior. Prog Neurobiol 76:279–327
Kuznetsov A, Bindokas VP, Marks JD, Philipson LH (2005) FRET-based voltage probes for confocal imaging: membrane potential oscillations throughout pancreatic islets. Am J Physiol Cell Physiol 289:C224–C229
Lakowicz JR (2006) Principles of fluorescence spectroscopy. Springer, New York, NY
Maher MP, Wu NT, Ao H (2007) pH-insensitive FRET voltage dyes. J Biomol Screen 12:656–667
Marin-Burgin A, Eisenhart FJ, Baca SM, Kristan WB Jr, French KA (2005) Sequential development of electrical and chemical synaptic connections generates a specific behavioral circuit in the leech. J Neurosci 25:2478–2489
Momose-Sato Y, Sato K et al (1999) Evaluation of voltage-sensitive dyes for long-term recording of neural activity in the hippocampus. J Membr Biol 172:145–157
Mutoh H, Perron A et al (2009) Spectrally-resolved response properties of the three most advanced FRET based fluorescent protein voltage probes. PLoS One 4:e4555
Palozza P, Krinsky NI (1992) Astaxanthin and canthaxanthin are potent antioxidants in a membrane model. Arch Biochem Biophys 297:291–295
Percival DB, Walden AT (1993) Spectral analysis for physical applications: multitaper and conventional univariate techniques. Cambridge University Press, New York, NY
Piston DW, Kremers GJ (2007) Fluorescent protein FRET: the good, the bad and the ugly. Trends Biochem Sci 32:407–414
Rink TJ, Montecucco C, Hesketh TR, Tsien RY (1980) Lymphocyte membrane potential assessed with fluorescent probes. Biochim Biophys Acta 595:15–30
Sacconi L, Dombeck DA, Webb WW (2006) Overcoming photodamage in second-harmonic generation microscopy: real-time optical recording of neuronal action potentials. Proc Natl Acad Sci U S A 103:3124–3129
Sakai R, Repunte-Canonigo V, Raj CD, Knopfel T (2001) Design and characterization of a DNA-encoded, voltage-sensitive fluorescent protein. Eur J Neurosci 13:2314–2318

Selvin PR (2000) The renaissance of fluorescence resonance energy transfer. Nat Struct Biol 7:730–734

Siegel MS, Isacoff EY (1997) A genetically encoded optical probe of membrane voltage. Neuron 19:735–741

Sjulson L, Miesenbock G (2008) Rational optimization and imaging in vivo of a genetically encoded optical voltage reporter. J Neurosci 28:5582–5593

Solly K, Cassaday J et al (2008) Miniaturization and HTS of a FRET-based membrane potential assay for K(ir) channel inhibitors. Assay Drug Dev Technol 6:225–234

Stryer L (1978) Fluorescence energy transfer as a spectroscopic ruler. Annu Rev Biochem 47:819–846

Taylor AL, Cottrell GW, Kleinfeld D, Kristan WB Jr (2003) Imaging reveals synaptic targets of a swim-terminating neuron in the leech CNS. J Neurosci 23:11402–11410

Thomson DJ (1982) Spectrum estimation and harmonic-analysis. Proc IEEE 70:1055–1096

Thomson DJ, Chave AD (1991) Jackknifed error estimates for spectra, coherences, and transfer functions. In: Haykin S (ed) Advances in spectrum analysis and array processing. Upper Saddle River, NJ, Prentice Hall

Tsau Y, Wu JY et al (1994) Distributed aspects of the response to siphon touch in *Aplysia*: spread of stimulus information and cross-correlation analysis. J Neurosci 14:4167–4184

Tsien RY, Gonzalez JE (2002) Detection of transmembrane potentials by optical methods. US patent # 6,342,379 B1

Tsutsui H, Karasawa S, Okamura Y, Miyawaki A (2008) Improving membrane voltage measurements using FRET with new fluorescent proteins. Nat Methods 5:683–685

Weinglass AB, Swensen AM et al (2008) A high-capacity membrane potential FRET-based assay for the sodium-coupled glucose co-transporter SGLT1. Assay Drug Dev Technol 6:255–262

Wu P, Brand L (1994) Resonance energy transfer: methods and applications. Anal Biochem 218:1–13

Wu JY, Cohen LB, Falk CX (1994a) Neuronal activity during different behaviors in *Aplysia*: a distributed organization? Science 263:820–823

Wu JY, Tsau Y et al (1994b) Consistency in nervous systems: trial-to-trial and animal-to-animal variations in the responses to repeated applications of a sensory stimulus in *Aplysia*. J Neurosci 14:1366–1384

Chapter 7
Monitoring Population Membrane Potential Signals from Neocortex

Jianmin Liang, Weifeng Xu, Xinling Geng, and Jian-young Wu

Abstract Voltage-sensitive dyes (VSDs) and optical imaging are useful tools for studying spatiotemporal patterns of population neuronal activity in cortex. Because fast VSDs respond to membrane potential changes with microsecond temporal resolution, these are better suited than calcium indicators for recording rapid neural signals. Here we describe methods for using a 464 element photodiode array and fast VSDs to record signals ranging from large scale network activity in brain slices and in vivo mammalian preparations with sensitivity comparable to local field potential (LFP) recordings. With careful control of dye bleaching and phototoxicity, long recording times can be achieved. Absorption dyes have less photo-toxicity than fluorescent dyes. In brain slices, the total recording time in each slice can be 1,000–2,000 s, which can be divided into hundreds of short recording trials over several hours. In intact brains when fluorescent dyes are used, reduced light intensity can also increase recording time. In this chapter, we will discuss technical details for the methods to achieve reliable VSD imaging with high sensitivity and long recording time.

Keywords Voltage-sensitive dyes • NK3630 • RH486 • RH1691 • Brain slice • Visual cortex • Propagating waves • Spirals • Coupled oscillators

---

J. Liang
Department of Neuroscience, Georgetown University Medical Center, 207A, Basic Science Building, 3900 Reservoir Road NW, Washington, DC 20057, USA

Department of Pediatrics, First Bethune Hospital of Jilin University, Changchun 130021, P. R. China

W. Xu • J.-y. Wu (✉)
Department of Neuroscience, Georgetown University Medical Center, 207A, Basic Science Building, 3900 Reservoir Road NW, Washington, DC 20057, USA
e-mail: wuj@georgetown.edu

X. Geng
Department of Neuroscience, Georgetown University Medical Center, 207A, Basic Science Building, 3900 Reservoir Road NW, Washington, DC 20057, USA

Shoool of Biomedical Engineering, Capital Medical University, Beijing 100069, P. R. China

M. Canepari et al. (eds.), *Membrane Potential Imaging in the Nervous System and Heart*, Advances in Experimental Medicine and Biology 859,
DOI 10.1007/978-3-319-17641-3_7

1 Introduction

Voltage-sensitive dye (VSD) imaging, an optical method of measuring transmembrane potential, has been undergoing constant development since pioneering work published about 40 years ago (Cohen et al. 1968; Tasaki et al. 1968; Chap. 1). With the development of better dyes and specialized apparatus, the method has gradually evolved from a technological demonstration of feasibility into a powerful experimental tool for imaging the activity of excitable tissues. After a heroic screening effort during which thousands of dye compounds were tested for voltage sensitivity (Gupta et al. 1981; Grinvald et al. 1982a, 1982b; Loew et al. 1992; Shoham et al. 1999), dozens of dyes were found useful, and many analogs became commercially available. These dyes all have very fast response times (<1 μs) and excellent linearity over the entire physiological range (Ross et al. 1977), but all show only small fractional changes in absorption or fluorescence. In biological tissues, these fractional changes amount to only about 10^{-2} to 10^{-5} of the resting light intensity per 100 mV of membrane potential change.

Until more sensitive probes become available, measurements of fast voltage activity are carried out using traditional organic dyes characterized by relatively low sensitivity in terms of fractional change in light intensity per unit change in membrane potential ($\Delta I/I$). However, a small fractional change in light intensity does not necessarily equate to poor signal-to-noise ratio. In this chapter, we will discuss methods to achieve high signal-to-noise ratio recordings of VSD activity in brain slices and in intact rodent cortex. If vibration noise is properly reduced, a signal-to-noise ratio of 5–30 can be achieved in single recording trials. This signal-to-noise ratio is sufficient to examine the initiation and propagation of wave activity in the cortical local circuits. Also, this high sensitivity is critical for examining the dynamics of population neuronal activities and activity patterns, such as spiral waves, with spatiotemporal structures that are too complex to be analyzed by averaging. Under *in vivo* conditions, new blue dyes (see Chap. 2) reduce the pulsation noise and allow us to achieve high sensitivity in single trials.

2 Cortical Slice, Solutions and Recording Conditions

The preparation of slices for imaging neuronal population activity is different from that for cellular electrophysiology. For imaging of population events, local circuits need to be well preserved with a larger proportion of viable neurons and functioning synapses. Thinner slices usually have poor signal-to-noise ratio, probably because of more extensive cell damage and poor cell recovery. We prefer slices with thickness of 400–600 μm, because local circuits can be better preserved. Thicker slices need to be well perfused from both sides with a fast perfusion rate.

In brief, following NIH animal care and use guidelines and institutional animal care committee protocols, the animals are deeply anesthetized with halothane and quickly decapitated (using a Stoelting small animal decapitator). The whole brain is carefully harvested and chilled for 90 s in cold (0–4 °C) artificial cerebrospinal fluid (ACSF) for slicing containing (in mM) 250 sucrose, 3 KCl, 2 $CaCl_2$, 2 $MgSO_4$, 1.25 NaH_2PO_4, 27 $NaHCO_3$, 10 dextrose, and saturated with 95 % O_2, 5 % CO_2. Then the brain is blocked and sliced at 400–600 μm thickness on a vibratome stage (e.g., 752 M Vibroslice, Campden Instruments, Sarasota, FL).

Brain slices can be cut in several directions to conserve different aspects of cortical circuitry. For example, coronal sections can preserve the cortical columns such as whisker barrels. Oblique directions can preserve the thalamocortical connections (Agmon and Connors 1992; MacLean et al. 2006). Slices cut tangentially can best preserve the horizontal connections in layer II–III to observe 2D waves such as spirals (Huang et al. 2004).

After slicing, the slices are transferred into a holding chamber with holding ACSF, containing (in mM): 126 NaCl, 2.5 KCl, 2 $CaCl_2$, 2 $MgSO_4$, 1.25 NaH_2PO_4, 26 $NaHCO_3$, 10 dextrose. We use a large holding chamber, containing about 500 ml of holding ACSF, which can keep slices viable for 8–12 h. The holding solution is bubbled with a mixture of 95 % O_2 and 5 % CO_2, and is slowly circulated by a magnetic stirring bar. Stirring provides a slow but steady convection around the slice for delivering oxygen and washing out noxious molecules produced by the cold shock and the trauma of slicing. Slices must be incubated for at least 2 h before the experiments to allow for recovery from the acute trauma.

3 Staining Slice with Voltage-Sensitive Dye

Slices are stained with the VSD before imaging experiment. We prefer lower dye concentrations and longer staining times of 1–2 h instead of shorter staining time under higher dye concentrations. Longer staining time allows the cells deep in the tissue to be evenly stained. Since staining time is long, aeration and convection are critical. The staining chamber contained about 50 ml of holding ACSF, bubbled slowly with a mixture of 95 % O_2 and 5 % CO_2 (about 1 bubble/s) and circulated by a stirring bar. Dyes have a tendency to be concentrated on the air-water interface and slower bubbling with large bubbles can reduce the depletion of the dye from the solution. For absorption measurements, we use 5–20 μg/ml of an oxonol dye, NK3630 (first synthesized by R. Hildesheim and A. Grinvald as RH482; available from Nippon Kankoh-Shikiso Kenkyusho Co.,Ltd., Japan, Charkit chemical Corp, Darien, CT). For fluorescence measurements, we use 10–50 μg/ml of fluorescent dye, RH414 or RH795 (first synthesized by R. Hildesheim and A. Grinvald; available from Molecular Probes/Invitrogen). After staining, the slices are transferred back to the holding chamber to rinse out excess dye and for incubation until recording.

4 Imaging Chamber for Brain Slice

During VSD imaging, the slice is mounted in a perfusion chamber on a microscope stage. The slice is set on a mesh in the center of the chamber so that both sides of the slice are well perfused. The perfusion speed is 6–10 ml per minute (80 drops/min). Perfusing ACSF contains (in mM) 126 NaCl, 2.5 KCl, 2 $CaCl_2$, 1.5 $MgSO_4$, 1.25 NaH_2PO_4, 26 $NaHCO_3$, and 10 dextrose. The solution is warmed prior to perfusion with fine Tygon tubing wound on a temperature-controlled heating block. The tubing in the heating block is about 100 cm long and the solution is heated from room temperature to 28–32 °C when entering the perfusion chamber.

5 Imaging Rat Cortex In Vivo

For rodent cortex in vivo we use the 'blue' dyes, RH1691 or RH1938, developed by Amiram Grinvald's group (Shoham et al. 1999), The blue dyes are excited by yellow light (630 nm), which does not overlap with the wavelengths absorbed by hemoglobin (510–590 nm). This virtually eliminates heart beat pulsation artifacts. By comparison, the traditional 'red' dyes, such as RH795 or Di-4-ANEPPS have pulsation artifacts that can sometimes exceed the evoked cortical signals by an order of magnitude (Orbach et al. 1985; London et al. 1989; Shoham et al. 1999; Grinvald and Hildesheim 2004; Ma et al. 2004).

Staining mammalian cortex in vivo requires great care with regard to the exposed cortex and the animal's physiological condition. An irritated cortex often leads to poor staining. Briefly, animals are pretreated with atropine sulfate (40 μg/kg IP) approximately 30 min prior to anesthetic induction. The animals are anesthetized with 1.5–2 % isoflurane in air via trachea catheter (16 G over-the-needle) and ventilated by a small animal respirator (Harvard Apparatus) at a rate of 60–100/min and volume of 2–4 ml. Rate and volume are adjusted to maintain an inspiratory pressure of ~5 cm H_2O and an end-tidal carbon dioxide value of 26–28 mmHg. A cranial window (~5 mm in diameter) is drilled and bone is carefully separated from the dura, which is left intact. Leaving the dura intact significantly reduces movement artifacts during optical recording (London et al. 1989; Lippert et al. 2007). Nontraumatic craniotomy is important for optimal staining. Irritated brain can appear reddish (due to increased blood flow), or CSF pressure can increase the potential subdural space, which leads to poor staining. Dexamethasone sulfate (1 mg/kg IP) can be given 24 h prior to the experiment, to reduce the inflammatory response of the dura. Voltage-sensitive dye RH 1691 (Optical Imaging) is dissolved at 2 mg/ml in artificial CSF solution, and staining is done through the intact dura mater. Drying dura before staining can increase the dural permeability to the dye. A temporary staining chamber is constructed around the craniotomy with silicone valve grease. Approximately 200 μl of dye solution will stain an area of 5 mm in diameter. During staining, the dye solution is continuously circulated by a custom-made perfusion pump (London et al. 1989). The pump has gear motor which gently presses the rubber nipple of a Pasteur pipette once every few seconds. The tip of this pipette is placed in the staining solution and performs a gentle, back-and-forth circulation of a small amount of dye (~100 μl). Staining lasts for 90 min, followed by washing with dye-free ACSF for >15 min. For further discussion of staining cortex in vivo see Lippert et al. 2007.

6 Imaging Device

A 469-element photodiode array (WuTech Instruments, Gaithersburg, MD) is used for VSD imaging (Fig. 7.1a). This array consists of 464 individual photodiodes, each glued to an end of an optical fiber (750 μm in diameter). All the optical fibers are bundled coherently into a hexagonal aperture (19 mm in diameter). With the hexagonal bundling and coupling of the optical fibers, the dead space on the imaging aperture is reduced (8 % of active/inactive space). The array can be mounted on standard camera port of a microscope (such as a "C" mount).

Photo-current from each diode is fed into a two stage amplifier circuit (Fig. 7.1b). The first stage converts the photocurrent into voltage-signal, about 5 V in our absorption measurements. This voltage signal has two components: one is a large DC voltage corresponding to resting light intensity. The second component is the dye signal, a small fractional change in the resting light. This signal ranges from 10^{-5} to 10^{-3} for absorption measurements and 10^{-3} to 10^{-2} for fluorescent measurements. For absorption measurements, the output of the first stage amplifier has a very small signal of 0.05–5 mV riding on top of a large 5 V resting light signal. If, for example, this 5 V + 0.5 mV is digitized directly by a 14 bit A/D converter, the 0.5 mV signal would be digitized by less than 1 bit, which is not sufficiently precise. In order to digitize 0.5 mV signal with at least 7 bit range, a DC subtraction circuit is used between the first and second stage amplifiers, for a hardware background subtraction (Fig. 7.1b).

The DC subtraction is done at the beginning of each recording trial when the illumination light is turned on. At this time the "reset" switch is closed and the capacitor C is charged to the output voltage of the resting light intensity. After about 80 ms when the C is fully charged, the reset switch is opened to allow recording to begin. Since the capacitor is charged to the same voltage as the resting light intensity at the input of the second stage, the voltage from the first stage and the voltage on the capacitor will be equal and have opposite polarities, resulting in a subtracted zero resting light intensity voltage. The second stage can then apply 100× to 1,000× voltage gain to the signal. As a result, a 100× voltage gain will extend the dynamic range of the imaging device by ~8 bit. Since resting light intensity on different detectors can differ largely, each diode needs a dedicated DC subtraction circuit between the first and second amplifier stage (Cohen and Lesher 1986; Wu and Cohen 1993). For this reason, the photodiode array with high spatial resolution is bulky and difficult to manufacture.

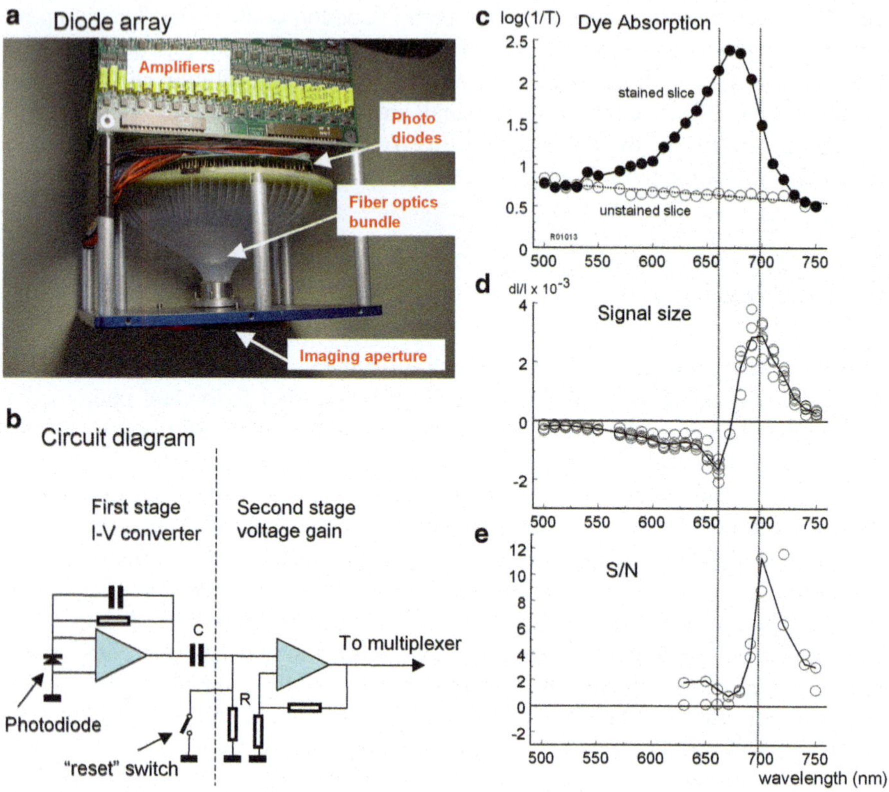

Fig. 7.1 Photodiode array, absorption spectra and signal-to-noise ratio. (**a**) WuTech 469 V photodiode array, a commercial version of the photodiode arrays developed by Cohen group (Cohen and Lesher 1986). WuTech 469 V has an integrated design with all 464 diode amplifiers housed together with the fiber optic bundles. This reduces both light and electrical interference. The hexagonal optic fiber bundle contains 469 optical fibers. Each of fibers is glued to a PIN photodiode and each diode is wired to a dedicated two-stage amplifier (**b**) Circuit diagram of one amplifier channel. The first stage converts the photocurrent outflow of the diode into voltage; the reset switch and the coupling capacitor C subtract the DC component in the light. The second stage provides a 100× voltage gain to boost the effective dynamic range. (**c**) The absorption spectrum of unstained slices (*unfilled circles*) and a slice stained with NK3630 (*solid dots*). (**d**) VSD signal amplitude at different wavelengths. Note that the maximum amplitude is not at the peak absorbance. (**e**) Signal-to-noise ratio at different wavelengths. (Modified from Jin et al. 2002)

We use a 14 bit data acquisition board (Microstar Laboratories, Bellevue, WA) installed in a desktop PC computer. The second stage amplifier also contains a low pass filter with 333 Hz corner frequency; this hardware filter will further improve the quality of the analog signal before digitizing. The sampling speed needs to be faster than any events to be measured, usually 1,000–2,000 frames/s are adequate for imaging brain slices.

7 Microscope and Light Source

For imaging with absorption dyes, inexpensive low NA objectives and a conventional 100 W tungsten halogen lamp from an ordinary research microscope can provide excellent results. 735 nm LED illuminator (ThorLabs) can also be used for illumination. LED illuminator runs cool, which appears to be better than the hot halogen lamp. However, LED can flicker at much higher frequency and so it is much more difficult to reduce the light fluctuation down to 10^{-5} of the illumination intensity (ordinary "super stable" LED driver uses switching circuit and only provides noise level down to 10^{-3}). Battery and a constant current source with linear (non-switching) circuits may be necessary for powering the LED.

For large imaging field such as that shown in Figs. 7.5, 7.6, and 7.7, we use a 5× 0.12 NA objective (Zeiss). When working with smaller imaging fields, 20× (0.6 NA, water immersion, Zeiss) or 40× (0.75 NA, water immersion, Zeiss) can be used accordingly. When a 5× objective is used, each photodetector of a 464-diode array receives light from a tissue area of 150 μm in diameter.

In absorption measurements in brain slice, the resting light intensity is usually about one billion photons per millisecond on each optical detector. The dye molecules absorb most of the light in the wavelength used for the measurement. Unstained tissue slices absorb an insignificant fraction of the light (Fig. 7.1c). The resting light intensity on each photodiode is about 10^9 photons per milliseconds.

For in vivo imaging of mammalian cortex with fluorescent dyes, high light gathering optics are necessary. We use a 5× "macroscope", which provides an imaging field of 4 mm in diameter (Kleinfeld and Delaney 1996; Prechtl, Cohen et al. 1997). The macroscope uses a commercial video camera lens (Navitar, 25 mm F0.95). This inexpensive camera lens provides a high numerical aperture of 0.45, compared to an ordinary 4× microscope objective (~0.12 NA). Such a high numerical aperture can gather about 100 times more light, significantly increasing the signal-to-noise ratio for in vivo fluorescent imaging. A tungsten halogen filament lamp (12 V, 100 W, Zeiss) or 630 nm LED (ThorLabs) can be used for illumination. Again the LED may need to be powered by a battery and a linear constant current source.

8 Optical Filters

Figure 7.1c shows the absorption spectrum of a cortical slice stained with dye NK3630 and an unstained slice. The unstained slice has a relatively even transmission with a tendency to absorb less at longer wavelengths (Fig. 7.1c, open circles). The stained slice has a peak of absorption around 670 nm (Fig. 7.1c, solid circles) and the peak stays the same after a long wash with dye-free ACSF. The light transmission at 670 nm (peak absorption) through a 400 μm well stained slice is reduced to 1/10–1/50 of that of unstained slices.

In Fig. 7.1d, e, the amplitudes of the dye signal (dI/I) and signal-to-noise ratio are plotted against wavelength. The dI/I and signal-to-noise ratio reaches a maximum at 705 nm. The signal decreases significantly at wavelengths shorter than 690 nm and reaches a minimum at ~675 nm. At wavelengths shorter than 670 nm the signal became larger but the polarity of the signal reverses. The signal reaches a second maximum at 660 nm. After the 660 nm peak, the signal decreases gradually and becomes undetectable at wavelengths of 550 nm or shorter. The signal-to-noise ratio at the 660 nm was not large in our actual measurement (Fig. 7.1e), probably due to less resting light at the wavelength. The plots in Fig. 7.1c–e indicate several useful wavelengths: for non-ratiometric measurement, band pass filtering around 705 nm should yield the largest signal; an alternating illumination at 700 and 660 nm may be used for ratiometric measurements; illumination at 675 nm should provide a minimum VSD signal, which can be used for measuring light scattering or for distinguishing VSD from broad-band intrinsic signals.

Optical filters for NK3630 should have a center wavelength around 705 nm. The bandwidth of the filter depends upon several factors: When the signal is small and shot noise is a limiting factor, a wider bandpass allows more light and thus increases the signal-to-noise ratio. In this case, the filter should have a center wavelength of 715 nm and a bandwidth of ± 30 nm. When measuring large signals and when a long total recording time is needed, a narrow band filter of 705 ± 15 nm should be used to reduce the illumination intensity. When the illumination light is strong enough, using narrow band filters helps to get larger signals and reduce dye bleaching and phototoxicity. Since the measurement wavelength is between 700 and 750 nm (Fig. 7.1d, e), the heat filter (infrared filter) in front of the light housing absorbs most energy in this range. For absorption measurements, this heat filter must be removed from the light path.

When the 735 nm LED illuminator is used for NK3630, the excitation filter may be omitted. This increases the light throughput significantly. However, the LED provides high intensity from 735 to 755 nm, which gives little VSD signal but large intrinsic optical signals (light scattering). The intrinsic optical signal appears as a large and slow drift of the baseline of the VSD recording traces (Fig. 7.3b-2, trace 2). In some recordings a flat baseline is preferred and a narrow band filter of 735 ± 10 nm can be used to reduce intrinsic signal.

For imaging mammalian cortex with the blue dyes, a filter cube (from Chroma or Semrock) is used, with a 630 ± 15 nm excitation filter, 655 nm dichroic mirror (Chroma Technology) and a 695 nm long pass filter. This filter cube fits well for both halogen lamp and LED. Köhler illumination is achieved with the macroscope.

9 Detecting Vibration Noise

Vibration noise is often the dominant noise source when recording absorption signals from brain slices. Vibration causes a movement of the image related to a photodetector and thus causes a fluctuation of the light intensity. Absorption measurement is characterized by small fractional changes and a high resting light intensity, so it is more sensitive to movement of the preparation compared to fluorescence measurements.

Vibration noise is proportional to the resting light intensity while shot noise is proportional to the square root of the light intensity. Thus when illumination intensity increases, vibration noise will ultimately become dominant. Above the level where vibration noise equals the shot noise, signal-to-noise ratio will no longer increase with higher illumination intensity. Therefore vibration control in many situations determines the ultimate signal-to-noise ratio.

In practice, we use the following method to evaluate the vibration noise on any given experimental set-up. (1) Under Koehler illumination, reduce the field diaphragm of the microscope so that a clear image of the diaphragm (as a round hole with a sharp edge) is formed in the field of view and projects the image onto the diode array. The sharp edge of the image generates a near maximum contrast on the imaging system. Due to vibration, the edge may move in the image plane so that the light intensity on some detectors recording the diaphragm edge will fluctuate between minimum and maximum. The intensity change on these photodetectors will reflect the worst possible scenario of vibration noise. The power spectrum of the signals from these noisy detectors viewing the edge of the field diaphragm can help to determine the vibration source. In many modern buildings, ~30 Hz floor vibrations are caused by ventilation fan motors attached to the building structure. Vibrations at less than 5 Hz sometimes come from movement of the building in wind, or resonance of the floor. When the experimental stage is properly isolated, noise at the edge of the field diaphragm should be no larger than that in the center of the aperture, indicating that the dominant noise is no longer vibration, but shot noise. With the chamber and preparation inserted into the light path, the vibration noise may also become larger because of the vibration of the water-air interface. A cover slip or water immersion lens can eliminate this vibration noise.

We find that many air tables and active isolation platforms perform poorly in filtering vibrations below 5 Hz. A novel isolation stage designed for atomic force microscopy, the "Minus K" table (www.minusk.com), is approximately ten times better at attenuating these frequencies than some standard air tables. With the Minus K table, the vibration noise can be reduced to below the level of shot noise at a resting light intensity of ~10^{11} photons/ms/mm^2.

10 Data Acquisition and Analysis

NeuroPlex, developed by A. Cohen and C. Bleau of RedShirtImaging, LLC, Decatur, GA, is a versatile program for acquiring and analyzing VSD data. The program can display the data in the form of traces for numerical analysis and pseudo color images for studying the spatiotemporal patterns. Data analysis can also be done with scripts written in MatLab (Mathworks). The singular value decomposition (SVD) method (Precthl et al. 1997) is a very effective way for removing random noise (e.g., shot noise) from the signal.

The figures in this chapter are examples of data display where the signal from the local field potential (LFP) electrode and that of an optical detector viewing the same location are plotted together. The spatiotemporal patterns of the activity are often presented with pseudo color maps (Senseman et al. 1990). To compose a pseudo color map, the signal from each individual detector is normalized to its own maximum amplitude (peak = 1 and baseline/negative peak = 0), and a scale of 16 or 256 colors is linearly assigned to the values between 0 and 1.

The imaging data are acquired at a high rate, usually at approximately 1,600 frames per second. Figures in published papers usually include only a few frames selected from this large imaging series. The signal-to-noise ratio is usually defined as the amplitude of the VSD signal divided by the root mean square (RMS) value of the baseline noise. Since most population neuronal activity in the neocortex is below 100 Hz, a low pass filter can be used to reduce RMS noise and improve the signal-to-noise ratio.

11 Sensitivity

The sensitivity of VSD measurement can be estimated by comparing the measurement with LFP recorded from the same tissue. Achieving a high sensitivity comparable to LFP recordings is important not only for detecting small population activity, but also for detecting spatiotemporal patterns when multiple trial averaging is not allowed.

In the experiment shown in Fig. 7.2, we used a rat neocortical slice stained with NK3630. Cortical layers II–III were monitored optically and a LFP electrode was placed inside of the imaging field of the array (Fig. 7.2a). Electrical stimulation is delivered in deep cortical layers and the evoked neuronal response in layers II–III is monitored both optically by the diode array and electrically by the local field potential electrode (Fig. 7.2b). At low stimulus intensities (0.8–2 V, 0.04 ms), the amplitude of the VSD and LFP is approximately linear with the stimulus intensity (Fig. 7.2d). Although the thresholds for VSD and LFP measurements varied in different slices, we found that the sensitivity of the VSD measurement is always higher than that of the LFP if the slice was properly stained (Jin et al. 2002).

In an ideal situation, the sensitivity of the measurement is limited only by the shot noise (the statistical fluctuations in the photon flux). In measurements using absorption dyes, the shot noise can be dominant when the resting light intensity is higher than 5×10^9 photons/ms/mm^2. This resting light intensity can be achieved with an ordinary 100 W tungsten-halogen filament lamp. Since the amplitude of the VSD signal is proportional to the resting light intensity while the shot noise is proportional to the square root of the resting light intensity, the signal-to-noise ratio is proportional to the square root of the resting light intensity (Cohen and Lesher 1986). Therefore increasing the illumination light can increase the signal-to-noise

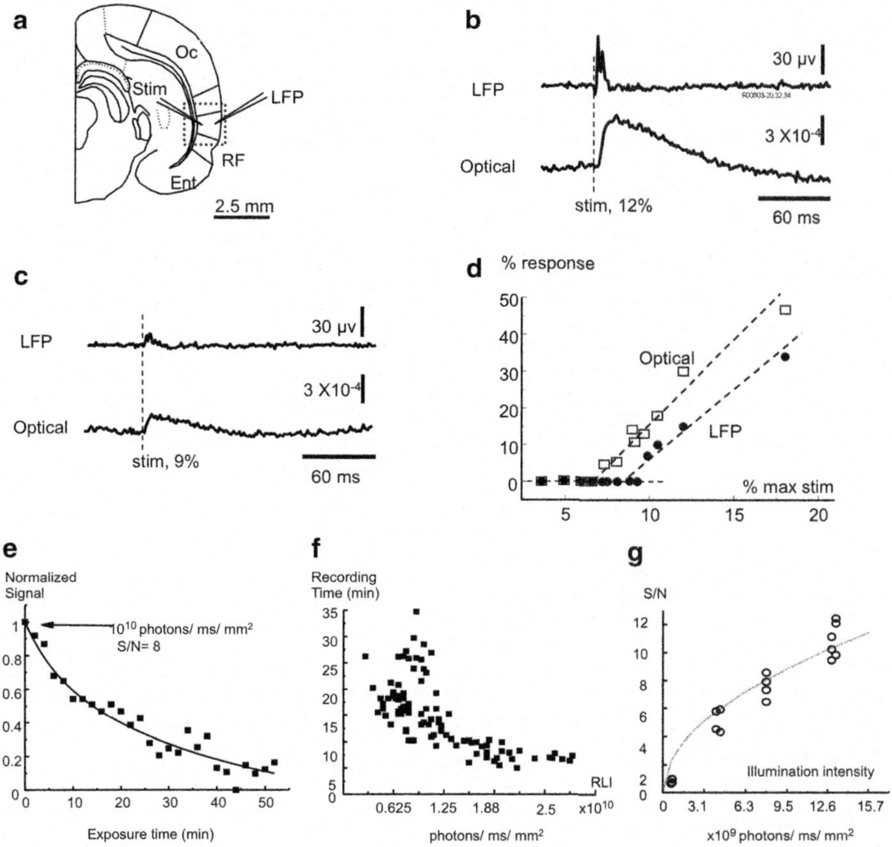

Fig. 7.2 Sensitivity of VSD imaging, dye bleaching and total recording time. (**a**) Experimental arrangement. Rat cortical slices are stained with NK3630 and imaged with a 5× 0.12 NA objective. The *broken line* marks the imaging field. Local field potential (LFP) is recorded from layer II–III, where a photo detector is viewing an area of 200 × 200 μm^2 of the tissue recorded by the electrode. Electrical stimulation (Stim) is delivered to deep layers. (**b**) At 12 % of the stimulation intensity that generated a maximal response, neuronal response is visible in both LFP and VSD signals. Note that stimulation artifact does not appear in the VSD signal. (**c**) A smaller stimulus evokes a reduced response in both VSD and LFP signals. A series of stimuli around the response threshold were applied to test the sensitivity of VSD recordings. (**d**) A plot of response amplitude at different stimulation intensities. Note that both VSD signals (*squares*) and LFP signals (*dots*) are linearly related to the stimulation intensity, and the sensitivity of VSD appears higher than that of the LFP. (**e**) Reduction of VSD signals due to light exposure. The slice was stained with NK3630 and exposed to an illumination intensity of 5×10^{11} photons/ms/mm^2 (resting light intensity was ~10^{10} photons/ms/mm^2). The exposures were 2 min trials with 2 min intermittent dark periods. (**f**) The total recording time is defined as the exposure time needed for the VSD signals to decrease to 50 % of the pre-exposure amplitude. The plot was composed of data from ~25 locations in four slices. (**g**) Signal-to-noise ratio of absorption measurement at different resting light intensities. The signal is the neuronal response in cortical layer II–III evoked by an electrical stimulation in deep layers. (Modified from Jin et al. 2002)

ratio. The diode array is a noise free imaging device in a wide range of light intensities when the shot noise dominates, because the dark noise of the system becomes negligible. However, the resting light intensity cannot be increased infinitely. Due to dye bleaching and phototoxicity, resting light intensity is a limiting factor for total recording time from a given field of view (see below).

12 Phototoxicity, Bleaching and Total Recording Time

When stained slices are continuously exposed to high intensity light (e.g., ~10^{12} photons/ms/mm^2), noticeable decline of both LFP and VSD signals is observed after 5–10 min. We refer to this reduction in electrical activity caused by light exposure as phototoxicity or photodynamic damage (Cohen and Salzberg 1978). Phototoxicity in cortical slices appears to be irreversible even after a long dark recovery period. Absorption dyes have much less phototoxicity than the fluorescent dye RH795 in brain slice experiments, probably because the latter produces free radicals when it is in the excited state.

Intermittent light exposure significantly reduces phototoxicity. With the same high intensity of 10^{12} photons/ms/mm^2, when the exposure was broken into 2 min sessions with 2 min dark intervals, the phototoxicity of NK3630 was smaller. The LFP signals decreased <30 % after 60 min of total exposure time. The optical signals, however, decreased ~90 % (Fig. 7.2e). We refer to the reduction in optical signal while LFP signals remain unchanged as "bleaching". After long light exposure, bleaching is visible even by eye, as loss of color in the exposed area. Optical signals can not be recovered in the bleached area even if the preparation is re-stained.

We define the total recording time as the exposure time needed to reduce the amplitude of the optical signal to 50 % of the pre-exposure level. Figure 7.2f shows the total recording time at different resting light intensities. Figure 7.2g shows the signal-to-noise ratio at four resting light intensities. Combining the results in Fig. 7.2f, g, we conclude that when illumination intensity is adjusted for a signal-to-noise ratio of 8–10, the total recording time is about 15–30 min. This time can be divided into hundreds of intermittent recording sessions (e.g., 100 of 8 s trials), enough for most of applications.

The total recording time of the fluorescent dye RH795 seems to be limited by phototoxicity. When slices stained with RH795 were illuminated with 520 nm light at an intensity of ~10^{12} photons/ms/mm^2, both LFP and optical signals disappeared in a few trials of 2 min exposures. After the signals disappeared, the fluorescence of the stained tissue remained at ~80 % of the pre-exposure level, suggesting that phototoxicity occurred and bleaching was insignificant.

13 Waveforms of the Population VSD Signal

When a single neuron was simultaneously monitored by a photo detector and an intracellular electrode, the waveforms of the two recordings were remarkably similar (Fig. 7.3a; Ross et al. 1977; Zecevic 1996). However, for population activity, the VSD signal may be greatly different from that recorded with an electrode.

Figure 7.3b compares the waveforms of LFP (top traces) and VSD (bottom traces) recordings during different types of neuronal activity in neocortex. Figure 7.3b-1 shows a short latency ("Fast") response to an electrical stimulus to the white matter. This response has a short latency and the amplitude of VSD signal is proportional to the stimulus intensity. Presumably the response is mainly based on depolarization of neurons stimulated either directly or via a few synapses. The LFP response had a short duration (<30 ms) but the VSD signal from the tissue surrounding the electrode had a much longer duration. This long-lasting optical signal did not appear at 520 nm, indicating that it was a wavelength dependent VSD signal, and not a wavelength-independent light-scattering ("intrinsic") signal. Dye molecules bound to the glial cells may also contribute to this long response due to glial membrane potential change caused by glutamate transporters (Kojima et al. 1999; Momose-Sato et al. 1999).

Figure 7.3b-2 shows the signals from an interictal-like spike (slice bathed in 10 μM bicuculline) evoked by weak electrical shock to the white matter, measured in cortical layers II–III. Interictal-like spikes are all-or-none population events; the amplitude of the dye signal is relatively large and independent of the stimulus intensity. Again, VSD signals had a long duration with a smooth waveform while LFP signals had a short duration and large fluctuations. The VSD signal during an interictal-like spike had an obvious undershoot (downward phase below the baseline following the positive peak). Part of this undershoot also occurred at 520 nm (Fig. 7.3b-2, trace 2), suggesting that it was caused by a light scattering signal. The sign of the light scattering signal under trans-illumination was opposite to that of the VSD signal of NK3630 at 705 nm.

Figure 7.3b-3 shows 40–80 Hz oscillations evoked by an electrical stimulus (Luhmann and Prince 1990; Metherate and Cruikshank 1999). The oscillations are only seen in LFP recordings but not in the VSD signals, suggesting that the local oscillatory populations are small and neurons with different phases are mixed (Wu et al. 2001). The VSD signal amplitude is only about 1/10–1/4 of that of the interictal-like spikes (Fig. 7.3b-2). This indicates that this type of ensemble activity involves a much smaller fraction of active neurons than do interictal-like spikes.

In Fig. 7.3b-4, the waveforms of VSD signals during the three types of population events depicted in Fig. 7.3b-1–b-3 are superimposed, illustrating differences in amplitude and time course. If the neurons are evenly stained, the amplitude of the VSD signal and the amount of synchrony in neuronal activity should have a liner relationship. The same should hold for the relationship between the area under the waveform envelope and the overall depolarization and spikes during the event. Figure 7.3b-4 suggests that the synchrony and duration during different cortical

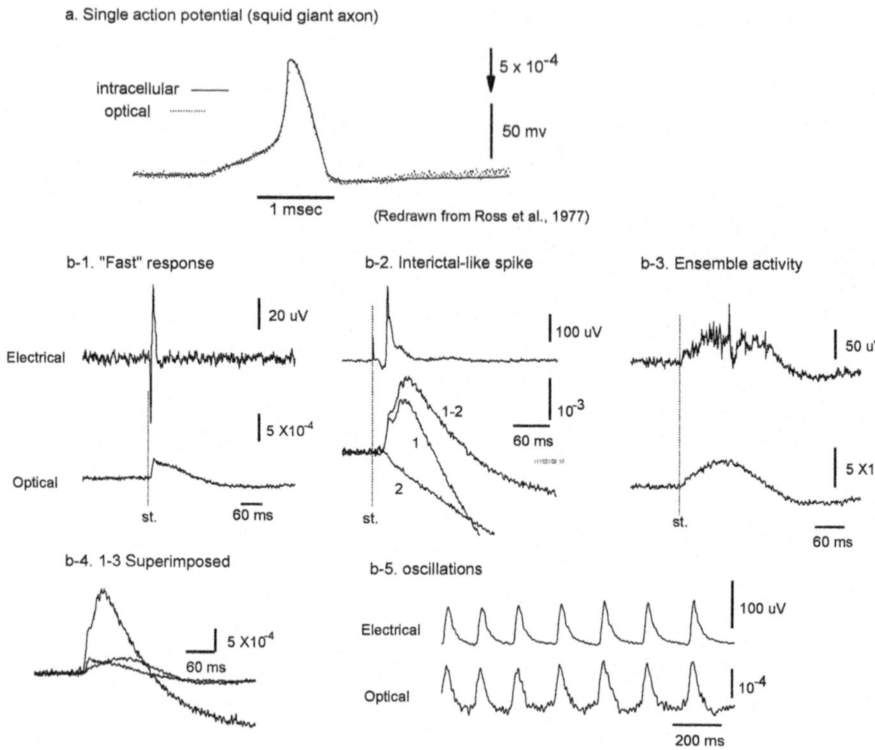

Fig. 7.3 Waveform of optical and electrical recordings. (**a**) VSD and intracellular recordings from a squid giant axon. The axon was stained with absorption dye XVII and the optical signal was measured at 705 nm (*dotted trace*). The actual membrane potential change (*smooth trace*) was simultaneously recorded by an intracellular electrode. The time course of the VSD signal mirrored precisely the time course of membrane potential change. (Redrawn from Ross et al. 1977, with permission from Springer). (**b**) LFP recordings (*top traces*) and VSD recordings (*bottom traces*) during different types of population activity. The recordings were simultaneously made from the same location in cortical layers II–III of rat somatosensory cortical slices. (**b-1**) A short latency activity (fast response) was evoked by an electrical shock to the white matter; the slice was bathed in normal ACSF. (**b-2**) An evoked interictal spike (Tsau et al. 1999); the slice was bathed in 10 μM bicuculline. During this activity, a large fraction of neurons were activated and changes in light scattering also contributed to the signal. Here, trace 1 is the light intensity change at 705 nm; trace 2 is the intensity change at 520 nm, presumably the light scattering signal; subtraction of the signal at the two wavelengths (trace 1–2) estimates the VSD signal without contamination from light scattering signal. (**b-3**) An evoked population oscillation (Wu et al. 2001). (**b-4**) VSD traces of **b-1**, **b-2** (trace 1–2), and **b-3** are superimposed, showing the difference in amplitude and time course. (**b-5**) A spontaneous 7–10 Hz oscillation (Wu et al. 1999) in which LFP and VSD signals are well correlated. (Modified from Jin et al. 2002)

events are two independent variables. Higher synchrony (a larger portion of neurons depolarizing in each time bin) are not always accompanied by longer duration.

Figure 7.3b-5 shows an oscillation around 10 Hz, which can be induced by bathing the slice in low Mg media (Flint and Connors 1996), or adding carbarchol to the media (Lukatch and MacIver 1997). The oscillations are seen in both VSD and LFP recordings and the VSD and LFP signals had very similar waveforms and similar frequency compositions (Wu et al. 1999; Bao and Wu 2003).

In general, low frequency (~10 Hz) oscillations can be seen in both LFP and VSD recordings (Wu et al. 1999; Bao and Wu 2003); High frequency oscillations (~40 Hz), can be seen in LFPs but cannot be seen in VSD recordings (However, see Mann et al. 2005). The waveform of the VSD and LFP signals can be largely different.

One explanation for why waveforms of VSD and LFP can be different is that the VSD and LFP record different signals from the active neurons in the cortex. VSD signals are linearly correlated to the potential changes of all stained membranes under one detector. LFP signals, on the other hand, are a non-linear summation of the current sources distributed around the tip of the electrode and thus its amplitude decreases with the square of the distance between the source and the electrode. When an oscillation contains synchronized neuronal populations, the VSD and LFP recordings would have similar waveforms. When oscillations are organized in local ensembles and the ensembles with different phases are overlapping spatially, the VSD signal may fail to see the oscillations while LFP electrode may still pick up oscillations from neurons near the electrode. The examples below further show that VSD signal may show no oscillation in a phase singularity of spiral waves when oscillations with all phases are mixed (Fig. 7.5).

14 Examples

Over the last 30 years, many authors have published VSD imaging studies of population activities in brain slices (just to list a few, Grinvald et al. 1982a, 1982b; Albowitz and Kuhnt 1993; Colom and Saggau 1994; Tanifuji et al. 1994; Hirota et al. 1995; Tsau et al. 1998; Demir et al. 1999; Tsau et al. 1999; Wu et al. 1999; Demir et al. 2000; Laaris et al. 2000; Mochida et al. 2001; Wu et al. 2001; Huang et al. 2004; Bai et al. 2006; Hirata and Sawaguchi 2008). Below, we will show three recent examples of spatiotemporal patterns in rodent neocortex. The signals in these examples must be measured with high sensitivity and in single trials, because averaging will obscure the dynamic spatiotemporal patterns.

14.1 Spiral Waves

When rodent cortical slices were perfused with carbachol and bicuculline, oscillations around 10 Hz develop (Lukatch and MacIver 1997) and such oscillations can be reliably observed in VSD signals (Fig. 7.4b). The spatiotemporal coupling of such oscillations manifests as propagating waves; at each cycle of oscillation a

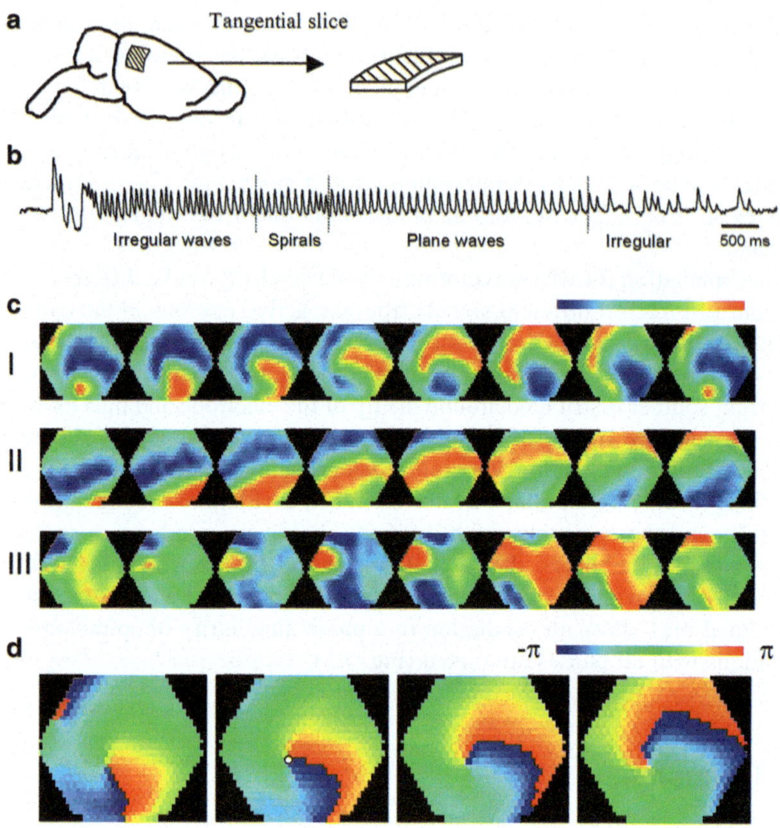

Fig. 7.4 Propagating waves in rat cortical slice. (**a**) The slice made as a $6 \times 6\text{-mm}^2$ tangential patch from rat visual cortex. (**b**) VSD signals of oscillations induced by bathing the slice in 100 μM carbachol and 10 μM bicuculline. Three types of wave patterns during one episode of oscillations (spirals, plane and irregular) can be identified from imaging data in (**c**). (**c**) Pseudo-color amplitude images during the oscillation epoch. In each row eight frames (12 ms inter-frame interval) are selected from thousands of consecutive frames take at a rate of 0.6 ms/frame. (**c-I**) Spirals, as a rotating wave around a center. (**c**-II) Plane waves, as a low curvature wave front moving across the field. (c-III) Irregular waves, as multiple wave fronts interacting with each other, sending waves with variety of velocities and directions. (d) Phase maps of a spiral wave. Color represents the oscillation phase between $-\pi$ and π according to a linear scale (*top right*). (Modified from data in Huang et al. 2004)

wave of activity propagates through the tissue (Bao and Wu 2003). When cortical slices are sectioned in a tangential plane (Fig. 7.4a), these waves are allowed to propagate in two-dimensions with a variety of patterns, such as spirals, target/ring, plane, and irregular waves (Fig. 7.4c). Plane waves have straight traveling paths across the tissue; target/ring waves develop from a center and propagate outward; Irregular waves have multiple simultaneous wave fronts with unstable directions and velocities. Spiral waves, likely arising out of the interactions of multiple plane waves, appeared as a wave front rotating around a center. Each cycle of the rotation was associated with a cycle of the oscillation. In experiments, spiral waves were observed in 48 % of the recording trials (Huang et al. 2004). Both clockwise and counter clockwise rotations are observed during different oscillation epochs from the same slice.

In the phase map of a spiral wave, the highest spatial phase gradient is observed at the pivot of the spiral (Fig. 7.4d, white dot), which is defined as a "phase singularity". The presence of phase singularity distinguishes spiral from other types of rotating waves and is the hallmark of a true spiral wave (Ermentrout and Kleinfeld 2001; Winfree 2001; Jalife 2003). Phase singularities in cortical slices have been hypothesized as a small region containing oscillating neurons with nearly all phases represented between $-\pi$ and π (Huang et al. 2004), with the phase mixing resulting in amplitude reduction in the optical signal. In Fig. 7.5 each detector covered a circular area of 128 μm in diameter (total field of view 3.2 mm in diameter). Signals from all detectors showed high amplitude oscillations before the formation of spirals (Fig. 7.5a, trace k–o before the first broken vertical line). During spiral waves, the phase singularity drifts across the tissue slowly. The four detectors, k–n, alternately recorded reduced amplitude as the spiral center approached each detector in turn. Such amplitude reduction was localized at the spiral center, and this reduced amplitude propagated with drift of the spiral center (Fig. 7.5a traces k–n). At locations distant from the spiral center (e.g., location o in Fig. 7.5a), the amplitude remained high during all rotations of the spiral. In plane or ring waves, no localized region of oscillatory amplitude reduction was seen.

To further confirm that amplitude reduction was caused by superposition of anti-phased oscillations, we examined signals surrounding a spiral center. Figure 7.5b shows the signals from a group of detectors when a spiral center drifted over the center detector (C). As the spiral center hovered briefly (~200 ms, two rotations, marked as p above the traces in Fig. 7.5b) over the center detector, the amplitude was reduced (trace C). Simultaneously, the six surrounding detectors (1–6) did not show amplitude reduction but the oscillation phases exactly opposed each other symmetrically across the center (traces 2 and 5 are shown). These results suggest that the area with amplitude reduction was less than or equal to the size of our optical detector field of view (128 μm diameter). When we added the signals from the six surrounding detectors, the averaged waveform showed a similar amplitude reduction as the center detector (Fig. 7.5b, trace AVG). This combined signal was nearly identical to the central signal, as demonstrated by the small residual when the two signals were subtracted (Fig. 7.7b, trace AVG-C). These findings strongly support the hypothesis that amplitude reduction is not caused by inactivity of the neurons, but rather by the superimposition of multiple widely distributed phases surrounding the singularity, and that the spiral center was fully confined within an area of ~100 μm in diameter.

This example also demonstrates that high sensitivity and single trial recording are essential for studying dynamic events such as spirals. At locations near the spiral singularity, for example, the VSD signal is reduced tenfold, and requires an order of magnitude better sensitivity than if one were observing simple oscillatory events without such dynamic spatiotemporal patterns (Fig. 7.5).

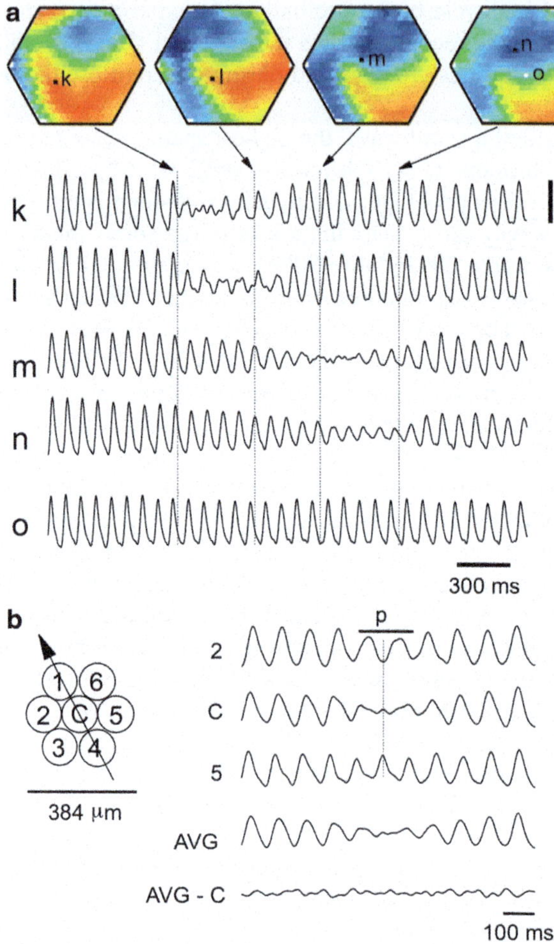

Fig. 7.5 Amplitude reduction at spiral singularity. (**a**) Color–coded representation of signal amplitude in four consecutive frames. The spirals started at about the time of the first image on the *left* and ended at the last image. Traces (k–o) Optical signals from five detectors indicated in individual frames. The *broken vertical lines* mark the times the images were taken. During spiral waves the four detectors, (k–n), sequentially recorded reduced amplitude as the spiral center approached each detector. Note that detector (o) has high amplitude signal throughout the entire period because the spiral center never reached that location. Calibration: 2×10^{-4} of rest light intensity. (**b**) *Left*: The trajectory (*arrow*) of a spiral center moving through a group of optical detectors (1–6 and C). *Right*: signals from the detectors mapped at left. *Black bar* (*p*) indicates the time when the spiral hovered over the center detector. The surrounding detectors (2, 5) did not show amplitude reduction but the oscillation phases exactly opposed each other. AVG, averaged signal of six surrounding detectors shows an amplitude reduction similar to that on the center detector C. AVG - C, numerical subtraction of trace C and AVG. These results indicate that the spiral center was fully confined within the area of the central detector. It also suggested that the amplitude reduction on trace C is caused by a superimposition of multiple phases surrounding the center detector. (Modified from Huang et al. 2004 with permission of the Society for Neuroscience)

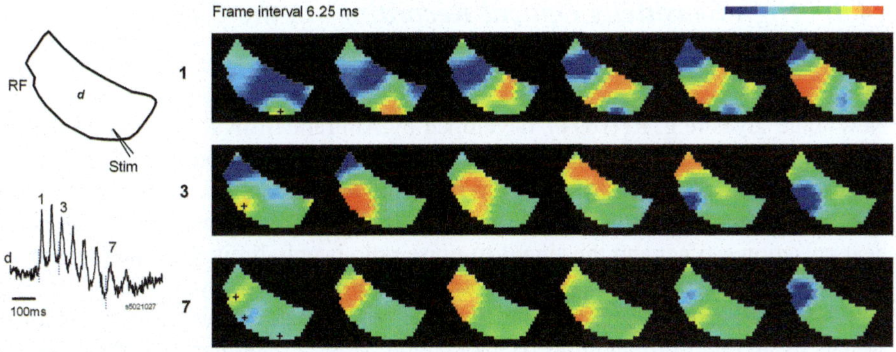

Fig. 7.6 Initiation foci of the ~25 Hz oscillation. *Left*: The orientation of the slice and stimulation sites (*top*). The VSD signal from one optical detector *d* (*bottom*). *Right*: Images of three oscillation cycles 1,3,7. Pseudo-color images generated according to a linear pseudo-color scale (*top right*, peak to *red* and valley to *blue*). All images are snapshots of 0.625 ms; the interval between images is 6.25 ms. Each row of images starts at the time marked by a *vertical broken line* in the trace display in the *left panel*. The initiation site was determined as the detector with earliest onset time and marked with *black crosses* in the first image of each row. The first spike (*1*) initiated at the location of stimulating electrode and propagated as a wave to the lateral (*top row*). Two following oscillation cycles (*3* and *7*) initiated from two different locations. All three initiation foci are marked at the first image of the *bottom row* for comparison. This optical recording trial contains 1,100 images (~0.7 s); only a few selected frames are shown for clarity. (Modified from Bai et al. 2006)

14.2 Locally Coupled Oscillations

VSD imaging has appropriate spatiotemporal resolution to reveal multiple local oscillations in brain slices. Figure 7.6 shows an evoked network oscillation (~25 Hz) in neocortical slices trigged by a single electrical shock (Bai et al. 2006). In a train of oscillations, for different cycles the waves may be initiated at different locations, suggesting local oscillators are competing for the pacemaker to initiate each oscillation cycle. The first spike initiated at the location of the stimulating electrode (Fig. 7.6, black cross in first top row image) and propagated across the slice. Two oscillation cycles, 3 and 7 initiated from two different locations (Fig. 7.6, black crosses in middle and bottom row images) and propagated in a concentric pattern surrounding their respective initiation foci. This pattern indicates that oscillations are not synchronized over space; multiple pacemakers exist and may lead different cycles alternately in turn. On the other hand, local oscillators are not completely independent, because there is a propagating wave accompanying each oscillation cycle. If the oscillators were completely independent, there would be no propagating waves. VSD imaging provides a powerful tool to study such spatial coupling of local oscillators in neocortical circuits.

14.3 High Sensitivity Optical Recordings in the Cortex In Vivo

New 'blue' dyes (e.g. RH1691), developed by Amiram Grinvald's group (Shoham et al. 1999; Chap. 2), have greatly advanced *in vivo* VSD imaging of mammalian cortex. The blue dyes are excited by red light (630 nm) that does not overlap with light absorption of hemoglobin (510–590 nm). This virtually eliminates the heart beat pulsation artifact and the signal-to-noise ratio is only limited by the movement noise and shot noise. By comparison, for the traditional 'red' dyes, such as RH795 or Di-4-ANEPPS, the pulsation artifact can sometimes exceed evoked cortical signals by an order of magnitude (Orbach et al. 1985; London et al. 1989; Shoham et al. 1999; Grinvald and Hildesheim 2004; Ma et al. 2004).

Methods for imaging mammalian cortex *in vivo* are described in detail in other chapters of this book. Some special experimental methods developed in our group have been discussed in Lippert et al. 2007. Below, we briefly demonstrate the sensitivity of *in vivo* VSD recordings.

When the cortex is properly stained, the sensitivity of VSD recording can be comparable to that of LFP recordings (Fig. 7.7). In this experiment, cortical activity was recorded with VSD imaging and LFP simultaneously from the same tissue. Comparing these two signals we found most of the peaks in the LFP showed a corresponding event in the VSD trace. Under 1.5 % isoflurane anesthesia, infrequent bursts of spontaneous activity were recorded on both electrical and optical recordings and most events within each burst correlated well between LFP and VSD signals (Fig. 7.7a). Adjusting the level of anesthesia allows us to further verify the correlation between LFP and the VSD signals. When the level of anesthesia was lowered, both VSD and LFP signals show continuous spontaneous fluctuations (Fig. 7.7b). The fluctuations in the VSD signals are unlikely to be baseline noise, since the baseline noise is much smaller in the quiescent segments of Fig. 7.7a, which were obtained from the same location in the same animal. Instead, these fluctuations are likely to be biological signals from local cortical activity, probably sleep-like oscillations or up/down states (Petersen et al. 2003a, 2003b).

LFP and VSD signals were frequently disproportionate in amplitude. In Fig. 7.7a, events labeled with dots had a higher amplitude in VSD trace while events labeled with triangles had a larger amplitude in the LFP trace. Events labeled with diamonds were seen in the LFP but not in the VSD signals. Under lower anesthesia, while the correlation between the LFP and VSD recordings decreased, many events in LFP were also seen in the VSD traces (exceptions are marked with squares in Fig. 7.7b). VSD recordings from two adjacent locations also show good correlation (bottom two traces in each panel of Fig. 7.7), further indicating that the fluctuations on the VSD traces are not noise. Since almost every event in the LFP can be recorded clearly in VSD imaging, the sensitivity of VSD measurement is comparable to that of LFP recordings. However, VSD and LFP signals have a low

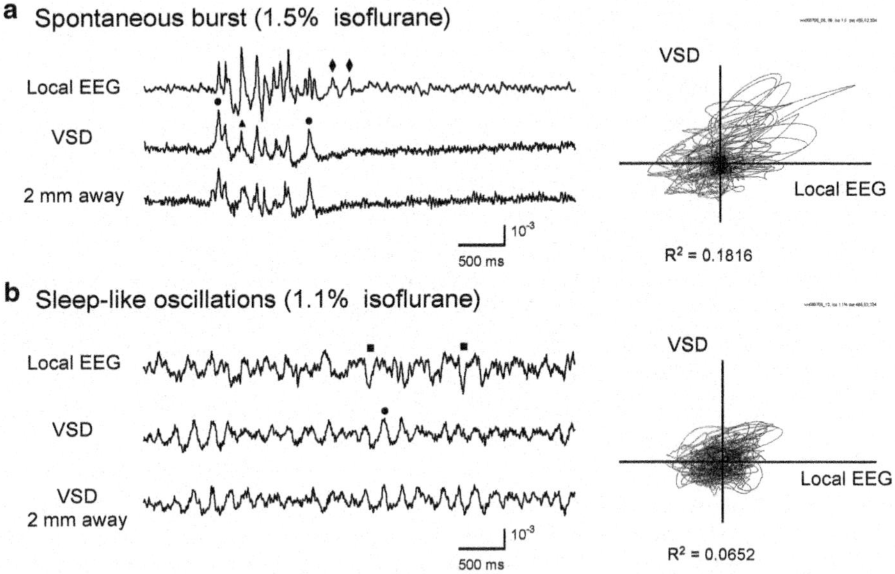

Fig. 7.7 Sensitivity of VSD recordings *in vivo*. The top two traces in each panel are LFP and VSD recordings from the same location in the visual cortex. The *bottom traces* in each panel are VSD recordings taken 2 mm away from this location. A correlation plot of the VSD and LFP of the raw data is presented on the *right*. (a) Under 1.5 % isoflurane anesthesia, the spontaneous spindle-like bursts were recorded. *Dots* mark events in which VSD has higher amplitude than LFP. *Triangles* mark events in which local field potential has higher amplitude than VSD. *Diamonds* mark events seen in local field potential but not in VSD. (b) Recordings from the same animal as in (a) about 5 min later, with isoflurane anesthesia lowered to 1.1 %. Most of the peaks in LFP are also seen in VSD, but the correlation between the two signals is lower. The baseline fluctuations cannot be attributed to noise, since recordings from the same preparation had almost no spontaneous fluctuation when anesthesia was elevated (a). (Modified from Lippert et al. 2007)

overall correlation of ~0.2 between 3 and 30 Hz (which includes 90 % of the power). This low overall correlation is probably due to some LFP peaks originating from deep cortical layers or subcortical sources.

The correlation between optical traces is much higher and each peak in the signal appears at many locations (Fig. 7.7a, b, bottom traces). The high correlation between optical traces is not due to light scattering or optical blurring between channels, because there is a small timing difference between locations, resulting in propagating waves in spatiotemporal patterns (Slovin et al. 2002; Petersen et al. 2003a, 2003b; Ferezou et al. 2006; Lippert et al. 2007; Petersen 2007; Xu et al. 2007, Mohajerani et al. 2010; Huang et al. 2010; Gao, et al. 2012). At a light intensity that results in the signal-to-noise ratio shown in Fig. 7.7, we can record up to 80–100 trials of 10 s, which is sufficient for many types of experiments. With longer recording time, the signal amplitude reduces and epileptiform spikes occasionally develop, probably due to phototoxicity.

14.4 Spatiotemporal Patters During Sleep-Like States

There are two main advantages that make high sensitivity in vivo VSD imaging unique: (1) Fast response time necessary to monitor wave dynamics which is too fast (e.g., 100 mm/ms waves) for Ca imaging. (2) High sensitivity which is mandatory for resolving patterns in single trial recordings (no averaging) to better reveal spatiotemporal characteristics of the signal. The example below demonstrates one application of the *in vivo* VSD imaging.

During sleep periods in the rat, propagating waves sweep across the cortex at a frequency of 1–4 Hz. VSD imaging provides a high spatial and temporal resolutions (464 detectors in a 5 mm diameter view field at 1,650 frames/s), sufficient for visualizing spiral routers and singularities (Huang et al. 2010). The sleep-like states were induced with slow withdrawal of pentobarbital anesthesia. The animals were first given an initial induction dose of pentobarbital (50 mg/kg), followed by tail vein infusion at a low rate of 6–9 mg/kg-h. When the initial anesthesia wore out after 2–3 h, an alternation of EEG patterns (theta/delta waves) occurred under the low maintenance infusion (Fig. 7.8a). The pentobarbital infusion rate (9–15 mg/kg-h) was then carefully adjusted for each individual animal to sustain the alternation of theta/delta wave patterns. The theta/delta alternations resemble the rapid eye movement (REM) state in rodent natural sleep (Montgomery et al., 2008). These alternations occur spontaneously, with clear difference in EEG and simultaneously recorded VSD signals (Fig. 7.8b). Spiral waves occurred frequently during both states (Examples shown in Fig. 7.8c). Double spirals with opposite rotation directions were often seen during sleep-like states (Fig. 7.8c right). The emergence of spiral waves corresponds with changes in the frequency of cortical activity. Frequency-time analysis showed that long-lasting spiral waves have prominent modulation effects on the oscillation frequencies (Fig. 7.8d). In this example frequency and power change can be clearly seen in raw trace and power spectrum, suggesting that established spiral waves have modulating effects on cortical dynamics.

Spiral waves in the intact cortex were not as stable as in slices. Most spiral waves during sleep-like states were shorter than two cycles. Ongoing background activity and long-range connections may contribute to this short duration and instability. Detailed discussion of spiral dynamics during sleep states can be found in Huang et al. 2010.

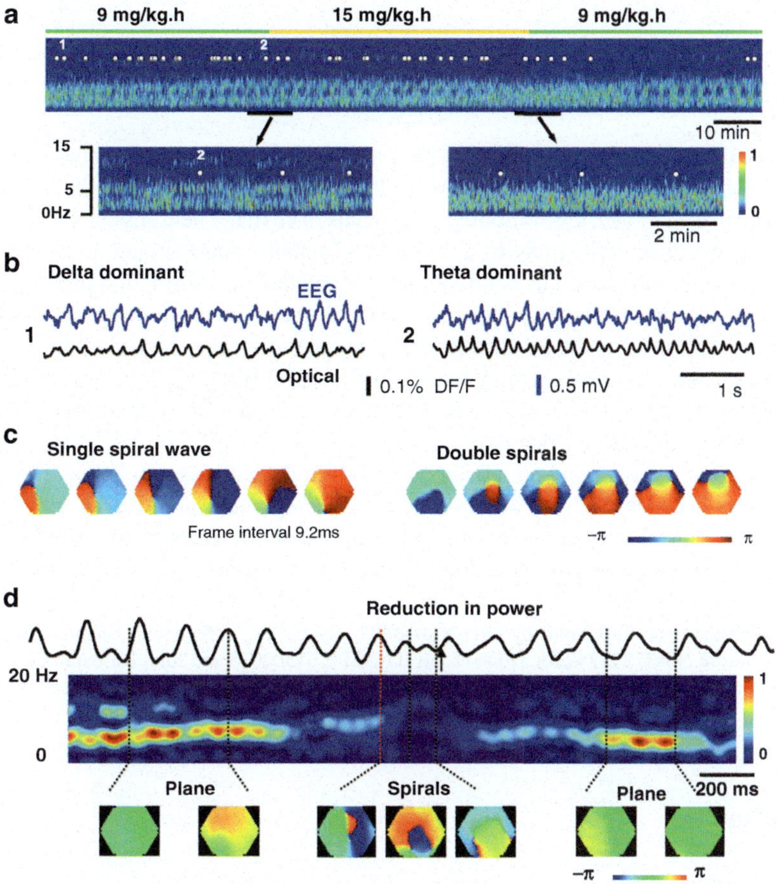

Fig. 7.8 Spatiotemporal dynamics in cortex during sleep-like states. (a) EEG is recorded by a *silver ball* electrode placed on the cortex at the corner of the imaging window. The image is the frequency-time map (frequency spectrum) made from the EEG signal. A low level of anesthesia is maintained by continuous tail vein infusion of pentobarbital at 9 mg/kg-h (*green bar*). The EEG power spectrum transitions periodically between theta (~6 Hz) and delta (1–4 Hz) dominant states when pentobarbital infusion is maintained at a low constant rate. With higher infusion rate (15 mg/kg-h) of pentobarbital (marked with *orange bar*), the alternating patterns in the EEG shifted to the delta-dominant state. *Note*: there are approximately 20 min delays of drug effects because of the slow infusion of pentobarbital. The EEG is continuously recorded and imaging trials are taken intermittently (marked by *white dots*) and two representative trials (*1*, *2*) are shown in (b). Simultaneous EEG (*blue*) and optical (*black*) traces recorded during the theta-dominant ((b), *trial 2*) and delta-dominant periods ((b), *trial 1*). Spiral waves are identified from the imaging trials (c). (d) *top trace* is raw VSD signal from a representing optical detector. *Middle image* is the frequency spectrum made from the detector. Fast Fourier transform was used to analyze the power spectrum of signals in a sliding window of 300 ms. A linear pseudocolor scale is used to represent the relative power of the frequency components, *red* for high power (*1*) and blue for low power (*0*). The *Bottom images* are selected from the phase movie for identifying wave patterns. *Broken vertical lines* indicate the times for these images. The *red broken lines* mark the first cycle of spiral waves and *black arrows* mark the end of spirals (b). Ongoing cortical activity during the theta-dominant period of sleep-like state, a pair of short-lived spiral waves was identified by phase maps (*bottom images*). During spiral waves the power of the activity is significantly reduced. (Modified from Huang et al. 2010)

15 Conclusion

Cortical tissue is rich in neuropil with a relatively large area of stained excitable membrane per unit volume. During population events, cortical neurons are usually activated in ensembles; the combined dye signals from many neurons can be very large. These factors make VSD imaging a useful tool for analyzing spatiotemporal dynamics in the local circuits as well as neuronal interactions during information processing.

Imaging with VSDs provides information about when and where activation occurs and how it spreads. In cortical tissue, oscillations recorded from a LFP electrode appear as propagating waves in the VSD imaging. The wave can have complex patterns such as spirals or irregular waves and these patterns cannot be seen by recording at any one site.

Using diode arrays and an advanced vibration control, we can achieve optimum conditions for measuring population activity in brain slices and *in vivo*. With this method, about 500 locations in 1 mm^2 of cortical tissue can be monitored simultaneously with a sensitivity better than or comparable to that of LFP electrode.

Acknowledgments Supported by NIH NS036477, NS059034, a Whitehall Foundation grant to JYW, Chinese NSFC 81171220, 31371125 to JL, NSFC 61302035 and Joint Research Fund for the Doctoral Program of Higher Education of China 20111107120018 to XG.

References

Agmon A, Connors BW (1992) Correlation between intrinsic firing patterns and thalamocortical synaptic responses of neurons in mouse barrel cortex. J Neurosci 12:319–329

Albowitz B, Kuhnt U (1993) Evoked changes of membrane potential in guinea pig sensory neocortical slices: an analysis with voltage-sensitive dyes and a fast optical recording method. Exp Brain Res 93:213–225

Bai L, Huang X, Yang Q, Wu JY (2006) Spatiotemporal patterns of an evoked network oscillation in neocortical slices: coupled local oscillators. J Neurophysiol 96:2528–2538

Bao W, Wu JY (2003) Propagating wave and irregular dynamics: spatiotemporal patterns of cholinergic theta oscillations in neocortex in vitro. J Neurophysiol 90:333–341

Cohen LB, Salzberg BM (1978) Optical measurement of membrane potential. Rev Physiol Biochem Pharmacol 83:35–88

Cohen LB, Lesher S (1986) Optical monitoring of membrane potential: methods of multisite optical measurement. Soc Gen Physiol Ser 40:71–99

Cohen LB, Keynes RD, Hille B (1968) Light scattering and birefringence changes during nerve activity. Nature 218:438–441

Colom LV, Saggau P (1994) Spontaneous interictal-like activity originates in multiple areas of the CA2-CA3 region of hippocampal slices. J Neurophysiol 71:1574–1585

Demir R, Haberly LB, Jackson MB (1999) Sustained and accelerating activity at two discrete sites generate epileptiform discharges in slices of piriform cortex. J Neurosci 19:1294–1306

Demir R, Haberly LB, Jackson MB (2000) Characteristics of plateau activity during the latent period prior to epileptiform discharges in slices from rat piriform cortex. J Neurophysiol 83:1088–1098

Ermentrout GB, Kleinfeld D (2001) Traveling electrical waves in cortex: insights from phase dynamics and speculation on a computational role. Neuron 29:33–44

Ferezou I, Bolea S, Petersen CC (2006) Visualizing the cortical representation of whisker touch: voltage-sensitive dye imaging in freely moving mice. Neuron 50:617–629
Flint AC, Connors BW (1996) Two types of network oscillations in neocortex mediated by distinct glutamate receptor subtypes and neuronal populations. J Neurophysiol 75:951–957
Gao X, Xu W, Wang Z, Takagaki K, Li B, Wu JY (2012) Interactions between two propagating waves in rat visual cortex. Neurosci 216:57–69
Grinvald A, Hildesheim R (2004) VSDI: a new era in functional imaging of cortical dynamics. Nat Rev Neurosci 5:874–885
Grinvald A, Manker A, Segal M (1982a) Visualization of the spread of electrical activity in rat hippocampal slices by voltage-sensitive optical probes. J Physiol 333:269–291
Grinvald A, Hildesheim R, Farber IC, Anglister L (1982b) Improved fluorescent probes for the measurement of rapid changes in membrane potential. Biophys J 39:301–308
Gupta RK, Salzberg BM, Grinvald A, Cohen LB, Kamino K, Lesher S, Boyle MB, Waggoner AS, Wang CH (1981) Improvements in optical methods for measuring rapid changes in membrane potential. J Membr Biol 58:123–137
Hirata Y, Sawaguchi T (2008) Functional columns in the primate prefrontal cortex revealed by optical imaging in vitro. Neurosci Res 61:1–10
Hirota A, Sato K, Momose-Sato Y, Sakai T, Kamino K (1995) A new simultaneous 1020-site optical recording system for monitoring neural activity using voltage-sensitive dyes. J Neurosci Methods 56:187–194
Huang X, Troy WC, Yang Q, Ma H, Laing CR, Schiff SJ, Wu JY (2004) Spiral waves in disinhibited mammalian neocortex. J Neurosci 24:9897–9902
Huang X, Xu W, Liang J, Takagaki K, Gao X, Wu JY (2010) Spiral wave dynamics in neocortex. Neuron 68:978–990
Jalife J (2003) Rotors and spiral waves in atrial fibrillation. J Cardiovasc Electrophysiol 14:776–780
Jin W, Zhang RJ, Wu JY (2002) Voltage-sensitive dye imaging of population neuronal activity in cortical tissue. J Neurosci Methods 115:13–27
Kleinfeld D, Delaney KR (1996) Distributed representation of vibrissa movement in the upper layers of somatosensory cortex revealed with voltage-sensitive dyes. J Comp Neurol 375:89–108
Kojima S, Nakamura T, Nidaira T, Nakamura K, Ooashi N, Ito E, Watase K, Tanaka K, Wada K, Kudo Y, Miyakawa H (1999) Optical detection of synaptically induced glutamate transport in hippocampal slices. J Neurosci 19:2580–2588
Laaris N, Carlson GC, Keller A (2000) Thalamic-evoked synaptic interactions in barrel cortex revealed by optical imaging. J Neurosci 20:1529–1537
Lippert MT, Takagaki K, Xu W, Huang X, Wu JY (2007) Methods for voltage-sensitive dye imaging of rat cortical activity with high signal-to-noise ratio. J Neurophysiol 98:502–512
Loew LM, Cohen LB, Dix J, Fluhler EN, Montana V, Salama G, Wu JY (1992) A naphthyl analog of the aminostyryl pyridinium class of potentiometric membrane dyes shows consistent sensitivity in a variety of tissue, cell, and model membrane preparations. J Membr Biol 130:1–10
London JA, Cohen LB, Wu JY (1989) Optical recordings of the cortical response to whisker stimulation before and after the addition of an epileptogenic agent. J Neurosci 9:2182–2190
Luhmann HJ, Prince DA (1990) Transient expression of polysynaptic NMDA receptor-mediated activity during neocortical development. Neurosci Lett 111:109–115
Lukatch HS, MacIver MB (1997) Physiology, pharmacology, and topography of cholinergic neocortical oscillations in vitro. J Neurophysiol 77:2427–2445
Ma HT, Wu CH, Wu JY (2004) Initiation of spontaneous epileptiform events in the rat neocortex in vivo. J Neurophysiol 91:934–945
MacLean JN, Fenstermaker V, Watson BO, Yuste R (2006) A visual thalamocortical slice. Nat Methods 3:129–134
Mann EO, Radcliffe CA, Paulsen O (2005) Hippocampal gamma-frequency oscillations: from interneurones to pyramidal cells, and back. J Physiol 562:55–63
Metherate R, Cruikshank SJ (1999) Thalamocortical inputs trigger a propagating envelope of gamma-band activity in auditory cortex in vitro. Exp Brain Res 126:160–174

Mochida H, Sato K, Arai Y, Sasaki S, Kamino K, Momose-Sato Y (2001) Optical imaging of spreading depolarization waves triggered by spinal nerve stimulation in the chick embryo: possible mechanisms for large-scale coactivation of the central nervous system. Eur J Neurosci 14:809–820

Mohajerani MH, McVea DA, Fingas M, Murphy TH (2010) Mirrored bilateral slow-wave cortical activity within local circuits revealed by fast bihemispheric voltage-sensitive dye imaging in anesthetized and awake mice. J Neurosci 30:3745–3751

Momose-Sato Y, Sato K, Arai Y, Yazawa I, Mochida H, Kamino K (1999) Evaluation of voltage-sensitive dyes for long-term recording of neural activity in the hippocampus. J Membr Biol 172:145–157

Montgomery SM, Sirota A, Buzsaki G (2008) Theta and gamma coordination of hippocampal networks during waking and rapid eye movement sleep. J Neurosci 28:6731–6741

Orbach HS, Cohen LB, Grinvald A (1985) Optical mapping of electrical activity in rat somatosensory and visual cortex. J Neurosci 5:1886–1895

Petersen CC (2007) The functional organization of the barrel cortex. Neuron 56:339–355

Petersen CC, Grinvald A, Sakmann B (2003a) Spatiotemporal dynamics of sensory responses in layer 2/3 of rat barrel cortex measured in vivo by voltage-sensitive dye imaging combined with whole-cell voltage recordings and neuron reconstructions. J Neurosci 23:1298–1309

Petersen CC, Hahn TT, Mehta M, Grinvald A, Sakmann B (2003b) Interaction of sensory responses with spontaneous depolarization in layer 2/3 barrel cortex. Proc Natl Acad Sci U S A 100:13638–13643

Prechtl JC, Cohen LB, Pesaran B, Mitra PP, Kleinfeld D (1997) Visual stimuli induce waves of electrical activity in turtle cortex. Proc Natl Acad Sci U S A 94:7621–7626

Ross WN, Salzberg BM, Cohen LB, Grinvald A, Davila HV, Waggoner AS, Wang CH (1977) Changes in absorption, fluorescence, dichroism, and Birefringence in stained giant axons: : optical measurement of membrane potential. J Membr Biol 33:141–183

Senseman DM, Vasquez S, Nash PL (1990) Animated pseudocolor activity maps PAMs: Scientific visualization of brain electrical activity. In: Schild D (ed) Chemosensory information processing, vol H39, NATO ASI series. Springer, Berlin, pp 329–347

Shoham D, Glaser DE, Arieli A, Kenet T, Wijnbergen C, Toledo Y, Hildesheim R, Grinvald A (1999) Imaging cortical dynamics at high spatial and temporal resolution with novel blue voltage-sensitive dyes. Neuron 24:791–802

Slovin H, Arieli A, Hildesheim R, Grinvald A (2002) Long-term voltage-sensitive dye imaging reveals cortical dynamics in behaving monkeys. J Neurophysiol 88:3421–3438

Tanifuji M, Sugiyama T, Murase K (1994) Horizontal propagation of excitation in rat visual cortical slices revealed by optical imaging. Science 266:1057–1059

Tasaki I, Watanabe A, Sandlin R, Carnay L (1968) Changes in fluorescence, turbidity, and birefringence associated with nerve excitation. Proc Natl Acad Sci U S A 61:883–888

Tsau Y, Guan L, Wu JY (1998) Initiation of spontaneous epileptiform activity in the neocortical slice. J Neurophysiol 80:978–982

Tsau Y, Guan L, Wu JY (1999) Epileptiform activity can be initiated in various neocortical layers: an optical imaging study. J Neurophysiol 82:1965–1973

Winfree AT (2001) The geometry of biological time. Springer, New York, NY

Wu JY, Cohen LB (1993) Fast multisite optical measurement of membrane potential. In: Mason WT (ed) Biological techniques: fluorescent and luminescent probes for biological activity. Academic, New York, NY, pp 389–404

Wu JY, Guan L, Tsau Y (1999) Propagating activation during oscillations and evoked responses in neocortical slices. J Neurosci 19:5005–5015

Wu JY, Guan L, Bai L, Yang Q (2001) Spatiotemporal properties of an evoked population activity in rat sensory cortical slices. J Neurophysiol 86:2461–2474

Xu W, Huang X, Takagaki K, Wu JY (2007) Compression and reflection of visually evoked cortical waves. Neuron 55:119–129

Zecevic D (1996) Multiple spike-initiation zones in single neurons revealed by voltage-sensitive dyes. Nature 381:322–325

Chapter 8
Voltage Imaging in the Study of Hippocampal Circuit Function and Plasticity

Brandon J. Wright and Meyer B. Jackson

Abstract Synaptic plasticity has the capacity to alter the function of neural circuits, and long-term potentiation (LTP) of synaptic transmission induced by high frequency electrical activity has the capacity to store information in neural circuits. The cellular and molecular mechanisms of LTP have been studied intensively for many years and much progress has been made on this front. By contrast, how synaptic plasticity alters circuit function has received much less attention and remains poorly understood. Voltage imaging provides a powerful general technique for the study of neural circuitry, and studies of synaptic plasticity with voltage imaging are beginning to reveal important aspects of how the function of a neural circuit can change when the strength of its synapses has been modified. The hippocampus has an important role in learning and memory and the plasticity of its synapses has received much attention. Voltage imaging with voltage sensitive dye in the CA1 region of a hippocampal slice has shown that spatial patterns of enhancement following LTP induction can diverge from the spatial patterns elicited by electrical stimulation, suggesting that LTP exhibits a distinct organizational structure. LTP can alter the throughput of electrical activity in the dentate gyrus of a hippocampal slice, to gate transmission on to the CA3 region. The spatial patterns evoked by complex electrical stimulation can be stored within the CA3 region in a hippocampal slice, allowing patterns to be reconstructed with simpler electrical stimulation. Thus, voltage imaging has demonstrated that the CA3 circuit has the capacity for pattern completion. These studies with voltage sensitive dye illustrate a range of interesting and novel questions that can be addressed at the population level. It is hoped that future imaging experiments with single-cell resolution using genetically-encoded voltage sensors will provide a more detailed picture of how synaptic plasticity modifies the information processing capabilities of neural circuits.

Keywords Synaptic plasticity • Long-term potentiation • Pattern completion • Pattern separation • Hippocampus • Dentate gyrus • Neural networks • Epilepsy

---

B.J. Wright • M.B. Jackson (✉)
Department of Neuroscience, University of Wisconsin, Madison, WI 53705, USA
e-mail: mbjackso@wisc.edu

M. Canepari et al. (eds.), *Membrane Potential Imaging in the Nervous System and Heart*, Advances in Experimental Medicine and Biology 859,
DOI 10.1007/978-3-319-17641-3_8

1 Introduction

Most of the synapses of the hippocampus exhibit plasticity and can change their efficacy as a result of various forms of use. The highly celebrated role of the hippocampus in learning and memory has naturally stimulated great interest in the plasticity of hippocampal synapses. Manipulation of the synaptic plasticity in the hippocampus either pharmacologically or genetically modifies the capacity of an animal to learn (Martin et al. 2000; Lynch 2004; Neves et al. 2008). However, although a large body of work has established a connection between behavioral plasticity and synaptic plasticity, this connection is at a very superficial level. Synapses exhibit many forms of plasticity, and behavior can be shaped and modified in many ways. The explicit relations between synaptic plasticity and behavior remain ill-defined. We do not know how the different types of synapses within and between hippocampal subfields contribute to specific forms of hippocampal information processing, or which of these synapses become modified when a specific kind of information is stored.

It is widely accepted that neural circuits store information as patterns of connectivity between neurons within the circuits that do some of the information processing. The nature of these patterns and their relation to the features of the environment they encode are completely unknown. Thus, place cells in the hippocampus, neurons that become active when an animal is in a particular location (Moser et al. 2008), may hook up in special ways depending on a larger structure in the environment or depending on some salient aspect of a location. The remapping of place cells following changes in environment may reflect changes in these higher order aspects of circuit structure (Colgin et al. 2008). Understanding the relation between synaptic plasticity and behavioral plasticity will require addressing questions about higher order circuit architecture and integrating synaptic plasticity into the picture. Thus, we can ask: can we find higher order connectivity patterns among the neurons exhibiting synaptic plasticity, and do these patterns represent a higher level of organization within established anatomical circuits? Are there subdivisions within neural circuits without visible structural distinctions, and do such subdivisions hold clues about assemblies that serve as units of information storage? Achieving a satisfactory understanding of mechanisms of hippocampal information storage and recall will almost certainly require answering such questions.

Voltage imaging provides a multisite view of electrical activity in brain slices with unique capabilities for addressing questions about neural circuitry. A number of studies have shown that imaging with voltage sensitive dyes reveals long term potentiation of evoked responses in hippocampal slices (Saggau et al. 1986; Momose-Sato et al. 1999; Hosokawa et al. 2003; Aihara et al. 2005; Chang and Jackson 2006a). This approach has the potential to address circuit level questions about synaptic plasticity, and offers an approach to exploring some of the issues confronting efforts to elucidate the relation between synaptic and behavioral plasticity. This chapter reviews work on the use of voltage imaging in the study of synaptic plasticity in the hippocampus. We go on to discuss how this approach has narrowed the gap between our understanding of synaptic plasticity and behavioral plasticity.

2 Heterogeneity

Voltage imaging provides a map of how different parts of a brain slice respond to electrical stimulation. These maps can be used to evaluate heterogeneity in the strength of specific pathways, as well as heterogeneity in the plasticity of these pathways. Electrical stimulation generally evokes responses that track the established anatomy of major pathways, although responses usually decline with distance from the site of stimulation. In an experiment on long-term potentiation (LTP) of synaptic transmission, one can take the difference between maps acquired before and after applying an LTP-inducing stimulus (typically theta bursts), and thus construct an LTP map. LTP maps from voltage imaging have revealed interesting features not evident in the response maps. In these cases LTP maps reveal structural elements that do not parallel response maps. LTP maps thus reveal a structure to the circuitry which arose from plasticity that does not follow the anatomy in a simple manner.

Voltage imaging in the CA1 region of rat hippocampal slices revealed optical responses over considerable distances, mainly in the stratum radiatum (Fig. 8.1A, Bi), with responses rising in a few msec and decaying in tens of milliseconds (Fig. 8.1Bii). Maps of response amplitude encoded as color show that the amplitudes and extent of spread increased following LTP induction by three theta bursts (Fig. 8.1Ci, Cii). The response amplitudes remained high for the duration of the experiment of 50 min, as is typical of LTP (Fig. 8.1D). Response maps were rather featureless (Fig. 8.1Ci, Cii), generally showing a maximum near the site of stimulation and falling off smoothly with distance. There were no peaks and valleys in these response maps (Fig. 8.1Ci, Cii). By contrast, the difference between response maps pre- and post-induction yielded a map that differed strikingly from the response maps and indicated a heterogeneous distribution of LTP (Fig. 8.1Ciii). The LTP maps had a great deal of structure, and often contained as many seven separate peaks (Chang and Jackson 2006b).

The basis for these features remains unclear, but the LTP maps bore a striking resemblance to the maps of action potential amplitude during the second and third theta burst used to induce LTP. Figure 8.2a displays the response amplitude map, along-side the spike amplitude maps during the three successive LTP-inducing theta bursts. There is a clear increase in spike map complexity with each theta burst, and the final spike map resembles the LTP map in the right panel. A plot of spike amplitude versus change in response amplitude shows a poor correlation for the first theta burst but a high correlation for the second two (Fig. 8.2b). The slope of these plots rose from ~0.35 to ~1 in the averages from eight such experiments (Fig. 8.2c). This experiment showed that regions where theta burst stimulation induced action potentials were subsequently found to be potentiated. This correlation reveals an important manifestation of the well-known Hebb condition of plasticity in which a stimulus can alter synaptic strength only when the stimulus drives action potentials in both the pre- and postsynaptic cells. Thus, this recasts the question of why LTP

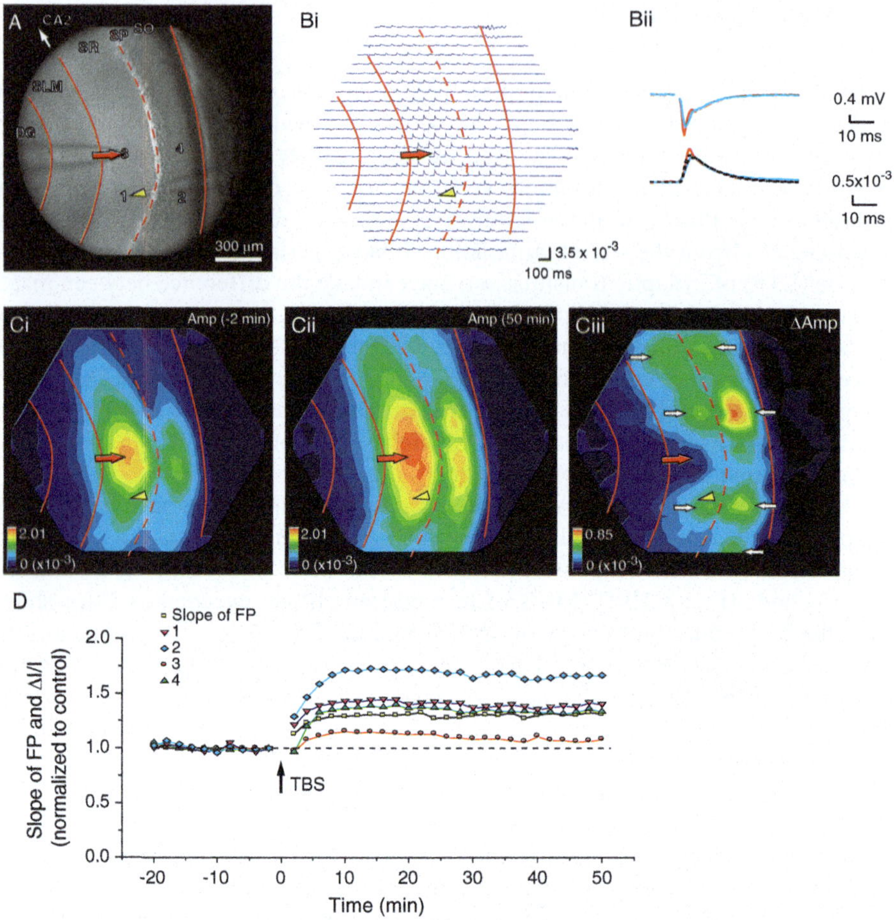

Fig. 8.1 Imaging LTP in the CA1 region. (**A**) A CCD image of the CA1 region in a hippocampal slice, with boundaries between layers highlighted in *red*. *SO* stratum oriens, *SP* stratum pyramidale, *SR* stratum radiatum, *DG* dentate gyrus. The CA2 region is above the field of view. The stimulating electrode is visible from the *left* (*red arrow*) and the field potential recording electrode is visible from the *right* (*white arrow head*). Four sites are indicated with *numbers*. (**Bi**). Optical traces are displayed from the 464 photodiodes that image the spatial distributions of responses to electrical stimulation (60 μA, 200 μs, each trace is an average of four trials). The *red curves* highlight boundaries in (**A**). (**Bii**). The field potential recording (above) and optical trace from the same location (site 1). The *dashed trace* represents the response 2 min before stimulation with three theta bursts and the *red trace* represents the response 50 min after. (**Ci**) The response map prior to theta burst stimulation, with amplitude encoded as color according to the scale in the lower *left* corner. (**Cii**) The response map after theta burst stimulation. (**Ciii**) The difference map created by subtracting (**Ci**) from (**Cii**) illustrates the spatial distribution of LTP. (**D**) The time course of response amplitude and field potential slope from the sites indicated in (**A**). Three theta bursts were applied at the *arrow* (TBS). (Modified from Chang and Jackson (2006b))

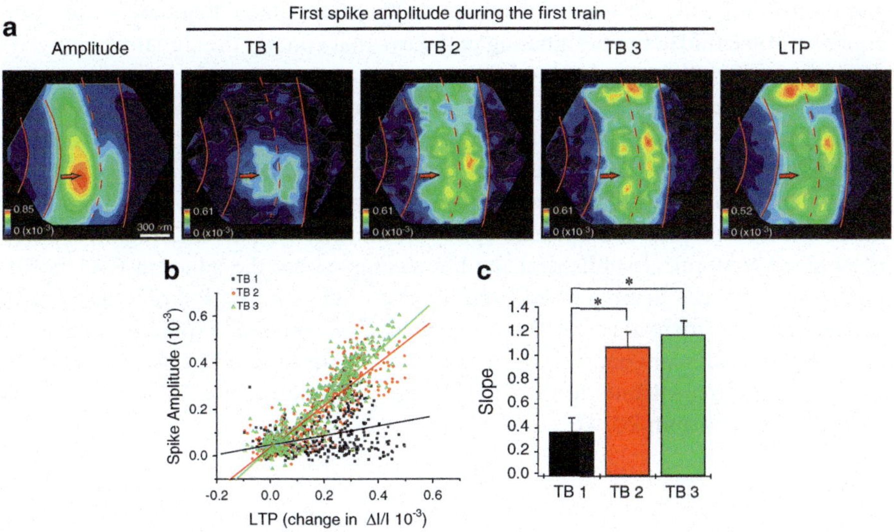

Fig. 8.2 Correlation between LTP and spike amplitude during induction. (**a**) Maps of response amplitude (*first panel*), spike amplitude during the first, second, and third theta bursts (next *three panels*), and LTP (*right panel*). *Red curves* highlight boundaries between layers as in Fig. 8.1a (Stimulation 80 μA). (**b**) Plot of spike amplitude during theta bursts versus the magnitude of LTP, with each point in the field of view contributing one point to the plots. The slopes increased with successive theta bursts. (**c**) The average slope from eight experiments as in (**b**) show statistically significant increases in slope after the first theta burst ($^*P < 0.005$). (Modified from Chang and Jackson (2006b))

induction is so heterogeneous. The question has been translated into a question about why postsynaptic spiking is so heterogeneous. Although we are in no better position to answer this question than the first question, knowing that the action potential map exhibits the same heterogeneity suggests that experiments to study the action potential map may ultimately shed light on the basis for the spatial heterogeneity of LTP.

One consistent feature in LTP maps was that they usually display a minimum around the site of stimulation. This was seen in both the CA1 region (Chang and Jackson 2006b) and the dentate gyrus (Wright and Jackson 2014). In the CA1 region this feature did not simply reflect saturation of responses. It was seen with a range of stimulus currents. Another possible explanation for the gap is that local activation of more inhibitory interneurons by the stimulating electrode could cause greater local release of GABA. The extent of the minimum around the stimulating electrode would then reflect the extent of directly activated GABA release. This may explain the results in the CA1 region as these experiments were performed in the absence of $GABA_A$ receptor blockade. But in the dentate gyrus experiments a $GABA_A$ receptor antagonist was present (and was necessary to obtain LTP with this induction protocol), so local inhibition cannot account for this particular aspect of heterogeneity. Near the stimulating electrode neurons are depolarized both directly, and indirectly by the activation of synapses. The more global depolarization of the entire neuron could activate signaling pathways that counter the signaling pathways activated by synapses that lead to LTP.

Another interesting feature of the LTP map was that a peak consistently appeared at the edge of the CA1 region bordering the CA2 region. This site was also shown to be a hot spot for epileptiform activity (Chang et al. 2007). Interestingly, the narrow CA2 region separating this hot spot from the CA3 region has much less LTP and lower amplitude epileptiform activity. The CA2 region is a well recognized anatomical subfield of the hippocampus, but the adjacent CA1 border region had not been distinguished from the rest of the CA1 region previously. This appears to be an interesting functional specialization, yet to be explained in terms of structure or molecular composition. These distinctions in terms of plasticity may be relevant to the recent finding that the CA2 region plays a role in social memory (Hitti and Siegelbaum 2014).

3 Gating

LTP directly increases the amplitude of responses to the activation of a targeted synapse. But whether this increase ends at that target or propagates further will be very relevant to the potential for plasticity at a given class of synapses to influence behavior. This issue is especially important in circuits with dense intrinsic connections. LTP at intrinsic synapses, as opposed to projection synapses, have the potential to alter how activity travels through a brain region and on to the next station. In this way LTP can modulate a gate that determines what passes through a circuit. Considering the plasticity of a synapse in isolation from its context ignores a potential for rich and interesting consequences that LTP can have on circuit function through complex polysynaptic interactions.

The dentate gyrus has been proposed to serve as the gateway to the hippocampus, and serves as a critical station in filtering and processing the inputs the hippocampus receives from the entorhinal cortex through the perforant path (Ang et al. 2006; Dudek and Sutula 2007; Heinemann et al. 1992; Lothman et al. 1992). Dentate granule cells serve as the principle relays in this pathway and project with axons known as mossy fibers to the CA3 region. However, the dentate gyrus performs a great deal of processing of cortical inputs. Granule cells innervate nearby mossy cells, an excitatory neuron located in the hilus. Mossy cells are thought to play an important role in the processing of information by the dentate gyrus (Buckmaster and Schwartzkroin 1994; Henze and Buzsaki 2007; Scharfman and Myers 2013). Mossy cells receive inputs from granule cells and reciprocate with synapses on granule cells in a powerful pathway of recurrent excitation (Jackson and Scharfman 1996; Scharfman 1995; Scharfman et al. 1990). This circuitry is illustrated in Fig. 8.3. Applying stimulation to the outer molecular layer of the dentate gyrus activates the extrinsic inputs from the entorhinal cortex,

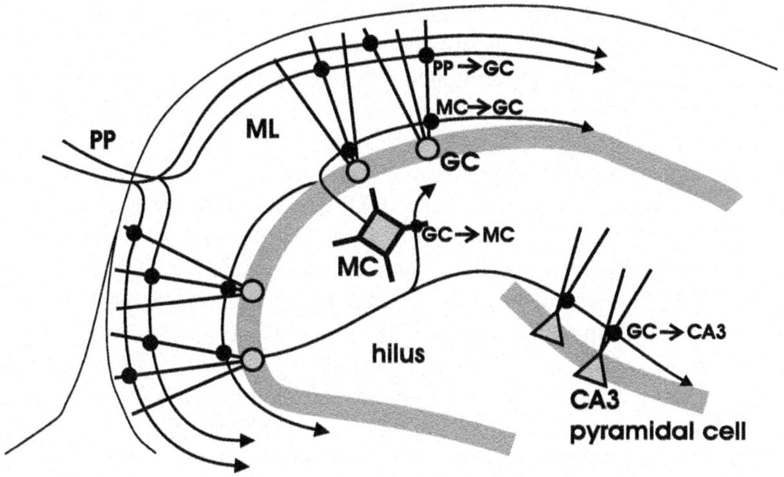

Fig. 8.3 Excitatory synapses of the dentate gyrus. The perforant path (PP) provides inputs from the entorhinal cortex to granule cell (GC) dendrites in the molecular layer (ML). GCs innervate mossy cells (MCs) in the hilus, which reciprocate with synapses on GCs. GCs also innervate pyramidal cells in the CA3 region (interneurons are not shown)

and the granule cell-mossy cell circuit distributes activity broadly throughout the dentate gyrus.

The phenomenon of LTP was first reported in the dentate gyrus of the rabbit at synapses between entorhinal cortex (perforant path) axons and granule cells (Bliss and Gardner-Medwin 1973). However, studies of the LTP of this pathway have yet to separate the contributions of cortical inputs from the mossy cell-granule cell circuit. This is especially difficult because the synapses are close together and the latencies are short. Synapses involving mossy cells do express LTP (Kleschevnikov and Routtenberg 2003; Lysetskiy et al. 2005) and strengthening these synapses would enhance responses to the activation of cortical inputs, thus creating the false impression of LTP at perforant path → granule cell synapses (Claiborne et al. 1993).

Voltage imaging has been used to evaluate responses and LTP throughout the dentate gyrus, and has provided some insight into the capacity of the different synapses for LTP as well as the impact of LTP at these synapses on the flow of activity through the dentate gyrus and on to the CA3 region. Applying theta bursts to the molecular layer potentiates responses throughout the molecular layer but does not potentiate responses in the CA3 region. However, applying theta bursts directly to granule cells potentiates responses in both the molecular layer and the CA3 region (Wright and Jackson 2014). The potentiation of molecular layer responses by granule cell theta bursts requires that LTP occur at either mossy cell → granule cell synapses, or granule cell → mossy cell synapses, or both (Fig. 8.3). Furthermore, it is intriguing that theta burst stimulation of inputs to granule cells in the molecular layer does not potentiate responses in the CA3 region but theta burst stimulation of granule cells does.

To identify the synapses potentiated by applying theta bursts to different sites, NMDA receptors were blocked with the NMDA receptor antagonist APV (Wright and Jackson 2014). NMDA receptor blockade should be very informative in this situation because the synapses formed by granule cells on CA3 pyramidal cells (mossy fiber synapses) express a form of LTP which is unusual in being resistant to NMDA receptor blockade. Thus, the block of LTP in the CA3 region induced by granule cell stimulation provides strong evidence against LTP of mossy fiber synapses. LTP of upstream synapses must be responsible for the robust potentiation in the CA3 region, and this implies LTP in dentate gyrus circuitry. It is further interesting that NMDA receptor blockade failed to inhibit LTP in the molecular layer induced by granule cell stimulation, but completely abolished LTP induced by theta burst stimulation of cortical inputs. Thus, synapses intrinsic to the dentate gyrus manifest an NMDA receptor independent form of LTP. This LTP most likely occurs in synapses formed by granule cell axons on mossy cells, as might be expected if NMDA receptor independent LTP is a general property of synapses formed by mossy cells.

Dual-site stimulation experiments (Fig. 8.4A1) provided an additional basis for evaluating the capacity of dentate gyrus synapses for LTP, and permitted unique tests of hypotheses that LTP at one set of synapses can enhance the response to other inputs. Thus, theta burst stimulation of cortical inputs potentiated responses to granule cell stimulation. This indicates that theta burst stimulation of cortical inputs does indeed induce LTP in downstream synapses between granule cells and mossy cells. Theta burst stimulation of granule cells potentiated responses in the molecular layer and CA3 regions (Fig. 8.4A3, B1, B2). Furthermore, this induction protocol not only potentiated responses to granule cells, but also potentiated the responses in the CA3 region to cortical inputs (Fig. 8.4A2, B1). This indicates that the mossy cell circuitry illustrated in Fig. 8.3 can amplify signals that enter the dentate gyrus through the perforant path from the cortex.

The results of experiments with NMDA receptor blockade together with dual site stimulation in the dentate gyrus created an interesting conundrum. They clearly show that very strong potentiation of responses in the CA3 region can result from LTP in upstream synapses in the dentate gyrus rather than LTP of local mossy fiber synapses. However, LTP of only mossy cell → granule cell synapses, as occurs with NMDA receptor blockade, or only of granule cell → mossy cells synapses, as occurs with theta burst stimulation of cortical inputs, fails to potentiate responses in the CA3 region. This issue is difficult to resolve but one intriguing possibility is that enhancement of transmission to the CA3 region requires LTP of both granule cell → mossy cell synapses and mossy cell → granule cell synapses. This suggests that the mossy cell circuitry can gate the flow of activity through the dentate gyrus, but only when a special condition has been met. Only when LTP occurs in these two distinct populations of synapses will transmission to the CA3 region be enhanced. Indeed this could require LTP in reciprocally connected pairs of granule cells and mossy cells. Granule cells have a high threshold for spiking and this has been

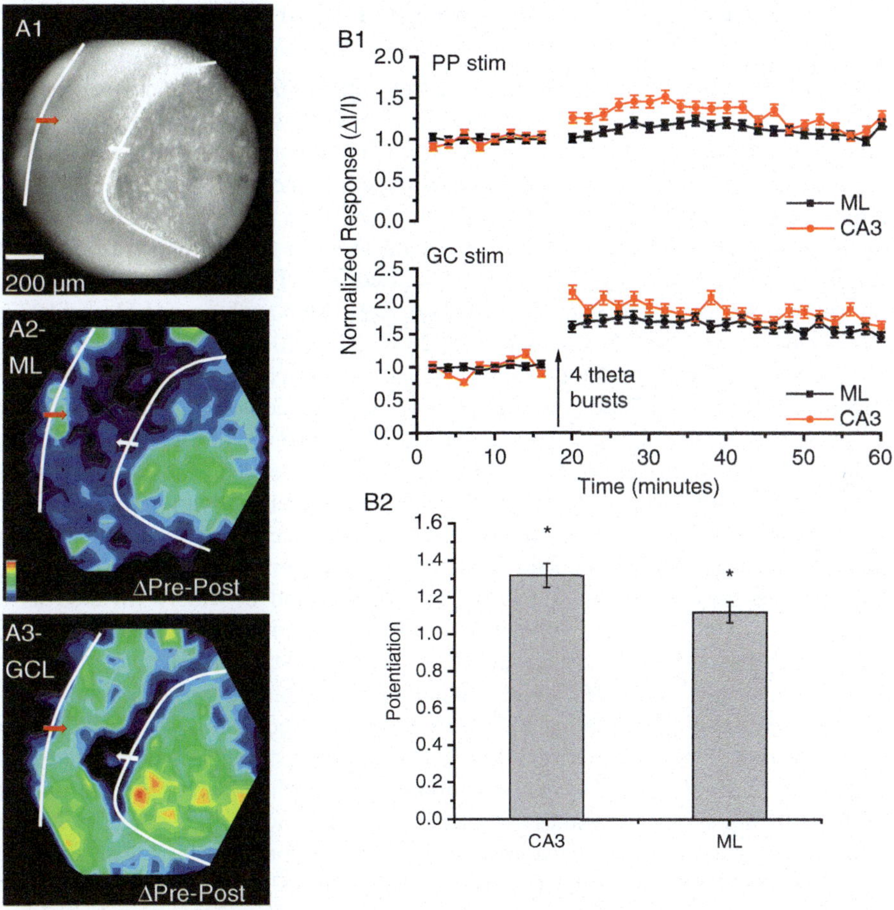

Fig. 8.4 Dual-site stimulation in the dentate gyrus. (**A**) A CCD image shows the dentate gyrus with stimulating electrodes positioned in the molecular layer to activate cortical inputs (*red arrow*) and in the granule cell layer to activate granule cells and MC inputs (*white arrow*). Four theta bursts were applied to the granule cell layer to induce LTP. (**A2**) The LTP map for responses to molecular layer stimulation shows that these responses were potentiated in the CA3 region. (**A3**) The LTP map for responses to granule cell layer stimulation shows that responses were potentiated in both the molecular layer and CA3 region. (**B1**) Time course plot of responses to stimulation of the molecular layer (PP, above) and GC layer (GC below). (**B2**) The heterosynaptic potentiation of responses to perforant path stimulation by theta burst stimulation of the granule cell layer. Responses in both the CA3 region and molecular layer were significantly potentiated (*$P < 0.05$)

proposed to play an important role in filtering inputs and selecting patterns of activity for further processing and transmission to the CA3 region (Hsu 2007; Leutgeb and Leutgeb 2007). This novel gating condition of reciprocal LTP may play an important role in how experience modifies the processing of information by the dentate gyrus, and alters the flow of information to the CA3 region.

4 Spatial Patterns and the Storage of Information

The nervous system takes in information from the environment and transforms it into patterns of activity distributed through neural networks, encoding the information as specific sets of activated neurons. If a pattern of activity leaves an imprint upon the network then some of the information will be stored. If the imprint enables a subsequent input to reactivate the original pattern, then the prior sensory input can be recalled. These ideas form the basis for many theories of information storage and recall by the nervous system, and in these theories LTP generally serves as the imprinting mechanism (Gerstner and Abbott 1997; McNaughton and Morris 1987; Rolls and Treves 1994; de Almeida et al. 2007; Rolls 2013).

The CA3 region of the hippocampus has an anatomical structure that should be capable of supporting these processes. CA3 pyramidal cells have a dense network of synapses with one another (Amaral and Lavenex 2007). When an event activates a set of CA3 pyramidal cells, LTP in the synapses between these co-activated neurons will store information about the event. Subsequent activation of a subset of these neurons would enable the strengthened synapses with other neurons to activate the full original set, and thus recall the information previously stored. This process is referred to as pattern completion, and is widely invoked as a mechanism of cortical and hippocampal storage (Leutgeb and Leutgeb 2007; Rolls and Treves 1998; Kesner et al. 2000; Yassa and Stark 2011).

Voltage imaging provided a test of the concept of pattern completion in hippocampal slices using stimulation by electrodes positioned in different locations to elicit different response patterns (Jackson 2013). Direct stimulation with electrodes is a far cry from activation by sensory inputs but nevertheless evokes patterns in a brain slice amenable to quantitative analysis and hypothesis testing. Electrical stimulation of two distinct locations in the CA3 region of a hippocampal slice evokes responses in two different overlapping regions which can be viewed as distinct response patterns. Stimulating both sites simultaneously activated what can be taken as a 'complete' spatial pattern that combined each of the two partial patterns (Fig. 8.5). When theta burst stimulation was applied to both sites simultaneously, LTP was induced in the synapses between the activated pyramidal cells. Subsequent electrical stimulation of only one of the original sites activated a larger spatial pattern than prior to induction, and this resembled the spatial pattern evoked by stimulation of the other site. Applying theta bursts to one site only did not increase the overlap between the two spatial response patterns. This increase in similarity between the distributions of responses through a slice represents the essence of the idea of information storage and recall by pattern completion. Pattern completion is evident in the examples of spatial maps presented in Fig. 8.5. The pre-induction maps in the left column of Fig. 8.5 are quite different and the post-induction maps in the middle column are quite similar. A quantitative analysis of pattern completion was conducted using the Euclidian distance as a measure of the similarity between the pre-theta burst and post-theta burst patterns. With data from experiments in ~20 slices, dual-site theta burst stimulation, but not single-site theta

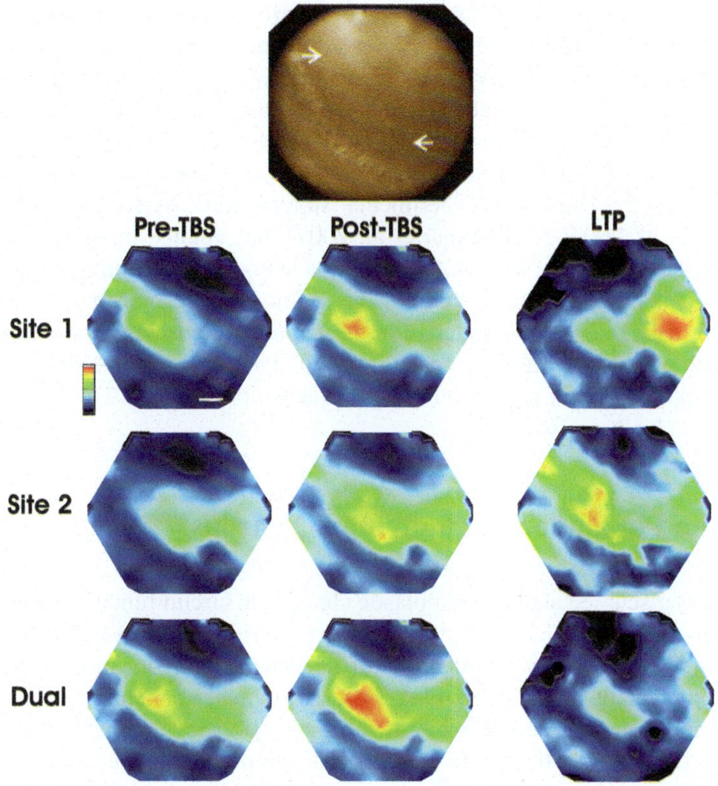

Fig. 8.5 Pattern completion in the CA3 region. Stimulating electrodes were positioned in two locations (*white arrows*) in the stratum radiatum of the CA3 region, as illustrated in the CCD image above. The *top row* presents response and LTP maps for stimulation of site 1. The *second row* presents response and LTP maps for stimulation of site 2. The *bottom row* presents response and LTP maps for stimulation of both sites. The *left column* presents these maps evoked by stimulation prior to the induction of LTP with four theta bursts (TBS) applied to both sites simultaneously. The *middle column* presents these maps after theta burst stimulation. The *right column* presents the LTP maps created by subtracting the pre-TBS map from the post-TBS map. Prior to TBS the three maps look quite different but after TBS the maps look more similar. The increased similarity indicates that TBS stored the information about the dual-site pattern in the slice so that stimulation of either site alone could recall the complete, dual-site pattern. All response maps were normalized to the maximum amplitude in the post-TBS dual-site response map. The LTP maps were normalized to the maximum LTP in the site 1 LTP map. The scale bar in the corner of the site 1 pre-TBS response map was 200 μm and applies to all the maps

burst stimulation, reduced the Euclidian distance between the response patterns evoked separately by each of the two sites. A similar analysis based on Hamming distance between response patterns yielded essentially the same result. These experiments demonstrated that the CA3 region has the capacity to store and recall information using LTP as the imprinting mechanism.

A concept closely related to pattern completion is pattern separation (Leutgeb and Leutgeb 2007; Rolls and Treves 1998; Kesner et al. 2000; Yassa and Stark 2011). In this process sensory inputs activate overlapping patterns of neurons and the neural network separates them into patterns that do not overlap. Such a process would enable the nervous system to classify objects or recognize two similar objects as distinct. The dentate gyrus has been proposed as a locus for pattern separation, and behavioral experiments support this assignment of function (McHugh et al. 2007; Niewoehner et al. 2007), but voltage imaging studies such as those described above in the CA3 region have not been carried out in the dentate gyrus. It will be important to address this point because more recent behavioral studies of tasks that test for pattern completion and pattern separation suggest that the dentate gyrus performs both operations, with pattern separation being carried by a distinct population of newborn neurons (Nakashiba et al. 2012).

5 From Populations to Individual Neurons

The studies discussed above about plastic changes in circuit function were based on measurements of population activity. Voltage sensitive dyes in brain slices generally provide signals from populations of neurons at each location. Signals from single cells cannot be resolved. One must ask whether conclusions about spatial heterogeneity, gating, and storage, at the population level can be related to the dynamics of circuits at the level of cells. If information is encoded at the population level then population experiments may provide the most useful insight into this class of problems. However, even if information is stored by mechanisms that are robust to variations in the activity of individual or small numbers of neurons, it seems likely that individual neurons connected in highly specific ways play an essential role in information processing and storage by the nervous system.

To bring the study of plasticity of circuit function to the cellular level will require technical improvements in imaging electrical activity in intact neural circuits. Voltage sensitive dyes stain all cells with very little specificity so an imaging experiment with one of these probes collects signals from an overwhelming number of neurons as well as glia. In intact circuits the processes of different neurons intermingle within volumes of the order of a micron. Even with highly improved performances, voltage sensitive dyes will likely fail to resolve signals from single cells under these circumstances.

With Ca^{2+} imaging the entire volume of a cell contributes so it is possible to obtain good single cell resolution of cell bodies through bulk filling of neurons. This does not work with voltage imaging because the dye must be on the cell surface, and cell bodies have much less surface area than neuropil. Ca^{2+} probes do not provide a readout of voltage; Ca^{2+} is an indirect measure of electrical activity and changes more slowly than voltage. Ca^{2+} imaging does not follow synchrony or timing on the msec time scale, and cannot detect electrical events that fail to elevate intracellular Ca^{2+}.

It seems unlikely at present that synthetic voltage sensitive dyes or Ca^{2+} probes will reach the point of providing satisfactory single cell resolution of rapid electrical signaling in large numbers of neurons in intact brain tissue. Genetically-encoded voltage sensors may offer the best hope of bringing the study of the plasticity of neural circuit function to the single cell level. Genetically encoded voltage sensors are still in an early stage of development and are likely to improve. The best of these provide good recordings of action potentials in brain slices as well as subthreshold synaptic potentials (Cao et al. 2013; Wang et al. 2010; Gong et al. 2014; St-Pierre et al. 2014; Ghitani et al. 2015). The trick will be to target them to desired populations in which separation is sufficient that signals can be clearly assigned to individual cells. If perhaps ~50 neurons are targeted among the ~thousands within a field of view, and manipulations can be performed to mimic sensory inputs, voltage imaging will reveal a fine structure to the functional connectivity that will illuminate the way in which circuits store and process information. At this level we will be able to ask how a neural circuit performs canonical computations such as normalization, pattern completion, and pattern separation. A detailed understanding of these elementary processes will serve as a significant step in bridging the gap between electrical signaling in the brain and behavior, as well as between synaptic plasticity and experience-induced modifications of behavior.

References

Aihara T, Kobayashi Y, Tsukada M (2005) Spatiotemporal visualization of long-term potentiation and depression in the hippocampal CA1 area. Hippocampus 15:68–78

Amaral DG, Lavenex P (2007) Hippocampal neuroanatomy. In: Andersen P, Morris R, Amaral DG, Bliss TV, O'Keefe J (eds) The hippocampus book. Oxford University Press, Oxford, pp 37–114

Ang CW, Carlson GC, Coulter DA (2006) Massive and specific dysregulation of direct cortical input to the hippocampus in temporal lobe epilepsy. J Neurosci 26:11850–11856

Bliss TV, Gardner-Medwin AR (1973) Long-lasting potentiation of synaptic transmission in the dentate area of the unanaestetized rabbit following stimulation of the perforant path. J Physiol 232:357–374

Buckmaster PS, Schwartzkroin PA (1994) Hippocampal mossy cell function: a speculative view. Hippocampus 4:393–402

Cao G, Platisa J, Pieribone VA, Raccuglia D, Kunst M, Nitabach MN (2013) Genetically targeted optical electrophysiology in intact neural circuits. Cell

Chang PY, Jackson MB (2006a) Heterogeneous spatial patterns of long-term potentiation in rat hippocampal slices. J Physiol 576:427–443

Chang PY, Jackson MB (2006b) Heterogeneous spatial patterns of long-term potentiation in hippocampal slices. Department of Biophysics. University of Wisconsin, Madison, WI, p 144

Chang PY, Taylor PE, Jackson MB (2007) Voltage imaging reveals the CA1 region at the CA2 border as a focus for epileptiform discharges and long-term potentiation in hippocampal slices. J Neurophysiol 98:1309–1322

Claiborne BJ, Xiang Z, Brown TH (1993) Hippocampal circuitry complicates analysis of long-term potentiation in mossy fiber synapses. Hippocampus 3:115–121

Colgin LL, Moser EI, Moser MB (2008) Understanding memory through hippocampal remapping. Trends Neurosci 31:469–477

de Almeida L, Idiart M, Lisman JE (2007) Memory retrieval time and memory capacity of the CA3 network: role of gamma frequency oscillations. Learn Mem 14:795–806
Dudek FE, Sutula TP (2007) Epileptogenesis in the dentate gyrus: a critical perspective. Prog Brain Res 163:755–773
Gerstner W, Abbott LF (1997) Learning navigational maps through potentiation and modulation of hippocampal place cells. J Comput Neurosci 4:79–94
Gong Y, Wagner MJ, Zhong Li J, Schnitzer MJ (2014) Imaging neural spiking in brain tissue using FRET-opsin protein voltage sensors. Nat Commun 5:3674
Ghitani N, Bayguinov BO, Ma Y, Jackson MB (2015) Single-trial imaging of spikes and synaptic potentials in single neurons in brain slices with genetically-encoded hybrid voltage sensor. J Neurophysiol 113:1249–1259
Heinemann U, Beck H, Dreier JP, Ficker E, Stabel J, Zhang CL (1992) The dentate gyrus as a regulated gate for the propagation of epileptiform activity. Epilepsy Res Suppl 7:273–280
Henze DA, Buzsaki G (2007) Hilar mossy cells: functional identification and activity in vivo. Prog Brain Res 163:199–216
Hitti FL, Siegelbaum SA (2014) The hippocampal CA2 region is essential for social memory. Nature 508:88–92
Hosokawa T, Ohta M, Saito T, Fine A (2003) Imaging spatio-temporal patterns of long-term potentiation in mouse hippocampus. Philos Trans R Soc Lond B Biol Sci 358:689–693
Hsu D (2007) The dentate gyrus as a filter or gate: a look back and a look ahead. Prog Brain Res 163:601–613
Jackson MB (2013) Recall of spatial patterns stored in a hippocampal slice by long-term potentiation. J Neurophysiol 110(11):2511–2519
Jackson MB, Scharfman HE (1996) Positive feedback from hilar mossy cells to granule cells in the dentate gyrus revealed by voltage-sensitive dye and microelectrode recording. J Neurophysiol 76:601–616
Kesner RP, Gilbert PE, Wallenstein GV (2000) Testing neural network models of memory with behavioral experiments. Curr Opin Neurobiol 10:260–265
Kleschevnikov AM, Routtenberg A (2003) Long-term potentiation recruits a trisynaptic excitatory associative network within the mouse dentate gyrus. Eur J Neurosci 17:2690–2702
Leutgeb S, Leutgeb JK (2007) Pattern separation, pattern completion, and new neuronal codes within a continuous CA3 map. Learn Mem 14:745–757
Lothman EW, Stringer JL, Bertram EH (1992) The dentate gyrus as a control point for seizures in the hippocampus and beyond. Epilepsy Res Suppl 7:301–313
Lynch MA (2004) Long-term potentiation and memory. Physiol Rev 84:87–136
Lysetskiy M, Foldy C, Soltesz I (2005) Long- and short-term plasticity at mossy fiber synapses on mossy cells in the rat dentate gyrus. Hippocampus 15:691–696
Martin SJ, Grimwood PD, Morris RG (2000) Synaptic plasticity and memory: an evaluation of the hypothesis. Annu Rev Neurosci 23:649–711
McHugh TJ, Jones MW, Quinn JJ, Balthasar N, Coppari R, Elmquist JK, Lowell BB, Fanselow MS, Wilson MA, Tonegawa S (2007) Dentate gyrus NMDA receptors mediate rapid pattern separation in the hippocampal network. Science 317:94–99
McNaughton BL, Morris RGM (1987) Hippocampal synaptic enhancement and information storage within a distributed memory system. Trends Neurosci 10:408–415
Momose-Sato Y, Sato K, Arai Y, Yazawa I, Mochida H, Kamino K (1999) Evaluation of voltage-sensitive dyes for long-term recording of neural activity in the hippocampus. J Membr Biol 172:145–157
Moser EI, Kropff E, Moser MB (2008) Place cells, grid cells, and the brain's spatial representation system. Annu Rev Neurosci 31:69–89
Nakashiba T, Cushman JD, Pelkey KA, Renaudineau S, Buhl DL, McHugh TJ, Rodriguez Barrera V, Chittajallu R, Iwamoto KS, McBain CJ, Fanselow MS, Tonegawa S (2012) Young dentate granule cells mediate pattern separation, whereas old granule cells facilitate pattern completion. Cell 149:188–201
Neves G, Cooke SF, Bliss TV (2008) Synaptic plasticity, memory and the hippocampus: a neural network approach to causality. Nat Rev Neurosci 9:65–75

Niewoehner B, Single FN, Hvalby O, Jensen V, Meyerzum Alten Borgloh S, Seeburg PH, Rawlins JN, Sprengel R, Bannerman DM (2007) Impaired spatial working memory but spared spatial reference memory following functional loss of NMDA receptors in the dentate gyrus. Eur J Neurosci 25:837–846
Rolls ET (2013) The mechanisms for pattern completion and pattern separation in the hippocampus. Front Syst Neurosci 7:74
Rolls ET, Treves A (1994) Neural networks in the brain involved in memory and recall. Prog Brain Res 102:335–341
Rolls ET, Treves A (1998) Neural networks and brain function. Oxford University Press, Oxford
Saggau P, Galvan M, ten Bruggencate G (1986) Long-term potentiation in guinea pig hippocampal slices monitored by optical recording of neuronal activity. Neurosci Lett 69:53–58
Scharfman HE (1995) Electrophysiological evidence that dentate hilar mossy cells are excitatory and innervate both granule cells and interneurons. J Neurophysiol 74:179–194
Scharfman HE, Myers CE (2013) Hilar mossy cells of the dentate gyrus: a historical perspective. Front Neural Circuits 6:106
Scharfman HE, Kunkel DD, Schwartzkroin PA (1990) Synaptic connections of dentate granule cells and hilar neurons: results of paired intracellular recordings and intracellular horseradish peroxidase injections. Neuroscience 37:693–707
St-Pierre F, Marshall JD, Yang Y, Gong Y, Schnitzer MJ, Lin MZ (2014) High-fidelity optical reporting of neuronal electrical activity with an ultrafast fluorescent voltage sensor. Nat Neurosci 17:884–889
Wang D, Zhang Z, Chanda B, Jackson MB (2010) Improved probes for hybrid voltage sensor imaging. Biophys J 99:2355–2365
Wright BJ, Jackson MB (2014) Long-term potentiation in hilar circuitry modulates gating by the dentate gyrus. J Neurosci 34:9743–9753
Yassa MA, Stark CE (2011) Pattern separation in the hippocampus. Trends Neurosci 34:515–525

Chapter 9
Monitoring Population Membrane Potential Signals During Development of the Vertebrate Nervous System

Yoko Momose-Sato, Katsushige Sato, and Kohtaro Kamino

Abstract The functional organization of the vertebrate central nervous system (CNS) during the early phase of development has long been unclear, because conventional electrophysiological means have several technical limitations. First, early embryonic neurons are small and fragile, and the application of microelectrodes is often difficult. Second, the simultaneous recording of electrical activity from multiple sites is limited, and as a consequence, response patterns of neural networks cannot be assessed. Optical recording techniques with voltage-sensitive dyes have overcome these obstacles and provided a new approach to the analysis of the functional development/organization of the CNS. In this review, we provide detailed information concerning the recording of optical signals in the embryonic nervous system. After outlining methodological considerations, we present examples of recent progress in optical studies on the embryonic nervous system with special emphasis on two topics. The first is the study of how synapse networks form in specific neuronal circuits. The second is the study of non-specific correlated wave activity, which is considered to play a fundamental role in neural development. These studies clearly demonstrate the utility of fast voltage-sensitive dye imaging as a powerful tool for elucidating the functional organization of the vertebrate embryonic CNS.

Keywords Optical recording • Voltage-sensitive dye • Development • Embryo • Functiogenesis • Neural network formation

Y. Momose-Sato (✉)
Department of Nutrition and Dietetics, College of Nutrition, Kanto Gakuin University, 1-50-1 Mutsuura-Higashi, Kanazawa-ku, Yokohama 236-8503, Japan
e-mail: yms@kanto-gakuin.ac.jp

K. Sato
Department of Health and Nutrition Sciences, Faculty of Human Health, Komazawa Women's University, Inagi-shi, Tokyo, Japan

K. Kamino
Tokyo Medical and Dental University, Bunkyo-ku, Tokyo, Japan

M. Canepari et al. (eds.), *Membrane Potential Imaging in the Nervous System and Heart*, Advances in Experimental Medicine and Biology 859,
DOI 10.1007/978-3-319-17641-3_9

Abbreviations

CNS	Central nervous system
DMNV	Dorsal motor nucleus of the vagus nerve
E	Embryonic day which shows days of incubation in chicks and days of pregnancy in rats and mice
EPSP	Excitatory postsynaptic potential
GABA	γ-Aminobutyric acid
NMDA	*N*-methyl-D-aspartate
NTS	Nucleus of the tractus solitarius
S/N	Signal-to-noise ratio
TTX	Tetrodotoxin

1 Introduction

Investigating the developmental organization of the embryonic nervous system has been one of the major challenges in neuroscience. It allows us to understand when and how the brain is functionally developed, and provides information about the onset of brain function. In addition, it is a powerful means of analyzing the complicated architecture of the adult central nervous system (CNS). Furthermore, understanding the developmental processes is crucial for treating diseases, because it is becoming increasingly apparent that the role of developmental signals is not over when embryogenesis has been completed, and the signals are continuously present in adults to maintain a physiological state. Therefore, exploring the functional organization of the embryonic nervous system would promote progress in developmental neuroscience, system neuroscience, and clinical neuroscience. Despite their significance, functional investigations of the vertebrate embryonic CNS have been hampered, since conventional electrophysiological means have some technical limitations. First, early embryonic neurons are small and fragile, and the application of microelectrodes is often difficult. Second, the simultaneous recording of electrical activity from multiple sites is limited, and as a consequence, spatio-temporal patterns of neural network responses cannot be assessed.

The advent of optical techniques using voltage-sensitive dyes has enabled the non-invasive monitoring of electrical activity in living cells and also facilitated the simultaneous recording of neural responses from multiple regions (for reviews see Cohen and Salzberg 1978; Salzberg 1983; Grinvald et al. 1988). Using optical recording techniques, it is now possible to follow the functional organization of the embryonic nervous system and to image the spatio-temporal dynamics of the neural network's formation. Here, we review the application of voltage-sensitive dyes to the brain and spinal cord of vertebrate embryos. The primary aim of this review is to provide detailed methodological information concerning the recording of optical signals in the embryonic nervous system. Thus, we first discuss several technical issues including preparations, voltage-sensitive dyes, recording systems, and optical signals. We then present two examples of optical studies in the embryonic nervous system. The first is the study of the synaptic network's formation in specific neuronal circuits. The second is the study of non-specific correlated activity that propagates widely like a wave. The application of voltage-sensitive dyes in post-natal/posthatched animals is a major branch of optical studies in developmental neuroscience. However, this issue is beyond the scope of this chapter and not discussed here.

2 Technical Considerations Regarding Optical Recording in the Embryonic Nervous System

2.1 Preparation: Advantages and Disadvantages of Optical Recording

Early embryonic brain tissue is small and thin. For example, in the chick embryo on E (embryonic day; days of incubation in chickens) 4, in which neural excitability has already been generated, the medulla is about 1.5 mm in width (right to left) and 150–200 μm in thickness (ventral surface to dorsal surface). By E7, when functional synaptic networks have formed, the chick medulla is about 2 mm in width and 800 μm in thickness. Furthermore, the early embryonic brain has a histologically loose structure with immature neurons/glia and undifferentiated connective tissues, and is relatively resistant to anoxia. As a consequence, a whole brain-spinal cord preparation is available in vitro, which makes it possible to investigate the functional organization of neural networks in the intact brain.

The immaturity of the embryonic nervous system has several advantages in optical recording. First, the loose cellular-interstitial structure of the embryonic preparation may allow the dye to diffuse readily from the surface to deeper regions, and consequently to stain neurons relatively well. Second, the brain tissue is highly translucent, allowing the incident light to pass easily through the preparation. Because of this, optical signals can be detected from deep in the brain. For example, it is possible to record neural responses in the dorsally located cranial nuclei from the ventral surface of the intact medulla. Third, the high translucency of the embryonic tissues gives a large signal-to-noise ratio (S/N) in absorption measurements, since the S/N is proportional to the square root of the transmitted or fluorescence background light intensity, if the dominant noise is shot noise (Grinvald et al. 1988). Fourth, the thinness and loose structure of the embryonic preparation may result in reduced light scattering that blurs the signals and limits the spatial resolution of the optical measurements. Finally, the simple structure and lower complexity of neural populations make it easier to analyze the optical signals. In optical recording systems, each detector receives light from many neurons and processes, and the signal detected by one detector is an integration of many signals from different cell populations. In early developmental stages, some neural populations are morphologically and/or functionally undifferentiated, and thus, the shape of the optical signals is relatively simple in some cases (also see Sect. 2.4 in this chapter).

The fragility and immaturity of embryonic neurons also have some disadvantages. First, it is difficult or impossible to compare directly the signals recorded optically with those measured electrophysiologically, because conventional electrophysiological means are difficult to employ. Second, when the meningial tissue surrounding the brain is removed to stain the tissue with the voltage-sensitive dye, the preparation (especially the root of the cranial and spinal nerves) is easily damaged. Third, postsynaptic potentials detected from the embryonic nervous system fatigue readily, so that signal averaging methods often cannot be employed. Finally, embryonic postsynaptic potentials, especially those in the CNS, usually have a slow time course (~1 s duration), so that the waveform of the optical signal is easily affected by various noise sources.

2.2 Voltage-Sensitive Dyes

Much progress has been achieved in the development of voltage-sensitive dyes for measuring rapid changes in membrane potential (for reviews see Dasheiff 1988; Loew 1988; Ebner and Chen 1995). Most voltage-sensitive dyes exhibit potential-dependent absorption and fluorescence changes. The high translucency of embryonic tissue gives a large S/N in absorption measurements, and thus, absorption, rather than fluorescence, voltage-sensitive dyes have usually been employed in embryonic preparations (Momose-Sato et al. 2001a). Absorption dyes have also been favored because they are less phototoxic than fluorescence dyes and can be used in experiments requiring prolonged imaging (O'Donovan et al. 2008).

The most commonly used voltage-sensitive absorption dye in embryonic tissue is a merocyanine-rhodanine dye, NK2761 (produced by Hayashibara Biochemical Laboratories Inc./Kankoh-Shikiso Kenkyusho, Okayama, Japan) (Fig. 9.1a). This dye was designed as an analogue of dye XXIII, in which the alkyl chain attached to the rhodanine nucleus is replaced by a butyl group (Fujii et al. 1981). It was first employed in experiments on embryonic hearts (for reviews see Kamino et al. 1989a; Kamino 1991) and subsequently used in embryonic nervous systems (Sakai et al. 1985; Kamino et al. 1989b). To look for additional new dyes to allow monitoring of embryonic neural activity, over 20 dyes including several newly synthesized absorption probes were screened in the chick embryo (Momose-Sato et al. 1995b). In this screening, signal size, S/N, photobleaching, and phototoxicity were examined, and several promising dyes for the embryonic nervous system were found among the merocyanine-rhodanine class of dyes. Large optical signals were obtained using NK2761, NK2776, NK3224, and NK3225, with no serious pharmacological or phototoxic effects on the evoked action potential. On the other hand, although an oxonol dye (RH155/NK3041) gave strong signals in other preparations, it was inferior to the merocyanine-rhodanine dyes for early embryonic preparations.

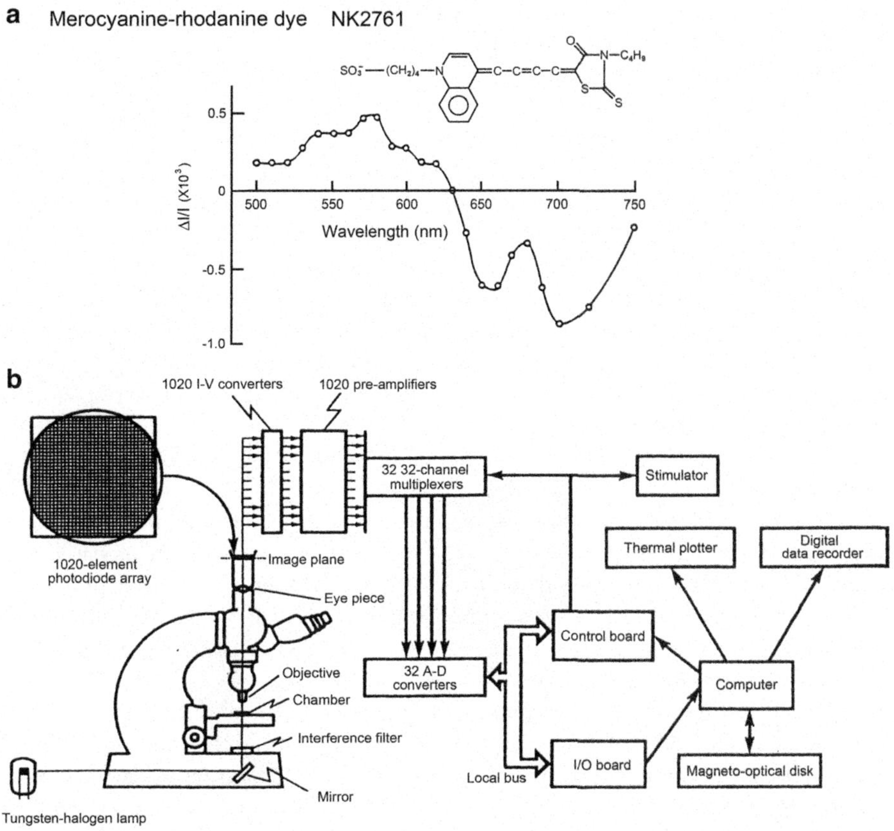

Fig. 9.1 Voltage-sensitive dye and optical recording system. (**a**) The chemical structure (*upper*) and wavelength dependence (*lower*) of a merocyanine-rhodanine dye NK2761. Fractional changes in transmitted light intensity ($\Delta I/I$: the change in light intensity divided by DC-background intensity) related to the action potential are plotted against wavelengths of incident light. (**b**) Schematic drawing of a 1,020-element optical recording system

Di-2-ANEPEQ had the largest S/N of the fluorescence dyes tested in the embryonic chick brain, while its photobleaching was faster and the recovery of neural responses after staining was slower than those of the other fluorescence dyes (Mullah et al. 2013). Although RH414 and RH795 were previously shown to be useful in other preparations (Grinvald et al. 1984, 1988; Obaid et al. 2004), these dyes exhibited very small or undetectable signals in the embryonic brain. The fractional change in fluorescence dyes was larger than that of absorption dyes, whereas its S/N was lower because of more noise. Even with di-2-ANEPEQ, which performed the best among the fluorescence dyes screened, its S/N was inferior (approximately 20 %) to that of NK2761 (Mullah et al. 2013). Nevertheless, fluorescence dyes may be a better option when optical recording is performed at a later developmental stage or when postnatal/posthatched animals, in which the translucency of tissue is low, are used.

Each voltage-sensitive dye has its own action spectrum, which is species- and tissue-specific. In the embryonic brain of chickens and rats, the action spectra of the merocyanine-rhodanine dyes including NK2761 are such that the transmitted light intensity increases (decrease in absorption) in the range of 500–620 nm, and decreases (increase in absorption) in the range of 630–750 nm, with the cross-over occurring at around 630 nm (Momose-Sato et al. 1995b) (Fig. 9.1a). The largest change in transmitted light intensity occurs at around 700 nm. This wavelength-dependency is useful because it allows one to distinguish membrane potential-related optical signals from intrinsic optical changes or other mechanical artifacts: voltage-dependent optical responses are detected at 700 nm, but not at 630 nm, while intrinsic signals and mechanical artifacts are observed at both wavelengths (also see Sect. 2.4).

A voltage-sensitive dye is usually applied to tissue by bathing the preparation to be recorded in a solution of the dye. To stain neurons well, it is often useful to remove meningial tissue surrounding the brain. Within a short time the dye binds to the cell membrane, after which excess dye is washed away with physiological solution. As an alternative approach to selectively labeling a specific neural population, relatively hydrophobic fluorescence styryl dyes (di-8-ANEPPS and di-12-ANEPPQ) have been employed in the embryonic nervous system (Tsau et al. 1996; Wenner et al. 1996). These dyes were injected into a nerve or axon tract and transported retrogradely to the parent neuronal somata by lateral diffusion, which enabled the selective labeling of a specific subpopulation of neurons. Recently, Ca^{2+}-indicators have been applied to embryonic and neonatal brains by electroporation (Bonnot et al. 2005; O'Donovan et al. 2008). This method has not yet been employed for voltage-sensitive dyes, but might be in the future.

When applying dyes to early embryonic preparations, certain physicochemical properties including the osmotic behavior of the dye should be considered, because the early embryonic cells are primitive and may be sensitive to small changes in the environment. To prevent toxic actions of the dye, a brief staining with low concentrations is preferable, but the staining conditions depend on several factors including the affinity of the preparation for the dye and tissue thickness. Usually, NK2761 is loaded for 5–20 min at a concentration of 0.1–0.4 mg/ml. Under these conditions, neither a pharmacological nor a phototoxic action of the dye is observed for evoked action potentials and postsynaptic potentials, while spontaneous activity is sometimes transiently suppressed by staining.

The fractional change in dye absorption ($\Delta A/A$) and fluorescence ($\Delta F/F$) is proportional to the magnitude of the changes in membrane potential in each cell and process, and to the membrane area of activated neural elements within the optically detected field, assuming that the amount of dye bound to the membrane is uniform (Orbach et al. 1985; Kamino et al. 1989a). In absorption measurements, $\Delta A/A$ is equal to $-\Delta I/(I_{before\ staining} - I_{after\ staining})$, where A and I are light absorbance of the dye and light intensity transmitted through the preparation, respectively, and ΔA and ΔI are their changes (Ross et al. 1977). In the majority of embryonic experiments, preparations have already been stained before being placed in the recording chamber. This is done to maximize diffusion of the dye into the tissue, but prevents the measuring of $I_{before\ staining}$ and $I_{after\ staining}$ from the same preparation. In embryonic preparations, variations in $I_{before\ staining}/I_{after\ staining}$ between regions are relatively small (Momose-Sato and Sato 2006), and thus, the optical signal has usually been expressed as $\Delta I/I_{after\ staining}$, assuming this to be linearly related to $\Delta A/A$. By measuring the size of the optical signals and examining their distribution, functional arrangements of the response cells can be assessed. This optical mapping method has been used to analyze the functional arrangement of the response area such as the cranial motor and sensory nuclei (see Sect. 3.2).

2.3 Optical Recording Systems

Several reviews concerning optical recording systems have been published (Salzberg 1983; Grinvald 1985; Cohen and Lesher 1986; Grinvald et al. 1988; Wu and Cohen 1993; Wu et al. 1998). Technical information is also available at http://www.redshirtimaging.com and described elsewhere in this issue. In optical recordings with voltage-sensitive dyes, high sensitivity and high temporal resolution together with an adequate spatial resolution are required to detect small optical signals (10^{-4}–10^{-2} as a fractional change) occurring with a time course of milliseconds. When designing and constructing recording systems for monitoring embryonic neural activity, the following basic requirements should be considered. First, the temporal resolution should be high enough to detect the action potential, which has a similar time course as that in adults. Second, slow changes in membrane potential, such as embryonic postsynaptic potentials that last more than a second, should be detectable. Third, the S/N should be as large as possible, hopefully sufficient to permit recording without averaging because the embryonic postsynaptic potential fatigues very rapidly. Finally, it is important to ensure that the dynamic range matches the fractional changes of the signals to be recorded, and that the system does not saturate. In absorption measurements in vitro embryonic preparations, the signals arise from a baseline of high intensity transmitted light. In this situation, a photodiode array or CMOS camera is preferable because other detectors such as a CCD camera would saturate at such high light intensities and the signal would be lost. In the case of fluorescence signals at low light levels, a cooled CCD camera might perform better. Overall, the choice of detection systems needs to be tailored to the preparation and the type of signals.

In our own studies using absorbance signals from merocyanine-rhodanine dyes, home-made optical recording systems with a 12 × 12 (144-elements) or 34 × 34 (1,020-elements) photodiode array have been used (Hirota et al. 1995; Momose-Sato et al. 2001a). In other laboratories, commercially available equipment (NeuroPDA, RedShirtImaging, Fairfield, CT; MiCAM, Brain Vision Inc., Tsukuba, Japan; ARGUS-50/PDA, Hamamatsu Photonics, Hamamatsu, Japan; Deltaron-1700, Fujifilm, Tokyo, Japan: the former two systems are now available) has been applied to the developing nervous system. In general, the systems are composed of three main parts: an optics system, a detection system, and a recording system with a computer. Figure 9.1b shows a schematic representation of the 1,020-element optical recording system, which is set for absorption measurements. The optics system is based on a biological microscope mounted on a vibration isolation table. Bright field illumination is provided by a 300 W tungsten-halogen lamp driven by a stable direct current (DC) power supply. In other laboratories, a 100–150 W tungsten-halogen lamp has also been used as a light source. Incident light is collimated, passed through a heat filter, rendered quasi-monochromatic with an interference filter, and focused onto the preparation, which is placed on the stage of a microscope. An interference filter should be selected depending on the wavelength dependency of the dye (see Sect. 2.2). For the merocyanine-rhodanine dye NK2761, an interference filter with a transmission maximum at 700 ± 11–15 nm (half width) has been used. An objective and a photographic eyepiece form a magnified real image of the preparation on the photodiode array. The focus is usually set on the surface of the preparation, but the optical signals seem to include activity from every depth, as it has been shown that the interior region of the embryonic brain is well stained with the dye (Sato et al. 1995), and that neuronal responses in the dorsally located cranial nerve nucleus can be detected from the ventral surface (Momose-Sato et al. 1991b).

The transmitted light intensity at the image plane is detected with a photosensitive device such as an array of silicon photodiodes. The spatial resolution in purely optical terms depends on the magnification of the microscope and the size of one photodetector. In our 1,020-element photodiode array system, each of the 1.35×1.35 mm^2 active elements of the array is separated by an insulating area 0.15 mm in width. Thus, each pixel of the array detects light from a region of 54×54 μm^2 of the preparation when a magnification of 25× (objective 10× and eyepiece 2.5×) is used to cover the entire region of a medulla slice. Using a 12×12-element photodiode array, in which each element has a 1.40×1.40 mm^2 active area, together with a water immersion 40× objective and a 6.7× photographic eyepiece, a spatial resolution of 5.2×5.2 μm^2 has been achieved, with which action potentials in the embryonic chick cervical vagus nerve bundle have been detected in single sweeps (K. Kamino, unpublished observation).

In the 1,020-element recording system shown in Fig. 9.1b, the outputs from the 1,020 elements of the photodiode array are fed into individual current-to-voltage converters followed by individual pre-amplifiers. The amplified outputs are fed into 32 sets of 32-channel analog multiplexers and then sent to a discretely designed subranging type analog-to-digital converter system. An advantage of this circuit is that the optical signal is detected with a resolution of 18 bits without AC-coupling, so that the embryonic slow responses can be recorded without a distortion of signal waveform. On the other hand, a drift in the DC baseline associated with dye bleaching causes some problems with long-term recordings. As an alternative method, in the 144-element optical recording system and currently modified 1,020-element system, the AC component of each output from individual pre-amplifiers is further amplified via AC-coupled circuits with low-cut filters (time constant, 3 s) and then digitally recorded. The frame rate provided by these systems is 1 kHz or more, which is sufficient to resolve individual action potentials and postsynaptic potentials.

2.4 Components of the Optical Signal

In optical recording systems, each element of the photodiode array detects optical signals from many neurons and processes when the dye is loaded by bath-application. Because the spatial resolution is not fine enough, optical signals often contain several components, such as antidromic action potentials, orthodromic action potentials, and postsynaptic potentials. One useful way to identify the origin of an optical signal is to analyze the waveform of the signal. Figure 9.2A, B show schematic representations of possible components of the optical signal. In Fig. 9.2A, examples are shown for optical responses evoked by stimulation of the preganglionic fibers in the embryonic chick superior cervical ganglion. Figure 9.2A(a) presents a case in which synaptic functions have not been generated or are blocked pharmacologically. Figure 9.2A(b) shows a case in which synaptic function is present. The upper, middle, and lower panels display possible structures included in the detected area, the observed optical signals, and possible components of the signal, respectively. The observed optical signals are composed of fast spike-like and slow signals. Physiological and pharmacological analyses have indicated that the fast signal corresponds to the presynaptic action potential (pre-AP) and postsynaptic firing (post-AP), and the slow signal, to the excitatory postsynaptic potential (EPSP) (also see Sect. 3.1). In the case of a mixed nerve such as the vagus nerve shown in Fig. 9.2B, the situation is more complex as the nerve contains motor and sensory nerve fibers that cannot be surgically separated. Therefore, stimulation simultaneously evokes antidromic action potentials in motoneurons and orthodromic action potentials in sensory nerve fibers, which are difficult to distinguish in the region where the motor and sensory nuclei functionally overlap.

Action potential-related fast optical signals are not clearly identified in some recordings although they are expected to be present. For example, dorsal root stimulation in the embryonic chick and neonatal rat spinal cords evoked action potential-related fast signals and EPSP-related slow signals in the dorsal horn, while only a slow component was detected in the ventral region (Arai et al. 1999; Mochida et al. 2001a; Ziskind-Conhaim and Redman 2005). In association with correlated wave activity, bursting discharges were electrically recorded in the cranial and spinal motor nerves, while the corresponding spike-like discharges were not relevant to optical recordings (Momose-Sato et al. 2005, 2007c, 2009, 2012a; Ren et al. 2006) (Fig. 9.2C). The amplitude of the optical signal is proportional to the magnitude of the change in membrane potential in each cell and process, and to the membrane area of activated neural elements within the receptive field of one photodiode. Thus, if the action potentials are asynchronous between

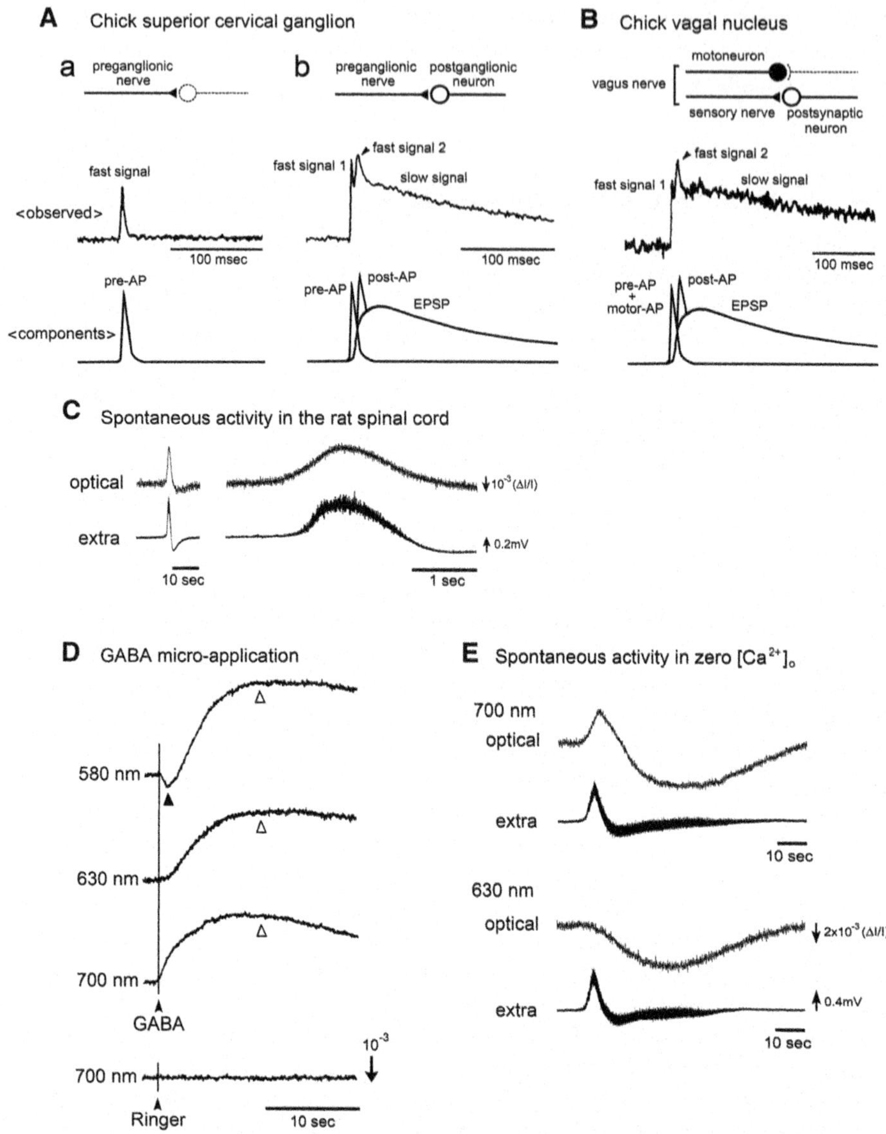

Fig. 9.2 Representation of different components of the optical signal. (**A–B**): Plausible origins of the fast and slow components of the optical signal evoked by stimulation of the preganglionic fibers in the chick superior cervical ganglion (**A**) and by stimulation of the vagus nerve in the chick brainstem (**B**). **A**(a) shows the case in which synaptic functions have not been generated or are blocked pharmacologically. **A**(b) and **B** show the case in which synaptic function is present. The *upper panels* show possible structures contributing to the optical signal. The *middle panels* display the detected optical signals. The *lower panels* present possible components of the optical signal. Pre-AP, post-AP, motor-AP, and EPSP indicate the presynaptic action potential, postsynaptic firing, action potential in motoneurons, and excitatory postsynaptic potential, respectively. (**C**) Optical (*upper traces*) and electrophysiological (*lower traces*) recordings of spontaneous activity from the lumbar spinal cord of an E17 rat embryo. The *right traces* are the expanded time base of the *left traces*. The electrophysiological recordings were made from the ventral root with a suction

neighboring neurons or originate from a small active membrane area, they are possibly undetected as clear spike-like signals.

Optical signals are classified into two groups: an extrinsic signal that depends on the absorption or fluorescence of the dye, and an intrinsic signal (Cohen and Salzberg 1978; Bonhoeffer and Grinvald 1995; Sato et al. 2004b). To check whether the detected signal contains any intrinsic optical change, the following procedures are useful. First, each voltage-sensitive dye has its own action spectrum, and thus, the wavelength dependence of the detected signals is one criterion to show that they are related to changes in membrane potential. Second, measurements from unstained preparations can be used to characterize the intrinsic signal.

Intrinsic light scattering changes are sometimes detected along with the voltage-dependent extrinsic signal in in vitro embryonic preparations. The micro-application of glutamate, γ-aminobutyric acid (GABA), and glycine to the embryonic brainstem slices induces a biphasic optical signal (Sato et al. 1997, 2001; Momose-Sato et al. 1998). An example is shown for GABA in Fig. 9.2D. The first component (indicated with a solid triangle) is wavelength-dependent and corresponds to membrane depolarization. The second, large slow component (indicated with open triangles) shows the same sign independent of wavelength and is observed in unstained preparations, indicating that this is an intrinsic optical change. Intrinsic signals with large amplitudes have also been reported in two studies in relation with correlated wave activity. One study found an increase in light transmission accompanying a non-synaptic wave recorded in a low or zero-$[Ca^{2+}]_o$ solution in the rat medulla-spinal cord (Ren et al. 2006) (Fig. 9.2E). In the other study, two types of intrinsic signals, a slow decrease and a fast rhythmic increase in light transmission, were observed in association with the spontaneous episode of activity in the chick spinal cord (Arai et al. 2007). Intrinsic signals in in vitro preparations have been attributed to several factors including alterations in cellular volume and extracellular space, organelle swelling or dendritic beading, and neurosecretion at the active terminal (Salzberg et al. 1985; Sato et al. 2004b). The intrinsic signals induced by the micro-application of glutamate and GABA/glycine seem to be associated with cell swelling and cell shrinkage,

Fig. 9.2 (continued) electrode. The optical signal exhibited a smooth waveform, which resembled the DC potential change of the electrical signal. (**D**) Optical signals evoked by micro-application of GABA in the chick brainstem. The recordings were made at 580 nm (*top trace*), 630 nm (*second trace*), and 700 nm (*third trace*). The *thin line* with an *arrowhead* shows the timing of GABA's application. The first component indicated by a *solid triangle* is the voltage-dependent extrinsic signal, and the second component indicated by *open triangles* is the intrinsic light scattering change. Note that the first extrinsic component is downward at 580 nm (decrease in absorption/increase in transmitted light intensity), upward at 700 nm, and null at 630 nm, which is the isosbestic wavelength of NK2761. The *lower trace* represents a control, which was obtained by micro-application of Ringer's solution. (**E**) Spontaneous activity recorded from the rat lumbar spinal cord in a zero-$[Ca^{2+}]_o$ solution. At 700 nm (*upper traces*), the optical signal showed a biphasic waveform, while at 630 nm (*lower traces*), only a long-lasting downward signal was observed respectively (Sato et al. 1997, 2001; Momose-Sato et al. 1998). The mechanisms underlying the intrinsic signals that accompany correlated activity are not well understood. Nevertheless, these signals can be used to complement the information obtained by imaging using extrinsic dyes, without the risk of phototoxicity (Arai et al. 2007).

Whether the optical signals detected from the embryonic brain contain a glial component is an important question; however, this has not yet been fully clarified. Glial differentiation generally occurs later in development after neuronal differentiation and migration are largely complete. The mature neuroglial form in the chick embryo has not been reported earlier than E8, while glial progenitor cells and radial glia-like structures appear earlier (Thomas et al. 2000; Korn and Cramer 2008). Optical signals similar to those recorded with NK2761 are observed using the voltage-sensitive dye, NK3630 (RH482), which is known to be relatively insensitive to changes in the membrane potential of glial cells (Konnerth et al. 1987). Although this result does not allow us to identify the relative contribution of glial components because the affinity of the dye for glial progenitor cells and radial glia remains unknown, it appears likely that adult-type astrocytes and oligodendrocytes are not the main contributors to the optical signal. It has been shown that Ca^{2+} waves propagate through radial glia in the developing neocortex (Weissman et al. 2004), and the contribution of these cells to the optical signal, especially to correlated wave activity discussed later in Sect. 4 in this chapter, should be examined further.

3 Application to the Study of Developing Neuronal Circuits

Optical recordings give dynamic spatio-temporal information on neural responses, and the visualization of neural circuits during their formative stages opens up many new avenues for experiments. Since the first application of voltage-sensitive dyes to the embryonic nervous system (Sakai et al. 1985; Kamino et al. 1989b), extensive investigations have been devoted to ontogenetic analyses of specific neuronal circuits in the brainstem, forebrain, spinal cord, and peripheral nervous system (Table 9.1). Some of these have been discussed previously in detail (Momose-Sato et al. 2001a; Momose-Sato and Sato 2006, 2011; Glover et al. 2008; O'Donovan et al. 2008), and here we briefly outline the key characteristics that have been revealed by optical studies.

Table 9.1 Application of voltage-sensitive dyes to the embryonic nervous system

Chick olfactory system (I)	Sato et al. (2007)
Chick visual system (II)	Miyakawa et al. (2004)
Chick oculomotor, trochlear and abducens nuclei (III, IV, VI)	Glover et al. (2003)[a]
Chick trigeminal nucleus (V)	Sakai et al. (1985) Momose-Sato and Sato (2014b) Sato et al. (1999)
Chick vestibulo-cochlear nucleus (VIII)	Asako et al. (1999) Glover et al. (2003)[a] Momose-Sato et al. (2006) Sato and Momose-Sato (2003)
Chick glossopharyngeal nucleus (IX)	Momose-Sato et al. (2007b) Sato and Momose-Sato (2004a, b) Sato et al. (1995, 2002a, b)
Chick vagal nucleus (X) (including pharmacological studies on NTS neurons)	Kamino et al. (1989b, 1990) Komuro et al. (1991) Momose-Sato and Sato (2005) Momose-Sato et al. (1991b, 1994, 1995a, 1997, 1998, 2007a, b) Sato and Momose-Sato (2004a) Sato et al. (1993, 1996, 1997, 2001, 2002b, 2004a)
Chick spinal cord	Arai et al. (1999, 2007) Mochida et al. (2001a)
Rat trigeminal nucleus (V)	Momose-Sato et al. (2004, 2005)
Rat facial nucleus (VII)	Momose-Sato et al. (2007c)
Rat glossopharyngeal nucleus (IX)	Momose-Sato et al. (2011)
Rat vagal nucleus (X)	Momose-Sato et al. (1999b, 2011, 2013) Sato et al. (1998, 2000)
Rat respiratory center	Onimaru and Homma (2005) Ikeda et al. (2004)
Rat spinal cord	Demir et al. (2002)
Peripheral nervous system	Momose-Sato et al. (1991a, 1999a) Sakai et al. (1985, 1991)
Correlated wave activity	
Chick	Arai et al. (2007) Komuro et al. (1993) Mochida et al. (2001b, 2009) Momose-Sato and Sato (2014a) Momose-Sato et al. (2001b, 2003a, b, 2009)
Rat	Momose-Sato et al. (2005, 2007c) Ren et al. (2006)
Mouse	Momose-Sato et al. (2012a, b)
Dye screening and improvements in optical recording systems	Hirota et al. (1995) Momose-Sato et al. (1995b) Mullah et al. (2013) Tsau et al. (1996) Wenner et al. (1996)

Roman numerals in parentheses indicate the number of cranial nerves, which were stimulated to induce the response, except for the oculomotor, trochlear, and abducens nuclei (III, IV, and VI), which were identified in the vestibulo-ocular reflex pathway by stimulation of the vestibular nerve

[a]Results are presented in abstract form

3.1 Identification of the Action Potential and Postsynaptic Response

The functional development of specific neuronal circuits has been most extensively studied in the chick vagal pathway (Table 9.1) (for reviews see Momose-Sato and Sato 2006, 2011), and the characteristics of optical signals as antidromic and orthodromic action potentials and postsynaptic responses were first established in this system. Following electrical stimulation of the vagus nerve, tetrodotoxin (TTX)-dependent fast spike-like signals were detected in the dorsal region of the brainstem, which correspond to antidromically evoked action potentials in vagal motoneurons and orthodromically propagated action potentials in vagal afferents (Kamino et al. 1989b). By comparing the spatial distribution of optical signals with anatomical information, the sources of the signals have been identified as the dorsal motor nucleus of the vagus nerve (DMNV) and the nucleus of the tractus solitarius (NTS), respectively. In addition to the TTX-dependent fast optical signals, slower and longer-lasting signals appear in the NTS at the later stage (Komuro et al. 1991). These signals were diminished by lowering the extracellular Ca^{2+} concentration and abolished by Mn^{2+} and Cd^{2+}, and were also sensitive to glutamate receptor antagonists, indicating that they represent excitatory postsynaptic potentials (EPSPs) in second-order neurons within the NTS. The early phase of these signals was mediated by the non-*N*-methyl-D-aspartate (NMDA) receptor, whereas the later phase was dependent on the NMDA receptor's function (Komuro et al. 1991; Momose-Sato et al. 1994). These EPSP-related optical signals were easily fatigued at the earliest stages of their appearance and increased in amplitude and robustness with development, demonstrating that voltage-sensitive dye recordings provide information about synaptic strength during the formation and later maturation of synapses.

Fast spike-like and slow optical signals with similar characteristics to those identified in the chick vagal pathway have also been observed in other cranial nuclei and spinal cord (Table 9.1). An exceptional finding was that Ca^{2+}-dependent action potentials were detected in the rat DMNV during particular developmental stages (Momose-Sato et al. 1999b). Ca^{2+} spikes were observed from E14 to a few days after birth (Fukuda et al. 1987; Momose-Sato et al. 1999b), suggesting that they are transiently expressed during development.

3.2 Functional Organization of the Motor and Sensory Nuclei

By combining voltage-sensitive dye recording and pharmacological manipulation, it is possible to chart the development of afferent and efferent projections and of central synaptic connections. The antidromic action potentials in motoneurons and orthodromic action potentials in afferent nerves can be detected at very early stages, as the motor and sensory axons are extending out of and into the brain and spinal cord. For example, action potentials in the chick glossopharyngeal and vagal pathways appear from Hamburger-Hamilton stages 23–24 (E3.5–E4) (Momose-Sato et al. 1991b, 2007b; Sato et al. 2002b), and those in the rat vagal pathway can be detected as early as E12 (Sato et al. 2000). Comparing the time of appearance of the optical signals with the birthdates of neural populations, it has been suggested that the motor and sensory neurons are electrically excitable soon after the final mitosis and as their axons are going to or from the periphery.

The time of appearance of slow optical signals indicates an expression of functional synaptic transmission. The onset of synaptic function in the chick embryo is earliest in the visual system (stage 27, E5.5) (Miyakawa et al. 2004), followed by the olfactory bulb, ophthalmic nucleus of the trigeminal nerve, spinal cord at stages 28–29 (E6) (Sato et al. 1999, 2007; Mochida et al. 2001a; Momose-Sato and Sato 2014b), and then the other systems at stage 30 (E7) (Momose-Sato et al. 1991b, 1994; Sato et al. 1999; Sato and Momose-Sato 2003, 2004b). In the rat embryo, the development of synaptic function has been studied in the trigeminal, glossopharyngeal, and vagal pathways, in which the EPSP-related slow optical signals appear at E14–E15 (Sato et al. 1998; Momose-Sato et al. 2004, 2011). As expected from NMDA receptor mediation of postsynaptic responses, the EPSP can be revealed at slightly earlier stages by eliminating Mg^{2+} from the bathing solution. This suggests that the NMDA receptor's function of postsynaptic neurons and glutamate-releasing activity in the sensory terminals have already been generated prior to the expression of non-NMDA receptors, and the onset of synaptic function is regulated by extracellular Mg^{2+}.

By measuring the size of the optical signals and by mapping their distribution, the functional architecture of the motor and sensory nuclei has been assessed. Contour line maps of the optical responses revealed that the number and/or activity of neurons in the motor and sensory nuclei are arranged in an orderly manner and change dynamically with development. A fine example is the three-dimensional maps of the vagal and glossopharyngeal responses in the chick embryo, which have been obtained by optical sectioning along the z-axis (Sato et al. 1995). The maps show a clear differential location of the motoneuron action potentials and the largest EPSP-signals related to the vagal and glossopharyngeal nerves, indicating that the nucleus cores of these nerves are separately located. A more detailed study of the developing vagal and glossopharyngeal motor nuclei (Sato et al. 2002b) showed an initial pattern of multiple small peaks of antidromic activation that eventually coalesced into a single large peak in each nucleus, suggesting spatial differences in the maturation of electrical excitability and developmental dynamics in these motor nuclei.

Another example that shows the utility of optical mapping is the study of the chick visual system (Miyakawa et al. 2004). Contour line maps of action potentials and EPSPs revealed distinct response areas in the diencephalon and mesencephalon, which were difficult to assess without optical mapping. By comparing the three-dimensional locations of optical response areas with morphologically identified visual nuclei, developmental time courses of functional synaptic connections in each nucleus have been determined.

The utility of voltage-sensitive dye recordings in assessments of the topographic specificity of synaptic connections has also been shown in the trigeminal system (Sato et al. 1999; Momose-Sato et al. 2004). Of particular interest in this system is the degree to which afferents in the three subdivisions of the trigeminal nerve segregate when terminating in the central sensory nuclei, in conjunction with somatotopy. Optical mapping of the responses to stimulation of the three trigeminal nerve branches in the rat embryo (Momose-Sato et al. 2004) showed that as soon as functional connections were made, there was already a hint of differential termination that may presage a somatotopic organization. Thus, voltage-sensitive dye recordings can be used to assess topographic relationships between and within brainstem nuclei with respect to functional differentiation and connectivity.

A trigeminal study previously demonstrated that postsynaptic function was generated well prior to the appearance of the morphological structure of conventional synapses (Momose-Sato et al. 2004). The discrepancy observed between morphological and functional findings suggests that communication between presynaptic and postsynaptic neurons at very early stages is not mediated by a typical "synapse".

3.3 Polysynaptic Neuronal Circuit Formation

In the context of sensorimotor integration exemplified by the brain and spinal cord, the voltage-sensitive dye recording technique clearly provides a way to rapidly assess the topographic specificity of synaptic connections and to follow information flow and integration in reflex pathways as those connections develop and mature. Polysynaptic neuronal circuits have been optically identified in several systems, including the chick olfactory system (Sato et al. 2007), auditory system (Momose-Sato et al. 2006), vestibulo-ocular reflex pathway (Glover et al. 2003), vagal system (Sato et al. 2004a), spinal reflex pathway (Arai et al. 1999; Mochida et al. 2001a), and rat vagal system (Momose-Sato et al. 2013). An example shown in Fig. 9.3 is the polysynaptic pathway in the chick auditory system (Momose-Sato et al. 2006). The high temporal resolution of the optical recording allows one to visualize how information flows from afferents to higher-order nuclei in a single sweep recording (Fig. 9.4). The development of the auditory system is particularly interesting because sensory information is split into two parallel channels that process different aspects of auditory stimuli, and thus, the system provides an excellent model for assessing the initial specificity of synaptic connections in a developing sensory pathway. The results of optical recording revealed that connections through third-order nuclei are made and become functional within a day after the afferents to first-order nuclei are established. Moreover, the basic adult connectivity pattern is apparent as soon as the connections become functional, indicating a high degree of specificity throughout.

Similar results have been obtained in the vagal and vestibulo-ocular reflex pathways (Glover et al. 2003; Momose-Sato et al. 2013; Sato et al. 2004a), and

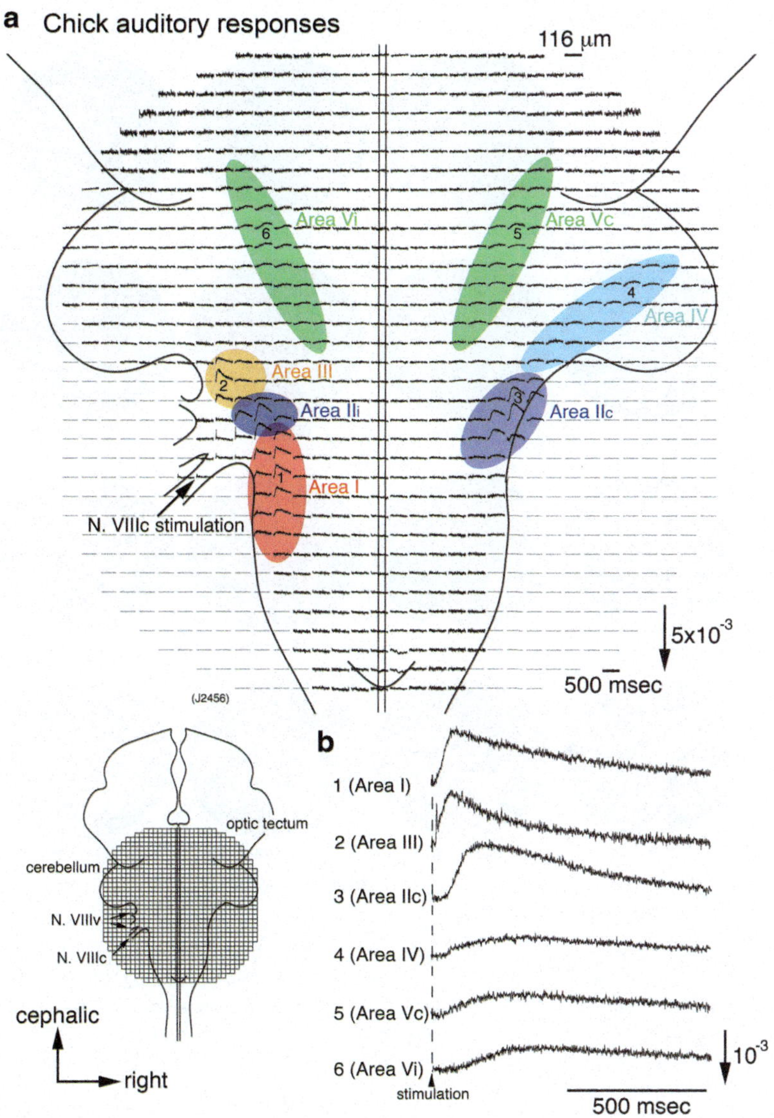

Fig. 9.3 Auditory responses in the embryonic chick brainstem. (**a**) Optical recording of neural responses to simulation of the auditory branch of the eighth cranial nerve (N. VIIIc) in an E8 chick brainstem. The recording was made in the region illustrated in the *bottom left inset*. *Area I*: ipsilateral nucleus magnocellularis; *Areas IIi* and *IIc*: ipsilateral and contralateral nucleus laminaris; *Area III*: ipsilateral nucleus angularis; *Area IV*: contralateral cerebellum; *Areas Vi* and *Vc*: ipsilateral and contralateral nucleus lemnisci lateralis. (**b**) Enlarged traces of the optical signals detected from six different regions indicated by the *numbers* in (**a**). *N.VIIIv* vestibular branch of the eighth cranial nerve

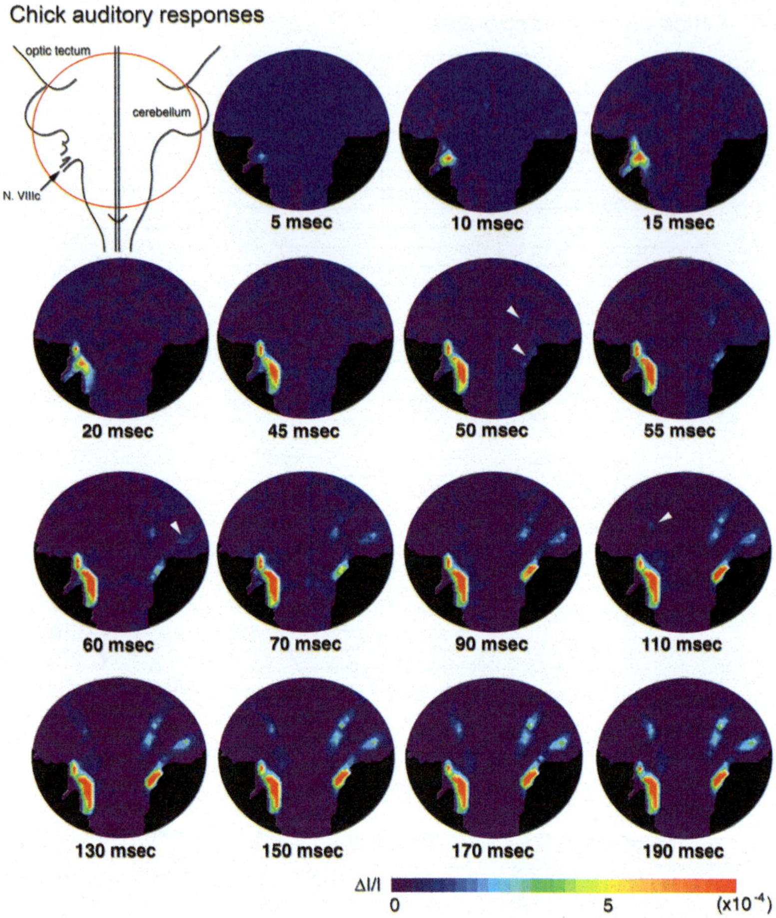

Fig. 9.4 Timelapse images of auditory responses in the chick embryo. Optical responses to stimulation of the auditory branch of the eighth cranial nerve (N. VIIIc) in an E8 chick brainstem are presented with pseudocolor images. The *circular* recorded region is shown in the *top left inset*, and the time after N. VIIIc stimulation is indicated below each image. Optical responses were first observed near the nerve entry zone in a region corresponding to the nucleus magnocellularis, nucleus angularis, and nucleus laminaris. Within 50 ms, responses appeared in contralateral areas, corresponding to the contralateral nucleus laminaris and nucleus lemnisci lateralis (*white arrowheads* in the sixth frame). Within 60 ms, responses were detected in the contralateral cerebellum (*white arrowhead* in the eight frame). Finally, at about 110 ms after stimulation, responses were detected in the ipsilateral nucleus lemnisci lateralis (*white arrowhead* in the 11th frame). The projection to the contralateral cerebellum had been noted previously in anatomical studies and considered to be a transient anomaly that is later eliminated (Young and Rubel 1986), but it had not been determined whether it ever became functional. As a demonstration of the power of the approach, the voltage-sensitive dye recording shows that the projection indeed generates a functional synaptic connection, although the function remains unclear

these observations raise the question of what role afferent activity may have during the formation and later maturation of the synaptic connections. Earlier electrophysiological studies in the chick auditory system had concluded that afferent activity is not important for initial specificity, because synaptic responses to afferent stimulation were not detected with microelectrodes until later (Rubel and Fritzsch 2002). This discrepancy can be ascribed to the invasiveness of microelectrode recordings, which must have been damaging enough to disrupt the early forming synapses and obscure their presence. Non-invasive optical imaging reveals functional connections as they appear. In this case it is now clear that as soon as afferents make contact with their first-order nuclear targets, functional synapses are established, and if the afferents are active at early stages, that activity may exert an influence on the development of the following pathway.

4 Application to the Study of Correlated Wave Activity

The developing nervous system generates spontaneous activity that is considered to play a fundamental role in neural development (Feller 1999; Ben-Ari 2001; Chatonnet et al. 2002; Moody and Bosma 2005). During the early phase of development, this activity recruits a large number of neurons, and thus, a wide region of the CNS is functionally correlated. A major challenge in the study of correlated activity is to identify the origin and spatio-temporal pattern of the activity's propagation, which is a prerequisite for resolving the functional significance of the activity. The optical imaging technique is an ideal tool to deal with this task. So far, several imaging studies have examined the spatio-temporal patterns of correlated activity in the embryo. However, most of these studies have used Ca^{2+}-indicators rather than voltage-sensitive dyes. This is probably because the integrated activity of neurons included in correlated activity has a slow time course, in the order of seconds, and thus, the low temporal resolution of Ca^{2+}-imaging does not place limits on the analysis of the activity. Nevertheless, application of voltage-sensitive dyes has provided a new avenue in the study of correlated wave activity. A typical example is the depolarization wave reported in the chick, rat, and mouse embryos (for a review see Momose-Sato and Sato 2013). Here we summarize the recent findings of the depolarization wave obtained with voltage-sensitive dye imaging and discuss them in relation to previously reported electrophysiological and Ca^{2+}-imaging studies.

4.1 Depolarization Wave

The first voltage-sensitive dye recording of correlated activity in embryos was made by Komuro et al. (1993), who detected spontaneous optical signals in slices of the embryonic chick brainstem. Spontaneous activity appeared with an episode interval of a few minutes and spread over the entire medulla slice. In this pioneering study, the spatio-temporal dynamics of the activity was not resolved in detail because of the limited spatial resolution of the recording apparatus with a 128-element photodiode array.

The next breakthrough was achieved by Momose-Sato et al. (2001b) by introducing a 1,020-element photodiode array system into the intact whole brain-spinal cord preparation and brainstem slice. In this study, the correlated wave was induced by electrical stimulation of the vagus nerve, which triggered a widespread depolarization via activation of postsynaptic neurons in the vagal sensory nucleus (Fig. 9.5). An outstanding feature of the wave, which has been termed the depolarization wave, is that a wide region of the CNS including the spinal cord, medulla, pons, cerebellum, midbrain, and part of the cerebrum is recruited across anatomical boundaries (Fig. 9.5). Such a broadly extended propagation of the wave has never been described in other studies, and this profile suggests that the depolarization wave may not serve as a simple regulator of the formation of specific neuronal circuits, but might play a more global role in the development of the CNS.

Consistent with this hypothesis, depolarization waves with similar spatial distribution patterns were induced by various types of stimulation of the cranial and spinal nerves including the trigeminal nerve, vestibulo-cochlear nerve, glossopharyngeal nerve, vagus nerve, and spinal nerve (Mochida et al. 2001b; Momose-Sato et al. 2003b). Furthermore, as was expected from the study of Komuro et al. (1993), the depolarization wave occurred spontaneously (Mochida et al. 2009; Momose-Sato et al. 2003b, 2009; Momose-Sato and Sato 2014a). These results show that the depolarization wave is triggered by multiple sources of external and endogenous activities, confirming that it is not specific to any neuronal circuit. In addition to chicks, the depolarization wave, either spontaneous or induced, was observed in rat embryos (Momose-Sato et al. 2005, 2007c) and mouse embryos (Momose-Sato et al. 2012a, b) (Fig. 9.6), suggesting that the widely-correlated depolarization wave is globally generated in different species including mammals.

The spatio-temporal characteristics of the depolarization wave are clearly different from those of the respiratory neuron network activity (Onimaru and Homma 2005) and spinal locomotor-like activity (Demir et al. 2002) detected in the rat embryo with voltage-sensitive dyes. In avian and mammalian embryos, spontaneous correlated activity has been electrically recorded from the hindbrain and spinal cord as episodic discharges of cranial and spinal motoneurons (Landmesser and O'Donovan 1984; Fortin et al. 1995; Milner and Landmesser 1999; Abadie et al. 2000; Hanson and Landmesser 2003; Ren and Greer 2003). These activities appear at a low frequency with an episode interval of a few minutes, and are considered to be primordial activity specifically observed during the early developmental stages (Chatonnet et al. 2002). Electrical discharges of vagal motoneurons detected in association with the spontaneous depolarization wave exhibit a similar pattern, suggesting that the depolarization wave is analogous to the previously reported primordial activity.

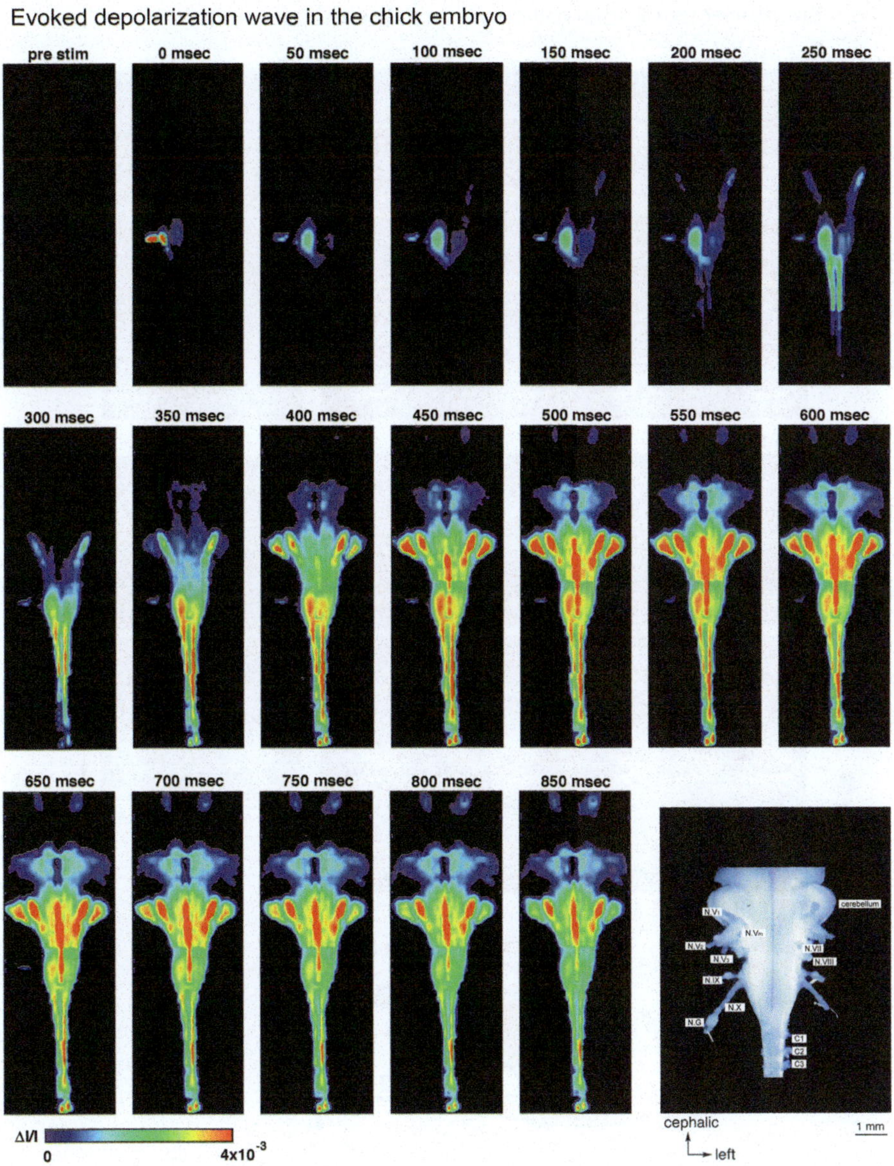

Fig. 9.5 Evoked depolarization wave in the chick embryo. The depolarization wave was induced by stimulation applied to the vagus nerve in an E8 chick embryo. The time after stimulation is indicated above each image. Activity first appeared in the vagal nuclei and then spread into the contralateral brainstem at 50–100 ms after stimulation. The wave bilaterally propagated to the spinal cord and cephalic region. At 350–400 ms, the wave activity appeared in the cerebellum, and at 450–500 ms, it reached the cerebral peduncle. *N.V*$_1$–*N.V*$_3$ ophthalmic, maxillar, and mandibular branches of the trigeminal nerve; *N.Vm* trigeminal motor nerve, *N.VII* facial nerve, *N.VIII* vestibulo-cochlear nerve, *N.IX* glossopharyngeal nerve, *N.X* vagus nerve, *N.G* nodose ganglion, *C* cervical nerve

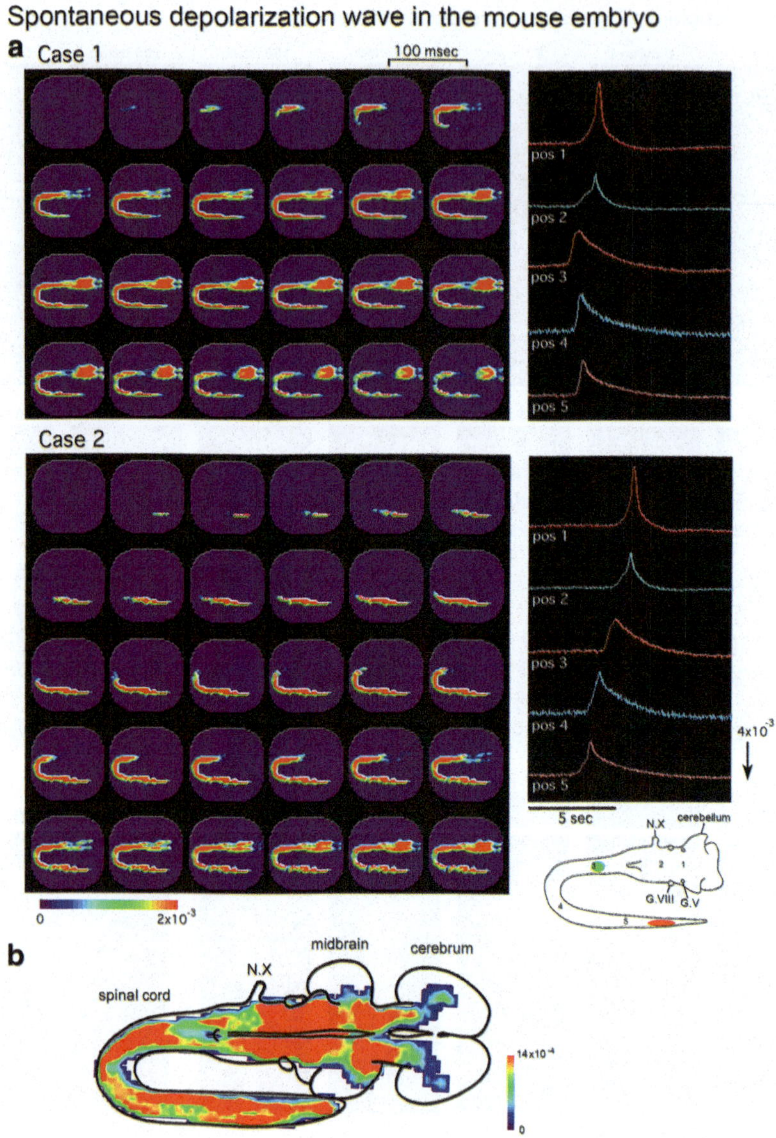

Fig. 9.6 Spontaneous depolarization wave in the mouse embryo. (**a**) Spatiotemporal patterns of the spontaneous depolarization wave in an E12 brainstem-spinal cord preparation. Images were acquired for case 1 and case 2 during two independent episodes in the same preparation. The signals on the *right* were detected from the five positions indicated in the *lower right inset*. In case 1, activity was initiated in the upper cervical cord (*green circle* in the *inset*), while the wave originated in the lumbosacral cord in case 2 (*red oval* in the *inset*). *G.V* trigeminal ganglion, *G.VIII* vestibulo-cochlear ganglion; *N.X* vagus nerve. (**b**) *Color-coded* representation of the maximum signal amplitudes of the spontaneous wave in an E13 mouse embryo

4.2 Common Characteristics

Voltage-sensitive dye imaging of the depolarization wave in different species together with electrophysiological studies has revealed that primordial-correlated activity in the embryonic brain-spinal cord has several common characteristics (for a review see Momose-Sato and Sato 2013). First, the origin of this activity is not fixed, but varies from event to event and depends on the developmental stage (Fig. 9.6) (Momose-Sato et al. 2007a, 2009, 2012a). Moreover, this activity was previously shown to be retained in transected segments at various levels of the brainstem and spinal cord (Fortin et al. 1995; Nakayama et al. 1999; Ren and Greer 2003; Thoby-Brisson et al. 2005). These properties indicate that this activity is generated by multiple regions rather than a single, fixed pacemaker.

Second, primordial-correlated activity recruits a large number of neurons and propagates like a wave over a wide region of the central nervous system (Figs. 9.5 and 9.6). The propagation velocity of the depolarization wave was too slow to be attributed to axonal conduction along unmyelinated fibers at the corresponding stage (Arai et al. 2007; Momose-Sato et al. 2007c). The mechanisms underlying the spread of this activity may include the sequential synaptic activation of adjacent regions coupled by short-range synaptic connections (Fortin et al. 1995; O'Donovan et al. 2008), the non-synaptic release of transmitters and paracrine-like intercellular communication (Demarque et al. 2002; Scain et al. 2010), and the coordination of chemical transmitters with gap junctions as well as electrical interactions between neighboring neurons (Hanson and Landmesser 2003; Ren et al. 2006).

Third, previous studies demonstrated that primordial-correlated activity was most commonly mediated by multiple neurotransmitters and also possibly by gap junctions, and its pharmacological nature was found to undergo developmental changes (Nakayama et al. 1999; Ren and Greer 2003; Myers et al. 2005; Ladle et al. 2007; Mochida et al. 2009; Momose-Sato et al. 2012b). Nicotinic acetylcholine receptors are known to play a significant role in mediating this activity during the early phase, but were shown to be dominated by glutamatergic transmission at the later stage. The second type of switching occurs in GABAergic and glycinergic systems, in which the neural response changes from depolarization/excitation to hyperpolarization/inhibition.

Fourth, primordial-correlated activity is expressed during a particular period of development. For example, the depolarization wave in the mouse embryo covers the spinal cord and brainstem at E11 and extends to the forebrain at E12–E13. This relationship between the brain and spinal cord declines at E14, and activity is localized to restricted regions of the medulla and lumbosacral cord (Momose-Sato et al. 2012a). Synchronization over the brain and spinal cord has also been observed during E4–E8 in the chick (Momose-Sato et al. 2009; Momose-Sato and Sato 2014a) and E13–E17 in the rat (Di Pasquale et al. 1992; Greer et al. 1992; Ren and Greer 2003). These profiles revealed widely propagating correlated activity during a specific time window, and also that the initially synchronized network segregates into more specialized sub-networks related to the adult rhythm generator. Several factors have been suggested as possible mechanisms underlying these changes, including a shift in GABA/glycine responses from depolarization to hyperpolarization, and the influence of neuromodulators such as serotonin (Whelan 2003; Marder and Rehm 2005; Momose-Sato et al. 2012b).

Finally, primordial-correlated activity is known to be under homeostatic control, which may regulate and maintain the overall excitability of the network. Previous studies reported that the removal of a crucial component of the circuit presenting spontaneous activity by the pharmacological blockade of neural transmission was often compensated for by the remaining components (Chub and O'Donovan 1998; Milner and Landmesser 1999). This compensation in network activity has not only been observed between transmitter systems, but also occurs in non-synaptic networks. The blockade of synaptic transmission by a zero-$[Ca^{2+}]_o$ solution abolished this spontaneous activity, but was followed by the emergence of a new form of activity mediated by non-synaptic mechanisms, such as gap junctions and electrotonic interactions between neighboring neurons (Ren et al. 2006). In addition to the compensatory regulation observed for pharmacological blockade, spontaneous activity is known to be homeostatically maintained following the deprivation of the primary rhythm generator. When the function of the primary pacing area or the pathway propagating from that area is disturbed, other regions become a new generator and produce activity with a similar pattern to the previous one (Momose-Sato et al. 2007c; Momose-Sato and Sato 2014a). This may be achieved because most regions of the brainstem and spinal cord have the ability to produce spontaneous activity, and the network is likely to behave as a self-distributing system.

Although much consensus has been obtained concerning the global features of primordial-correlated activity, several questions need to be addressed in future investigations. The most important issue is the functional significance of this activity, especially the necessity of its large-scale distribution over a wide region of the central nervous system. Primordial activity with characteristics similar to those described here has been observed in several other structures of the developing nervous system (Wong 1999; Moody and Bosma 2005; Torborg and Feller 2005; Ben-Ari et al. 2007; Allene and Cossart 2010; Blankenship and Feller 2010), and investigations from multiple perspectives will improve our understanding of this activity and its role in the development of the nervous system.

5 Conclusion

In the past two decades, optical recording techniques with voltage-sensitive dyes have facilitated the characterization of neural responses in the embryonic nervous system. The signals obtained with photodiode array detectors provide a millisecond time resolution with a large field of view, and are particularly useful for identifying action potentials and postsynaptic responses and for visualizing the dynamics of neural responses in real time. The ability to identify and follow the functional development of sequential synaptic relays covering a large area allows us to generate four-dimensional maps of developing neuronal circuits, which cannot be obtained by electrophysiological means. Recently, voltage-sensitive dye imaging has also proved to be a powerful tool for analyzing spontaneous correlated wave activity. Studies on depolarization waves would open the way to further works focusing on the respective role of correlated activity during the development of the central nervous system.

Acknowledgements The present study was supported by grants from the Ministry of Education-Science-Culture of Japan and the Institute of Human Environmental Studies of Kanto Gakuin University.

References

Abadie V, Champagnat J, Fortin G (2000) Branchiomotor activities in mouse embryo. NeuroReport 11:141–145

Allene C, Cossart R (2010) Early NMDA receptor-driven waves of activity in the developing neocortex: physiological or pathological network oscillations? J Physiol (Lond) 588:83–91

Arai Y, Momose-Sato Y, Sato K et al (1999) Optical mapping of neural network activity in chick spinal cord at an intermediate stage of embryonic development. J Neurophysiol 81:1889–1902

Arai Y, Mentis GZ, Wu J-Y et al (2007) Ventrolateral origin of each cycle of rhythmic activity generated by the spinal cord of the chick embryo. PLoS One 2(5), e417

Asako M, Doi T, Matsumoto A et al (1999) Spatial and temporal patterns of evoked neural activity from auditory nuclei in chick brainstem detected by optical recording. Acta Otolaryngol 119:900–904

Ben-Ari Y (2001) Developing networks play a similar melody. Trend Neurosci 24:353–359

Ben-Ari Y, Gaiarsa J-L, Tyzio R et al (2007) GABA: a pioneer transmitter that excites immature neurons and generates primitive oscillations. Physiol Rev 87:1215–1284

Blankenship AG, Feller MB (2010) Mechanisms underlying spontaneous patterned activity in developing neural circuits. Nat Rev Neurosci 11:18–29

Bonhoeffer T, Grinvald A (1995) Optical imaging based on intrinsic signals: the methodology. The Weizmann Institute of Science, Rehovot

Bonnot A, Mentis GZ, Skoch J et al (2005) Electroporation loading of calcium-sensitive dyes into the CNS. J Neurophysiol 93:1793–1808

Chatonnet F, Thoby-Brisson M, Abadie V et al (2002) Early development of respiratory rhythm generation in mouse and chick. Respir Physiol Neurobiol 131:5–13

Chub N, O'Donovan MJ (1998) Blockade and recovery of spontaneous rhythmic activity after application of neurotransmitter antagonists to spinal networks of the chick embryo. J Neurosci 18:294–306

Cohen LB, Lesher S (1986) Optical monitoring of membrane potential: methods of multisite optical measurement. In: De Weer P, Salzberg BM (eds) Optical methods in cell physiology. Wiley, New York, NY, pp 71–99

Cohen LB, Salzberg BM (1978) Optical measurement of membrane potential. Rev Physiol Biochem Pharmacol 83:35–88

Dasheiff RM (1988) Fluorescent voltage-sensitive dyes: applications for neurophysiology. J Clin Neurophysiol 5:211–235

Demarque M, Represa A, Becq H et al (2002) Paracrine intercellular communication by a Ca^{2+}- and SNARE-independent release of GABA and glutamate prior to synapse formation. Neuron 36:1051–1061

Demir R, Gao B-X, Jackson MB et al (2002) Interactions between multiple rhythm generators produce complex patterns of oscillation in the developing rat spinal cord. J Neurophysiol 87:1094–1105

Di Pasquale E, Monteau R, Hilaire G (1992) In vitro study of central respiratory-like activity of the fetal rat. Exp Brain Res 89:459–464

Ebner TJ, Chen G (1995) Use of voltage-sensitive dyes and optical recordings in the central nervous system. Prog Neurobiol 46:463–506

Feller MB (1999) Spontaneous correlated activity in developing neural circuits. Neuron 22:653–656

Fortin G, Kato F, Lumsden A et al (1995) Rhythm generation in the segmented hindbrain of chick embryos. J Physiol (Lond) 486:735–744

Fujii S, Hirota A, Kamino K (1981) Action potential synchrony in embryonic precontractile chick heart: optical monitoring with potentiometric dyes. J Physiol (Lond) 319:529–541

Fukuda A, Nabekura J, Ito C et al (1987) Development of neuronal properties of rat dorsal motor nucleus of the vagus (DMV). J Physiol Soc Jpn 49:395

Glover JC, Momose-Sato Y, Sato K (2003) Functional visualization of emerging neural circuits in the brain stem of the chicken embryo using optical recording. In: Abstracts of the 33rd annual meeting Society for Neuroscience, pp 148.7

Glover JC, Sato K, Momose-Sato Y (2008) Using voltage-sensitive dye recording to image the functional development of neuronal circuits in vertebrate embryos. Dev Neurobiol 68:804–816

Greer JJ, Smith JC, Feldman JL (1992) Respiratory and locomotor patterns generated in the fetal rat brain stem-spinal cord in vitro. J Neurophysiol 67:996–999

Grinvald A (1985) Real-time optical mapping of neuronal activity: from single growth cones to the intact mammalian brain. Annu Rev Neurosci 8:263–305

Grinvald A, Anglister L, Freeman JA et al. (1984) Real-time optical imaging of naturally evoked electrical activity in intact frog brain. Nature 308:848–850

Grinvald A, Frostig RD, Lieke E et al (1988) Optical imaging of neuronal activity. Physiol Rev 68:1285–1366

Hanson MG, Landmesser LT (2003) Characterization of the circuits that generate spontaneous episodes of activity in the early embryonic mouse spinal cord. J Neurosci 23:587–600

Hirota A, Sato K, Momose-Sato Y et al (1995) A new simultaneous 1020-site optical recording system for monitoring neural activity using voltage-sensitive dyes. J Neurosci Meth 56:187–194

Ikeda K, Onimaru H, Yamada J et al (2004) Malfunction of respiratory-related neuronal activity in Na^{+}, K^{+}-ATPase-α subunit-deficient mice is attributable to abnormal Cl^{-} homeostasis in brainstem neurons. J Neurosci 24:10693–10701

Kamino K (1991) Optical approaches to ontogeny of electrical activity and related functional organization during early heart development. Physiol Rev 71:53–91

Kamino K, Hirota A, Komuro H (1989a) Optical indications of electrical activity and excitation-contraction coupling in the early embryonic heart. Adv Biophys 25:45–93

Kamino K, Katoh Y, Komuro H et al (1989b) Multiple-site optical monitoring of neural activity evoked by vagus nerve stimulation in the embryonic chick brain stem. J Physiol (Lond) 409:263–283

Kamino K, Komuro H, Sakai T et al (1990) Optical assessment of spatially ordered patterns of neural response to vagal stimulation in the embryonic chick brainstem. Neurosci Res 8:255–271

Komuro H, Sakai T, Momose-Sato Y et al (1991) Optical detection of postsynaptic potentials evoked by vagal stimulation in the early embryonic chick brain stem slice. J Physiol (Lond) 442:631–648

Komuro H, Momose-Sato Y, Sakai T et al (1993) Optical monitoring of early appearance of spontaneous membrane potential changes in the embryonic chick medulla oblongata using a voltage-sensitive dye. Neuroscience 52:55–62

Konnerth A, Obaid AL, Salzberg BM (1987) Optical recording of electrical activity from parallel fibres and other cell types in skate cerebellar slices in vitro. J Physiol (Lond) 393:681–702
Korn MJ, Cramer KS (2008) Distribution of glial-associated proteins in the developing chick auditory brainstem. Dev Neurobiol 68:1093–1106
Ladle DR, Pecho-Vrieseling E, Arber S (2007) Assembly of motor circuits in the spinal cord: driven to function by genetic and experience-dependent mechanisms. Neuron 56:270–283
Landmesser LT, O'Donovan MJ (1984) Activation patterns of embryonic chick hind limb muscles recorded in ovo and in an isolated spinal cord prepraration. J Physiol (Lond) 347:189–204
Loew LM (1988) How to choose a potentiometric membrane probe. In: Loew LM (ed) Spectroscopic membrane probes, vol 2. CRC Press, Boca Raton, FL, pp 139–151
Marder E, Rehm KJ (2005) Development of central pattern generating circuits. Curr Opin Neurobiol 15:86–93
Milner LD, Landmesser LT (1999) Cholinergic and GABAergic inputs drive patterned spontaneous motoneuron activity before target contact. J Neurosci 19:3007–3022
Miyakawa N, Sato K, Momose-Sato Y (2004) Optical detection of neural function in the chick visual pathway in the early stages of embryogenesis. Eur J Neurosci 20:1133–1149
Mochida H, Sato K, Arai Y et al (2001a) Multiple-site optical recording reveals embryonic organization of synaptic networks in the chick spinal cord. Eur J Neurosci 13:1547–1558
Mochida H, Sato K, Arai Y et al (2001b) Optical imaging of spreading depolarization waves triggered by spinal nerve stimulation in the chick embryo: Possible mechanisms for large-scale coactivation of the CNS. Eur J Neurosci 14:809–820
Mochida H, Sato K, Momose-Sato Y (2009) Switching of the transmitters that mediate hindbrain correlated activity in the chick embryo. Eur J Neurosci 29:14–30
Momose-Sato Y, Sato K (2005) Primary vagal projection to the contralateral non-NTS region in the embryonic chick brainstem revealed by optical recording. J Membr Biol 208:183–191
Momose-Sato Y, Sato K (2006) Optical recording of vagal pathway formation in the embryonic brainstem. Auton Neurosci 126–127:39–49
Momose-Sato Y, Sato K (2011) The embryonic brain and development of vagal pathways. Resp Physiol Neurobiol 178:163–173
Momose-Sato Y, Sato K (2013) Large-scale synchronized activity in the embryonic brainstem and spinal cord. Front Cell Neurosci 7(Article 36):1–15
Momose-Sato Y, Sato K (2014a) Maintenance of the large-scale depolarization wave in the embryonic chick brain against deprivation of the rhythm generator. Neuroscience 266:186–196
Momose-Sato Y, Sato K (2014b) Optical survey of initial expression of synaptic function in the embryonic chick trigeminal sensory nucleus. Neurosci Lett 570:92–96
Momose-Sato Y, Komuro H, Sakai T et al (1991a) Optical monitoring of cholinergic postsynaptic potential in the embryonic chick ciliary ganglion using a voltage-sensitive dye. Biomed Res 12 (Suppl 2):139–140
Momose-Sato Y, Sakai T, Komuro H et al (1991b) Optical mapping of the early development of the response pattern to vagal stimulation in embryonic chick brain stem. J Physiol (Lond) 442:649–668
Momose-Sato Y, Sakai T, Hirota A et al (1994) Optical mapping of early embryonic expressions of Mg^{2+}-/APV-sensitive components of vagal glutaminergic EPSPs in the chick brainstem. J Neurosci 14:7572–7584
Momose-Sato Y, Sato K, Sakai T et al (1995a) A novel γ-aminobutyric acid response in the embryonic brainstem as revealed by voltage-sensitive dye recording. Neurosci Lett 191:193–196
Momose-Sato Y, Sato K, Sakai T et al (1995b) Evaluation of optimal voltage-sensitive dyes for optical monitoring of embryonic neural activity. J Membr Biol 144:167–176
Momose-Sato Y, Sato K, Hirota A et al (1997) Optical characterization of a novel GABA response in early embryonic chick brainstem. Neuroscience 80:203–219

Momose-Sato Y, Sato K, Hirota A et al (1998) GABA-induced intrinsic light-scattering changes associated with voltage-sensitive dye signals in embryonic brain stem slices: coupling of depolarization and cell shrinkage. J Neurophysiol 79:2208–2217
Momose-Sato Y, Komuro H, Hirota A et al (1999a) Optical imaging of the spatiotemporal patterning of neural responses in the embryonic chick superior cervical ganglion. Neuroscience 90:1069–1083
Momose-Sato Y, Sato K, Kamino K (1999b) Optical identification of calcium-dependent action potentials transiently expressed in the embryonic rat brainstem. Neuroscience 90:1293–1310
Momose-Sato Y, Sato K, Kamino K (2001a) Optical approaches to embryonic development of neural functions in the brainstem. Prog Neurobiol 63:151–197
Momose-Sato Y, Sato K, Mochida H et al (2001b) Spreading depolarization waves triggered by vagal stimulation in the embryonic chick brain: optical evidence for intercellular communication in the developing central nervous system. Neuroscience 102:245–262
Momose-Sato Y, Miyakawa N, Mochida H et al (2003a) Optical analysis of depolarization waves in the embryonic brain: a dual network of gap junctions and chemical synapses. J Neurophysiol 89:600–614
Momose-Sato Y, Mochida H, Sasaki S et al (2003b) Depolarization waves in the embryonic CNS triggered by multiple sensory inputs and spontaneous activity: optical imaging with a voltage-sensitive dye. Neuroscience 116:407–423
Momose-Sato Y, Honda Y, Sasaki H et al (2004) Optical mapping of the functional organization of the rat trigeminal nucleus: Initial expression and spatiotemporal dynamics of sensory information transfer during embryogenesis. J Neurosci 24:1366–1376
Momose-Sato Y, Honda Y, Sasaki H et al (2005) Optical imaging of large-scale correlated wave activity in the developing rat CNS. J Neurophysiol 94:1606–1622
Momose-Sato Y, Glover J, Sato K (2006) Development of functional synaptic connections in the auditory system visualized with optical recording: afferent-evoked activity is present from early stages. J Neurophysiol 96:1949–1962
Momose-Sato Y, Kinoshita M, Sato K (2007a) Development of vagal afferent projections circumflex to the obex in the embryonic chick brainstem visualized with voltage-sensitive dye recording. Neuroscience 148:140–150
Momose-Sato Y, Kinoshita M, Sato K (2007b) Embryogenetic expression of glossopharyngeal and vagal excitability in the chick brainstem as revealed by voltage-sensitive dye recording. Neurosci Lett 423:138–142
Momose-Sato Y, Sato K, Kinoshita M (2007c) Spontaneous depolarization waves of multiple origins in the embryonic rat CNS. Eur J Neurosci 25:929–944
Momose-Sato Y, Mochida H, Kinoshita M (2009) Origin of the earliest correlated neuronal activity in the chick embryo revealed by optical imaging with voltage-sensitive dyes. Eur J Neurosci 29:1–13
Momose-Sato Y, Nakamori T, Sato K (2011) Functional development of the vagal and glossopharyngeal nerve-related nuclei in the embryonic rat brainstem: optical mapping with a voltage-sensitive dye. Neuroscience 192:781–792
Momose-Sato Y, Nakamori T, Sato K (2012a) Spontaneous depolarization wave in the mouse embryo: origin and large-scale propagation over the CNS identified with voltage-sensitive dye imaging. Eur J Neurosci 35:1230–1241
Momose-Sato Y, Nakamori T, Sato K (2012b) Pharmacological mechanisms underlying the switching from the large-scale depolarization wave to segregated activity in the mouse CNS. Eur J Neurosci 35:1242–1252
Momose-Sato Y, Nakamori T, Mullah SH-E-R et al (2013) Optical survey of vagus nerve-related neuronal circuits in the embryonic rat brainstem. Neurosci Lett 535:140–145
Moody WJ, Bosma MM (2005) Ion channel development, spontaneous activity, and activity-dependent development in nerve and muscle cells. Physiol Rev 85:883–941

Mullah SH-E-R, Komuro R, Yan P et al (2013) Evaluation of voltage-sensitive fluorescence dyes for monitoring neuronal activity in the embryonic central nervous system. J Membr Biol 246:679–688

Myers CP, Lewcock JW, Hanson MG et al (2005) Cholinergic input is required during embryonic development to mediate proper assembly of spinal locomotor circuits. Neuron 46:37–49

Nakayama K, Nishimaru H, Iizuka M et al (1999) Rostrocaudal progression in the development of periodic spontaneous activity in fetal rat spinal motor circuits in vitro. J Neurophysiol 81:2592–2595

Obaid AL, Loew LM, Wuskell JP et al. (2004) Novel Naphthylstyryl-pyridinium potentiometric dyes offer advantages for neural network analysis. J Neurosci Meth 134:179–190

O'Donovan MJ, Bonnot A, Mentis GZ et al (2008) Imaging the spatiotemporal organization of neural activity in the developing spinal cord. Dev Neurobiol 68:788–803

Onimaru H, Homma I (2005) Developmental changes in the spatio-temporal pattern of respiratory neuron activity in the medulla of late fetal rat. Neuroscience 131:969–977

Orbach HS, Cohen LB, Grinvald A (1985) Optical mapping of electrical activity in rat somatosensory and visual cortex. J Neurosci 5:1886–1895

Ren J, Greer JJ (2003) Ontogeny of rhythmic motor patterns generated in the embryonic rat spinal cord. J Neurophysiol 89:1187–1195

Ren J, Momose-Sato Y, Sato K et al (2006) Rhythmic neuronal discharge in the medulla and spinal cord of fetal rats in the absence of synaptic transmission. J Neurophysiol 95:527–534

Ross WN, Salzberg BM, Cohen LB et al (1977) Changes in absorption, fluorescence, dichroism, and birefringence in stained giant axons: optical measurement of membrane potential. J Membr Biol 33:141–183

Rubel EW, Fritzsch B (2002) Auditory system development: primary auditory neurons and their targets. Annu Rev Neurosci 25:51–101

Sakai T, Hirota A, Komuro H et al (1985) Optical recording of membrane potential responses from early embryonic chick ganglia using voltage-sensitive dyes. Dev Brain Res 17:39–51

Sakai T, Komuro H, Katoh Y et al (1991) Optical determination of impulse conduction velocity during development of embryonic chick cervical vagus nerve bundles. J Physiol (Lond) 439:361–381

Salzberg BM (1983) Optical recording of electrical activity in neurons using molecular probes. In: Barker JL, McKelvy JF (eds) Current methods in cellular neurobiology, vol 3, electrophysiological techniques. Wiley, New York, NY, pp 139–187

Salzberg BM, Obaid AL, Gainer H (1985) Large and rapid changes in light scattering accompany secretion by nerve terminals in the mammalian neurohypophysis. J Gen Physiol 86:395–411

Sato K, Momose-Sato Y (2003) Optical detection of developmental origin of synaptic function in the embryonic chick vestibulo-cochlear nuclei. J Neurophysiol 89:3215–3224

Sato K, Momose-Sato Y (2004a) Optical detection of convergent projections in the embryonic chick NTS. Neurosci Lett 371:97–101

Sato K, Momose-Sato Y (2004b) Optical mapping reveals developmental dynamics of Mg^{2+}-/APV-sensitive components of glossopharyngeal glutamatergic EPSPs in the embryonic chick NTS. J Neurophysiol 92:2538–2547

Sato K, Momose-Sato Y, Sakai T et al (1993) Optical assessment of spatial patterning of strength-duration relationship for vagal responses in the early embryonic chick brainstem. Jpn J Physiol 43:521–539

Sato K, Momose-Sato Y, Sakai T et al (1995) Responses to glossopharyngeal stimulus in the early embryonic chick brainstem: spatiotemporal patterns in three dimensions from repeated multiple-site optical recording of electrical activity. J Neurosci 15:2123–2140

Sato K, Momose-Sato Y, Hirota A et al (1996) Optical studies of the biphasic modulatory effects of glycine on excitatory postsynaptic potentials in the chick brainstem and their embryogenesis. Neuroscience 72:833–846

Sato K, Momose-Sato Y, Arai Y et al (1997) Optical illustration of glutamate-induced cell swelling coupled with membrane depolarization in embryonic brain stem slices. NeuroReport 8:3559–3563

Sato K, Momose-Sato Y, Hirota A et al (1998) Optical mapping of neural responses in the embryonic rat brainstem with reference to the early functional organization of vagal nuclei. J Neurosci 18:1345–1362

Sato K, Momose-Sato Y, Mochida H et al (1999) Optical mapping reveals the functional organization of the trigeminal nuclei in the chick embryo. Neuroscience 93:687–702

Sato K, Yazawa I, Mochida H et al (2000) Optical detection of embryogenetic expression of vagal excitability in the rat brainstem. NeuroReport 11:3759–3763

Sato K, Mochida H, Sasaki S et al (2001) Optical responses to micro-application of GABA agonists in the embryonic chick brain stem. NeuroReport 12:95–98

Sato K, Mochida H, Sasaki S et al (2002a) Developmental organization of the glossopharyngeal nucleus in the embryonic chick brainstem slice as revealed by optical sectioning recording. Neurosci Lett 327:157–160

Sato K, Mochida H, Yazawa I et al (2002b) Optical approaches to functional organization of glossopharyngeal and vagal motor nuclei in the embryonic chick hindbrain. J Neurophysiol 88:383–393

Sato K, Miyakawa N, Momose-Sato Y (2004a) Optical survey of neural circuit formation in the embryonic chick vagal pathway. Eur J Neurosci 19:1217–1225

Sato K, Momose-Sato Y, Kamino K (2004b) Light-scattering signals related to neural functions. In: Fagan J, JN D, Shimizu N (eds) Recent research developments in membrane biology, vol 2. Research Signpost, Trivandrum, pp 21–45

Sato K, Kinoshita M, Momose-Sato Y (2007) Optical mapping of spatiotemporal emergence of functional synaptic connections in the embryonic chick olfactory pathway. Neuroscience 144:1334–1346

Scain A-L, Le Corronc H, Allain A-E et al (2010) Glycine release from radial cells modulates the spontaneous activity and its propagation during early spinal cord development. J Neurosci 30:390–403

Thoby-Brisson M, Trinh J-B, Champagnat J et al (2005) Emergence of the pre-Bötzinger respiratory rhythm generator in the mouse embryo. J Neurosci 25:4307–4318

Thomas J-L, Spassky N, Perez Villegas EM et al (2000) Spatiotemporal development of oligodendrocytes in the embryonic brain. J Neurosci Res 59:471–476

Torborg CL, Feller MB (2005) Spontaneous patterned retinal activity and the refinement of retinal projections. Prog Neurobiol 76:213–235

Tsau Y, Wenner P, O'Donovan MJ et al (1996) Dye screening and signal-to-noise ratio for retrogradely transported voltage-sensitive dyes. J Neurosci Meth 70:121–129

Weissman TA, Riquelme PA, Ivic L et al (2004) Calcium waves propagate through radial glial cells and modulate proliferation in the developing neocortex. Neuron 43:647–661

Wenner P, Tsau Y, Cohen LB et al (1996) Voltage-sensitive dye recording using retrogradely transported dye in the chicken spinal cord: staining and signal characteristics. J Neurosci Meth 70:111–120

Whelan PJ (2003) Developmental aspects of spinal locomotor function: insights from using the *in vitro* mouse spinal cord preparation. J Physiol (Lond) 553:695–706

Wong ROL (1999) Retinal waves and visual system development. Annu Rev Neurosci 22:29–47

Wu J-Y, Cohen LB (1993) Fast multisite optical measurement of membrane potential. In: Mason WT (ed) Fluorescent and luminescent probes for biological activity. Academic, Boston, MA, pp 389–404

Wu J-Y, Lam Y-W, Falk CX et al (1998) Voltage-sensitive dyes for monitoring multineuronal activity in the intact central nervous system. Histochem J 30:169–187

Young SR, Rubel EW (1986) Embryogenesis of arborization pattern and topography of individual axons in N. laminaris of the chicken brain stem. J Comp Neurol 254:425–459

Ziskind-Conhaim L, Redman S (2005) Spatiotemporal patterns of dorsal root-evoked network activity in the neonatal rat spinal cord: optical and intracellular recordings. J Neurophysiol 94:1952–1961

Chapter 10
Imaging the Dynamics of Mammalian Neocortical Population Activity In-Vivo

Amiram Grinvald, David Omer, Shmuel Naaman, and Dahlia Sharon

Abstract Neural computations underlying sensory perception, cognition and motor control are performed by populations of neurons at different anatomical and temporal scales. Few techniques are currently available for exploring dynamics of local and large range populations. Voltage-sensitive dye imaging (VSDI) reveals neural population activity in areas ranging from a few tens of microns to a couple of centimeters, or two areas up to ~10 cm apart. VSDI provides a sub-millisecond temporal resolution, and a spatial resolution of about 50 μm. The dye signal emphasizes subthreshold synaptic potentials. VSDI has been applied in the mouse, rat, gerbil, ferret, tree shrew, cat and monkey cortices, in order to explore lateral spread of retinotopic or somatotopic activation, the dynamic spatiotemporal pattern resulting from sensory activation, including the somatosensory, olfactory, auditory, and visual modalities, as well as motor preparation and the properties of spontaneously-occurring population activity. In this chapter we focus on VSDI *in-vivo* and review results obtained mostly in the visual system in our laboratory.

Keywords Cat • Cortical response field • Evoked response • Horizontal connections • Lateral spread • Monkey • Orientation map • Orientation selectivity • Spatiotemporal response • Spikes • Spontaneous activity • Subthreshold response • V1 • V2 • Visual cortex

A. Grinvald (✉)
Department of Neurobiology, Weizmann Institute of Science,
P.O. Box 26, Rehovot 76100, Israel

The Grodetsky Center for Research of Higher Brain Functions,
The Weizmann Institute of Science, Rehovot 76100, Israel
e-mail: Amiram.grinvald@weizmann.ac.il

D. Omer • S. Naaman
Department of Neurobiology, Weizmann Institute of Science,
P.O. Box 26, Rehovot 76100, Israel

D. Sharon
Department of Neurobiology, Weizmann Institute of Science,
P.O. Box 26, Rehovot 76100, Israel

Department of Psychology, Stanford University, Stanford, CA 94305, USA

M. Canepari et al. (eds.), *Membrane Potential Imaging in the Nervous System and Heart*, Advances in Experimental Medicine and Biology 859,
DOI 10.1007/978-3-319-17641-3_10

1 Introduction

The activity of highly distributed neural networks is thought to underlie sensory processing, motor coordination and higher brain functions. These intricate networks are composed of large numbers of individual neurons, which interact through synaptic connections in complex, dynamically regulated spatiotemporal patterns. To understand network properties and functions, it is helpful to study the ensemble activity of neuronal populations because coherent activity of many neurons is often responsible for performing the function rather than individual cells. Functionally related sub-networks of neurons are often spatially segregated, making imaging techniques ideal for monitoring population activity. Our understanding of the contribution of single neurons for generating percepts and controlling behavior can be improved within the context of the relationship between individual-level and population-level activity.

The remarkable performance of the mammalian brain arises from computations taking place in the neocortex. The neocortex is organized into cortical columns (Mountcastle 1957; Hubel and Wiesel 1962), which have lateral dimensions of a few hundred microns. Interactions within and between cortical columns occur on the time scale of milliseconds. To follow neuronal computations at the fundamental level of cortical columns in real-time therefore requires a spatial resolution of ~100 μm and a temporal resolution of ~1 ms. *In vivo* Voltage-Sensitive Dye Imaging (VSDI) (Grinvald et al. 1984; Orbach et al. 1985) fulfils these technical requirements and should help resolve many fundamental questions.

Electrical communication in cortical networks comprises two basic signals: subthreshold potentials (reflecting synaptic input onto dendrites) and suprathreshold action potentials (forming the neuronal output). VSDI appears to relate primarily to spatiotemporal patterns of the subthreshold synaptically-driven membrane potentials simply because of the relatively larger area of neocortical dendrites. Therefore, VSDI allows monitoring the fluctuations of the membrane potential of the population away from and towards the threshold for action potential. Single unit and multi-unit recordings cannot monitor this activity, reflecting only spiking activity. The local field potential, which does reflect synaptic potentials, changes its polarity depending on the activity source and therefore provides ambiguous information about the sign of recorded activity (inhibition versus excitation). Furthermore, its spatial resolution is far lower. Therefore VSDI has a unique place among available techniques for measuring neural activity.

Here we describe how *in-vivo* VSDI (Grinvald et al. 1984; Shoham et al. 1999; Grinvald and Hildesheim 2004) can be applied to anesthetized mammals. We describe previous findings and technical advances, mostly from our lab, and how VSDI has been combined with intracortical microstimulation, single-unit recording, local-field-potential recording and targeted injections. We conclude by discussing further technical developments to overcome current limitations.

2 Principles of Voltage-Sensitive Dye Imaging In-Vivo

VSDI *in-vivo* is a form of functional optical imaging of neuronal activity that utilizes extrinsic fluorescence probes to monitor membrane potential changes in real-time. It thus differs from optical imaging based on intrinsic signals in two important ways. The first, its biggest advantage, is that it offers a millisecond temporal resolution without compromising spatial resolution. The second, its biggest disadvantage, is that it relies on minimally invasive staining of the cortex with vital voltage-sensitive dyes (VSDs).

To perform optical imaging of electrical activity *in-vivo*, the preparation under study is first stained with a suitable voltage-sensitive dye. The dye molecules bind to the external surface of excitable membranes and act as molecular transducers that transform changes in membrane potential into optical signals. These optical signals are observed as changes in absorption or emitted fluorescence, and respond to membrane potential changes in microseconds. The amplitude of the VSD signal varies linearly with both the membrane potential changes and the membrane area of the stained neuronal elements. These optical changes are monitored with light imaging devices positioned in a microscope image plane. Optical signals using voltage-sensitive dyes were first recorded by Tasaki et al. (1968) in the squid giant axon and by Salzberg et al. (1973) in the leech ganglia. *In-vivo* experiments in rat visual cortex began in 1982 (Grinvald, Orbach and Cohen, unpublished results) and revealed several complications that had to be overcome. One complication was the large amount of noise caused by respiratory and heartbeat pulsations. In addition, the relative opacity and packing density of the cortex limited the penetration of the excitation light and the ability of dyes to stain deep layers of the cortex. Subsequently, new, improved dyes that overcame these problems were developed (e.g. RH-414) and an effective remedy for the heartbeat and respiratory noise was found by synchronizing data acquisition and the respiration with the electrocardiogram and subtracting a no-stimulus trial. These improvements facilitated the *in-vivo* imaging of several different sensory systems, including the retinotopic responses in the frog optic tectum (Grinvald et al. 1984), and the whisker barrels in rat somatosensory cortex (Orbach et al. 1985), and experiments in the salamander olfactory bulb (Orbach and Cohen 1983). The development of more hydrophilic dyes improved the quality of the results obtained in cat and monkey visual cortex (e.g. RH-704 and RH-795, Grinvald et al. 1994).

The newest generation of voltage-sensitive dyes, introduced in the late 1990s, offers a 30-fold improvement in signal-to-noise ratio over the above-mentioned early dyes. This was accomplished by designing oxonol dyes, which are excited outside the absorption band of hemoglobin, thus minimizing pulsation and hemodynamic noise (Fig. 10.1).

With this advance it has become possible to reveal the dynamics of cortical information processing and its underlying functional architecture at the necessary spatial and temporal resolution in both anesthetized and behaving animals. Furthermore, it finally became possible to obtain good signals without trial averaging,

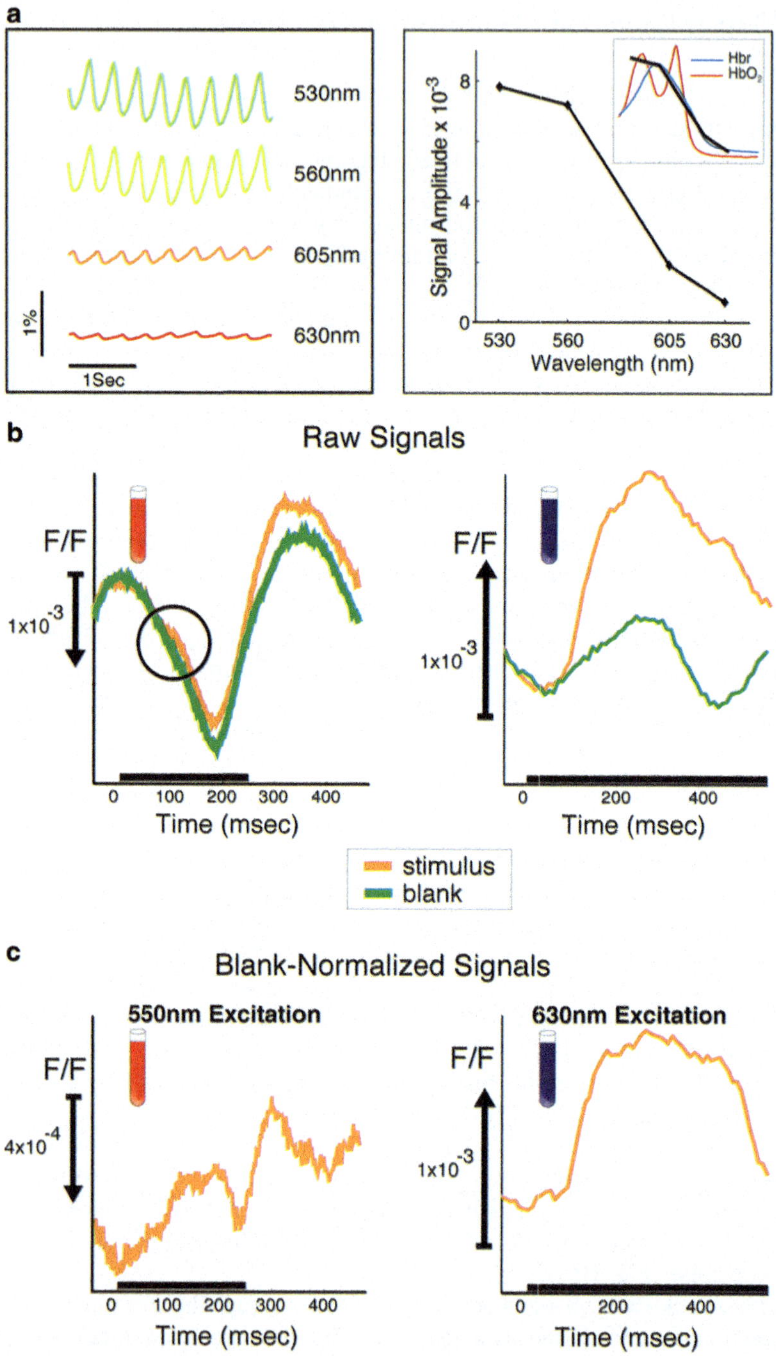

Fig. 10.1 "Blue dyes": Thirty-fold improvement in the signal-to-noise ratio. (**a**) Wavelength dependency of heart-beat and respiration signals. The *left panel* illustrates optical reflectance signals from cat visual cortex measured at several wavelengths. The *color* of the *line* schematically represents the wavelength of the illuminating light. The signals were averaged over the entire

thus facilitating the exploration of neocortical dynamics including spontaneous on-going activity. Additional advances related to the implantation of transparent artificial dura by Arieli et al. (2002) now allow chronic recordings to be taken over a long period of time. Optical imaging has also been performed simultaneously with intracellular recording, extracellular recording, microstimulation and tracer injection due to the development of an electrode assembly attached to a cranial window (Arieli and Grinvald 2002).

2.1 Relationship Between Voltage-Sensitive Dye Signals and Intracellular Recordings In-Vivo

In simple preparations, when single cells or their processes can be visualized, the dye signal looks just like an intracellular recording. Controversies about what the dye signal reflects during in vivo measurements were resolved by combining VSDI with intracellular recordings in vivo, first in cat visual cortex (Sterkin et al. 1998; Grinvald et al. 1999) and in the rat somatosensory cortex (Petersen et al. 2003a). The results established that the dye signal precisely reports changes in membrane potential (red and green traces in Fig. 10.2a, red and yellow in Fig. 10.2b and red and black in Figs 10.2c–f). The tight correlation between the intracellular neuronal recording and the VSD signal also indicates that the contribution of slow glial depolarization to the VSD signal is minimal. The linearity is shown in Fig. 10.2f.

During in vivo imaging of the neocortex, a single pixel contains the blurred images of various neuronal compartments—including the dendrites, axons and somata of a population of neurons—rather than a single cell. The VSD signal is linearly related to the stained membrane area, and most of the dye signal originates

Fig. 10.1 (continued) imaged area. The *right panel* plots the peak to peak amplitude as a function of wavelength, illustrating the fall off with increasing wavelength in this range. The *inset* shows the amplitude plot (normalized) on the same axes as the textbook absorption of oxy- and deoxy-hemoglobin, illustrating the close relationship between the two. (**b**) Time course of the evoked voltage-sensitive dye signals. The *left* and *right top panels* illustrate the signals from different experiments using the red dye RH-795 and the blue dye RH-1692 respectively. The *curves* show the raw fluorescence signals, without subtracting the blank, integrated over the imaged area (cat visual area 18) and averaged over trials. The *orange curves* are from the stimulated condition and the *green lines* are the blank data (no stimulus). Data acquisition was synchronized to the heart-beat. The *solid black line* marks the stimulus period. The *circle* in the *left panel* (RH-795) points to the time of the fast evoked signal showing its minute magnitude relative to the heartbeat signal. Note the large amplitude of the evoked signal relative to the heart-beat signal for the blue dye (*right*). This is true despite the larger total raw signal for the red dye. Note also that the sign of the heartbeat signals from the two dyes are opposite. (**c**) The same data as (**b**) but with the heartbeat signal removed by subtracting the blank condition. Note the large heart-beat artifact with the red dye at the time of the second heart-beat. Again the blue dye produces much better signal to noise ratio, allowing examination of small components in the response and longer recording interval that are still reliable. In both cases 16 presentations were averaged

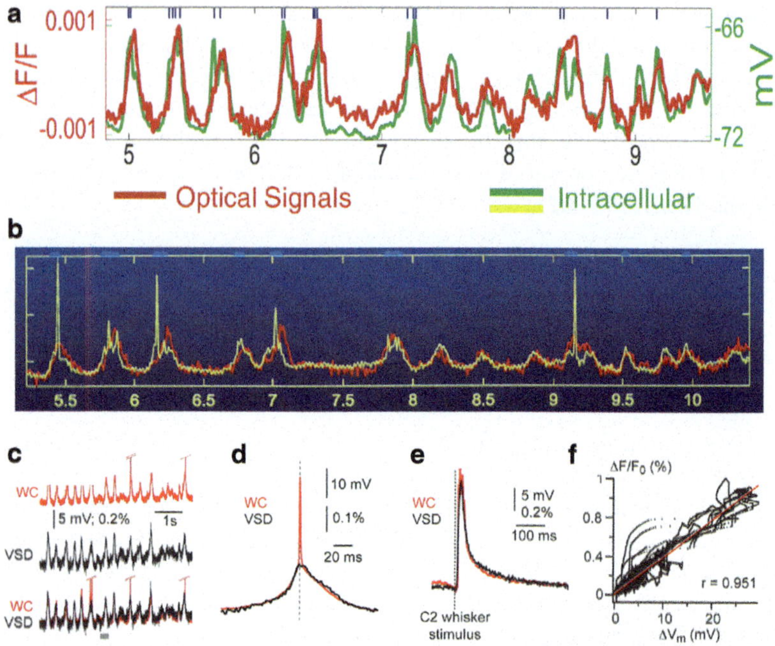

Fig. 10.2 The similarity between the cortical dye signal in-vivo from a small population of neurons and intracellular recording. (a) Two traces showing simultaneous intracellular and optical recording for 5 s, performed in a deeply anesthetized cat, a condition in which spontaneous changes in membrane potential are highly synchronized in a large population of neuron. The intracellular recording is depicted by the *green trace* and the population signal by the *red trace*. The action potentials were truncated and occurred at times marked by the *short blue lines* at the *top*. (**b**) The intracellular recording is depicted by the *yellow trace*. The action potentials were not truncated. One can see that the optical signal from the population next to the electrode is not picking up action potentials, presumably because the membrane area of the dendrites monitored by a single pixel is much larger than the membrane area of the non myelinated axons originating from that cell (the action potentials are not synchronized whereas the synaptic potentials are). (**a, b**) Modified from Sterkin et al. (1998). (**c**) Similar simultaneous recording from the rat somatosensory cortex showing once again that the population signal is similar to the synaptic potential but not to the action potential. *WC* whole cell recordings. (**d**) Same but on an expanded time scale, showing the resemblance between the dye signal and the intracellular recording. (**e**) Simultaneous recordings of an evoked response resulting from sensory stimulation of moving whisker C2. (**f**) The linearity of the Dye signal vs. the membrane potential change. (**c–f**) Modified from Petersen et al. (2003b)

from cortical dendrites and non-myelinated axons rather than cell bodies, because their membrane areas are orders of magnitude larger than that of neuronal somata or confined non myelinated axons. The dendrites of cortical cells are often far more confined than the axons, so most of the signals in a given pixel originate from the dendrites of nearby cortical cells. Therefore, the existence of a dye signal in a particular cortical site does not necessarily imply that cortical neurons at that site are generating action potentials. However, the peak/s of the VSD response to a suprathreshold stimulus corresponds to the spiking zone, which is to be expected from the fact that synaptic activity peaks in this zone too (Jancke et al. 2004; Sharon et al. 2007). *In-vivo* the VSD signal thus reflects mainly dendritic activity, making VSDI optimal for exploring population dynamics of local subthreshold neuronal activity. However, the VSD signal also contains some spiking information (output), which remains to be exploited.

2.2 *Imaging from the Mammalian Brain In Vivo*

A major technical impediment to imaging anesthetized or awake animals is movement of the brain, A critical aspect of imaging neocortical activity *in-vivo* is therefore to obtain a stable view of the cortex. The obvious solution is to fix the head relative to the imaging apparatus as done for electrical or intracellular recordings. Head-fixation bars can be attached to the skull of the animal. Next, a sealed recording chamber should be used to minimize brain pulsations relative to the skull. A tandem-lens macroscope (Ratzlaff and Grinvald 1991) can then be mounted in a stable location relative to the fixed head position for epifluorescent VSDI with suitable illumination and camera equipment.

2.3 *Spatial and Temporal Resolution*

As already mentioned, voltage-sensitive dyes respond to membrane potential changes in microseconds at a spatial resolution of ~0.5 μm. However, the practical spatiotemporal resolution that can be obtained by *in-vivo* imaging depends on factors other than the resolution of the dye molecule's response, such as light scattering within the imaged tissue, photodynamic damage, and optics of the data acquisition apparatus including the detector resolution. Below we discuss some of these factors. Empirically, we find that current dyes and equipment allow excellent signal-to-noise at a sampling rate of 5–10 ms with each pixel looking at a 50×50 μm area of cortex. Light scattering is a major determinant of spatial resolution, and photodynamic damage is an important determinant of signal-to-noise ratio, both spatially and temporally because it limits the light intensity used to reduce shot noise or the extent of signal averaging used to improve the signal to noise ratio.

Additional concerns include limited depth of penetration into the cortex, and possible pharmacological side effects. The new oxonol dyes have largely alleviated the problems of pharmacological side-effects and photodynamic damage. First, intracellular recordings *in-vivo* have directly confirmed that stained cortical cells maintain their response properties. (Ferster, Lampl, Arieli and Grinvald, unpublished results). Furthermore, long-term VSDI in awake monkeys have indicated that, even after a year of imaging, monkey visual cortex continued to function normally: the animal maintained normal performance in tasks that required the normal functioning of the cortical area which was repeatedly imaged for up to a year.

The primarily dendritic origin of the VSD signal (see above) also affects the spatial resolution of VSDI, because the dendrites of cells in a given cortical column typically cross cortical areas that correspond to adjacent cortical columns, which have different functional properties. Therefore, the dendritic view of activity offered by VSDI is in effect somewhat blurred relative to the somatic activity. Differential imaging provides some remedy to this problem as well as the light scattering problem. Using differential imaging whenever applicable may provide even 1 μm resolution and depends mostly on the signal to noise ratio which can be obtained.

Improvements in the dyes and in the spatial resolution of fast cameras have thus made it possible to obtain high resolution functional maps of somatosensory whisker barrels, olfactory glomeruli and visual orientation columns, "lighting up" in milliseconds with a signal-to-noise ratio even better than that obtained with the slow intrinsic signals. Additional developments will undoubtedly introduce further improvement.

2.4 *Long-Term Repeated VSDI*

To facilitate repeated imaging sessions, the resected dura matter is substituted with artificial dura mater. A careful cleaning procedure has allowed repeated VSDI sessions in monkeys 2–3 times a week over a period longer than a year (Arieli et al. 2002). Long term imaging allows exciting new type of experiments related to development, plasticity, learning and memory, and recovery after trauma or stroke.

2.5 *Integrating VSDI with Electrode Techniques*

VSDI can easily be combined with standard electrode-based techniques. Whereas optimal mechanical stability for imaging is provided by sealed cranial windows, these do not allow electrodes to be introduced. One approach is to cover the cortical surface with agarose and stabilise it by a cover slip. A small gap between the edge of the cover slip and the wall of the recording chamber allows oblique entry of electrodes into the rodent cortex (Petersen et al. 2003a, b; Berger et al. 2007; Ferezou et al. 2006, 2007). For cats and monkeys a sliding-top cranial window was developed with a removable microdrive-positioned electrode (Arieli and Grinvald 2002). VSDI can then be combined with microstimulation, extracellular recording (single- and multiple-unit recording and local field potential), intracellular recording, patch recordings and targeted injection of tracers. Recordings can either be made simultaneously and/or can be targeted to specific regions based on the functional imaging data. (Arieli and Grinvald 2002)

3 Explorations of Population Dynamics in Visual Cortex

3.1 *Lateral Extent of Focal Activation and the Lateral Spread Dynamics*

Retinotopic VSDI experiments in monkey have been used to investigate how far across the cortical surface synaptic activation spreads from a sensory point stimulus—the cortical point-spread function (Fig. 10.3a). Activity spreads over a cortical area much larger than predicted on the basis of standard retinotopic measurements in layer 4, but is consistent both with the anatomical finding of long-range horizontal connections in visual cortex and with intracellular recordings. The cortical point spread function was calculated from these experiments and projected onto a histological section of cytochrome oxidase blobs (Fig. 10.3b) to show its relationship to individual cortical modules.

The stimulus caused spiking activity only in neurons in the marked small square, which contains four cortical modules. However, more than 250 blobs had access to the information carried by the signal spread, at a signal amplitude of at least $1/e^2$ of maximum. The apparent space constant (signal decrement to 1/e of peak) for the spread was 1.5 mm along the cortical axis parallel to the ocular dominance (OD) columns, and 3 mm along the perpendicular axis. The spread velocity was 0.1–0.2 m/s. Much higher spatial resolution data showing similar spread in the somatosensory cortex has been first reported by Petersen and colleagues (2003a, b; Fig. 10.3c) and several subsequent papers from the same group.

These results indicate extensive distributed processing, and were further investigated also in cat visual cortex with the new generation of dyes and an imaging system offering subcolumnar resolution, taking a more detailed look at the "inverse" of the receptive field—the region of cortical space whose spatiotemporal pattern of electrical activity is influenced by a given sensory stimulus. The spatiotemporal properties of this activated area, which we refer to as the cortical response field, were studied using VSDI for responses to small, local drifting oriented gratings. One might expect a smooth decline in activity level from the cortical response field peak to periphery, as the overlap between the stimulus and the aggregate receptive field (RF) at each cortical location decreases. However, this is not the case, and as shown in Fig. 10.4a, the average cortical response field during a late time window (~140–515 ms after stimulus onset) has a distinctive spatial structure: an initial rapid drop from the peak is followed by a slower-sloped region, the plateau, beyond the rim of which (black contour line) a rapid decline in activity level occurs in the periphery. Plateau rim location was largely independent of

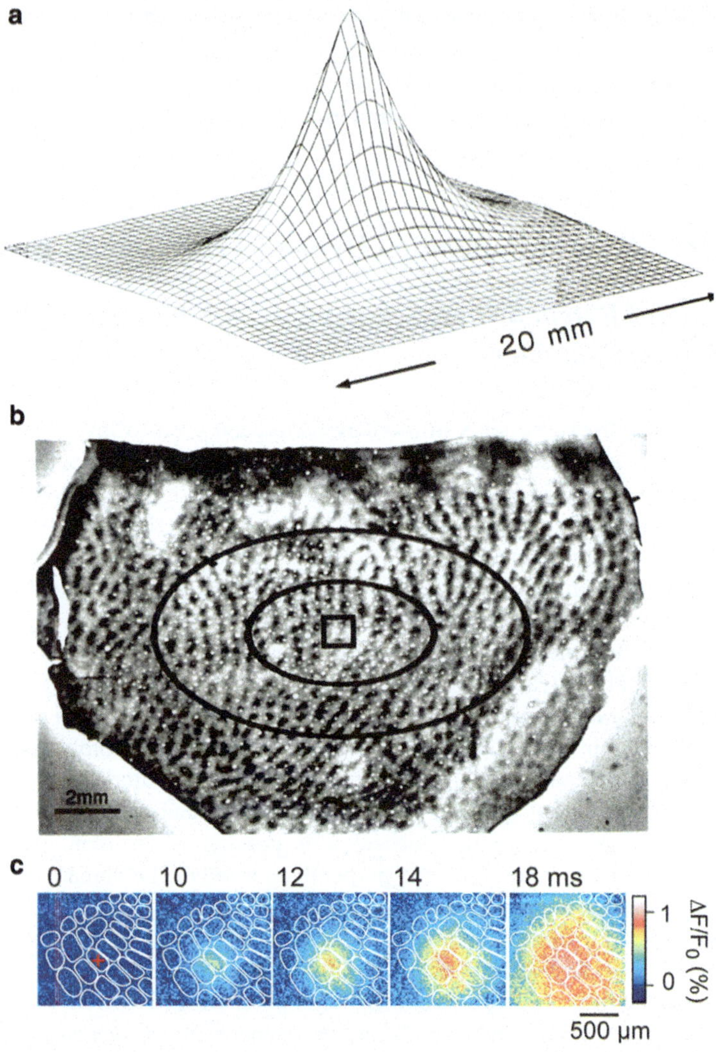

Fig. 10.3 Many functional domains are activated during the processing of a small retinal image or a single whisker. (**a**) Calculation of the activity spread from a small patch in layer 4 (a 1×1 mm square) within the upper cortical layers of a macaque. This cortical activation was produced by a retinal image of approximately $0.50° \times 0.25°$ that was presented to both eyes. The 'space constants' for the exponential-activity-spread measured with the dye experiments were 1.5 and 2.9 mm in the cortical axes perpendicular and parallel to the vertical meridian representation, respectively. (**b**) Direct activation in layer 4 (of the area within the square; calculated from retinotopic extracellular recordings) and the spread in layers 2/3 (elliptical contours; calculated from VSDI) are shown superimposed on a histological section showing the mosaics of cytochrome oxidase blobs, close to the border between cortical areas V1 and V2. The *center ellipse* shows the contour at which the amplitude of cortical activity drops to 37 % of its peak. The *larger ellipse* shows the contour at which the spread amplitude drops to 14 %. More than 10,000,000 neurons reside in the cortical area bounded by the large ellipse containing a regular mosaic of about 250 blobs. Adapted from Grinvald et al. (1994) with permission from the Society for Neuroscience. (**c**) The exact spatio-temporal pattern of the spread in the somato-sensory cortex in response to a single barrel stimulation. The temporal resolution was 2 ms/frame. Modified from Petersen et al. (2003b) and from Ferezou et al. (2006)

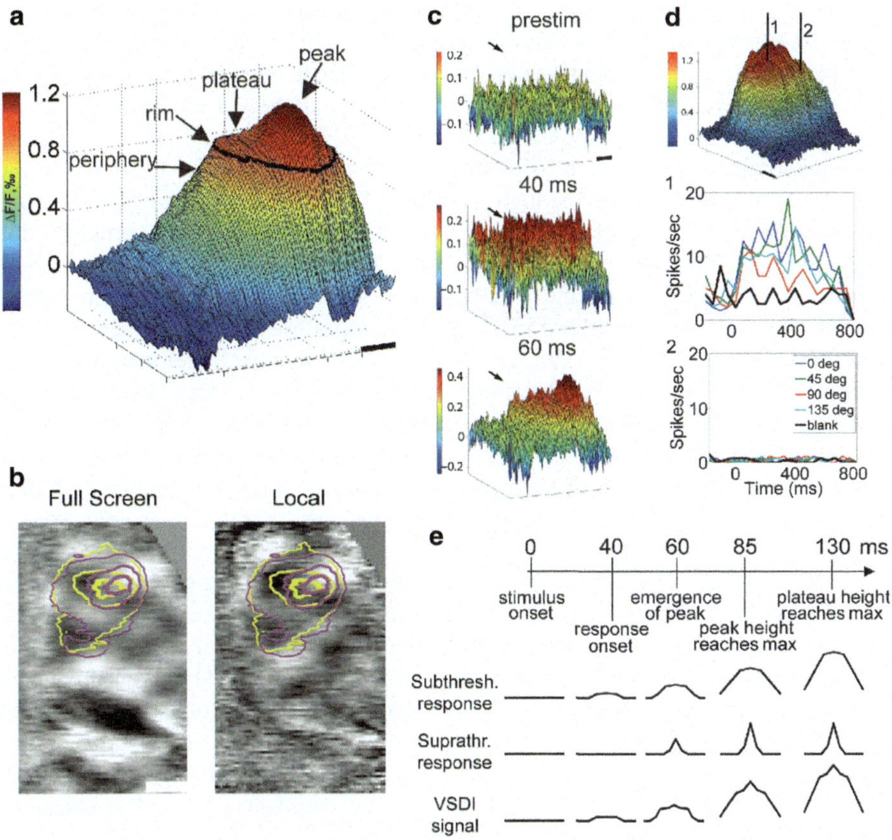

Fig. 10.4 Cortical response fields in response to local visual stimuli: structure and dynamics. (**a**) The cortical response field averaged over 140–515 ms after presentation of a local drifting grating exhibits the characteristic plateau + peak spatial structure. Beyond plateau rim, responses in the periphery decay quickly. (**b**) Relationship between response field to local grating stimuli of two orthogonal orientations (*yellow* and *magenta contour lines*) and differential orientation map from full-screen (*left*) and local (*right*) stimuli. Plateau rim (*outer contour lines*) is similar among the two orientations, whereas the peak zone (*inner contour lines*) corresponds better to the underlying orientation map. (**c**) Early dynamics: the initial response (40 ms) is a flat-topped plateau, with the peak emerging upon it only later (60 ms). (**d**) Spiking is restricted to the peak zone. (*Top*) A cortical response field with the locations of two extracellular electrodes. (*Middle*) Spiking responses are evoked in the peak zone (location 1) by the local stimulus. (*Bottom*) Spiking is absent in response to the same stimulus in the plateau away from the peak (location 2). (**e**) Timeline of the population response (*top*) and the proposed underlying subthreshold and suprathreshold responses (*middle*). The VSDI signal (*bottom*) is a sum of these two components, with the subthreshold response contributing a larger component due to relative membrane areas

stimulus orientation (Fig. 10.4b). The task of obtaining single-condition responses to these small stimuli at 10 ms resolution, especially demanding during early activation when response amplitude is low, was undertaken to determine the dynamics and significance of this spatial structure.

Results show that initially the response consists of a flat-topped plateau (Fig. 10.4c, 40 ms), and that the peak emerges upon it only about 20 ms later (Fig. 10.4c, 60 ms). The peak region amplifies relative to plateau amplitude for an average of 25 ms, and after this time only the plateau's height above the periphery continued to increase, reaching a maximum at 130 ms. The qualitatively different response dynamics suggested to us that spiking activity may be restricted to the peak zone, and that despite the high amplitude VSD signal at the plateau, neurons are not spiking. Extracellular recording showed that indeed spiking is evoked in the peak zone, but not at the plateau (Fig 10.4d). These results can be explained by a simple model which views the VSD signal as a weighted sum of all membrane potential changes: subthreshold responses and suprathreshold responses, with the former supplying the larger contribution to the signal due the larger membrane area involved (Fig 10.4e). The subthreshold response is wider in spatial extent because subthreshold receptive fields are larger than suprathreshold receptive fields. In this model, the sharp drop in activity occurring at the rim of the plateau and beyond is a result of reaching the edge of the area where neurons' subthreshold receptive field fully contains the stimulus. Smaller decrements are exhibited as long as the stimulus is fully contained within neurons' receptive field, i.e. within plateau rim.

The view that arises from these results is that, while unsurprisingly spiking activity occurs only for neurons with receptive fields at the correct location and orientation, a near-maximal subthreshold response is evoked in an area spanning several square millimeters, plunging quickly only once subthreshold receptive field overlap with the stimulus starts to decrease. Plateau activation to just below spiking threshold thus encompasses all orientations in an area representing a region of visual space larger than the stimulus. This interesting neuronal strategy primes the cortex for processing of subsequent stimuli, and could be involved in stimulus-driven attentional capture and in motion processing.

3.2 *Dynamics of Shape Processing*

Although orientation selectivity in the visual cortex has been explored for decades, the question of the interplay between feedforward processes and intracortical connections underlying this property has remained open. Two families of mechanisms have been proposed to play a role in the emergence of the highly orientation-selective responses of cortical visual neurons, a property not shared by their thalamic inputs. Feedforward-only models suggest appropriate alignment of thalamic input as the mechanism, whereas recurrent models suggest that intracortical interactions are more important. The response dynamics in visual cortex are fundamental to answering this question, because the feedforward explanation predicts that orientation selectivity should remain constant with time from stimulus onset, whereas if recurrent interactions are important then selectivity should change as the cortical network performs its processing. Traditional electrophysiological techniques have not been able to give a conclusive result regarding this issue. VSDI is ideally-suited to resolve it, being a high spatiotemporal resolution population imaging technique that emphasizes synaptic input.

To investigate the dynamics of orientation selectivity, high-quality single-condition maps were obtained in cat visual cortex. A time-series of the initial response is shown in Fig. 10.5a for two orthogonal orientations. As soon as the response can be observed (30–40 ms), the two orthogonal stimuli preferentially

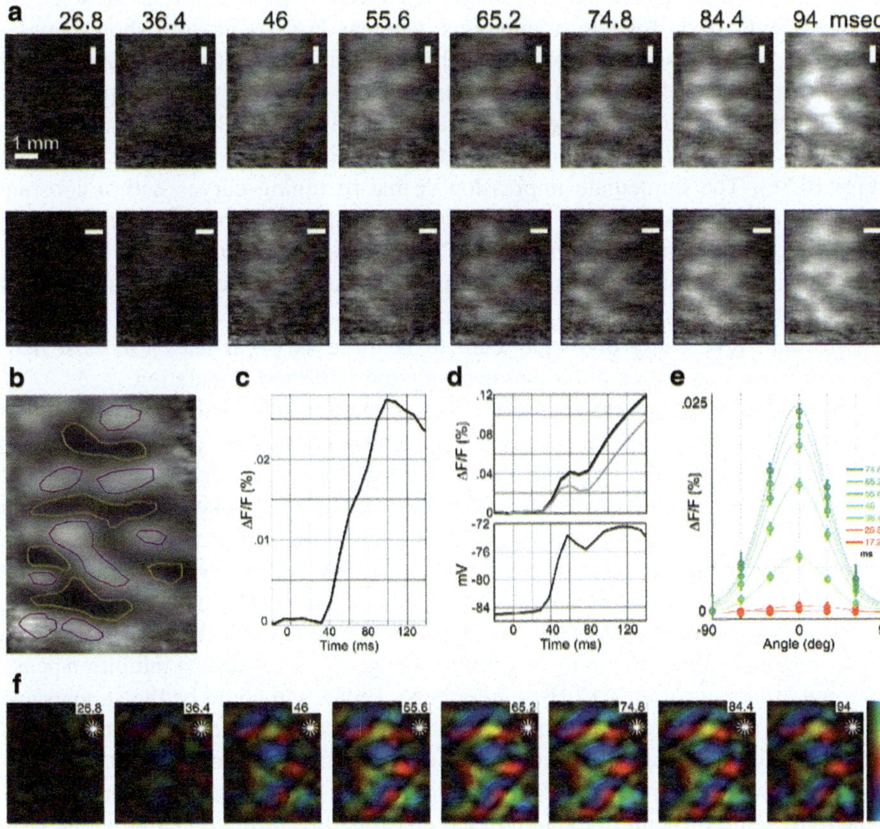

Fig. 10.5 Width of the orientation tuning curve is constant, but its amplitude increases dynamically during the first 40–60 ms of the response. (**a**) Time-series of single-condition orientation maps in cat produced in response to two orthogonal visual stimuli (*top* and *bottom rows*). (**b**) Spatial pattern revealed by differential imaging. Neurons within the *yellow lines* are selective for vertical; *red*—horizontal. (**c**) Temporal pattern of differential activity. The evoked DA (deceleration–acceleration) notch is barely visible. (**d**) *Top*, time-course of the evoked response to preferred (*blue*) and orthogonal (*green*) orientations. The two responses have the same onset latency but the response to the preferred orientation is larger from the beginning. A striking feature of the evoked responses is the DA notch, confirmed with intracellular recordings (*bottom*). (**e**) Orientation tuning curves at different times after stimulus onset. The curves after response onset (*green*) have the same shape but different amplitudes. The baseline of the curves was shifted to facilitate comparison. (**f**) Time-series of orientation-preference 'polar maps': *color* represents preferred orientation (*top* to *bottom* of color scale on right is 0–180°) and *brightness* represents the amplitude of differential response. Preferred orientation is steady from response onset. Adapted from Sharon and Grinvald (2002) with permission from the American Association for the Advancement of Science

activate complementary patches of cortex, seen in the differential map (Fig. 10.5b).

The modulation depth, i.e. the amplitude of the difference between the responses to the two stimuli, increases at the beginning of the response, and peaks at about 100 ms (Fig. 10.5c). In order to calculate orientation tuning curves, however, differential maps and time-courses are of no use, and single-condition responses to several orientations are needed. Single-condition maps at different latencies are shown in Fig. 10.5a for two orientations, whereas Fig. 10.5d (top) shows single-condition timecourses averaged over the imaged pixels. The average evoked response to the preferred and orthogonal stimuli are plotted (blue and yellow, respectively). The tuning curve at each time point was calculated from the single-condition responses corresponding to the six different orientations presented (Fig. 10.5e). The immediate impression is that of tuning curves with a constant shape but changing amplitude. The map of preferred orientation does not change as time from stimulus onset progresses, as shown in Fig. 10.5f. Indeed, the half-width at half height of tuning curves, as well as preferred orientation, were steady right from response onset (30–40 ms after stimulus onset). Therefore, sustained intracortical processing does not seem to be necessary—at least for most neurons—to determine orientation tuning width and preferred orientation.

Other aspects of the response also affect orientation selectivity. The modulation depth of the response changes over time, decreasing after a peak at 100 ms (Fig. 10.5c). An intriguing phenomenon is nearly undetectable in the differential time-course, but becomes obvious when considering the evoked single-condition responses (Fig. 10.5d). Between 50 and 80 ms after stimulus onset, there is a notch (Fig. 10.5d) in the evoked response, equivalent to a deceleration followed by acceleration in the rise-time—this is termed the 'evoked DA notch' (deceleration-acceleration). Furthermore, the evoked DA notch is more pronounced in response to the orthogonal stimulus than to the preferred (Fig. 10.5d). Shunting inhibition peaks at about 70 ms in cat area 17 (Bringuier et al. 1998), and could be the suppressive mechanism underlying the DA notch. These response properties—the increase and subsequent decrease of modulation depth (dynamic amplification of the orientation-selective response component), the existence of the DA notch and its orientation-selectivity—are difficult to explain based on properties of the thalamic input.

These results suggest that thalamic input may be the major determinant of orientation tuning width for most cortical neurons, but that intracortical processing is critical in amplifying the orientation selective component of the response. They also suggest that intracortical suppression contributes to this process by preventing the orthogonal response from increasing as rapidly as the response to the preferred orientation.

3.3 Long-Range Horizontal Spread of Orientation Selectivity Is Controlled by Intracortical Cooperativity

During the last three decades, the intra-cortical anatomical connectivity made by the horizontal connections has been extensively explored in multiple species. It has generally been concluded that horizontal axons in visual cortex bind distant columns sharing similar orientation preference. However, the functional selectivity of the horizontal spread has never been measured directly. One reason is that the connections made by the horizontal axons are difficult to record since they only have a subthreshold impact on their post-synaptic targets. Therefore, to unveil the

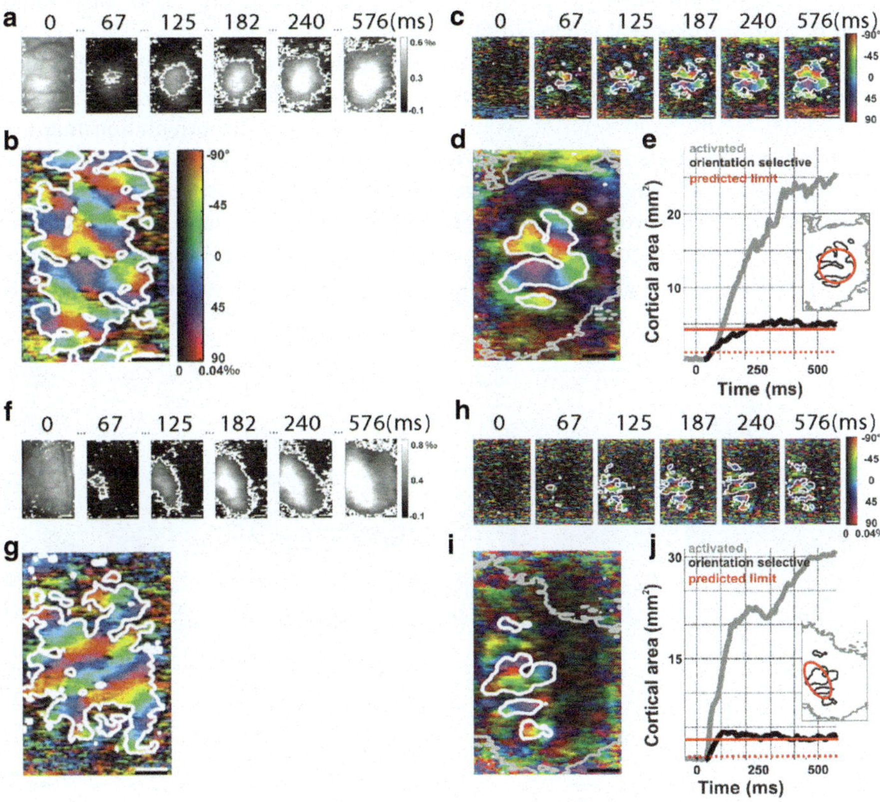

Fig. 10.6 The horizontal spread away from the feedforward imprint is loosing its orientation tuning. (**a, f**) Time-series of the cortical response propagation evoked by a local stimulus (averaged over four orientations). (**a**) is an example from area 17 (stimulus diameter 2°; eccentricity of 4.1°; average of 28 trials) and (f) is from area 18 (stimulus diameter 4°; eccentricity of 7.6°; average of 32 trials). The imaged cortical area is shown in the first frame. (**b, g**) The polar maps obtained with full field stimulation using all orientations. Color hue and brightness code respectively for the preferred orientation and the strength of the orientation bias. (**c, h**) and (**d, i**) Polar maps of the orientation tuning in response to the local small stimulus (**c, h** time series **d, i** polar maps at late times). The *white contour* delineates the region significantly activated. (**e, j**) Superimposition over the late activation map is the feedforward imprint (*red ellipse*) limit of the stimulus

functional expression of the horizontal connectivity, it is mandatory to have access to the activation of the horizontal network at the subthreshold, postsynaptic integration, level (Chavane et al. 2009) reported a multi-scale analysis of visually driven horizontal network activation, using direct population and intracellular measures of post-synaptic integration. Voltage-sensitive dye imaging shows that while global activation in response to a local stimulus exhibits long-range horizontal spread (Fig. 10.6a, f), the orientation-selective component of this response does not spread beyond the feedforward cortical imprint of the stimulus (Fig. 10.6c, h). Orientation selectivity decreases exponentially with horizontal distance. Therefore, beyond a distance of one hypercolumn, horizontal functional connectivity no longer obeys a binding rule preserving iso-orientation preference with this type of stimulus. Intracellular recordings show that this loss of orientation selectivity arises from the diversity of convergence patterns of intracortical synaptic input originating from beyond the classical receptive field.

In contrast, when increasing the spatial summation evoked by the stimulus—for example by presenting annular stimuli, orientation-selective activation spreads beyond the feedforward imprint (compare Fig. 10.7a, b: the orientation map). It was therefore concluded that stimulus-induced cooperativity at the network level is needed for the emergence of long-range orientation-selective spread.

This study shows two different dynamic behaviors of the same network for two distinct stimulus configurations: a single local stimulus does not propagate orientation preference through the long-range horizontal cortical connections, whereas

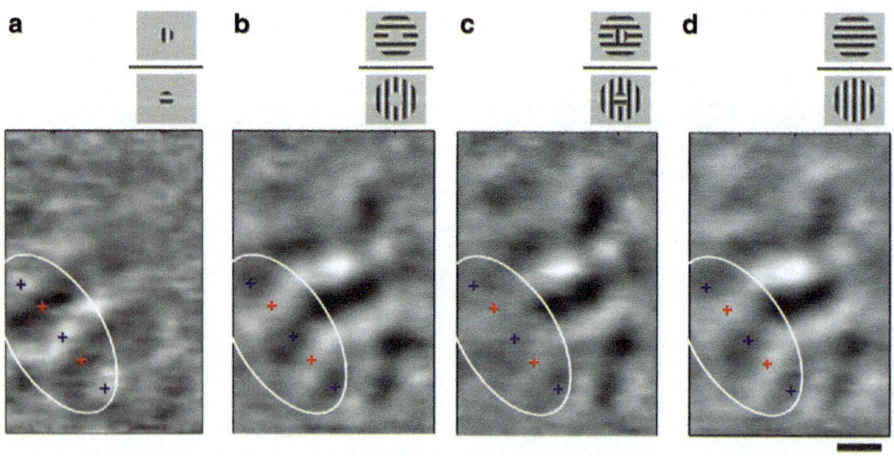

Fig. 10.7 Horizontal spread and center-surround competition. (**a**) Area 18 differential map obtained by dividing the responses to horizontal vs. vertical gratings presented through a central disk aperture (stimulus diameter 4° at an eccentricity of 7.6°; see cartoons). Orientation map is confined locally. (**b**) In comparison, displaying a larger grating through an annular aperture without stimulating the central disk (stimulus inner diameter, 4°, outer diameter, 12°, same eccentricity) results in the propagation of an oriented-selective signal within the cortical representation of the central disk region. (**c**) When both stimuli compete in a composite cross-oriented configuration, the orientation map disappears within the retinotopic representation of the central disk. (**d**) Control condition with iso-oriented center-surround configurations. The *white contour* encircles the expected retinotopic representation of the central disk. Scale bar is 1 mm

stimulation imposing spatial summation and temporal coherence facilitates the build-up of propagating activity exhibiting a strong orientation preference. These observations do not necessarily contradict each other. They point to the possibility that complex stimulus configurations are allowed and expected based on activation within a locally-confined region in visual space, whereas an iso-oriented stimulus with a large spatial extent supports the expectation of that specific orientation throughout a yet-larger area.

3.4 Dynamics of Transitions Between Different Cortical Representations of Visual Inputs

The previous series of experiments dealt with questions regarding the representation of a single stimulus. In a different series of experiment, Na'aman and Grinvald set out to characterize the temporal transition between stimulus representations. In one example switches between full-field oriented gratings were studied, with different conditioning stimuli presented for 500 ms followed by a constant test stimulus (Fig. 10.8). Transitions between representation pairs took 30–100 ms,

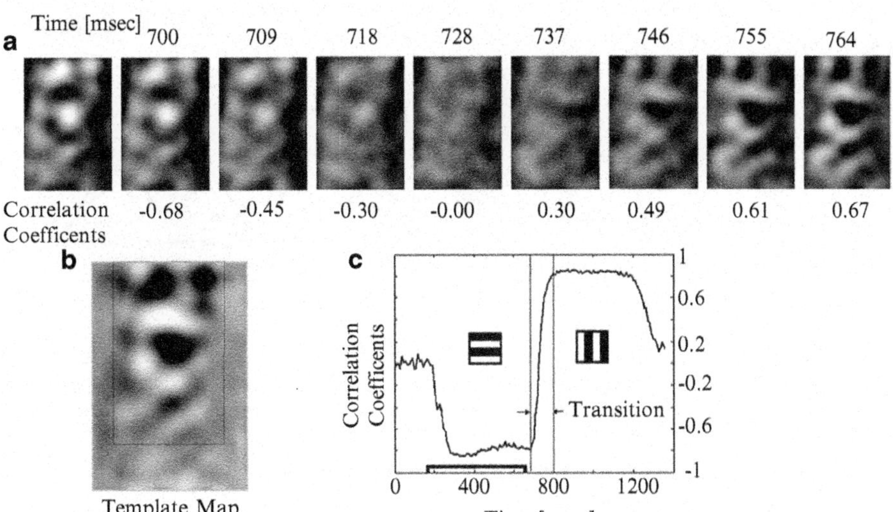

Fig. 10.8 Quantifications of the transition dynamics. (**a**) Differential orientation maps dynamics just before, during and after transition. The acquisition time of each frame is 9.1 [ms] shown at the top, Here time 0 is defined as start of data acquisition. Correlation coefficient values of each frame with the template map (**b**) appear at the bottom of each frame. (**b**) The average response to the test stimuli the *square gray* region used as the template map. (**c**) Correlation coefficients as a function of time calculated between the template map to each frame during the presentation of the conditioning and the test stimuli. These data represent and average of 32 repetitions

depending on the spatial frequency, drift velocity, orientation and contrast of the two stimuli. Additional details are still unpublished.

3.5 Neural Mechanisms of the Line-Motion Illusion

In another experiment we explored a temporal transition between two stimuli that gives rise to a visual illusion. The illusion of motion that can be produced by transition between two non-moving stimuli was described almost a century ago by Gestalt psychologists. In one variant of this phenomenon, the 'line-motion' illusion, motion perception can be induced by a stationary square immediately followed by a long, stationary bar. The percept is that of the bar 'sweeping' away from the square

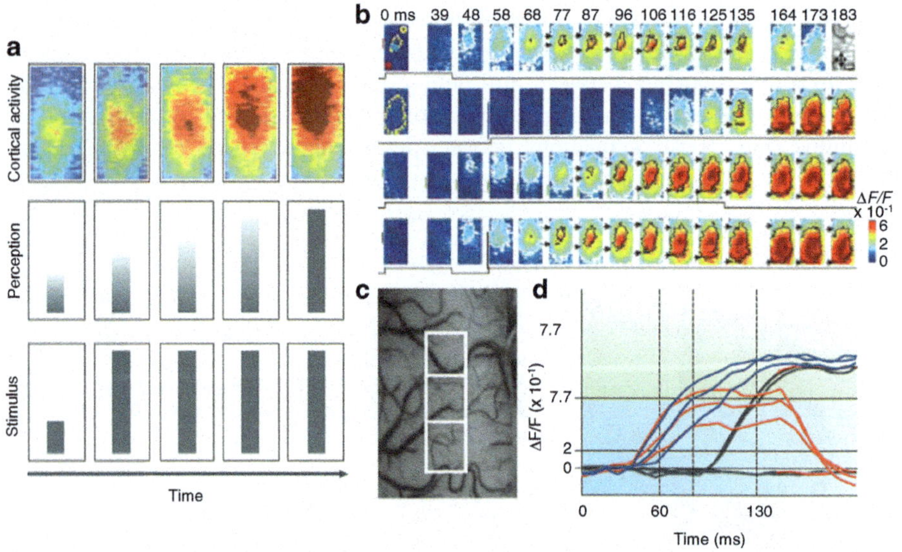

Fig. 10.9 Priming by subthreshold spread of synaptic potential can account for the illusion of motion. (**a**) The line-motion stimulus (*bottom*), the perceived stimulus (*middle*) and the evoked cortical activity (*top*) over time. (**b**) Time series of the cortical representations of (from top to bottom) the stationary *square* and *bars* alone, the small moving *square* and the line-motion stimulus. *Yellow dotted contours* shown at 0 ms approximate the retinotopic representation of the stimuli; *white contours* delimit low-amplitude activity ($p < 0.05$). *Colored vertical lines* shown to the left of individual panels after stimulus onset indicate estimated sizes (or trajectory) of the stimuli along a posterior-anterior axis. Time in milliseconds after stimulus onset is indicated at the *top*. The time courses of stimulation are outlined at the bottom of each row. (**c**) The cortical area imaged with three contiguous regions of interest (ROIs, *white squares*) from which the time courses of responses shown in (**d**) were derived. (**d**) Responses to the *small square* (*red*), the line-motion condition (delayed bar after the square; *blue*) and identical flashed *bar alone* (*black*) within the three ROIs shown in (**c**). Note the large effect of the priming by the square on the latency of the response to the flashed bar (*blue traces* compared with *black traces*). Adapted from Jancke et al. (2004) with permission from Nature Publishing Group

(Fig. 10.8a). The neuronal underpinnings of this illusion remain unknown, and VSDI was used to explore the associated subthreshold responses, combined with extracellular recordings to monitor spiking responses. We tested the hypothesis that the preceding square sets up a subthreshold activity gradient, such that when the stationary bar appears its response is superimposed on the preceding gradient, resulting in an increasing delay for reaching spike threshold with distance from the representation of the square stimulus.

The responses of anaesthetized cat visual cortex to five stimuli were imaged using VSDI: a stationary small square; a stationary long bar; a moving square; a drawn-out bar; and the line-motion. Flashing the bar alone evoked the expected localized, short-latency and high-amplitude activity patterns (Fig. 10.9b second row). Three regions of interest (ROIs) were defined (Fig. 10.9c), and responses to the bar alone simultaneously crossed threshold at all three ROIs (Fig. 10.9d, black traces). However, presenting a square 60–100 ms before the bar (line-motion stimulus) induced dynamic activity patterns (Fig. 10.9b fourth row), which resembled those produced by fast movement (Fig. 10.9b third row). ROI timecourses confirmed that a critical threshold VSD signal amplitude was reached at successive times for the line-motion stimulus (Fig. 10.9d, blue traces). The spatiotemporal activity gradient evoked by the square (Fig. 10.9b, top) set up a propagating cortical response to the subsequently-presented bar that correlated with illusory motion because it was indistinguishable from cortical representations of real motion in this area. These findings emphasize the effect of spatiotemporal patterns of subthreshold synaptic potentials on cortical processing and the shaping of perception.

4 Spontaneous Cortical Activity

Another series of studies was aimed at uncovering the principles underlying cortical activity in the absence of sensory stimulation, both that of single cells and that of the neuronal population.

4.1 On-Going Coherent Activity Is Large and Accounts for the Variability of Evoked Responses

Hebb suggested that neurons operate in assemblies—networks of neurons, local or widespread, that communicate coherently to perform the computations required for various behavioural tasks (Hebb 1949). A significant contribution of VSDI has been the visualization of the dynamics of coherent neuronal assemblies, i.e., neuronal assemblies in which the activity of cells is time-locked, across the entire imaged area. The firing of a single neuron, recorded with an extracellular electrode, is used as a time reference for spike-triggered averaging of the VSD signal of each imaged pixel. With a sufficient number of spikes, any neuronal activity not time-locked to the reference neuron is averaged out, enabling selective visualization of those cortical locations in which activity consistently occurred coherently with the firing of the reference neuron. A series of studies in anesthetized cat visual cortex explored the relationship of coherent spontaneous cortical states to evoked stimulus responses and to single-cell spikes.

An early surprising finding was that the amplitude of coherent, spontaneous, ongoing activity in neuronal assemblies was over half the amplitude of evoked activity. This suggested that spontaneous fluctuations in assembly activity affects evoked responses to sensory stimuli.

Indeed, while non averaged evoked responses display a high degree of variability, this variability can be accounted for by the spatiotemporal patterns of ongoing activity (Fig. 10.10). The variable single-trial evoked response is well-predicted by summing the average evoked response (averaged over many stimulus presentations; Fig. 10.10a top) with the ongoing activity immediately preceding stimulus onset

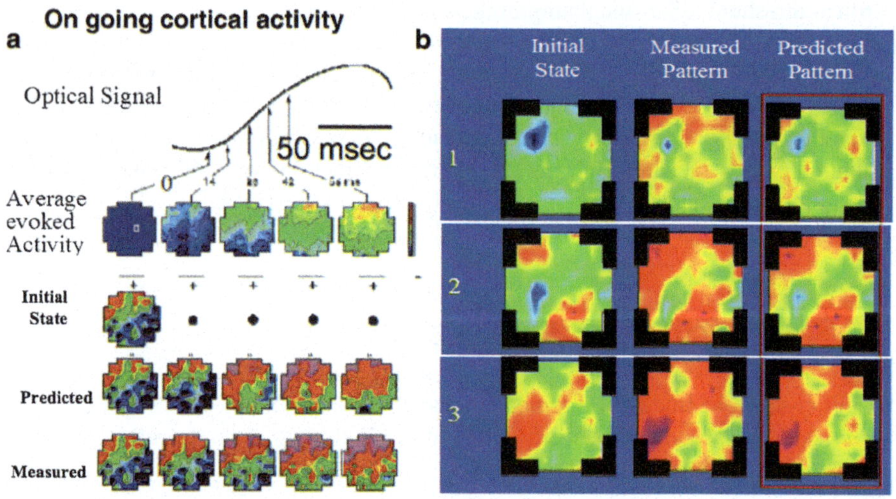

Fig. 10.10 The impact of on-going activity; predicting cortical evoked responses in spite of their large variability. (**a**) In spite of the large variability of the evoked response, it can be predicted as a linear combination of the fixed pattern of an evoked response, and the instantaneous, constantly varying on-going activity pattern. Both patterns can be measured. The *top trace* shows the evoked response after averaging dozens of evoked trials to remove the variability. The average series of temporal patterns of the developing evoked response is shown in the *upper row* of frames. The *second row* shows the instantaneous on-going pattern which is assumed to be changing rather slowly and therefore assumed fixed for less than 50 ms. It is thus approximated by the initial state, i.e. the on-going pattern just prior to the onset of the evoked response). The *third row* (Predicted) shows the sum of the pattern of the initial state and the temporal sequence of the averaged pattern of evoked activity. The *bottom row* (Measured) shows that this predicted response and the measured response are very similar. (**b**) Three examples of comparing predicted and measured responses are shown. The predicted response was calculated as shown in (**a**). *Left columns*: Three very different Initial states. *Middle column*: Measured responses. *Right columns*: Predicted responses, obtained by adding the initial state in each case. Modified from Arieli et al. (1996)

(the initial state), provided that for the duration of the response the ongoing state of the assembly remains similar to the initial state. This result suggests that ongoing activity is not merely "noise" and may provide the neuronal substrate for the dependence of sensory information processing on context, attention, behavioral and consciousness states, memory retrieval etc.

4.2 *Spontaneous Cortical States Are Internal Representations of Visual Attributes*

A subsequent study explored the relationship between the spatial pattern of population activity coherent with a single neuron's spikes during spontaneous and evoked activity (Fig. 10.11). They were found to be very similar. Furthermore, the instantaneous firing rate of a spontaneously active single neuron strongly depends on the spatial pattern of ongoing population activity in a large cortical area. During spontaneous activity, whenever the instantaneous spatial population pattern correlated highly with the spatial population pattern evoked by the preferred stimulus of the recorded neuron (Fig. 10.12a), its firing rate increased (Fig. 10.12c).

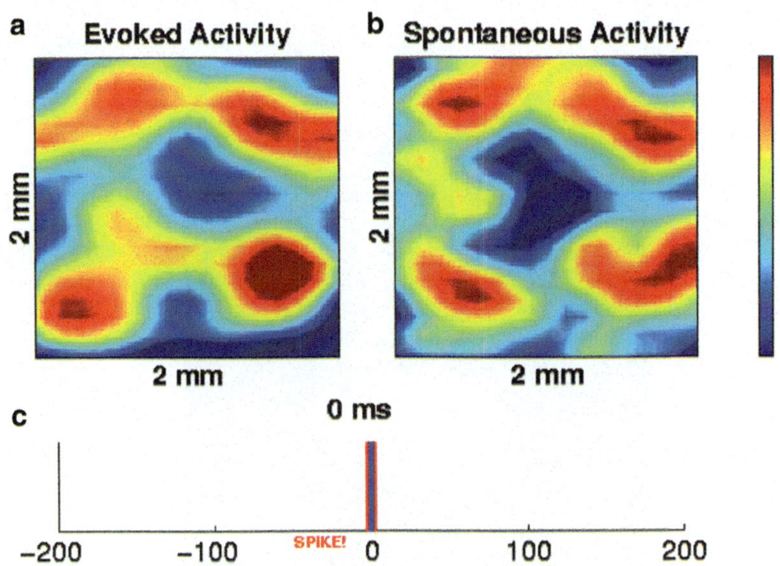

Fig. 10.11 The preferred cortical state of a reference tuned neuron is similar to the related global functional architecture. (**a**) Orientation map of a 2 mm neighbourhood of the recorded neuron. This map represents the orientation-selective functional architecture of this neuron's environment. (**b**) The cortical state corresponding to spontaneous action potentials obtained by spike triggered averaging of the spontaneous activity. The two patterns are nearly identical at the time of action potential occurrence. (**c**) The timing of the spike. In panels (**a, b**) it is stationary at time 0. This result suggests that a neuron tends to fire spontaneously whenever a pattern corresponding to its parent functional architecture appears spontaneously

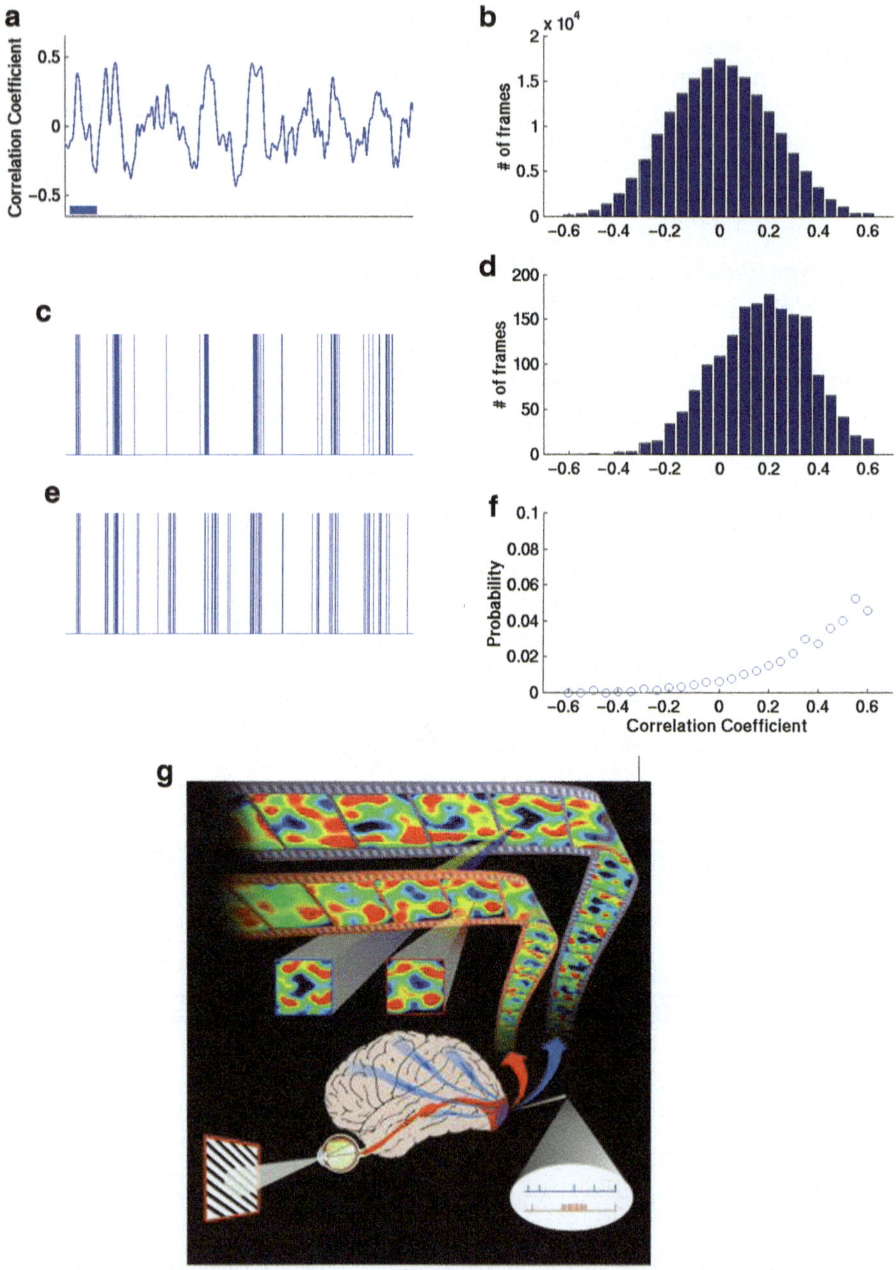

Fig. 10.12 A spontaneous spike train of a single neuron can be predicted from the similarity of the instantaneous population activity pattern to the functional architecture. (**a**) The correlation coefficient between the instantaneous snapshot of population activity and the relevant functional architecture. (**b**) Histogram of the correlation coefficient for 20,480 frames. (**c**) The observed spike train: a strong upswing in the values of the similarity index is seen in *panel* (**a**) each time the neuron emits a burst of action potentials. Every strong burst is followed by a marked downswing in the values of the correlation coefficient. (**d**) Histogram of correlation coefficients for 338 frames

Compared to the zero-centered distribution of correlation coefficients between each instantaneous recorded frame and the evoked orientation map (Fig. 10.12b), correlation coefficients during spikes had a strong bias towards positive values (Fig. 10.12d, and normalized in Fig. 10.12f). Strikingly, the dependence of single neurons' spiking on the global spontaneous activity is strong enough to allow reconstruction of the instantaneous spontaneous activity of single neurons (Fig. 10.12e).

In addition to revealing network influences in spontaneous spiking activity of individual neurons, the latest improvements in VSDI enabled exploration of spontaneous population dynamics and their relation to internal representations of sensory attributes. Ongoing activity in the visual cortex of the anaesthetized cat was shown to be comprised of dynamically switching intrinsic cortical states, many of which closely correspond to orientation maps (Fig. 10.13). Cortical states that correspond to each and all orientations emerge spontaneously. When such an orientation state emerged spontaneously, it spanned several cortical hypercolumns and was usually followed by a state corresponding to a proximal orientation, or dissolved into an unrecognized state or noisy pattern after a few tens of milliseconds.

Such dynamically switching cortical states could reflect the brain's internal context. Exploring cortical states is likely to reveal new fundamental principles about neural strategies for cortical processing, representations of objects, memories, context, expectations, and particularly about the interplay between internal cortical representations and the sensory input in primary sensory areas.

4.3 Ongoing Activity in Anesthetized Monkeys

The previous studies using voltage sensitive dye imaging (VSDI) were carried out on anesthetized cats (Grinvald et al. 1989; Arieli et al. 1995, 1996; Tsodyks et al. 1999; Kenet et al. 2003; Ringach 2003). This has raised the question whether these results are relevant to behavior. Therefore, Omer and coworkers (Omer and Grinvald 2004; Omer and Rom (2008) preformed VSDI of ongoing cortical activity in the visual cortices of awake monkeys simultaneously with measurements of single unit activity and the local-field potential. Coherent activity was found,

Fig. 10.12 (continued) during spike occurrences is highly skewed towards larger values. (**e**) Predicted instantaneous firing rate of a neuron, determined from the instantaneous similarity index, showing marked similarity to the bursting features in the observed spike train. (**f**) Same as (**d**), but normalized to the entire distribution shown in (**b**). (**g**) Dynamic brain activity in a given region is a combination of spontaneous internal activity in many cortical areas (*blue* stream) and the evoked responses (*red*). The spatial pattern that corresponds to spontaneous action potentials (*blue frame*) obtained by spike-triggered averaging and the evoked activity (*red frame*) are nearly identical. Adapted from Tsodyks et al. (1999) with permission from the American Association for the Advancement of Science

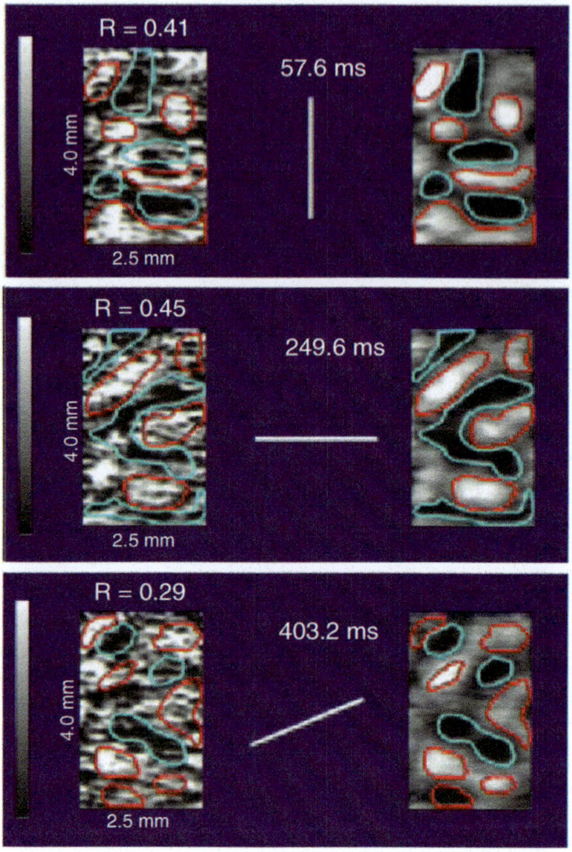

Fig. 10.13 Snapshot of the spontaneous appearance of orientation representations-orientation maps. The *three panels* compare instantaneous patterns of spontaneous (*left side*) and evoked activity to the averaged functional map (*right side*). The orientation maps were obtained from responses to full-field gratings (at the orientation indicated in the *middle*). The maps on the *left* are 10 ms snapshots of spontaneously appearing patterns; i.e. a single frame from a spontaneous recording session. R is the correlation value between the cortical representation of a given orientation and the snapshot. The number at the *middle*, *top*, denotes the time after recording onset at which the instantaneous pattern appeared

however distinct cortical states reminiscent of orientation maps have so far remained undetectable. To rule out species difference Omer and his colleagues explored the anesthetized monkey, where the results were found to be similar to those in anesthetized cat. However, in the anesthetized monkey the spontaneous cortical activity shows a larger repertoire of cortical states; not surprisingly the two orthogonal OD maps were also spontaneously represented (Fig. 10.14), and to a larger extent than that of orientation.

Similarly, spike triggered averaging (STA) using a sharply monocular reference neuron revealed a spatial pattern essentially identical to the evoked OD map (Fig. 10.15). It was noted the often both the orientation map and the OD map appeared spontaneously only on a small cortical area, unlike the findings in cat (Fig. 10.12) when the spontaneous pattern covered the entire imaged area.

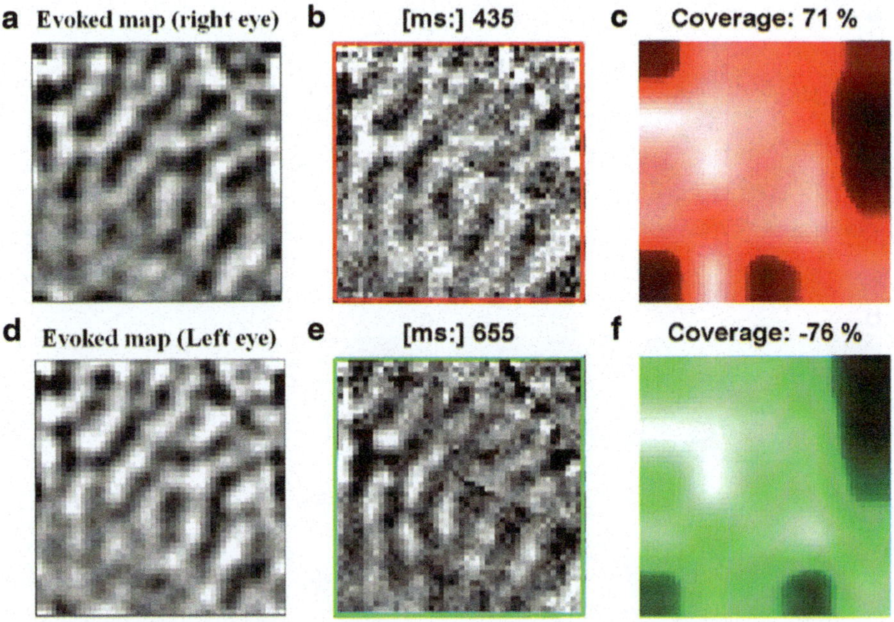

Fig. 10.14 Spontaneous appearance of OD maps in anesthetized monkey. (**a**) OD map obtained by signal averaging of alternating stimulation of the right eye and the left eye each of them separately. (**b**) Snapshot of the spontaneous appearance of right eye map at 435 ms after the imaging onset. (**c**) The spontaneous map in (**b**) is highly correlated with the OD map in (**a**) in the *red-white* region the coverage was 71 %. During the session the coverage of the cortical area with a functional map varied from 0 to ~80%. (**d**) The reversed differential OD map. (**e**) Snapshot of the spontaneous map for the contralateral-eye. (**f**) Coverage in *green* for the contralateral eye. Similar maps were observed spontaneously also for orientation. Note the excellent signal to noise ratio for the 10 ms snap shot seen without any signal averaging

5 Outlook

Several groups, including those of Waggoner (Waggoner and Grinvald 1977; Waggoner 1979), Loew (Loew 1987) and Fromherz (Kuhn and Fromherz 2003), as well as our own, have made significant progress in developing VSDI during the last three decades by designing and synthesizing of new VSDs. This chapter suggests that the technique has matures for shedding new light on neocortical function. Nevertheless, the signal-to-noise ratio that can be obtained with dyes, and the photodynamic damage they can cause, are limiting factors for yet better VSDI. Furthermore, different preparations often require dyes with different properties (Ross and Reichardt 1979; Cohen and Lesher 1986; Grinvald et al. 1988). Even different cortical areas in the same species can require different dyes (for example, one dye provided a high-quality signal in the rat somatosensory cortex but not in the rat olfactory bulb, although the bulb was well stained (Spors and Grinvald 2002). Therefore, continuing efforts to make better dyes and new biophysical and

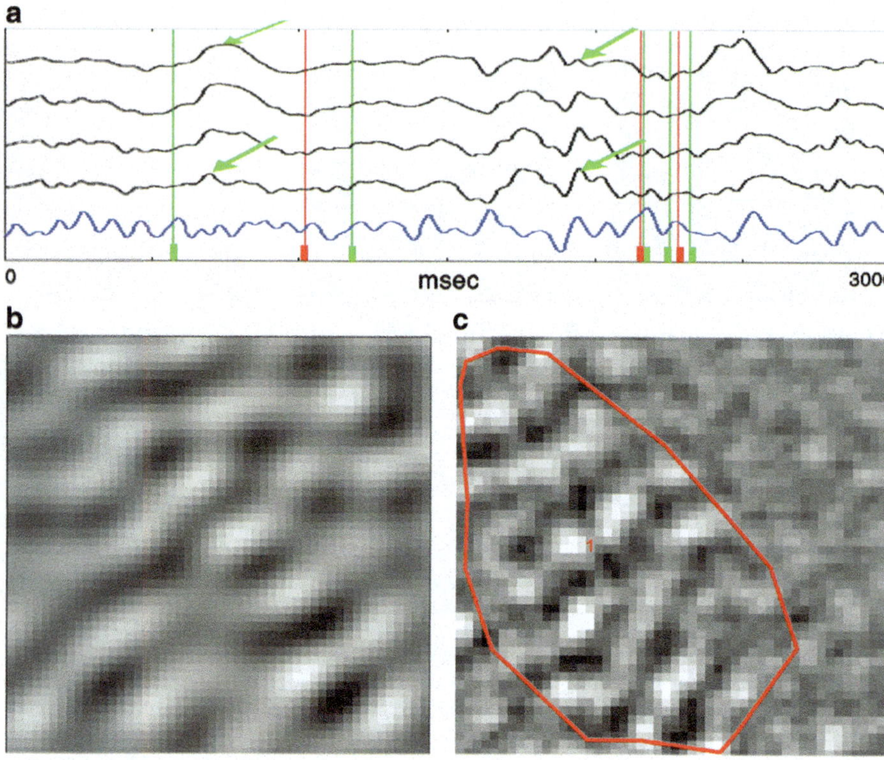

Fig. 10.15 Spike triggered averaging of spontaneous firing reveals that the cell fired when the entire population looked like its parent functional architecture. (**a**) Three seconds of an imaging session. The *top four traces* are optical signals from 4 pixels out of 1,000,000 showing mostly slow waves. The simultaneously recorded spikes and the local field potential (LFP) are shown in *blue*. The *red bars* show the timing of single unit activity detected by the same electrode. (**b**) OD map obtained by differential imaging and signal averaging. (**c**) The same map was obtained by spike triggered averaging of a single cell (*red* "1" in the center). The spatial pattern enclosed by the *red contour* exhibits a spatial pattern that corresponds to the OD map shown in (**b**). Note that the map has not spread to the entire imaged region

optical innovations are important. Similarly, the development of new genetically engineered *in vivo* probes (Miyawaki et al. 1997; Siegel and Isacoff 1997) will make experiments easier and improve the results. Of particular importance are probes that will stain only specific cell types and/or specific cellular compartments (axons, somata or dendrites).

Additional technological developments using multi-photon imaging, adaptive optics or other optical innovations to achieve three dimensional imaging are desirable (for example, using special lenses that offer a shallow field of view; Rector et al. 1999). The combination of existing VSD probes with fluorescence resonance energy transfer (FRET) to detect electrical activity is also promising (Cacciatore et al. 1999). Similarly, excitation at the red edge of existing VSDs has been shown to provide up to ten-fold larger fractional change in the VSD signal size (Kuhn and Fromherz 2003; Kuhn et al. 2004), and imaging at the second harmonic frequency of new VSDs (Millard et al. 2003; Nemet et al. 2004) might also provide larger signals, enhancing the signal-to-noise ratio. The amount and nature of the data that have been and will be accumulated necessitate further methods of analysis, modelling and theoretical research that should lead to new conceptual frameworks regarding cortical function.

To increase the scope of meaningful neurophysiological data obtained from the same patch of cortex, VSDI should be combined with targeted-tracer injections, retrograde labelling, microstimulation and intracellular and extracellular recording. VSDI combined with electrical recordings enables neuroscientists to obtain information about spatiotemporal patterns of activity in coherent neuronal assemblies in the neocortex, at sub-columnar resolution. At present, no alternative imaging technique for visualizing organization and function in-vivo provides a comparable spatial and temporal resolution. This level of resolution allows us to address questions of both *where* and *when* processing takes place, providing great promise that this technique, integrated with other approaches, will contribute to the study of *how* processing occurs.

6 Conclusion

With the advent of the latest generation of VSDs and imaging equipment, a new era of exploration of cortical dynamics has ushered in. New-generation VSDI studies in cat and monkey visual cortex have revealed intricate spatiotemporal activity patterns underlying a variety of perceptual phenomena, underscoring the importance of large-scale subthreshold population responses and spontaneous states.

References

Arieli A, Grinvald A (2002) Combined optical imaging and targeted electrophysiological manipulations in anesthetized and behaving animals. J Neurosci Methods 116:15–28

Arieli A, Shoham D, Hildesheim R, Grinvald A (1995) Coherent spatio-temporal pattern of on-going activity revealed by real time optical imaging coupled with single unit recording in the cat visual cortex. J Neurophysiol 73:2072–2093

Arieli A, Sterkin A, Grinvald A, Aertsen A (1996) Dynamics of ongoing activity: explanation of the large variability in evoked cortical responses. Science 273:1868–1871

Arieli A, Grinvald A, Slovin H (2002) Dural substitute for longterm imaging of cortical activity in behaving monkeys and its clinical implications. J Neurosci Methods 114:119–133

Berger T, Borgdorff AJ et al (2007) Combined voltage and calcium epifluorescence imaging in vitro and in vivo reveals subthreshold and suprathreshold dynamics of mouse barrel cortex. J Neurophysiol 97:3751–3762

Bringuier V, Chavane F, Glaeser L, Frégnac Y (1998) Horizontal propagation of visual activity in the synaptic integration field of area 17 neurons. Science 283:695–699

Cacciatore TW, Brodfuehrer PD et al (1999) Identification of neural circuits by imaging coherent electrical activity with FRET-based dyes. Neuron 23:449–459

Chavane F, Sharon D, Jancke D, Marre O, Frégnac Y, Grinvald A (2011) Lateral spread of orientation selectivity in V1 is controlled by intracortical cooperativity. Front Syst Neurosci. doi:10.3389/fnsys.2011.00004

Cohen LB, Lesher S (1986) Optical monitoring of membrane potential: methods of multisite optical measurement. Soc Gen Physiol Ser 40:71–99

Ferezou I, Bolea S, Petersen CCH (2006) Visualizing the cortical representation of whisker touch: voltage-sensitive dye imaging in freely moving mice. Neuron 50:617–629

Ferezou I, Haiss F et al (2007) Spatiotemporal dynamics of cortical sensorimotor integration in behaving mice. Neuron 56:907–923

Grinvald A, Hildesheim R (2004) VSDI: a new era in functional imaging of cortical dynamics. Nat Rev Neurosci 5:874–885

Grinvald A, Anglister L, Freeman JA, Hildesheim R, Manker A (1984) Real-time optical imaging of naturally evoked electrical activity in intact frog brain. Nature 308:848–850

Grinvald A, Frostig RD, Lieke E, Hildesheim R (1988) Optical imaging of neuronal activity. Physiol Rev 68:1285–1366

Grinvald A, Frostig RD et al (1989) Optical imaging of activity in the visual cortex. In: Lam D, Gilbert CD (eds) Neurall mechanism of visual perception. Portfolio Publishing Company, Austin, TX

Grinvald A, Lieke EE, Frostig RD, Hildesheim R (1994) Cortical point-spread function and long-range lateral interactions revealed by real-time optical imaging of macaque monkey primary visual cortex. J Neurosci 14:2545–2568

Grinvald A, Shoham D et al (1999) In-vivo optical imaging of cortical architecture and dynamics. In: Johansson H, Windhorst U (eds) Modern techniques in neuroscience research. Springer, New York, NY

Hebb DO (1949) The organization of behavior. Wiley, New York, NY

Hubel DH, Wiesel TN (1962) Receptive fields, binocular interactions and functional architecture in the cat's visual cortex. J Physiol 160:106–154

Jancke D, Chavane F, Grinvald A (2004) Imaging cortical correlates of a visual illusion. Nature 428:424–427

Kenet T, Bibitchkov D, Tsodyks M, Grinvald A, Arieli A (2003) Spontaneously occurring cortical representations of visual attributes. Nature 425:954–956

Kuhn B, Fromherz P (2003) Anellated hemicyanine dyes in neuron membrane: molecular stark effect and optical voltage recording. J Phys Chem B 107:7903–7913

Kuhn B, Fromherz P, Denk W (2004) High sensitivity of Stark-shift voltage-sensing dyes by one- or two-photon excitation near the red spectral edge. Biophys J 87:631–639

Loew LM (1987) Optical measurement of electrical activity. CRC press, Boca Raton (FL)

Millard AC, Jin L, Lewis A, Loew LM (2003) Direct measurement of the voltage sensitivity of second-harmonic generation from a membrane dye in patch-clamped cells. Opt Lett 28:1221–1223

Miyawaki A, Llopis J et al (1997) Fluorescent indicators for Ca^{2+} based on green fluorescent proteins and calmodulin. Nature 388:834–835

Mountcastle VB (1957) Modality and topographic properties of single neurons of cat's somatic sensory cortex. J Neurophysiol 20:408–434

Nemet BA, Nikolenko V, Yuste R (2004) Second harmonic imaging of membrane potential of neurons with retinal. J Biomed Opt 9:873–881

Omer DB, Grinvald A (2004) The dynamics of evoked and ongoing activity in the behaving monkey. Rev Neurosci 19:S50

Omer DB, Rom L (2008) The dynamics of ongoing activity in awake and anesthetized monkey are significantly different. Annual meeting of the Society of neuroscience abstract, Washington, DC

Orbach HS, Cohen LB (1983) Simultaneous optical monitoring of activity from many areas of the salamander olfactory bulb. A new method for studying functional organization in the vertebrate CNS. J Neurosci 3:2251–2262
Orbach HS, Cohen LB, Grinvald A (1985) Optical mapping of electrical activity in rat somatosensory and visual cortex. J Neurosci 5:1886–1895
Petersen CCH, Grinvald A, Sakmann B (2003a) Spatiotemporal dynamics of sensory responses in layer 2/3 of rat barrel cortex measured in vivo by voltage-sensitive dye imaging combined with whole-cell voltage recordings and neuron reconstructions. J Neurosci 23:1298–1309
Petersen CCH, Hahn TTG, Mehta M, Grinvald A, Sakmann B (2003b) Interaction of sensory responses with spontaneous depolarization in layer 2/3 barrel cortex. Proc Natl Acad Sci U S A 100:13638–13643
Ratzlaff EH, Grinvald A (1991) A tandem-lens epifluorescence macroscope: hundred-fold brightness advantage for wide-field imaging. J Neurosci Methods 36:127–137
Rector DM, Rogers RF, George JS (1999) A focusing image probe for assessing neural activity in vivo. J Neurosci Methods 91:135–145
Ringach DL (2003) Neuroscience: states of mind. Nature 425(6961):912–913
Ross WN, Reichardt LF (1979) Species-specific effects on the optical signals of voltage sensitive dyes. J Membr Biol 48:343–356
Salzberg BM, Davila HV, Cohen LB (1973) Optical recording of impulses in individual neurons of an invertebrate central nervous system. Nature 246:508–509
Sharon D, Grinvald A (2002) Dynamics and constancy in cortical spatiotemporal patterns of orientation processing. Science 295:512–515
Sharon D, Jancke D, Chavane F, Na'aman S, Grinvald A (2007) Cortical response field dynamics in cat visual cortex. Cereb Cortex 17:2866–2877
Shoham D, Glaser DE et al (1999) Imaging cortical dynamics at high spatial and temporal resolution with novel blue voltage-sensitive dyes. Neuron 24:791–802
Siegel MS, Isacoff EY (1997) A genetically encoded optical probe of membrane voltage. Neuron 19:735–741
Spors H, Grinvald A (2002) Temporal dynamics of odor representations and coding by the mammalian olfactory bulb. Neuron 34:1–20
Sterkin A, Lampl I, Ferster D, Grinvald A, Arieli A (1998) Real time optical imaging in cat visual cortex exhibits high similarity to intracellular activity. Neurosci Lett 51:S41
Tasaki I, Watanabe A, Sandlin R, Carnay L (1968) Changes in fluorescence, turbidity, and birefringence associated with nerve excitation. Proc Natl Acad Sci U S A 61:883–888
Tsodyks M, Kenet T, Grinvald A, Arieli A (1999) The spontaneous activity of single cortical neurons depends on the underlying global functional architecture. Science 286:1943–1946
Waggoner AS (1979) Dye indicators of membrane potential. Annu Rev Biophys Bioeng 8:47–63
Waggoner AS, Grinvald A (1977) Mechanisms of rapid optical changes of potential sensitive dyes. Ann N Y Acad Sci 303:217–241

Chapter 11
Imaging the Dynamics of Neocortical Population Activity in Behaving and Freely Moving Mammals

Amiram Grinvald and Carl C.H. Petersen

Abstract The development of functional imaging techniques applicable to neuroscience and covering a wide range of spatial and temporal scales has greatly facilitated the exploration of the relationships between cognition, behaviour and electrical brain activity. For mammals, the neocortex plays a particularly profound role in generating sensory perception, controlling voluntary movement, higher cognitive functions and planning goal-directed behaviours. Since these remarkable functions of the neocortex cannot be explored in simple model preparations or in anesthetised animals, the neural basis of behaviour must be explored in awake behaving subjects. Because neocortical function is highly distributed across many rapidly interacting regions, it is essential to measure spatiotemporal dynamics of cortical activity in real-time. Extensive work in anesthetised mammals has shown that *in vivo* Voltage-Sensitive Dye Imaging (VSDI) reveals the neocortical population membrane potential dynamics at millisecond temporal resolution and subcolumnar spatial resolution. Here, we describe recent advances indicating that VSDI is also already well-developed for exploring cortical function in behaving monkeys and mice. The first animal model, the non-human primate, is well-suited for fundamental exploration of higher-level cognitive function and behavior. The second animal model, the mouse, benefits from a rich arsenal of molecular and genetic technologies. In the monkey, imaging from the same patch of cortex, repeatedly, is feasible for a long period of time, up to a year. In the rodent, VSDI is applicable to freely moving and awake head-restrained mice. Interactions between different cortical areas and different cortical columns can therefore now be dynamically mapped through VSDI and related to the corresponding behaviour. Thus by applying VSDI to mice and monkeys one can begin to explore

A. Grinvald
Department of Neurobiology, Weizmann Institute of Science,
P.O. Box 26, Rehovot 76100, Israel

C.C.H. Petersen (✉)
Laboratory of Sensory Processing, Brain Mind Institute, Faculty of Life Sciences,
École Polytechnique Fédérale de Lausanne (EPFL), Lausanne CH-1015, Switzerland
e-mail: carl.petersen@epfl.ch

M. Canepari et al. (eds.), *Membrane Potential Imaging in the Nervous System and Heart*, Advances in Experimental Medicine and Biology 859,
DOI 10.1007/978-3-319-17641-3_11

how behaviour emerges from neuronal activity in neuronal networks residing in different cortical areas.

Keywords Cat • Cortical response field • Evoked response • Horizontal connections • Lateral spread • Monkey • Mouse • Orientation map • Orientation selectivity • Sensorimotor cortex • Spatiotemporal response • Spikes • Spontaneous activity • Subthreshold response • V1 • V2 • Visual cortex • Whisker

1 Introduction

The activity of highly distributed neural networks is thought to underlie sensory processing, motor coordination and higher brain functions. These intricate networks are composed of large numbers of individual neurons, which interact through synaptic connections in complex dynamically regulated spatiotemporal patterns. To understand network properties and functions, it is helpful to begin by studying the ensemble activity of neuronal populations. Functionally related networks of neurons are often spatially organised and imaging techniques are therefore ideally suited to monitoring neural network activity. Having defined the spatiotemporal dynamics of the ensemble population network activity, it may then be possible to begin to understand the relationships and contributions of single neurons and their synaptic connections to the larger networks that are essential for generating percepts and controlling behaviour.

The high performance of the mammalian brain is thought to result, in part, from computations taking place in the neocortex. The neocortex is organized into cortical columns (Mountcastle 1957; Hubel and Wiesel 1962), which have lateral dimensions of a few hundred microns. Interactions within and between cortical columns occur on the time scale of milliseconds. To follow neuronal computations at the fundamental level of cortical columns in real-time therefore requires a spatial resolution of ~100 μm and a temporal resolution of ~1 ms. *In vivo* Voltage-Sensitive Dye Imaging (VSDI) (Grinvald et al. 1984; Orbach et al. 1985) fulfils these technical requirements and should help resolve many outstanding fundamental questions. Electrical communication in cortical networks comprises two basic signals: subthreshold potentials (reflecting dendritic synaptic integration) and suprathreshold action potentials (forming the neuronal output). VSDI appears to relate primarily to spatiotemporal patterns of the subthreshold synaptically-driven membrane potentials. VSDI in behaving animals therefore allows a correlation of subthreshold synaptic processing with perception and other higher brain functions.

Sensory perception is an active process, in which information is often acquired by self-generated movements of sensors. Multiple brain regions concerning motor control and different sensory modalities must cooperate to generate the mental construct of the world around us. Obvious examples that we are familiar with from our own experiences are vision and touch. In vision we actively move our eyes to examine visual targets of interest. For touch, we move our fingers into contact with objects to obtain tactile information about shape and texture. In order to understand sensory perception it is therefore of critical importance to investigate the neural activity of highly distributed sensorimotor networks in awake behaving animals. Here we describe how *in vivo* VSDI (Grinvald et al. 1984; Shoham et al. 1999; Grinvald and Hildesheim 2004) can be applied to provide such information on cortical spatiotemporal dynamics during behaviour. We describe technical advances and first results from VSDI of behaving animals and how VSDI can be combined with intracortical microstimulation, single-unit recording and local-field-potential recording from awake animals. We conclude by discussing further technical developments to overcome current limitations and additional questions that are likely to be explored.

2 Voltage-Sensitive Dyes

To explore the neuronal basis of behavior it is important to be able to image cortical dynamics without signal averaging. This is now possible through the development of new voltage-sensitive dyes (Shoham et al. 1999) with excitation and emission spectra in the far red. These new dyes reduce pulsation artefacts and haemodynamic noise, since they operate beyond the major absorption band of haemoglobin. This advance has enabled the functional architecture and dynamics of cortical information processing to be revealed at subcolumnar spatial and millisecond temporal resolution in behaving animals.

Intracellular recordings *in vivo* under anesthesia combined with simultaneous VSDI, both in the cat visual cortex (Grinvald et al. 1999) and in the rodent somatosensory cortex (Petersen et al. 2003a, b; Ferezou et al. 2006; Berger et al. 2007) indicate that the VSD signal correlates closely with changes in subthreshold membrane potential (red and blue traces in Fig. 11.1). *In vitro* measurements show microsecond VSD response times and, *in vivo,* VSD signals correlate on the millisecond timescale with membrane potential recorded in individual neurons. However, it should be noted that these simultaneous recordings have so far only been carried out under anesthesia and in the future it will be important to repeat these experiments combining VSDI and intracellular electrophysiology in awake behaving animals.

The spatial resolution of VSDI in awake animals is primarily limited by light scattering, movement artefacts and signal-to-noise ratio. Each camera pixel will receive fluorescence originating from dendrites, axons and somata belonging to a population of neurons. Since the VSD stains membranes, dendrites and axons will dominate the total observed fluorescence. Dendrites and axons can extend for considerable lateral distances, which could therefore strongly affect spatial resolution of VSDI. In practice, however, VSDI has been shown to provide high resolution maps clearly differentiating activity in nearby cortical columns (Petersen et al. 2003a; Ferezou et al. 2006) and offering a spatial resolution better than 50 μm (Shoham et al. 1999; Grinvald et al. 1999). One possible explanation may

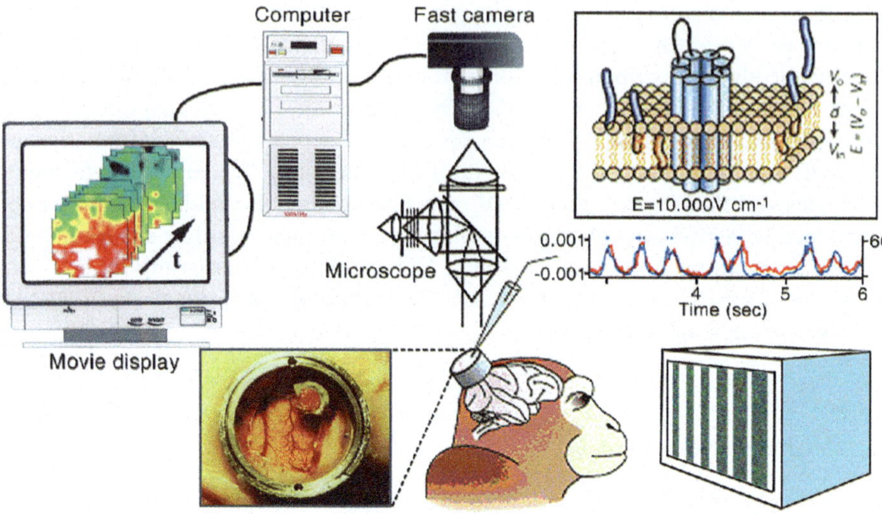

Fig. 11.1 Voltage-sensitive dye imaging of behaving monkeys. Setup for VSD imaging of cortical dynamics in behaving monkeys. The exposed monkey cortex is covered with a sealed cranial window (*bottom left inset*), which acts as if the monkey had a transparent skull. After the cortex is stained with a suitable voltage-sensitive dye (*blue*/*orange* molecules in *top right inset*), it is illuminated with excitation light of the appropriate wavelength (630 nm; 30 nm bandwidth, in this case). A sequence of images of the fluorescing cortex is taken with a fast camera (100–10,000 Hz) using a macroscope offering a numerical aperture of ~0.4. During image acquisition the monkey fixates on a visual stimulus or the eyes are covered with shutters in order to monitor ongoing activity. The acquired images are digitized and transferred to the computer controlling the experiment. Functional maps or movies of the cortical activity are analysed and displayed on a colour monitor. Various types of electrical recording, microstimulation or tracer injection are often carried out simultaneously with the imaging. The similarity between the two traces above the electrode, comparing intracellularly recorded electrical activity (*blue*) from one cell and VSD population activity (*red*), indicate that, *in vivo*, VSD measures mostly subthreshold synaptic membrane potential changes. Spikes recorded in the intracellular recording are truncated to increase vertical scale, but spikes are indicated by *blue lines* at the *top*. Spikes were not detected in the optical signal

be that synaptic processing on different dendritic branches is spatially mapped with respect to cortical columns.

3 Imaging from Awake Behaving Animals

A major technical impediment to imaging in awake animals is movement of the brain, as is the case for all forms of *in vivo* brain imaging. Movements are of course particularly large during wakefulness and overt movements are often an integral part of behavioural paradigms in animal experiments. A critical aspect of imaging brain function during behaviour is therefore to obtain a stable view of the brain.

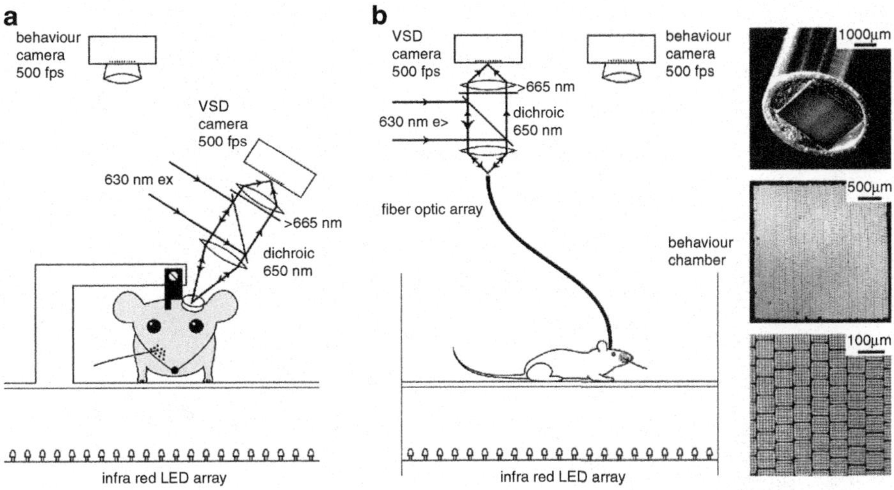

Fig. 11.2 Voltage-sensitive dye imaging of awake and freely moving mice. (**a**) Setup for voltage-sensitive dye imaging of the sensorimotor cortex in an awake head-restrained mouse. An implanted metal head-fixation post is attached to a fixed mechanical arm. The tandem lens optics are turned to include as large an area of sensorimotor cortex as possible in the focal depth of the epifluorescence tandem lens optics. With the cortex stained with RH1691, the excitation light (630 nm) is reflected by a dichroic mirror (650 nm) and focused on the brain through an objective. Emitted fluorescence is collected by the same lens and imaged on to a fast camera with large well-depth. Simultaneous filming of behaviour enables correlation of neocortical activity with sensorimotor behavioural events. (**b**) For imaging freely moving mice, the experimental setup is almost identical, except that a flexible fiber optic image bundle relays light from the focal plane of the tandem lens imaging system to the mouse brain. The fibers are arranged in a orderly fashion at each end of the fiber (*right*). Each individual fiber measures 10 μm in diameter and they are assembled in a multistep process first involving 6×6 multifibers that are then wound to form a 300×300 array of fibers extending approximately 3×3 mm

The most obvious paradigm is to fix the head relative to the imaging apparatus. Head-fixation is routinely applied in neurophysiological animal experiments and has proven to be extremely useful both in constraining behavioural variability and in facilitating measurements. Head-fixation bars can be attached to the skull of monkeys (Fig. 11.1) and mice (Fig. 11.2a). The animals rapidly learn to accept head-restraint during a period of behavioural habituation. As for imaging in anesthetised animals, a sealed recording chamber should be used to minimise brain movements relative to the skull. Furthermore, to facilitate staining in monkeys, the resected dura matter is substituted with artificial dura mater (Fig. 11.3). A careful cleaning procedure allowed repeated VSDI sessions in monkeys 2–3 times each week over a period longer than a year (Arieli et al. 2002).

A tandem-lens macroscope (Ratzlaff and Grinvald 1991) can then be mounted in a stable location relative to the fixed head position for epifluorescent voltage-sensitive dye imaging with suitable illumination and camera equipment (similar to that used for imaging anesthetised animals, except for additional equipment

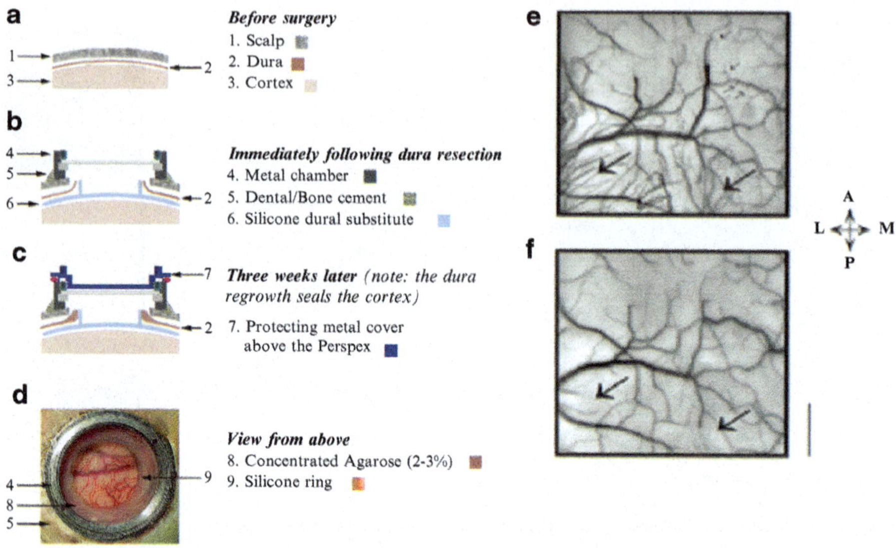

Fig. 11.3 The silicone dural substitute for long-term chronic recordings. (**a**) Scalp, dura and cortex prior to the surgery. (**b**) When the dural substitute was positioned in its proper position, the cut edge of the natural dura was enfolded and touched the silicon ring in the middle of the sheet, in order to stabilize it and prevent the natural dura from growing back into the central area used for imaging. (**c**) Within a few weeks after the operation the dura had grown around the ring, thus creating a good seal of the cortex. (**d**) View from above 2 weeks after the artificial dural transplantation. (**e**, **f**) Cortical reaction to dural implant. The cortical vascularisation 12 months (**e**) and 3 months (**f**) after implantation of artificial dura. Note the new growth of blood vessels originating from the Lunate sulcus and other sites (*big arrows*). This neovascularization is different from the blood vessels over the cortical tissue since it has the shape of 'hair like' growth. Note that cortical blood vessels also react: the veins show new 'growth cones' which were directed towards the new microvascularization (*small arrows*)

necessary for controlling and monitoring behaviour). Voltage-sensitive dye imaging of awake head-restrained mice (Petersen et al. 2003b; Ferezou et al. 2007) and awake head-restrained monkeys (Slovin et al. 2002; Seidemann et al. 2002; Omer and Grinvald 2008; Fekete et al. 2009) has begun to reveal cortical correlates of sensorimotor processing.

An alternative approach, which has been successfully tested in mice (Ferezou et al. 2006), is to image voltage-sensitive dye signals via flexible fiber optics (Fig. 11.2b). Fiber optic bundles containing well-ordered arrays of individual fibers are able to efficiently transmit images from one end of the fiber to the other. At each end, the fibers are held rigidly in place, but in between the ends, the fibers can run separately from each other allowing high flexibility and a considerable range of rotation. Under these conditions the animal is therefore relatively free to move around within the tether limits of the light-weight fiber optics. Excitation light can be passed in one direction through the same fibers that also relay emitted fluorescence back from the animal. One end of the fiber bundle can be placed in the location normally imaged by the tandem lens epifluorescence macroscope. The other end of the fiber optic image bundle can be placed in direct contact with the brain surface and firmly anchored on the skull. Movement artefacts are surprisingly small in this imaging configuration (Ferezou et al. 2006; Flusberg et al. 2008), perhaps because of correlated movement of the fiber bundle and the brain.

4 Integrating VSDI with Electrode Techniques

VSDI can easily be combined with standard electrode-based techniques. Whereas optimal mechanical stability for imaging is provided by sealed cranial windows, these do not allow electrodes to be introduced. One approach is to cover the cortical surface with agarose and stabilise it by a cover slip. A small gap between the edge of the cover slip and the wall of the recording chamber allows oblique entry of electrodes into the rodent cortex (Petersen et al. 2003a, b; Berger et al. 2007; Ferezou et al. 2006, 2007). For cats and monkeys a 'sliding-top cranial window' was developed with a removable microdrive-positioned electrode (Arieli and Grinvald 2002). VSDI can then be combined with microstimulation, extracellular recording (single- and multiple-unit recording and local field potential), or intracellular recording, and the targeted injection of tracers. Recordings can either be made simultaneously and/or can be targeted to specific regions based on the functional imaging data.

5 Imaging the Monkey Visual Cortex During Behaviour

5.1 Dynamics of the Spread of Retinotopically Evoked Activity in Macaque Visual Cortex

Using VSDI in awake behaving monkeys, Slovin et al. (2002) imaged cortical dynamics with high spatial and temporal resolution from the same patch of cortex for up to 1 year (Fig. 11.4). In the original study, the visual cortices of trained macaques were stained 1–3 times a week. Immediately after each staining session, the monkey started to carry out a simple behavioural task, while the visual cortex was imaged using VSDI. The first important task of this study was to show that such repeated use of VSDI does not interfere with normal cortical function. The functional maps obtained by VSDI were confirmed by imaging based on intrinsic signals (Grinvald et al. 1986) over a period of up to a year (Fig. 11.4b). Orientation preference maps can be used as a sensitive assay to assess any potential cortical damage since orientation preference is formed in the cortex itself rather than lower brain areas. The imaging results, as well as the behavioural performance, indicated that pharmacological side-effects or photodynamic damage were negligible.

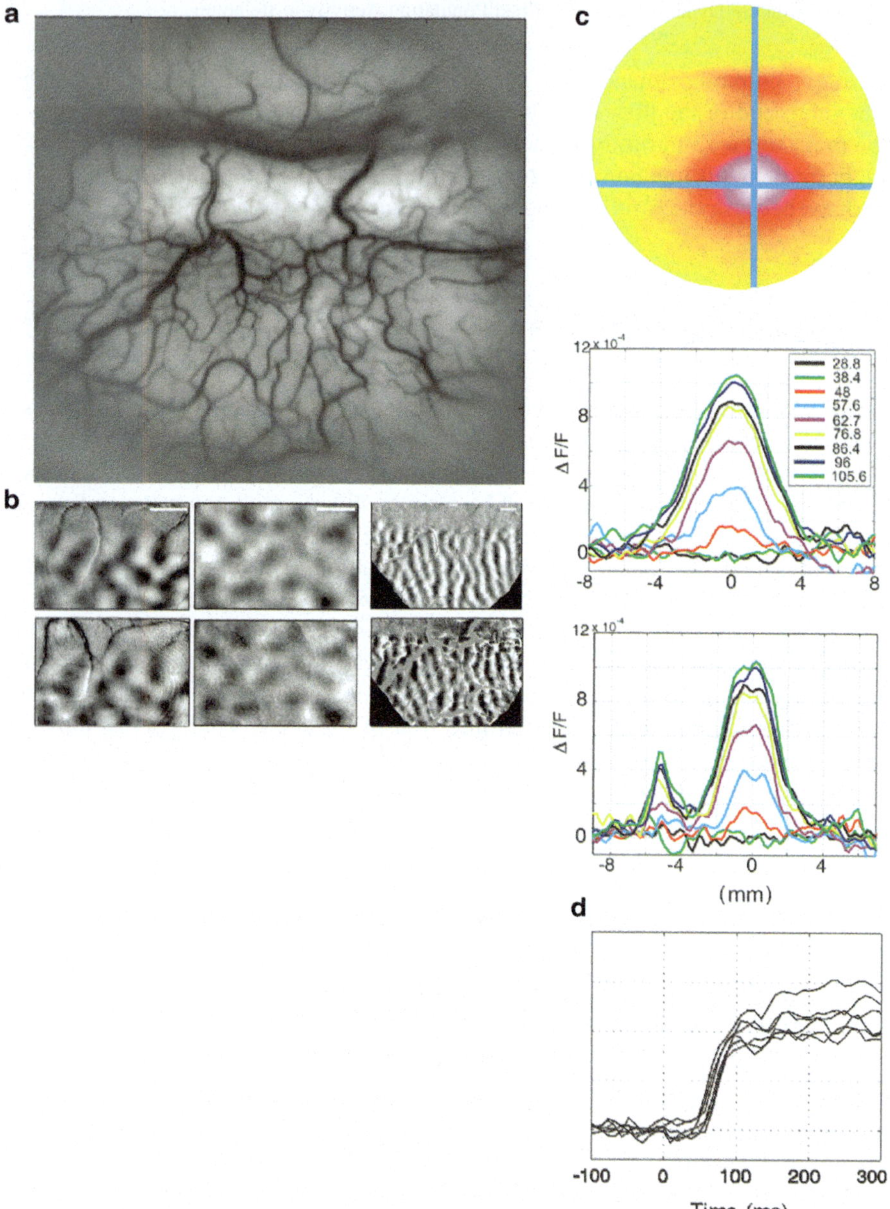

Fig. 11.4 Normal cortical architecture and function is preserved throughout long-term chronic VSDI for up to a year. (**a**) A large area of exposed cortex in excellent condition after almost a year of repeated VSDI, intrinsic imaging and electrical recording in this monkey. (**b**) The functional VSDI maps of ocular dominance in two sites (*left* and *right*) and orientation (*middle*), which were recorded from the same patch of cortex in recording sessions separated by up to a year, are virtually identical (compare *top* and *bottom rows* of images). (**c**) Dynamics of retinotopic activation. *Top*: retinotopic activation of V1 and V2 (width of cortical area about 14 mm). *Bottom*: the spread of cortical dynamics along horizontal (*upper*) and vertical (*lower*) axes as a function of time. Time is marked by different *colours*. (**d**) Single-trial evoked responses in the awake monkey. As in the anaesthetized animal, the response is variable. Scale bar: 1 mm

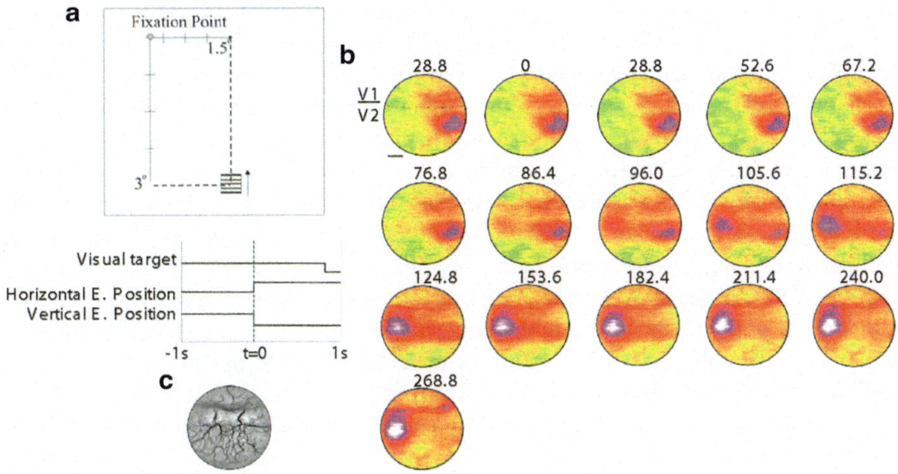

Fig. 11.5 Fast saccadic eye movement generates a cortical response along its track. (**a**) Schematic representation of the visual stimulus. (**b**) Time series of the average optical signal triggered by the onset of a saccade to the new visual stimulus averaged over 17 trials. The first few frames show the fully developed evoked response to the small (0.5°) single isoluminant drifting grating, which was turned on 500–800 ms earlier. After a saccadic eye movement to the stimulus ($t = 0$), the activity on the cortex is shifted to a more foveal location (lateral direction). The *thin black line* in the first frame denotes approximately the V1/V2 border. Scale bar 3 mm. (**c**) *Black-and-white* image of the blood vessel pattern from the imaged cortical areas, the V1/V2 border was detected by obtaining an ocular-dominance map with intrinsic imaging. *A* anterior, *P* posterior, *L* lateral, *M* medial

Having established this point, Slovin et al. (2002) studied the spatiotemporal representation of a point stimulus in primary and secondary visual cortex (Fig. 11.4c), quantifying the latency and time course of the cortical response (Fig. 11.4d). The individual traces indicate that evoked activity can be detected without signal averaging in the awake monkey, offering VSDI as an important tool to explore real-time cortical dynamics. This is beneficial because averaging may mask important variability among the instantaneous responses. In an example behavioural experiment, the cortical dynamics as the monkey makes a saccade are depicted in Fig. 11.5. Evidently despite the high speed of the eye movements, an evoked response was detected also along the cortical representation of the fast eye movement.

5.2 *Visualization of Neuronal Assemblies*

With the improvements in the signal-to-noise ratio of *in vivo* VSDI offered by the blue dyes, evoked activity can be detected without signal averaging in the awake monkey (Fig. 11.4d; Slovin et al. 2002), as previously found in anaesthetized cats (Arieli et al. 1995, 1996; Tsodyks et al. 1999; Kenet et al. 2003). This finding indicates that cortical spatiotemporal dynamics *per se* can now be explored in awake behaving monkeys. Hebb (1949) suggested that neocortical neurons operate in assemblies, defined as networks of neurons, local or widespread, that communicate coherently to carry out the computations that are required for various behavioral tasks. Neurons located within the same pixel of a VSDI movie might belong to the same or different neuronal assemblies. It is therefore of great interest to monitor the spiking activity of a single neuron in the context of larger network dynamics, to reveal the co-active nodes in the assembly during a given computation. This can be achieved by combining single unit (or multiple-unit recording) with real-time optical VSD imaging of the dynamics of coherent neuronal assemblies. First, a single unit with the desired tuning properties is selected. Second, VSDI and unit recordings are conducted for a long time without any signal averaging. During off-line analysis, which involves performing spike-triggered averaging on the optical data, the firing of a single neuron serves as a time reference to selectively visualize only the population activity that was synchronized with it. With a sufficient number of spikes, any neuronal activity not time-locked to these spikes will be averaged out and the net, clean spatiotemporal pattern of coherent activity will be thus obtained.

Such spike-triggered averaging of VSD signals can be performed on both evoked and spontaneous activity. Even in primary sensory areas, there is a large amount of spontaneous on-going electrical activity, primarily subthreshold activity, in the absence of any sensory input. Some of this activity is coherent over large cortical areas. Is this spontaneous activity just network noise? How large is it relative to evoked activity? Does it affect evoked activity? And does it have an important functional role? VSDI is ideally suited to exploring such questions at the level of neuronal assemblies. To compare the amplitude of coherent, ongoing activity with that of evoked activity (both representing large subthreshold activity, to which VSDI is sensitive), spike-triggered averaging was used during both spontaneous- and evoked-activity imaging sessions with anaesthetized cats (Grinvald et al. 1991; Arieli et al. 1995, 1996; Tsodyks et al. 1999; Kenet et al. 2003). Surprisingly, the amplitude of coherent, spontaneous, ongoing activity in neuronal assemblies was nearly as large as that of evoked activity. Large, coherent, ongoing activity was also found in the somatosensory cortex (Petersen et al. 2003b; Ferezou et al. 2006, 2007). The large fluctuations of ongoing activity affect how far cortical neurons are from their firing threshold. Therefore, the amplitude of ongoing activity indicates that it might have an important role in shaping spatiotemporal patterns evoked by sensory input. The idea that network activity can converge to cortical states is a central concept in theoretical brain research. What are the dynamics of intrinsic cortical states? It is now possible to visualize certain aspects of cortical networks and their states in action, at a high spatiotemporal resolution. Such studies can provide insights into the dynamic interplay between activated internal cortical representations and incoming sensory input. The previous studies on anesthetized cats are reviewed in the previous chapter. In those VSDI studies we reported that spontaneous ongoing cortical activity in areas 17 and 18 represented dynamic spatial patterns, and about 20 % of these patterns resembled the functional architecture of orientation domains. These patterns covered

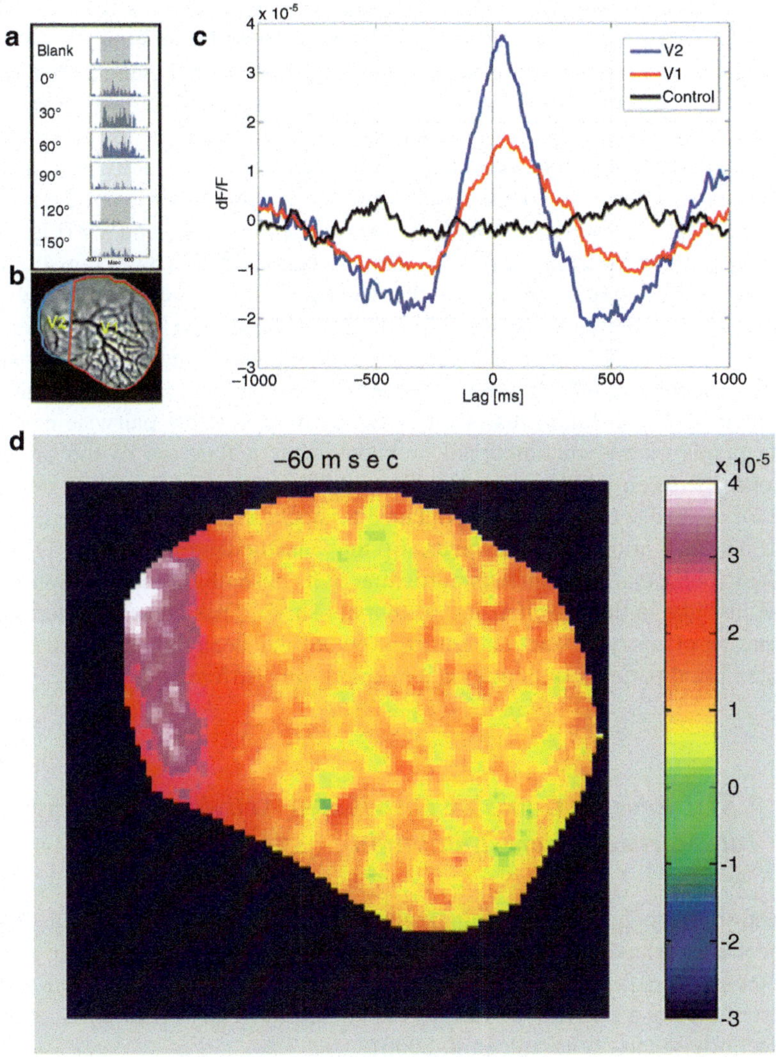

Fig. 11.6 Visualization of coherent neuronal assemblies across visual areas. (**a**) The tuning properties of a single cell in V1 of primary visual cortex of the awake monkey, which was used for the spike triggered averaging. (**b**) Cartoon of the imaged area showing the V1/V2 border that was identified by mapping the ocular dominance columns in V1. (**c**) The time course of spike triggered average (STA) of the optical signal from V2 (*blue curve*) as well as the optical signal from V1 (*red curve*) on the time of the firing of the V1 reference neuron. Evidently, whenever this cell fired, there was activity over a large area in V1 as well as V2. In fact, in this case, whenever this cell in V1 fired there was larger coherent activity in V2 than in V1. (**d**) The average spatial pattern 60 ms before the V1 action potential occurred is shown. Courtesy of David Omer and Amiram Grinvald; unpublished results

large cortical areas—up to the entire imaged areas of up to 6×6 mm (Grinvald et al. 1989; Arieli et al. 1995, 1996; Tsodyks et al. 1999; Kenet et al. 2003).

To find whether these results are relevant to the awake behaving primate, Omer and Grinvald (2008) performed VSDI of ongoing cortical activity in the visual cortices of awake monkeys simultaneously with measurements of single unit activity and the local-field potential. They found coherent activity also in the awake monkey (Fig. 11.6). This figure shows the tuning properties of the reference cell (Fig. 11.6a), the border between the two imaged areas V1 and V2 (Fig. 11.6b), time course of the coherent activity in area V1 (red) and V2 (blue; Fig. 11.6c) and the instantaneous spatial pattern of the assembly activity just before the action potential of the reference neuron (Fig. 11.6d). Overall, the dynamics were very different from that found in anesthetized cats and in a single epoch it was not possible to recognize any pattern that looked like the known functional architecture. Nevertheless, it was found that such patterns do appear: the pairwise correlation among pixels of known functional architecture as a function of the orientation difference between these pixels or their ocularity index had the same stereotypic shape for evoked activity and for spontaneous epochs. Similar results were obtained in anesthetized monkeys. However, in the anesthetized monkeys, Omer and Grinvald found that spontaneous cortical activity resembled the functional architecture just like in the anesthetized cat (unpublished). These results underscore the importance of carrying out experiments on awake behaving animals since the spatiotemporal patterns of activity are so drastically different.

5.3 VSDI Reveals Complex Distributed Activity Evoked by Intracortical Microstimulation

Microstimulation in various brain regions has been used extensively to affect neuronal and behavioural responses. However, many questions remain about the spatiotemporal patterns of activity that are evoked by microstimulation and their characteristics as a function of various parameters such as the current strength and the electrode shape (Tehovnik et al. 2006).

Intracortical stimulation has been used as a tool to discover the pattern of muscle and/or movement representation in motor areas of the cortex (Slovin et al. 2003). The direct and synaptic spread of activation induced by stimulation has been the subject of considerable uncertainty and controversy. To help resolve some of these issues we used voltage sensitive-dye (VSD) imaging to examine the patterns of activation evoked by different stimulation parameters in the primary motor cortex (M1) of an awake monkey. Single cathodal pulses in M1 (15–30 μA, 0.2 ms duration) evoked short latency VSD responses that continued for 40–50 ms after the onset of stimulation. These responses spread in an asymmetric manner from the stimulation site in M1; the width of the activated area was about 1.5–3 mm. Single pulses larger than 15 μA also evoked activity at spatially separate sites within M1. Short trains of pulses in M1 (2–20 pulses, 330–500 Hz, 30–100 μA) produced activation that spread from the electrode site to adjacent premotor areas and to the contralateral hemisphere. A distinct hyperpolarization followed the activation evoked by a high current (>60 μA), short stimulus train (2–20 pulses). Finally, long stimulus trains (70 μA, 200 pulses, 330 Hz) evoked activation that spread over large portions of M1 and the adjacent premotor areas. These responses outlasted the stimulus train by several hundred milliseconds. Our results emphasize the importance of considering the spread of activation when evaluating the results of M1 stimulation.

VSDI studies on the effect of microstimulation in the frontal and motor cortices of awake monkeys have also recently shed light on the evoked spatiotemporal dynamics. The frontal eye field and neighbouring area '8Ar' of the primate cortex are involved in the programming and execution of fast eye movements. Electrical microstimulation in these regions elicits short-latency contralateral saccades. A combination of VSDI and microstimulation has been used to determine how the spatiotemporal dynamics of microstimulation-evoked activity are converted into saccade plans. Microstimulation was shown to elicit neural activity with complex spatiotemporal dynamics, both inhibitory and excitatory; these dynamics depend on the stimulated area, and have important behavioural correlates. The observed spread was large and depended on the amplitude of the microstimulation (Seidemann et al. 2002). These results emphasize the importance of further characterization of microstimulation-evoked activity for the interpretation of its behavioural effects and what it has revealed about the neural basis of cognition (Newsome et al. 1989; Cohen and Newsome 2004). VSDI seems to be one of the most appropriate tools to nail down these issues, which also have important implications for neural prosthesis and brain machine interfaces.

5.4 Imaging the Correlates of Go/No-Go Delayed Response Task in the Motor Cortex

Electrical recording studies during Go/No-Go tasks (Strick et al. 2003) have found that M1 neurons display changes in activity, which begin primarily after a Move cue, whereas PMd neurons display changes in activity, which can begin during Instructed Delay periods. We used both intrinsic imaging and VSD imaging in a trained monkey to examine the spatiotemporal patterns of subthreshold population activity, which underlies these neuronal responses.

One monkey was trained to perform a Go/No-Go task. The monkey began a trial by placing his hand over a small photodiode placed in front of him. This caused a small white square to appear on a computer screen. After a variable precue period (3,000–4,000 ms), the white square was changed to red (Go mode) or green (No-Go mode). Then, after a variable Instructed Delay period (300–1,500 ms), a luminance change of the square told the monkey to follow the prior instruction: uncover (Go) or continue to cover the photodiode (No-Go).

Intrinsic imaging showed patches of activation in PMd and in M1 for the Go mode. In the No-Go mode, small amplitude responses were present in the PMd, but not in M1. Changes in VSD signals were present in both PMd and M1 during the Go mode. However, changes in VSD signals were much smaller in M1 during the No-Go mode. In the PMd, the VSD signal changes began shortly after the onset of the Go or No-Go Instructions (i.e., at the start of the Instructed Delay periods). In M1, the signal changes appeared to be locked to movement onset as confirmed by single unit recordings. These observations suggest that the preparation to perform or withhold a motor response is associated mainly with marked subthreshold activity in the PMd.

6 Imaging the Mouse Sensorimotor Cortex During Behaviour

Mice are nocturnal animals living in tunnels. For much of their lives they therefore receive impoverished visual input, and instead they rely heavily upon olfactory and somatosensory information. The prominent mystacial whiskers surrounding the snout form a sensitive array of tactile detectors, allowing texture discrimination (von Heimendahl et al. 2007) and object location (Knutsen et al. 2006; Mehta et al. 2007). Sensory information is signalled to the primary somatosensory cortex via a first synapse in the brainstem and a second synapse in the thalamus. The thalamus in turn directly signals to the neocortex. Whisker-responsive thalamic neurons mainly project to the primary somatosensory cortex. Each whisker is represented in the primary somatosensory cortex by an anatomically identifiable structure, termed a barrel. These barrels are somatotopically arranged in a stereotypical manner matching the layout of the whiskers on the snout (Woolsey and van der Loos 1970). As one might expect from such an anatomical map, the functional cortical sensory processing evoked by a single whisker deflection indeed begins within the related barrel column, i.e. if the C2 whisker is deflected, then the neurons in the C2 barrel column are the first to respond (reviewed by Petersen 2007). Many of the pyramidal neurons located in a given barrel column project dense axonal arborisations into neighbouring cortical columns (Petersen et al. 2003a) and indeed also to other cortical regions such as secondary somatosensory cortex and motor cortex (Chakrabarti and Alloway 2006; Ferezou et al. 2007). Consistent with such anatomical data, sensory signals can propagate extensively across the sensorimotor cortex, which can be directly visualised by voltage-sensitive dye imaging (Orbach et al. 1985; Kleinfeld and Delaney 1996; Derdikman et al. 2003; Petersen et al. 2003a; Ferezou et al. 2006, 2007; Berger et al. 2007). Here, we will focus on how different behavioural states in awake mice affect both sensory processing and spontaneous cortical activity.

6.1 State-Dependent Processing in Mouse Sensorimotor Cortex

In order to deliver a well-controlled whisker deflection in awake mice, we attach a small metal particle to the C2 whisker and evoke movement of this whisker by driving a brief (1–2 ms) current pulse through an electromagnetic coil to generate a magnetic field which exerts a force on the metal particle attached to the whisker. Such a stimulus evokes a small, brief and reproducible whisker deflection (Ferezou et al. 2006). The evoked sensory response was imaged in head-fixed mice, whose sensorimotor cortex had been stained with voltage-sensitive dye RH1691 through a large craniotomy (Ferezou et al. 2007). When the mouse was not spontaneously moving, defining a period of quiet wakefulness, deflection of the C2 whisker evoked a sensory response, which at first was localised to the C2 barrel column (Fig. 11.7a, upper). In the following milliseconds the depolarisation spread across the barrel field, presumably through local excitatory synaptic connections between neighbouring cortical columns. In addition, a second localised spot of activity was evoked in the whisker motor cortex, after a delay of ~8 ms following the first signals in S1. This initially localised secondary sensory response in motor cortex also spread over the following milliseconds and within 50 ms of the stimulus a surprisingly large cortical area had depolarised in response to a single whisker deflection. Interestingly, the mouse often began to actively whisk following such a sensorimotor response. The propagating sensory response may serve as a wake-up call, bringing the mouse from an initially quiet state into one optimised for actively processing subsequent sensory stimuli. Rapid signalling of sensory information to motor cortex may be important in general for optimising motor control. For example when we touch an object with our hands, we change our movements appropriately as informed by sensory feedback. Rodents also often change their pattern of whisker movements following whisker-object contact, perhaps as a way to improve the quality of sensory information.

During exploratory behaviour, mice move their whiskers back and forth at high frequencies (typically around 10 Hz) scanning the environment surrounding their snout. Mice also spontaneously whisk during head-restraint and by chance some of our magnetic whisker deflections occurred during such active whisking bouts. Interestingly, the sensory-evoked responses were dramatically different during active states. The sensory responses were again initially localised to the C2 barrel column, but instead of spreading across large cortical areas, the response remained localised and had only small amplitude and a brief duration (Fig. 11.7a, lower). The ongoing behaviour of the mouse therefore makes a dramatic difference to sensory processing. Here in this example, it is again clear that neocortical spatiotemporal patterns of activity must be directly explored during behaviour, since brain states and behavioural states play such profound roles.

In order to investigate if a similar state-dependence of sensory processing occurs in freely moving mice, we used fiber optic image bundles to visualise the membrane potential dynamics of the primary somatosensory barrel cortex (Ferezou

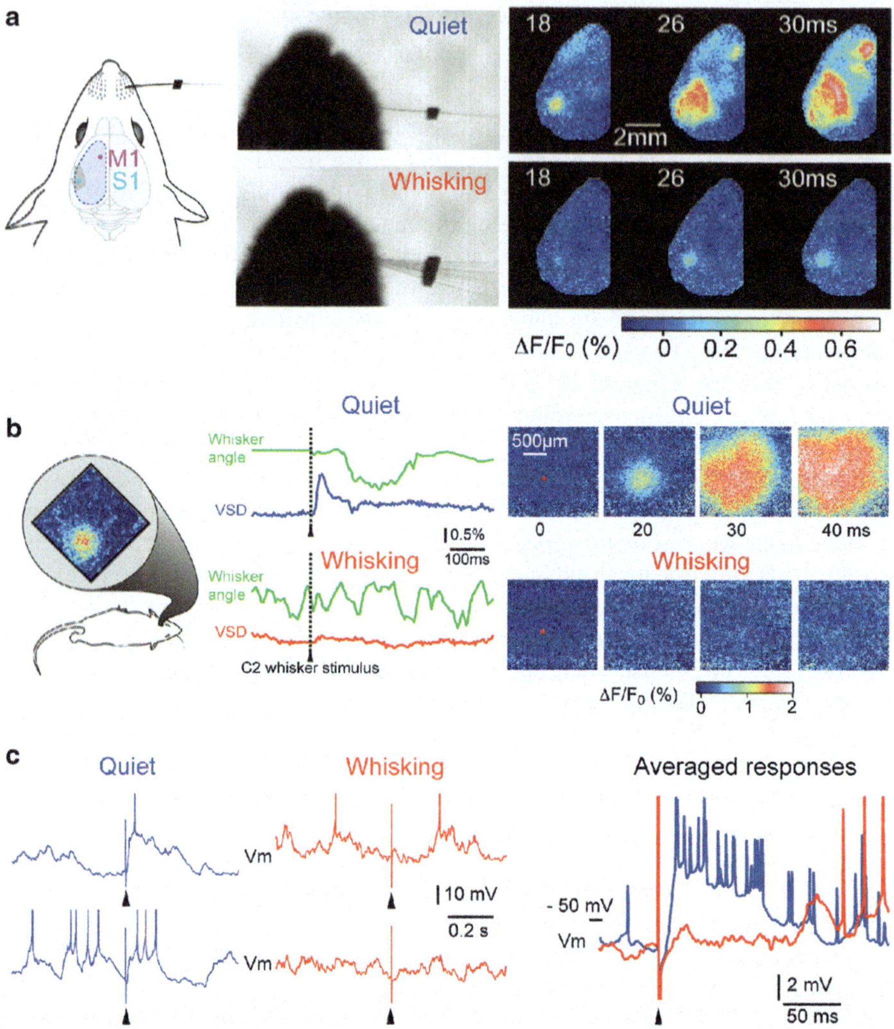

Fig. 11.7 State-dependent processing of sensory information in the sensorimotor cortex of the awake mouse. (**a**) A large fraction of the sensorimotor cortex was stained with voltage-sensitive dye including primary somatosensory barrel cortex (S1) and the whisker motor cortex (M1). A small piece of metal was attached to the whisker allowing brief magnetic pulses to evoke a whisker deflection and subsequent sensory processing. The whisker-related behaviour was filmed simultaneously with voltage-sensitive dye imaging allowing direct correlation of cortical activity with behaviour. The image of the mouse nose and whisker is a superposition of many frames. The *upper image* extends over a quiet prestimulus period. The *lower image* extends over a period of spontaneous whisking in the prestimulus period. Deflection of the whisker during these two different behavioural periods evokes dramatically different sensory responses. During quiet wakefulness, the response is initially localised but rapidly spreads across S1 and a secondary sensory response is observed in motor cortex. The same stimulus applied during a period of spontaneous whisking, evokes a small amplitude sensory response that remains localised to S1. (**b**) Voltage-sensitive dye imaging of barrel cortex in freely moving mice through fiber optics

et al. 2006). The fiber array we used only extended across somatosensory barrel cortex and did not include motor cortex. Similar to the observations made in head-restrained mice, we found that a brief whisker stimulus delivered during quiet wakefulness evoked a large, spreading sensory response (Fig. 11.7b, upper panel) whereas the stimulus delivered during active whisking was strongly suppressed (Fig. 11.7b, lower panel). The propagating sensory responses evoked by a single whisker stimulation observed during anaesthesia (Orbach et al. 1985; Kleinfeld and Delaney 1996; Derdikman et al. 2003; Petersen et al. 2003a) are therefore not artefacts of anaesthesia, *per se*, but appear prominently during some behavioural states of wakefulness (Ferezou et al. 2006, 2007). That a large part of the cortex can be informed about a single whisker deflection might allow integration of sensory information from different whiskers and even different modalities, which might be crucial for associative learning and sensory perception.

In separate experiments, we examined whether we could find single cell correlates of the voltage-sensitive dye imaging data. In head-fixed mice, we obtained whole-cell recordings from excitatory neurons and again stimulated the C2 whisker by magnetic pulses (Crochet and Petersen 2006). In good agreement with the voltage-sensitive dye imaging experiments, we found that brief, small-amplitude sensory responses were evoked during active whisking periods whereas larger amplitude sensory responses were evoked during quiet wakefulness (Fig. 11.7c). Similar to observations from anesthetised animals (Petersen et al. 2003a, b; Ferezou et al. 2006; Berger et al. 2007), in awake mice there is therefore also good agreement between membrane potential measurements made at the single cell level through whole-cell recordings and membrane potential measurements made at the ensemble level with voltage-sensitive dye.

6.2 *State-Dependence of Spontaneous Dynamics in Mouse Sensorimotor Cortex*

Having identified a profound role of the prestimulus behaviour in governing sensory processing, we were obviously interested to examine what changes in brain states accompany the changes in behavioural states. During quiet wakefulness in head-fixed mice, using voltage-sensitive dyes, we imaged propagating waves of depolarisation

Fig. 11.7 (continued) provides further evidence for state-dependent sensory processing of tactile information. During a quiet period the deflection of a whisker evoked a large-amplitude propagating sensory response. During active whisking, the same stimulus barely evoked a measurable response. (**c**) State-dependent processing is also a prominent feature of whole-cell recordings of awake head-restrained mice. Whisker deflection during quiet wakefulness evoked a large-amplitude depolarisation, whereas during active whisking periods the same stimulus evoked only a very small and brief sensory response. There is therefore good agreement between voltage-sensitive dye imaging data and whole-cell recording data

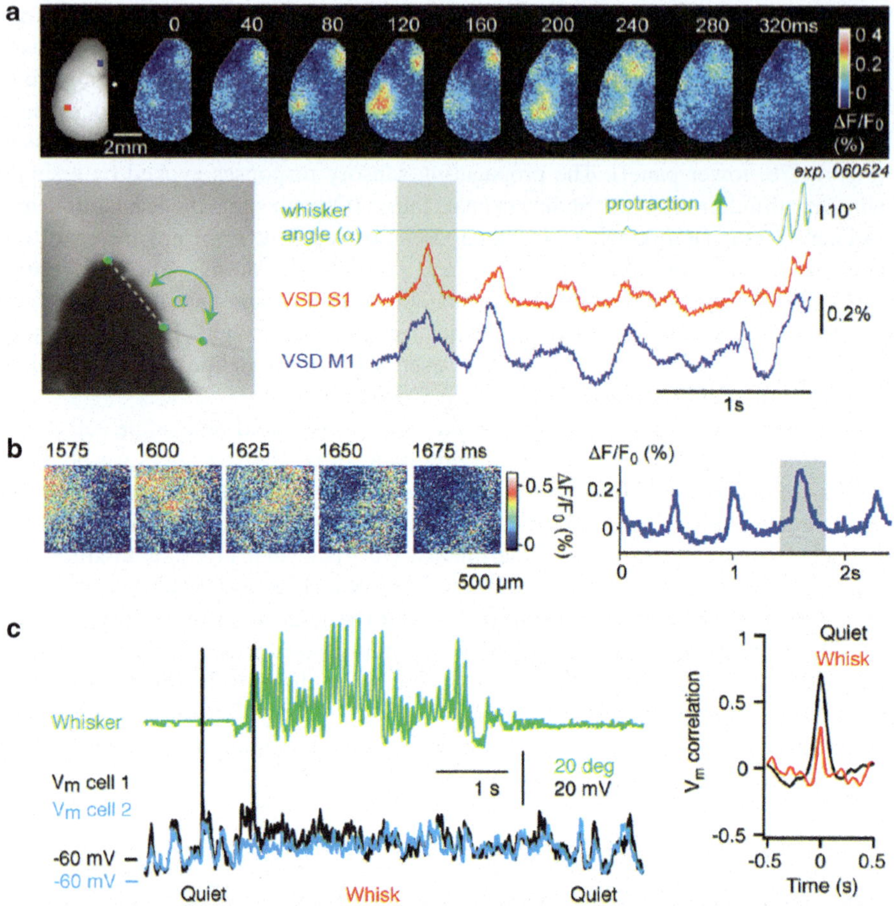

Fig. 11.8 State-dependent spontaneous activity in the sensorimotor cortex of the awake mouse. (**a**) Spontaneous activity was imaged across the sensorimotor cortex in awake head-restrained mice. During quiet wakefulness (when the whisker was not moving), propagating waves of depolarisation can be observed in complex spatiotemporal patterns. Quantified across regions of interest located in S1 (*red*) and M1 (*blue*), the membrane potential changes appear as slow oscillations (*grey* shading indicates the period shown in the above sequence of images). (**b**) In a freely moving mouse, spontaneous waves of activity can also be imaged during quiet wakefulness through fiber optics. The signal changes were quantified across a region of interest in the C2 barrel column and period shown in the image sequence (*left*) is shaded in *grey*. (**c**) Dual whole-cell recordings from nearby neurons in the barrel cortex indicate that correlated, slow, large-amplitude membrane potential changes dominate during quiet wakefulness. During active whisking periods, the membrane potential fluctuations show lower variance and are less correlated. Both of these factors will decrease the amplitude of spontaneous voltage-sensitive dye signals during whisking. During quiet wakefulness, it is likely that the strong correlations of membrane potential changes in nearby neurons forms the basis of the travelling waves of spontaneous activity imaged with voltage-sensitive dye

spreading across the sensorimotor cortex in complex patterns (Fig. 11.8a). Quantified over time at a given location on the cortical surface, these appear as slow oscillations in the local ensemble membrane potential. Although a detailed analysis of these spontaneous activity patterns has not yet been carried out, in some cases we found correlated depolarisations in primary somatosensory cortex and motor cortex (Fig. 11.8a), which was not very different from the sensory-evoked response. Spontaneous activity patterns in sensorimotor cortex may, in part, reflect major signalling pathways between different cortical areas, as already discussed for the occurrence of spontaneous orientation maps in the visual cortex (Kenet et al. 2003). Analysis of these spontaneous patterns of activity might therefore give insight into the functional connectivity between cortical areas, something that has also already been studied for spontaneous dynamics recorded through functional magnetic resonance imaging (Vincent et al. 2007).

In freely moving mice, we also observed waves of activity propagating across the primary somatosensory barrel cortex during quiet wakefulness (Fig. 11.8b). The waves of activity during rest are therefore not imposed by head-restraint, but appear to be a normal physiological pattern of cortical activity during rest. However, during active whisking we did not observe such large-amplitude propagating waves of spontaneous activity.

In order to understand why spontaneous activity was not resolved during active periods, we obtained dual whole-cell recordings from pyramidal neurons in the barrel cortex of awake head-fixed mice (Poulet and Petersen 2008). During quiet wakefulness, we observed slow, highly-correlated, large-amplitude subthreshold membrane potential changes, which qualitatively resemble those observed under anaesthesia (Lampl et al. 1999; Petersen et al. 2003b). Action potential activity occurred independently in each neuron across our dual recording data set, despite remarkably high correlations in subthreshold membrane potential changes. These highly correlated subthreshold membrane potential changes are likely to provide the basis for the large-amplitude propagating waves of depolarisation imaged with voltage-sensitive dye. Each depolarised oscillation recorded in an individual neuron is therefore likely to reflect a propagating wave of activity travelling across the location of the recorded neuron.

Interestingly, the spontaneous membrane potential fluctuations differ dramatically comparing periods of quiet wakefulness to periods of active whisking. The slow large-amplitude membrane potentials are suppressed during active periods (Crochet and Petersen 2006; Poulet and Petersen 2008). The remaining higher frequency membrane potential changes have lower amplitude giving an overall lower variance to the membrane potential fluctuations. This reduction in the amplitude of spontaneous membrane potential fluctuations therefore provides one reason that spontaneous activity is less obvious during active periods in the voltage-sensitive dye images. A second and equally important factor is that the remaining membrane potential fluctuations are more independent in nearby neurons during active periods. The correlation coefficients of membrane potentials recorded in nearby neurons indicates that synchrony is approximately halved during active periods (Poulet and Petersen 2008). Since our method of voltage-sensitive dye imaging measures ensemble membrane potential, we need correlated changes in membrane potentials of nearby neurons in order to measure a signal. Both the decorrelation and the reduction in amplitude of the membrane potential fluctuations during active periods will reduce voltage-sensitive dye signals relating to spontaneous activity. Nonetheless, even during active behaviours and active brain states, it is likely that useful smaller amplitude signals will be uncovered in the future by more detailed analysis of spontaneous activity in mice imaged with voltage-sensitive dye.

7 Conclusions and Future Perspectives

Voltage-sensitive dye imaging of awake animals is beginning to provide a unique insight into the spatiotemporal dynamics of cortical processing during behaviour. VSDI provides information on cortical dynamics with millisecond temporal resolution and subcolumnar spatial resolution. Correlations of spatiotemporal dynamics of VSDI with behaviour indicate sequential activation of different cortical areas. Combining VSDI with extracellular recordings has begun to shed light on the larger neuronal network dynamics that through concerted interactions drive spiking in individual neurons. Combining VSDI with intracortical microstimulation reveals the spatiotemporal extent of microstimulation, which is important for the interpretation of motor mapping experiments. The ease, with which VSDI can be combined with other techniques, makes it broadly applicable to a wide range of neurophysiological experiments. There are of course also many limitations to what has so far been accomplished with VSDI and there is a great need for further technical development.

1. Perhaps most importantly, *in vivo* VSDI is currently limited to the exposed areas of the cortical surface. In the future, use of fiber optic microendoscopes may provide a way to image from deep structures in the brain (Flusberg et al. 2005). In this context it is interesting to note that the results of VSDI and functional MRI in behaving monkeys (Logothetis et al. 2001) may complement each other in a very helpful manner. VSDI can provide temporal dynamics and higher spatial resolution to the static images obtained by fMRI. On the other hand, fMRI can contribute details about activity in the entire brain including subcortical areas.
2. The application of VSD to the cortical surface ensures even staining of the underlying neuronal networks, but the approach inherently lacks cell-type specificity. High resolution *in vivo* imaging using two photon microscopy (Denk et al. 1990) of sparsely labeled individual neurons would allow investigation of membrane potential dynamics in single cells, without however providing information on the population activity. Genetically encoded fluorescent functional

reporter proteins (Miyawaki et al. 1997) would provide a possible alternative to synthesised dyes allowing the targeting to specific cell types. There has been progress in developing voltage-sensitive fluorescent proteins (Siegel and Isacoff 1997; Sakai et al. 2001; Ataka and Pieribone 2002; Dimitrov et al. 2007; Lundby et al. 2008; Akemann et al., 2010; Kralj et al., 2011; Jin et al., 2012; St-Pierre et al., 2014) giving hope for future prospects of cell-type specific VSDI. Transgenic animals expressing voltage-sensitive fluorescent proteins now begin to allow simple repetitive imaging of cortical activity in preparations with sealed cranial windows (Trachtenberg et al. 2002; Madisen et al. 2015). This is particularly important since, so far, repetitive staining and imaging of the same cortical area with VSDs has not yet been accomplished in mice (although as described above this works well in monkeys).

3. VSDI appears to be dominated by subthreshold membrane potential dynamics. On the other hand calcium signals from neuronal networks are dominated by suprathreshold action potential activity (Stosiek et al. 2003). It would therefore be of great interest to combine VSDI with calcium imaging in order to simultaneously image sub- and supra-threshold activity (Berger et al. 2007). In future experiments, such combined imaging techniques should be applied to awake behaving animals.
4. So far, only very simple behaviours have been imaged with VSD. It will be of great interest to image more complex behaviours in order to study the spatio-temporal dynamics of learned behaviours.

Acknowledgements AG's research was funded by the Weizmann Institute of Science, The Grodetsky Center, the Goldsmith, Glasberg, Heineman and Korber foundations, BMBF, ISF grants, Ms. Enoch and the EU daisy grants. CCHP's research was funded by grants from the Swiss National Science Foundation and SystemsX.ch.

References

Akemann W, Mutoh H, Perron A, Rossier J, Knöpfel T (2010) Imaging brain electric signals with genetically targeted voltage-sensitive fluorescent proteins. Nat Methods 7:643–649

Arieli A, Grinvald A (2002) Combined optical imaging and targeted electrophysiological manipulations in anesthetized and behaving animals. J Neurosci Methods 116:15–28

Arieli A, Shoham D, Hildesheim R, Grinvald A (1995) Coherent spatio-temporal pattern of on-going activity revealed by real time optical imaging coupled with single unit recording in the cat visual cortex. J Neurophysiol 73:2072–2093

Arieli A, Sterkin A, Grinvald A, Aertsen A (1996) Dynamics of ongoing activity: explanation of the large variability in evoked cortical responses. Science 273:1868–1871

Arieli A, Grinvald A, Slovin H (2002) Dural substitute for longterm imaging of cortical activity in behaving monkeys and its clinical implications. J Neurosci Methods 114:119–133

Ataka K, Pieribone VA (2002) A genetically targetable fluorescent probe of channel gating with rapid kinetics. Biophys J 82:509–516

Berger T, Borgdorff AJ et al (2007) Combined voltage and calcium epifluorescence imaging in vitro and in vivo reveals subthreshold and suprathreshold dynamics of mouse barrel cortex. J Neurophysiol 97:3751–3762

Chakrabarti S, Alloway KD (2006) Differential origin of projections from SI barrel cortex to the whisker representations in SII and MI. J Comp Neurol 498:624–636
Cohen MR, Newsome WT (2004) What electrical microstimulation has revealed about the neural basis of cognition. Curr Opin Neurobiol 14:169–177
Crochet S, Petersen CCH (2006) Correlating whisker behavior with membrane potential in barrel cortex of awake mice. Nat Neurosci 9:608–610
Denk W, Strickler JH, Webb WW (1990) Two-photon laser scanning fluorescence microscopy. Science 248:73–76
Derdikman D, Hildesheim R, Ahissar E, Arieli A, Grinvald A (2003) Imaging spatiotemporal dynamics of surround inhibition in the barrels somatosensory cortex. J Neurosci 23:3100–3105
Dimitrov D, He Y et al (2007) Engineering and characterization of an enhanced fluorescent protein voltage sensor. PLoS One 2, e440
Fekete T, Pitowsky I, Grinvald A, Omer DB (2009) The representational capacity of cortical tissue. Comput Neurosci 178:31–39
Ferezou I, Bolea S, Petersen CCH (2006) Visualizing the cortical representation of whisker touch: voltage-sensitive dye imaging in freely moving mice. Neuron 50:617–629
Ferezou I, Haiss F et al (2007) Spatiotemporal dynamics of cortical sensorimotor integration in behaving mice. Neuron 56:907–923
Flusberg BA, Cocker ED et al (2005) Fiber-optic fluorescence imaging. Nat Methods 2:941–950
Flusberg BA, Nimmerjahn A et al (2008) High-speed, miniaturized fluorescence microscopy in freely moving mice. Nat Methods 5:935–938
Grinvald A, Hildesheim R (2004) VSDI: a new era in functional imaging of cortical dynamics. Nat Rev Neurosci 5:874–885
Grinvald A, Anglister L, Freeman JA, Hildesheim R, Manker A (1984) Real-time optical imaging of naturally evoked electrical activity in intact frog brain. Nature 308:848–850
Grinvald A, Lieke E, Frostig RD, Gilbert CD, Wiesel TN (1986) Functional architecture of cortex revealed by optical imaging of intrinsic signals. Nature 324:361–364
Grinvald A, Frostig RD et al (1989) Optical imaging of activity in the visual cortex. In: Lam D, Glibert CD (eds). MIT Press, Cambridge, MA
Grinvald A, Bonhoeffer T et al (1991) Optical imaging of architecture and function in the living brain. In: Squire L (ed) Memory, organization and locus of change. Oxford University Press, Oxford
Grinvald A, Shoham D et al (1999) In-vivo optical imaging of cortical architecture and dynamics. In: Windhorst U, Johansson H (eds) Modern techniques in neuroscience research. Springer, Berlin
Hebb D (1949) The organization of behavior. Wiley, New York, NY
Hubel DH, Wiesel TN (1962) Receptive fields, binocular interactions and functional architecture in the cat's visual cortex. J Physiol 160:106–154
Jin L, Han Z, Platisa J, Wooltorton JR, Cohen LB, Pieribone VA (2012) Single action potentials and subthreshold electrical events imaged in neurons with a fluorescent protein voltage probe. Neuron 75:779–785
Kenet T, Bibitchkov D, Tsodyks M, Grinvald A, Arieli A (2003) Spontaneously occurring cortical representations of visual attributes. Nature 425:954–956
Kleinfeld D, Delaney KR (1996) Distributed representation of vibrissa movement in the upper layers of somatosensory cortex revealed with voltage-sensitive dyes. J Comp Neurol 375:89–108
Kralj JM, Douglass AD, Hochbaum DR, Maclaurin D, Cohen AE (2011) Optical recording of action potentials in mammalian neurons using a microbial rhodopsin. Nat Methods 9:90–95
Knutsen PM, Pietr M, Ahissar E (2006) Haptic object localization in the vibrissal system: behavior and performance. J Neurosci 26:8451–8464
Lampl I, Reichova I, Ferster D (1999) Synchronous membrane potential fluctuations in neurons of the cat visual cortex. Neuron 22:361–374

Logothetis NK, Pauls J, Augath M, Trinath T, Oeltermann A (2001) Neurophysiological investigation of the basis of the fMRI signal. Nature 412:150–157

Lundby A, Mutoh H, Dimitrov D, Akemann W, Knöpfel T (2008) Engineering of a genetically encodable fluorescent voltage sensor exploiting fast Ci-VSP voltage-sensing movements. PLoS One 3, e2514

Madisen L, Garner AR, Shimaoka D, Chuong AS, Klapoetke NC, Li L, van der Bourg A, Niino Y, Egolf L, Monetti C, Gu H, Mills M, Cheng A, Tasic B, Nguyen TN, Sunkin SM, Benucci A, Nagy A, Miyawaki A, Helmchen F, Empson RM, Knöpfel T, Boyden ES, Reid RC, Carandini M, Zeng H (2015) Transgenic mice for intersectional targeting of neural sensors and effectors with high specificity and performance. Neuron 85:942–958

Mehta SB, Whitmer D, Figueroa R, Williams BA, Kleinfeld D (2007) Active spatial perception in the vibrissa scanning sensorimotor system. PLoS Biol 5, e15

Miyawaki A, Llopis J et al (1997) Fluorescent indicators for Ca^{2+} based on green fluorescent proteins and calmodulin. Nature 388:834–835

Mountcastle VB (1957) Modality and topographic properties of single neurons of cat's somatic sensory cortex. J Neurophysiol 20:408–434

Newsome WT, Britten KH, Movshon JA (1989) Neuronal correlates of a perceptual decision. Nature 341:52–54

Omer DB, Grinvald A (2008) The dynamics of evoked and ongoing activity in the behaving monkey. Rev Neurosci 19:S50

Orbach HS, Cohen LB, Grinvald A (1985) Optical mapping of electrical activity in rat somatosensory and visual cortex. J Neurosci 5:1886–1895

Petersen CCH (2007) The functional organization of the barrel cortex. Neuron 56:339–355

Petersen CCH, Grinvald A, Sakmann B (2003a) Spatiotemporal dynamics of sensory responses in layer 2/3 of rat barrel cortex measured in vivo by voltage-sensitive dye imaging combined with whole-cell voltage recordings and neuron reconstructions. J Neurosci 23:1298–1309

Petersen CCH, Hahn TTG, Mehta M, Grinvald A, Sakmann B (2003b) Interaction of sensory responses with spontaneous depolarization in layer 2/3 barrel cortex. Proc Natl Acad Sci U S A 100:13638–13643

Poulet JFA, Petersen CCH (2008) Internal brain state regulates membrane potential synchrony in barrel cortex of behaving mice. Nature 454:881–885

Ratzlaff EH, Grinvald A (1991) A tandem-lens epifluorescence macroscope: hundred-fold brightness advantage for wide-field imaging. J Neurosci Methods 36:127–137

Sakai R, Repunte-Canonigo V, Raj CD, Knöpfel T (2001) Design and characterization of a DNA-encoded, voltage-sensitive fluorescent protein. Eur J Neurosci 13:2314–2318

Seidemann E, Arieli A, Grinvald A, Slovin H (2002) Dynamics of depolarization and hyperpolarization in the frontal cortex and saccade goal. Science 295:862–865

Shoham D, Glaser DE et al (1999) Imaging cortical dynamics at high spatial and temporal resolution with novel blue voltage-sensitive dyes. Neuron 24:791–802

Siegel MS, Isacoff EY (1997) A genetically encoded optical probe of membrane voltage. Neuron 19:735–741

Slovin H, Arieli A, Hildesheim R, Grinvald A (2002) Long-term voltage-sensitive dye imaging reveals cortical dynamics in behaving monkeys. J Neurophysiol 88:3421–3438

Slovin H, Strick PL, Hildesheim R, Grinvald A (2003) Voltage sensitive dye imaging in the motor cortex I. Intra- and intercortical connectivity revealed by microstimulation in the awake monkey. Soc Neurosci Abstr 554(8)

Stosiek C, Garaschuk O, Holthoff K, Konnerth A (2003) In vivo two-photon calcium imaging of neuronal networks. Proc Natl Acad Sci U S A 100:7319–7324

Strick P, Grinvald A, Hildesheim R, Slovin H (2003) Voltage sensitive dye imaging in the motor cortex II. Cortical correlates of Go/No-Go delayed response task. Soc Neurosci Abstr 918(8)

St-Pierre F, Marshall JD, Yang Y, Gong Y, Schnitzer MJ, Lin MZ (2014) High-fidelity optical reporting of neuronal electrical activity with an ultrafast fluorescent voltage sensor. Nat Neurosci 17:884-889

Tehovnik EJ, Tolias AS, Sultan F, Slocum WM, Logothetis NK (2006) Direct and indirect activation of cortical neurons by electrical microstimulation. J Neurophysiol 96:512–521
Trachtenberg JT, Chen BE et al (2002) Long-term in vivo imaging of experience-dependent synaptic plasticity in adult cortex. Nature 420:788–794
Tsodyks M, Kenet T, Grinvald A, Arieli A (1999) The spontaneous activity of single cortical neurons depends on the underlying global functional architecture. Science 286:1943–1946
Vincent JL, Patel GH et al (2007) Intrinsic functional architecture in the anaesthetized monkey brain. Nature 447:83–86
von Heimendahl M, Itskov PM, Arabzadeh E, Diamond ME (2007) Neuronal activity in rat barrel cortex underlying texture discrimination. PLoS Biol 5, e305
Woolsey TA, Van der Loos H (1970) The structural organization of layer IV in the somatosensory region (SI) of mouse cerebral cortex. The description of a cortical field composed of discrete cytoarchitectonic units. Brain Res 17:205–242

Part IV
Monitoring Membrane Potential in the Heart

Chapter 12
Optical Imaging of Cardiac Action Potential

Arkady Pertsov, Richard D. Walton, and Olivier Bernus

Abstract This chapter reviews the major milestones and scientific achievements facilitated by optical imaging of the action potential in the heart over more than four decades since its introduction. We discuss the limitations of this technique, which sometimes are not fully recognized; the unresolved issues, such as motion artifacts, and the newest developments and future directions.

Keywords Optical imaging • Heart • Voltage-sensitive dyes • Arrhythmias • Motion artifact

1 History of Cardiac Membrane Potential Imaging

The optical mapping technique utilizing voltage-sensitive dyes migrated to cardiac electrophysiology from neurophysiology. The first successful optical recordings of cardiac action potentials using voltage-sensitive dyes have been reported in 1976 by (Salama and Morad) just 3 years after pioneering work in neurons by Davila et al. (1973) and Salzberg et al. (1973). These early optical recordings were obtained using the voltage-sensitive dye merocyanine 540. The magnitude of the optical signal in heart experiments, measured as voltage-dependent change in fluorescence ΔF over the background ΔF/F exceeded 1 %. This was an order of magnitude greater than could be achieved in neurons. However, at that time, the new technique did not provide immediate and significant advantages over conventional microelectrode recordings, which significantly delayed its adoption in cardiac electrophysiology research. It took more than a decade before voltage-sensitive dyes have made their way into mainstream cardiac electrophysiology to become a viable alternative to electrode array-based systems and a major tool for mapping cardiac excitation.

A. Pertsov (✉)
Department of Pharmacology, SUNY Upstate Medical University, Syracuse, NY, USA
e-mail: pertsova@upstate.edu

R.D. Walton • O. Bernus
L'Institut de rythmologie et modélisation cardiaque LIRYC, Centre de recherche cardio-thoracique de Bordeaux, Inserm U1045, Université de Bordeaux, Bordeaux, France

M. Canepari et al. (eds.), *Membrane Potential Imaging in the Nervous System and Heart*, Advances in Experimental Medicine and Biology 859,
DOI 10.1007/978-3-319-17641-3_12

The first optical mapping system for monitoring electrical excitation in the heart was built by Dillon and Morad in 1981. In their setup, the dye was excited via a laser beam which rapidly scanned preselected points on the heart surface. The fluorescence was continuously recorded with a single photo-detector. The data stream from the detector was de-multiplexed to produce time-dependent signals from individual locations, which were subsequently processed to construct activation maps. While this first study showed promising results, this experimental approach did not spread outside the inventors' laboratory.

The penetration of optical mapping into cardiac electrophysiology accelerated after the introduction of large photodiode arrays and fast CCD cameras and their successful application for multi-site optical recordings of electrical activity in neurons. In 1987 Salama et al. reported the first photodiode system for mapping cardiac action potentials, which was soon reproduced in other laboratories (Davidenko et al. 1990). These setups were similar to a previously developed multi-site neuronal mapping system developed by Grinvald et al. 1981.

The use of CCD cameras for cardiac mapping was pioneered by Davidenko et al. (1992) following the lead by Blasdel and Salama (1986), and Kauer (1988) who were the first to use CCD technology for recording neuronal activity in the visual cortex and olfactory bulb, respectively. CCD-based cardiac mapping systems often contain several synchronized cameras, which enables simultaneous imaging of the endocardial and epicardial surfaces (Zaitsev et al. 2000), different cardiac compartments, and almost the entire myocardial surface (panoramic imaging) (Bray et al. 2000; Kay et al. 2004).

For more than a decade the photodiode and CCD-based systems for cardiac electrophysiological mapping were used concurrently. The systems based on photodiode arrays provided greater dynamic range, higher acquisition speeds, and signal-to-noise ratio, whereas CCD-based setups had an advantage of higher pixel resolution, lower cost, and simplicity of use. However, with the advancement of CCD and CMOS technology, the custom-built photodiode-based systems in the majority of applications were gradually replaced by commercial systems utilizing fast, high-resolution cooled CCD and CMOS cameras. One of the remaining niche applications for photodiode-based systems is the mapping of electrical activity in monolayers of cultured cardiac myocytes (Tung and Zhang 2006). Photodiode arrays are also used in combination with fiber-optics-based optrodes for recording electrical activity in the thickness of the myocardial wall (Hooks et al. 2001; Kong et al. 2011).

The key application of cardiac optical mapping has been the characterization of cardiac arrhythmias and fibrillation. The optical mapping has provided the first experimental demonstration of spiral waves in the heart and has elucidated their role in cardiac arrhythmias (Davidenko et al. 1990, 1992; Gray et al. 1995), which to date represents one of the most striking achievements of this technology.

High-resolution spatial and temporal information about the electrical activity in the heart, generated during optical mapping experiments, brought to life fundamentally new approaches to the analysis of the ventricular and atrial fibrillation and characterization of their complexity. The most prominent of those analytical tools

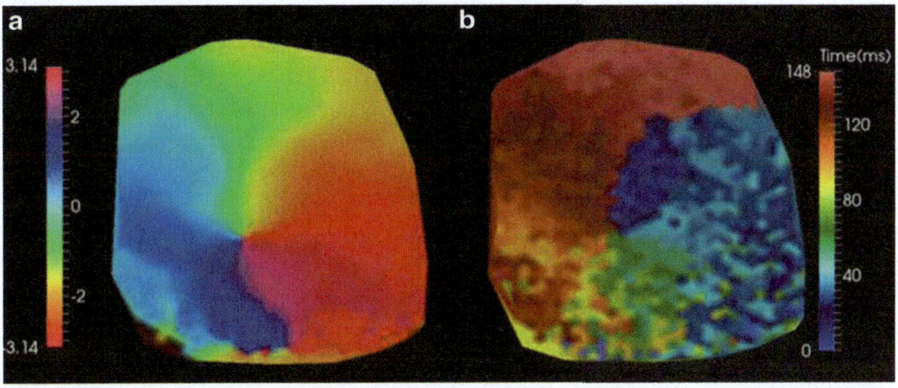

Fig. 12.1 (**a**) Phase-map of a re-entrant arrhythmia in a perfused sheep left atrium. A phase singularity (*circle*) can be observed at the center of spiral wave. (**b**) Corresponding activation map obtained directly from the raw optical signals over one activation cycle (courtesy of R. Dubois)

have become the so-called phase singularities (Gray et al. 1998) and dominant frequency maps (Zaitsev et al. 2000), which are now extensively used both in basic and clinical studies to elucidate the mechanisms responsible for the maintenance of atrial and ventricular fibrillation. Recently, mapping of phase singularities or so-called rotors have been adopted in clinical cardiac electrophysiology for guiding electrical ablation of atrial fibrillation (Haissaguerre et al. 2014; Schricker et al. 2014). Figure 12.1 shows an example of phase mapping applied to optical imaging of atrial fibrillation and allowing to identify and localize a phase-singularity and re-entrant spiral wave.

The ability to monitor electrical activity without direct contact with myocardium makes optical mapping particularly useful for the analysis of the mechanism of electrical defibrillation. Strong electrical shocks utilized for defibrillation can cause severe interference with conventional electrode recordings making such recordings technically challenging. The lack of electrical contact of the optical detector with the tissue completely eliminates this problem.

Among the great successes of optical mapping in defibrillation studies were the experimental demonstration of a "dog-bone" polarization pattern around the unipolar electrode (Wikswo et al. 1995) and quatrefoil reentry patterns (Lin et al. 1999), which were predicted by the bidomain models of electrical defibrillation, but were not previously observed experimentally. The confirmation of theoretical predictions enabled by optical mapping was a triumph of bidomain theory and contributed to its acceptance by the scientific community (see also Chap. 14).

On the other hand, optical mapping studies of defibrillation have also produced multiple unexpected observations, which cannot be readily explained by the conventional bidomain theory. One such effect is the bulk tissue hyperpolarization during strong electrical shocks (Fast et al. 2002). This effect contradicts the observations of symmetric polarization in isolated cardiac myocytes (Knisley et al. 1993) and stirred controversy, which still remains unresolved

(Plank et al. 2008). Another paradoxical observation was the anomalously low surface polarization, and the absence of cathodal activation during near-threshold field stimuli (Zemlin et al. 2006; Caldwell et al. 2010). These findings also challenge the conventional bidomain model and await a mechanistic explanation. Additional information on the topic of this chapter can be found in recent reviews (Herron et al. 2012; Efimov and Salama 2012).

2 Optical Mapping of Calcium Transients

An important advantage of optical mapping is that one can use the same experimental setup for monitoring not only the transmembrane potential but also other physiological parameters as long as respective fluorescent indicators are available (Salama et al. 1987). The most common parameter that is being measured concurrently with the transmembrane voltage is the intracellular calcium transient. Effective fluorescent molecular probes sensitive to free intracellular calcium (Tsien 1980) were developed in the early eighties and are being used to study intracellular calcium transients in intact hearts since 1987 (Lee et al. 1987). It is interesting that one of the first applications of the calcium transient mapping was the visualization of excitation propagation in monolayers of cardiac myocytes (Bub et al. 1998). In the monolayers, the maximum ratio of ΔF/F and the signal-to-noise ratio that can be achieved using voltage-sensitive dyes are usually much lower than those with the calcium indicators which makes the latter more convenient to use. It should be noted however that while normally the calcium wave follows the propagation of the depolarization front with a fixed delay, this is not always the case. There are situations when calcium transients are not triggered by the depolarization front and occur independently. This uncertainty with the interpretation of the "propagation patterns" obtained using calcium transient mapping can be readily resolved using simultaneous Voltage (Vm) and calcium mapping.

Simultaneous optical mapping of voltage and calcium was introduced independently by two groups in 2000 (Choi and Salama 2000; Fast and Ideker 2000). Currently such dual mapping is being used extensively to investigate the link between the regulation of intracellular calcium and cardiac arrhythmias. Over time the technical implementation of dual Vm and calcium mapping underwent significant modification. Originally, simultaneous Vm and calcium imaging required two separate photodiode arrays for measuring voltage and calcium. Recently, a new technical approach has been proposed that achieves the same goal with just one high-resolution, fast CCD camera. The new approach utilizes alternating multiple wavelength excitation, custom-designed double-band fluorescence filters, with subsequent de-multiplexing (Lee et al. 2011).

3 Interpretation of Optical Mapping Signals and Limitations of Cardiac Optical Imaging

It is well established that for the majority of voltage sensitive probes currently in use that the fractional fluorescence change is proportional to the transmembrane voltage. Consequently, the optical recordings of voltage-dependent fluorescence look similar to microelectrode recordings of the transmembrane action potential. However, this perception can be often deceptive in particular when the recordings are taken from intact myocardial wall.

Figure 12.2a compares upstrokes of the optically and electrically recorded action potentials recorded from the same spot in a normally beating rat heart. One can see that the electrical recording has a significantly steeper upstroke as compared to that of the optical recording. Another example of an optical recording which would be difficult to interpret in terms of transmembrane potential is illustrated in Fig. 12.2b. It shows a representative single pixel optical recording of the onset of ventricular fibrillation induced by electrical stimulation. The recording was taken from the epicardial surface of a coronary-perfused sheep right ventricular wall. Note a significant upwards shift of the zero line after the onset of arrhythmia. If this shift would have represented the actual membrane depolarization then during the entire duration of the arrhythmia the transmembrane voltage would remain near 0, which would prevent reactivation of the majority of sodium channels as well as a significant portion of calcium channels. As a result one would not be able to observe any stable propagating electrical waves. However, this was not the case. The analysis of the entire data set from all pixels in this particular experiment revealed a well defined propagation pattern consistent with a 3D reentry. It is interesting that the baseline rapidly returns to normal after application of a defibrillation shock suggesting that the baseline shift is unrelated with changes in intracellular/

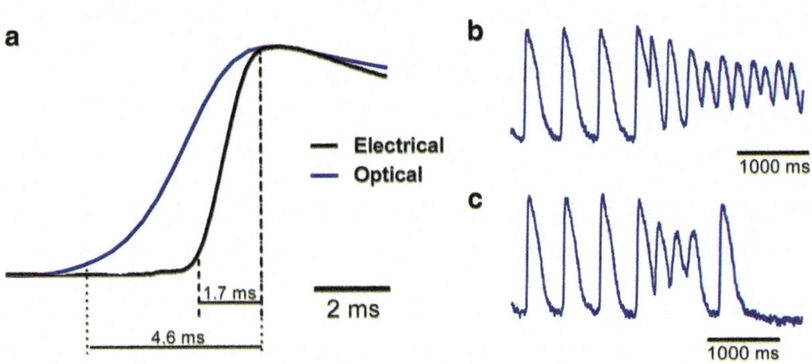

Fig. 12.2 (**a**) Optical and electrical action potential upstrokes measured at the same location in an isolated rat heart. The optical upstroke is significantly longer than the electrical upstroke. (**b**) Optical trace of the induction and maintenance ventricular fibrillation in a coronary-perfused sheep right ventricular preparation. (**c**) Optical trace of a non-sustained ventricular arrhythmia in a coronary perfused sheep right ventricular preparation

extracellular ionic compositions. An example showing a rapid return to baseline from a different experiment is illustrated in panel c.

The unusual shapes of the optically recorded action potentials illustrated in Fig. 12.2 are the result of the blurring effect caused by light scattering in the tissue. Any optical recordings, even those obtained with infinitely small sensors represent a weighted sum of signals originating in a certain tissue volume, which we call below “integration volume”. The integration volume depends on many factors including the wavelengths of the excitation and fluorescent light, the types of illumination (broad field versus point), the numerical aperture of the lens in front of the detector, the thickness of the myocardial wall, the type of perfusion (blood versus saline) etc. For different experimental settings it can vary from a fraction of 1 mm (Salama and Morad 1976) to as large as ~50 mm^3. The largest value corresponds to experiments in large animal hearts (pig, dog) utilizing near-infrared voltage sensitive dyes and broad field illumination.

As the excitation front passes through the interrogation volume, the latter contains cells at different stages of depolarization. The larger the volume, the greater is the range of differences; the greater is the blurring effect. The reduced conduction velocity increases voltage gradient over the unit length increasing the degree of blurring. The latter factor is particularly relevant during high frequency arrhythmia which is usually characterized by significant reduction in conduction velocity. The long duration of the optical upstroke and shift of the zero level during arrhythmias in Fig. 12.2 are both the result of the blurring effect discussed above. In the latter case, in any moment of time during arrhythmia the integration volume contains cells at different stages of depolarization which shifts the average and prevents the signal from going all the way down. The effect is most pronounced in experiments with near infrared voltage sensitive dyes and thick tissue samples characterized by large integration volumes. More detailed discussions on this topic can be found in Chaps. 15 and 16.

3.1 Quantification of Volume Averaging Effects

The interest towards volume averaging effects has emerged with the development of transillumination optical mapping when the light source and the photodetector face opposite surfaces of the myocardial wall (Baxter et al. 2001). The purpose of such transillumination recordings was to glean information about three-dimensional propagation inside ventricular myocardium during reentrant arrhythmias (see Sect. 4.4 below as well as dedicated Chap. 16 in this book). The important feature of fluorescence signals recorded in transillumination mode is that they represent weighted contributions through the entire thickness of the ventricular wall. This study motivated the development of detailed computational models that enabled the prediction of the shape of the optical action potential using a combination of light transport equations with three-dimensional models of electrical propagation (Hyatt et al. 2003; Bernus et al. 2005; Bishop et al. 2006).

One of the major parameters which determines the integration volume is the attenuation length (δ), which characterizes the rate of light decay in the tissue with the distance from the illuminated surface. The value of δ depends on the wavelength of the light, ranging from <0.5 mm for blue-green light to >2 mm for red light (see also Chaps. 15 and 16). Thus, the integration volume is smaller when voltage sensitive dyes with excitation in green part of the spectrum are used as compared to near-infrared voltage-sensitive dyes. The differences are caused by more rapid attenuation of light intensity inside the tissue in the green as opposed to red part of the spectrum.

3.2 Spatial and Temporal Resolution

The blurring caused by volume averaging effect determines the fundamental physical limits of the spatial and temporal resolution of the optical mapping technology. It serves as a low pass filter that suppresses high-frequency components of the voltage sensitive fluorescent signals to a point beyond which it cannot be restored by increasing pixel resolution or frame rate of the detectors. Contemporary optical mapping systems often have sufficiently high pixel resolution and frame rates to reproduce the spectral properties of such, devoid of high-frequency content, fluorescent signals. In many cases one can use heavy spatial and temporal filtering without causing any additional signal blurring (Mironov et al. 2006). Some publications even attempt to propose universal, one size fit all filter parameters (Laughner et al. 2012). It is important, however, that any such recommendation also informs the readers about the limitations and the range of applicability of a given recommendation. The readers should be aware that the appropriate degree of filtering is dependent on specifics of the experimental setup, the spectral characteristics of the dyes etc and should be determined separately when the experimental parameters fall outside the recommended range.

3.3 Motion Artifact

One of the key limitations of optical mapping is the motion artifact caused by myocardial contraction, which can cause significant distortion of the optically recorded action potential. There are rare situations when the motion artifact is not an issue. One such case is optical mapping of long duration ventricular fibrillation (Venable et al. 2010) when generalized contractions are not present and local contractions are very weak. However, the majority of optical mapping studies would not be possible without suppressing the motion artifact.

The most common approach, which is being used to mitigate the motion artifact problem, is the pharmacological elimination of mechanical contractions while preserving the electrical activity. Among those drugs known as electromechanical uncouplers most prevalent are BDM (Liu et al. 1993) Cytochalasin D (Wu et al. 1998) and Blebbistatin (Fedorov et al. 2007). Such drugs can have electrophysiological effects and other limitations (Brack et al. 2013; Lou et al. 2012; Baker et al. 2004; Lee et al. 2001; Swift et al. 2012) which should be taken into consideration when designing the experiments and interpreting the experimental results. There have been multiple attempts to eliminate motion artifact without electromechanical uncouplers by mechanical heart immobilization, ratiometric approaches (Knisley et al. 2000), computational means (Rohde et al. 2005; Khwaounjoo et al. 2015), or a combination of the above. While showing some success, these approaches so far proved having only a narrow range of applicability and have not yet been universally adopted. Resolving the motion artifact problem at the whole heart level remains a major challenge. Finding an effective solution to this problem will remove a significant obstacle to the use of optical mapping in pharmacological studies, where chemical uncouplers are unacceptable due to their potential interference with the tested drug. It may also open the way towards clinical applications of optical mapping such as interventional cardiac electrophysiology.

4 Novel Approaches in Cardiac Optical Imaging

4.1 New Voltage-Sensitive Probes

Since the first pioneer studies utilizing voltage-sensitive dyes (Salama and Morad 1976) there has been slow but significant progress in achieving greater fractional fluorescence, reduced toxicity and phototoxicity, as well as reduced bleaching and internalization rates compared to the original probes. The $\Delta F/F$ of the newest styryl dyes (Matiukas et al. 2007) di-4-ANBDQPQ and di-4-ANBDQBS reach 20 %, which is an order of magnitude greater than that of merocyanine-540 used in the original work by Morad and Salama (1976). The new dyes remain effective for hours without a major decay of the voltage-sensitive signal. The time it takes for the voltage-sensitive signal to drop 50 % from its maximum value reaches and sometimes exceeds 2 h (Matiukas et al. 2007) compared to several minutes for early dyes. Importantly, when used in intact tissues, the new dyes show no detectable internalization, photo-bleaching or photo-toxicity (Kanaporis et al. 2012) which significantly simplifies their use. They can be equally well excited in the green and the near-infrared (NIR) parts of the spectrum (Matiukas et al. 2007). The NIR excitation allows the new dyes to be effectively utilized in blood perfused tissues removing one of the important obstacles to clinical applications of optical mapping (Matiukas et al. 2007). This represents a major advantage over dyes with green-blue excitation which cannot be efficiently used in blood perfused tissues because of strong hemoglobin absorption, which significantly reduces the signal in this spectral range.

4.2 Optogenetics

Rather than introducing voltage-sensitive dyes externally, one can make the cell to make its own recombinant proteins that are sensitive to changes in the transmembrane potential. This idea was inspired by the success of GFP-based protein expression markers, which have become indispensible in contemporary cell biology. The first success in the development of voltage-sensitive proteins was achieved in neurobiology (Sakai et al. 2001). However, the voltage response of the protein-based voltage sensors is relatively slow. The frequency response of the fastest and the most recent indicators is limited to 200 Hz, which makes them less attractive for cardiac optical mapping.

4.3 Optical Mapping as a Drug Screening Tool

Since 1993 optical mapping is being used to study electrical propagation and patterned strands of cultured cardiac myocytes (Fast and Kléber 1993), and monolayers of cultured cardiac myocytes (Fast and Kleber 1994). This application proved particularly useful for the analysis of propagation at a single cell level, which would be difficult using conventional electrical recordings. More recently, optical mapping of cardiac propagation in monolayers of cardiac myocytes is being utilized for the analysis of electrophysiological manifestations of the genetic modifications of connexins (Beauchamp et al. 2006) and ion channels (Campbell et al. 2012) as well as for the physiological characterization of human embryonic stem cells (Lee et al. 2012). In the future, optical mapping of monolayers of cultured stem cells-derived human cardiac myocytes may become a useful tool for screening cardiac toxicity of potential pharmacological agents.

4.4 Three-Dimensional Optical Mapping

The heart tissue is essentially three-dimensional (3D). In larger animals such as pigs and dogs the thickness of the ventricular wall can easily exceed 2 cm. Would it be possible to glean 3D information about the electrical activity deep inside the ventricular wall using optical mapping? As was discussed above, the light can penetrate deep into the tissue, in particularly in the red and near-infrared parts of the spectrum. However, significant light scattering in the tissue and significant signal blurring significantly complicates the efforts to extract useful 3D information. It is interesting that in early studies, a “2D heart” model was developed, in which all intramural tissue was cryoablated leaving intact only thin viable epicardial layer (Girouard et al. 1996), which largely eliminated the complications caused by 3D effects.

The first successful attempts to extract 3D information from optical signals were reported by Efimov and Mazgalev who were able to detect activation of the AV nodal cells buried under the thin layer of atrial tissue in the rabbit (Efimov and Mazgalev 1998). Their method takes advantage of significant differences in the activation time of the surface layer of atrial cells and underlying layer of the AV nodal cells. The optical summation of signals from these two layers produces specific double-hump signals, which can be readily related to one or the other layer. A similar approach was recently used to study electrical propagation in human SA-node (Fedorov et al. 2010). In humans, the layer of atrial myocytes over the AV node is significantly thicker than in the rabbit. To obtain strong enough signal from the AV-nodal layer the authors utilized the novel NIR dyes, which enabled the signal detection from significantly greater depth compared to a dye with green-blue excitation used in the rabbit studies.

An important step towards the development of 3D optical mapping was the introduction of the alternating transillumination technique (Baxter et al. 2001). The transillumination method enabled interrogation of the entire thickness of the ventricular wall and was not limited to specific anatomical structures such as AV-node and SA-node. This technique in combination of NIR voltage-sensitive dyes has enabled the analysis of 3D reentrant activity or so-called scroll waves (Mitrea et al. 2009) as well as the detection and 3D localization of intramyocardial sources of excitation (Mitrea et al. 2011).

The use of Diffusive Optical Tomography allows in principle reconstructing not only the centers of activation and one-dimensional objects such as filaments, but the entire excitation front. The first successful attempt of such reconstruction was reported by Hillman et al. (Hillman et al. 2007). We would like to refer the readers to Chap. 16 on 3D optical mapping which describes in great detail the major achievements, technical challenges and limitations of 3D optical mapping which is still in its infancy.

References

Baker LC, Wolk R, Choi BR, Watkins S, Plan P, Shah A, Salama G (2004) Effects of mechanical uncouplers, diacetyl monoxime, and cytochalasin-d on the electrophysiology of perfused mouse hearts. Am J Physiol Heart Circ Physiol 287:H1771–H1779

Baxter WT, Mironov SF, Zaitsev AV, Jalife J, Pertsov AM (2001) Visualizing excitation waves inside cardiac muscle using transillumination. Biophys J 80:516–530

Beauchamp P, Yamada KA, Baertschi AJ, Green K, Kanter EM, Saffitz JE, Kleber AG (2006) Relative contributions of connexins 40 and 43 to atrial impulse propagation in synthetic strands of neonatal and fetal murine cardiomyocytes. Circ Res 99:1216–1224

Bernus O, Wellner M, Mironov SF, Pertsov AM (2005) Simulation of voltage-sensitive optical signals in three-dimensional slabs of cardiac tissue: application to transillumination and coaxial imaging methods. Phys Med Biol 50:215–229

Bishop MJ, Rodriguez B, Eason J, Whiteley JP, Trayanova N, Gavaghan DJ (2006) Synthesis of voltage-sensitive optical signals: application to panoramic optical mapping. Biophys J 90:2938–2945

Blasdel GG, Salama G (1986) Voltage-sensitive dyes reveal a modular organization in monkey striate cortex. Nature 321:579–585

Brack KE, Narang R, Winter J, Ng GA (2013) The mechanical uncoupler blebbistatin is associated with significant electrophysiological effects in the isolated rabbit heart. Exp Physiol 98:1009–1027

Bray MA, Lin SF, Wikswo JP Jr (2000) Three-dimensional surface reconstruction and fluorescent visualization of cardiac activation. IEEE Trans Biomed Eng 47:1382–1391

Bub G, Glass L, Publicover NG, Shrier A (1998) Bursting calcium rotors in cultured cardiac myocyte monolayers. Proc Natl Acad Sci U S A 95:10283–10287

Caldwell BJ, Wellner M, Mitrea BG, Pertsov AM, Zemlin CW (2010) Probing field-induced tissue polarization using transillumination fluorescent imaging. Biophys J 99:2058–2066

Campbell K, Calvo CJ, Mironov S, Herron T, Berenfeld O, Jalife J (2012) Spatial gradients in action potential duration created by regional magnetofection of herg are a substrate for wavebreak and turbulent propagation in cardiomyocyte monolayers. J Physiol 590:6363–6379

Choi BR, Salama G (2000) Simultaneous maps of optical action potentials and calcium transients in guinea-pig hearts: mechanisms underlying concordant alternans. J Physiol 529 (Pt 1):171–188

Davidenko JM, Kent PF, Chialvo DR, Michaels DC, Jalife J (1990) Sustained vortex-like waves in normal isolated ventricular muscle. Proc Natl Acad Sci 87:8785–8789

Davidenko JM, Pertsov AV, Salomonsz R, Baxter W, Jalife J (1992) Stationary and drifting spiral waves of excitation in isolated cardiac muscle. Nature 355:349–351

Davila HV, Salzberg BM, Cohen LB, Waggoner AS (1973) A large change in axon fluorescence that provides a promising method for measuring membrane potential. Nat New Biol 241:159–160

Dillon S, Morad M (1981) A new laser scanning system for measuring action potential propagation in the heart. Science 214:453–456

Efimov IR, Mazgalev TN (1998) High-resolution, three-dimensional fluorescent imaging reveals multilayer conduction pattern in the atrioventricular node. Circulation 98:54–57

Efimov I, Salama G (2012) The future of optical mapping is bright: Re: Review on: "Optical imaging of voltage and calcium in cardiac cells and tissues" by Herron, Lee, and Jalife. Circ Res 110:e70–e71

Fast VG, Ideker RE (2000) Simultaneous optical mapping of transmembrane potential and intracellular calcium in myocyte cultures. J Cardiovasc Electrophysiol 11:547–556

Fast VG, Kléber AG (1993) Microscopic conduction in cultured strands of neonatal rat heart cells measured with voltage-sensitive dyes. Circ Res 73:914–925

Fast VG, Kleber AG (1994) Anisotropic conduction in monolayers of neonatal rat heart cells cultured on collagen substrate. Circ Res 75:591–595

Fast VG, Sharifov OF, Cheek ER, Newton JC, Ideker RE (2002) Intramural virtual electrodes during defibrillation shocks in left ventricular wall assessed by optical mapping of membrane potential. Circulation 106:1007–1014

Fedorov VV, Lozinsky IT, Sosunov EA, Anyukhovsky EP, Rosen MR, Balke CW, Efimov IR (2007) Application of blebbistatin as an excitation–contraction uncoupler for electrophysiologic study of rat and rabbit hearts. Heart Rhythm 4:619–626

Fedorov VV, Glukhov AV, Chang R, Kostecki G, Aferol H, Hucker WJ, Wuskell JP, Loew LM, Schuessler RB, Moazami N, Efimov IR (2010) Optical mapping of the isolated coronary-perfused human sinus node. J Am Coll Cardiol 56:1386–1394

Girouard SD, Pastore JM, Laurita KR, Gregory KW, Rosenbaum DS (1996) Optical mapping in a new guinea pig model of ventricular tachycardia reveals mechanisms for multiple wavelengths in a single reentrant circuit. Circulation 93:603–613

Gray RA, Jalife J, Panfilov AV, Baxter WT, Cabo C, Davidenko JM, Pertsov AM (1995) Mechanisms of cardiac fibrillation. Science 270:1222–1223, author reply 1224–1225

Gray RA, Pertsov AM, Jalife J (1998) Spatial and temporal organization during cardiac fibrillation. Nature 392:75–78

Grinvald A, Cohen LB, Lesher S, Boyle MB (1981) Simultaneous optical monitoring of activity of many neurons in invertebrate ganglia using a 124-element photodiode array. J Neurophysiol 45:829–840

Haissaguerre M, Hocini M, Denis A, Shah AJ, Komatsu Y, Yamashita S, Daly M, Amraoui S, Zellerhoff S, Picat MQ, Quotb A, Jesel L, Lim H, Ploux S, Bordachar P, Attuel G, Meillet V, Ritter P, Derval N, Sacher F, Bernus O, Cochet H, Jais P, Dubois R (2014) Driver domains in persistent atrial fibrillation. Circulation 130:530–538

Herron TJ, Lee P, Jalife J (2012) Optical imaging of voltage and calcium in cardiac cells & tissues. Circ Res 110:609–623

Hillman EM, Bernus O, Pease E, Bouchard MB, Pertsov A (2007) Depth-resolved optical imaging of transmural electrical propagation in perfused heart. Opt Express 15:17827–17841

Hooks DA, LeGrice IJ, Harvey JD, Smaill BH (2001) Intramural multisite recording of transmembrane potential in the heart. Biophys J 81:2671–2680

Hyatt CJ, Mironov SF, Wellner M, Berenfeld O, Popp AK, Weitz DA, Jalife J, Pertsov AM (2003) Synthesis of voltage-sensitive fluorescence signals from three-dimensional myocardial activation patterns. Biophys J 85:2673–2683

Kanaporis G, Martisiene I, Jurevicius J, Vosyliute R, Navalinskas A, Treinys R, Matiukas A, Pertsov AM (2012) Optical mapping at increased illumination intensities. J Biomed Opt 17:96007-1

Kauer JS (1988) Real-time imaging of evoked activity in local circuits of the salamander olfactory bulb. Nature 331:166–168

Kay MW, Amison PM, Rogers JM (2004) Three-dimensional surface reconstruction and panoramic optical mapping of large hearts. IEEE Trans Biomed Eng 51:1219–1229

Khwaounjoo P, Rutherford SL, Svrcek M, LeGrice IJ, Trew ML, Smaill BH (2015) Image-based motion correction for optical mapping of cardiac electrical activity. Ann Biomed Eng 43(5):1235–1246

Knisley S, Blitchington T, Hill B, Grant A, Smith W, Pilkington T, Ideker R (1993) Optical measurements of transmembrane potential changes during electric field stimulation of ventricular cells. Circ Res 72:255–270

Knisley SB, Justice RK, Kong W, Johnson PL (2000) Ratiometry of transmembrane voltage-sensitive fluorescent dye emission in hearts. Am J Physiol Heart Circ Physiol 279:H1421–H1433

Kong W, Pollard AE, Fast VG (2011) A new optrode design for intramural optical recordings. IEEE Trans Biomed Eng 58:3130–3134

Laughner JI, Ng FS, Sulkin MS, Arthur RM, Efimov IR (2012) Processing and analysis of cardiac optical mapping data obtained with potentiometric dyes. Am J Physiol Heart Circ Physiol 303: H753–H765

Lee HC, Smith N, Mohabir R, Clusin WT (1987) Cytosolic calcium transients from the beating mammalian heart. Proc Natl Acad Sci U S A 84:7793–7797

Lee MH, Lin SF, Ohara T, Omichi C, Okuyama Y, Chudin E, Garfinkel A, Weiss JN, Karagueuzian HS, Chen PS (2001) Effects of diacetyl monoxime and cytochalasin d on ventricular fibrillation in swine right ventricles. Am J Physiol Heart Circ Physiol 280: H2689–H2696

Lee P, Bollensdorff C, Quinn TA, Wuskell JP, Loew LM, Kohl P (2011) Single-sensor system for spatially resolved, continuous, and multiparametric optical mapping of cardiac tissue. Heart Rhythm 8:1482–1491

Lee P, Klos M, Bollensdorff C, Hou L, Ewart P, Kamp TJ, Zhang J, Bizy A, Guerrero-Serna G, Kohl P, Jalife J, Herron TJ (2012) Simultaneous voltage and calcium mapping of genetically purified human induced pluripotent stem cell-derived cardiac myocyte monolayers. Circ Res 110:1556–1563

Lin SF, Roth BJ, Wikswo JP Jr (1999) Quatrefoil reentry in myocardium: an optical imaging study of the induction mechanism. J Cardiovasc Electrophysiol 10:574–586

Liu Y, Cabo C, Salomonsz R, Delmar M, Davidenko J, Jalife J (1993) Effects of diacetyl monoxime on the electrical properties of sheep and guinea pig ventricular muscle. Cardiovasc Res 27:1991–1997
Lou Q, Li W, Efimov IR (2012) The role of dynamic instability and wavelength in arrhythmia maintenance as revealed by panoramic imaging with blebbistatin vs. 2,3-butanedione monoxime. Am J Physiol Heart Circ Physiol 302:H262–H269
Matiukas A, Mitrea BG, Qin M, Pertsov AM, Shvedko AG, Warren MD, Zaitsev AV, Wuskell JP, Wei MD, Watras J, Loew LM (2007) Near-infrared voltage-sensitive fluorescent dyes optimized for optical mapping in blood-perfused myocardium. Heart Rhythm 4:1441–1451
Mironov SF, Vetter FJ, Pertsov AM (2006) Fluorescence imaging of cardiac propagation: spectral properties and filtering of optical action potentials. Am J Physiol Heart Circ Physiol 291: H327–H335
Mitrea BG, Wellner M, Pertsov AM (2009) Monitoring intramyocardial reentry using alternating transillumination. Conf Proc IEEE Eng Med Biol Soc 2009:4194–4197
Mitrea BG, Caldwell BJ, Pertsov AM (2011) Imaging electrical excitation inside the myocardial wall. Biomed Opt Express 2:620–633
Plank G, Prassl A, Hofer E, Trayanova NA (2008) Evaluating intramural virtual electrodes in the myocardial wedge preparation: simulations of experimental conditions. Biophys J 94:1904–1915
Rohde GK, Dawant BM, Lin SF (2005) Correction of motion artifact in cardiac optical mapping using image registration. IEEE Trans Biomed Eng 52:338–341
Sakai R, Repunte-Canonigo V, Raj CD, Knopfel T (2001) Design and characterization of a DNA-encoded, voltage-sensitive fluorescent protein. Eur J Neurosci 13:2314–2318
Salama G, Morad M (1976) Merocyanine 540 as an optical probe of transmembrane electrical activity in the heart. Science 191:485–487
Salama G, Lombardi R, Elson J (1987) Maps of optical action potentials and NADH fluorescence in intact working hearts. Am J Physiol 252:H384–H394
Salzberg BM, Davila HV, Cohen LB (1973) Optical recording of impulses in individual neurones of an invertebrate central nervous system. Nature 246:508–509
Schricker AA, Lalani GG, Krummen DE, Narayan SM (2014) Rotors as drivers of atrial fibrillation and targets for ablation. Curr Cardiol Rep 16:509
Swift LM, Asfour H, Posnack NG, Arutunyan A, Kay MW, Sarvazyan N (2012) Properties of blebbistatin for cardiac optical mapping and other imaging applications. Pflugers Arch 464:503–512
Tsien RY (1980) New calcium indicators and buffers with high selectivity against magnesium and protons: design, synthesis, and properties of prototype structures. Biochemistry 19:2396–2404
Tung L, Zhang Y (2006) Optical imaging of arrhythmias in tissue culture. J Electrocardiol 39:S2–S6
Venable PW, Taylor TG, Shibayama J, Warren M, Zaitsev AV (2010) Complex structure of electrophysiological gradients emerging during long-duration ventricular fibrillation in the canine heart. Am J Physiol Heart Circ Physiol 299:H1405–H1418
Wikswo JP Jr, Lin SF, Abbas RA (1995) Virtual electrodes in cardiac tissue: a common mechanism for anodal and cathodal stimulation. Biophys J 69:2195–2210
Wu J, Biermann M, Rubart M, Zipes DP (1998) Cytochalasin d as excitation-contraction uncoupler for optically mapping action potentials in wedges of ventricular myocardium. J Cardiovasc Electrophysiol 9:1336–1347
Zaitsev AV, Berenfeld O, Mironov SF, Jalife J, Pertsov AM (2000) Distribution of excitation frequencies on the epicardial and endocardial surfaces of fibrillating ventricular wall of the sheep heart. Circ Res 86:408–417
Zemlin CW, Mironov S, Pertsov AM (2006) Near-threshold field stimulation: intramural versus surface activation. Cardiovasc Res 69:98–106

Chapter 13
Optical Mapping of Ventricular Fibrillation Dynamics

Sarah A. Park and Richard A. Gray

Abstract There is very limited information regarding the dynamic patterns of the electrical activity during ventricular fibrillation (VF) in humans. Most of the data used to generate and test hypotheses regarding the mechanisms of VF come from animal models and computer simulations and the quantification of VF patterns is non-trivial. Many of the experimental recordings of the dynamic spatial patterns of VF have been obtained from mammals using "optical mapping" or "video imaging" technology in which "phase maps" are derived from high-resolution transmembrane recordings from the heart surface. The surface manifestation of the unstable reentrant waves sustaining VF can be identified as "phase singularities" and their number and location provide one measure of VF complexity. After providing a brief history of optical mapping of VF, we compare and contrast a quantitative analysis of VF patterns from the heart surface for four different animal models, hence providing physiological insight into the variety of VF dynamics among species. We found that in all four animal models the action potential duration restitution slope was actually negative during VF and that the spatial dispersion of electrophysiological parameters were not different during the first second of VF compared to pacing immediately before VF initiation. Surprisingly, our results suggest that APD restitution and spatial dispersion may not be essential causes of VF dynamics. Analyses of electrophysiological quantities in the four animal models are consistent with the idea that VF is essentially a two-dimensional phenomenon in small rabbit hearts whose size are near the boundary of the "critical mass" required to sustain VF, while VF in large pig hearts is three-dimensional and exhibits the maximal theoretical phase singularity density, and thus will not terminate spontaneously.

S.A. Park
F.M. Kirby Research Center for Functional Brain Imaging, Kennedy Krieger Institute, Russell H. Morgan Department of Radiology, The Johns Hopkins University School of Medicine, Baltimore, MD 21205, USA

R.A. Gray (✉)
Division of Physics, Office of Science and Engineering Laboratories, Center for Devices and Radiological Health, Food and Drug Administration, 10903 New Hampshire Ave, WO62 Room 1114, Silver Spring, MD 20993, USA
e-mail: Richard.Gray@fda.hhs.gov

M. Canepari et al. (eds.), *Membrane Potential Imaging in the Nervous System and Heart*, Advances in Experimental Medicine and Biology 859,
DOI 10.1007/978-3-319-17641-3_13

Keywords Fibrillation • Optical mapping • Spiral • Action potential • Conduction velocity

1 Ventricular Fibrillation Dynamics

Ventricular fibrillation (VF) is often a precursor to sudden cardiac death, the leading cause of mortality in the United States. It is generally believed that multiple unstable reentrant waves sustain VF (Jalife et al. 1998; Nanthakumar et al. 2004; Nash et al. 2006). Reentrant waves are typically visible on the epicardial and/or endocardial surfaces as two-dimensional (2D) spiral waves rotating about their organizing centers (i.e., phase singularities) (Jalife et al. 1998). In three-dimensions (3D), reentrant waves are scroll waves that rotate about string-like filaments (singularity filaments), the ends of which, when they exist, are phase singularities on the heart surface (Wellner et al. 2002a). When reentrant waves are stable (i.e., stationary) without "fibrillatory conduction", the electrocardiogram (ECG) resembles monomorphic ventricular tachycardia (VT). When reentrant waves are unstable, the ECG resembles polymorphic VT or VF, because of the spatiotemporal disorder that arises from the presence of a single moving or multiple rotors (Jalife et al. 1998; Gray et al. 1995a, b; Chen et al. 2000a; Moe 1975; Riccio et al. 1999). The ECG can resemble VF or polymorphic VT even when only a single rotor is present due to the temporal disorder resulting from the "Doppler" effect.

Several factors are believed to contribute to the initiation and maintenance of the complex spatio-temporal patterns observed during VF. Three such factors are: (1) action potential duration (APD) restitution (Riccio et al. 1999; Karma 1994; Qu et al. 1999; Cao et al. 1999); (2) spatial heterogeneity of APD and conduction velocity (CV) (Han et al. 1966; Han and Moe 1964; Kuo et al. 1983; Lammers et al. 1990; Girouard et al. 1996); and (3) heart geometry (Mines 1913; Garrey 1914; Winfree 1994; Kavanagh et al. 1992; Fenton and Karma 1998).

APD restitution, and the corresponding presence of APD alternans, is considered a mechanism by which reentrant waves during VT break up to become VF (*i.e.*, spiral wave breakup). Alternans, which are characterized by a sequence of alternating APDs and may occur in tissue that is stimulated at a fast rate (Riccio et al. 1999; Cranefield et al. 1972), are a purported cause of spiral wave break up (Karma 1994; Qu et al. 1999). Prior theoretical work shows that the slope of the APD, diastolic interval (DI) relationship is predictive of the presence of alternans (Nolasco and Dahlen 1968). According to this hypothesis, the development of alternans and the promotion of VF induction and maintenance via wavebreak require a slope >1 whereas, alternans and wavebreak will not occur when the slope <1 (Riccio et al. 1999; Karma 1994; Courtemanche et al. 1993). However, Banville et al. (2004) showed that APD, DI pairs during the maintenance of VF and VT did not lie on the restitution curve but clustered significantly below the curve; furthermore, APD, DI pairs during the initial beats of VF fell above the curve.

Therefore, it may not be appropriate to use the APD restitution slope to explain wavebreak and reentry formation during VF (Cytrynbaum and Keener 2002).

Spatial heterogeneity of APD and conduction velocity (CV) have been shown to increase the heart's vulnerability to VF induction (Han and Moe 1964; Kuo et al. 1983; Lammers et al. 1990; Girouard et al. 1996). In general, if the refractoriness of two adjacent regions of the myocardium is different, the region with the longer refractory period may block an oncoming wave but the neighboring region with a shorter refractory period will allow the wavefront to propagate through it. Similarly, if a wave travels at different speeds, the region of the wave that travels faster will approach refractory tissue faster to potentially cause conduction block. Both situations may produce wavebreak and may result in reentry. Girouard et al. (1996) demonstrated that wavelength (λ), calculated as the product of APD and CV, continuously changes with time and space during established reentrant VT around an anatomic obstacle due to abrupt changes in loading during wavefront pivoting. The changes in λ were attributed to heterogeneities in CV rather than to heterogeneities in APD. Spatial heterogeneities can be either functional or inherent/structural. Functional heterogeneities can result from the effects of: wavefront curvature on CV (Cabo et al. 1994); discordant alternans (Fenton et al. 2002); DI gradients; etc. Examples of inherent heterogeneities are: ion channel distribution (Viswanathan et al. 1999; Samie et al. 2001a); tissue anisotropy (Spach et al. 1987); fiber rotation (Fenton and Karma 1998); and cell-to-cell coupling (Choi et al. 2003).

The relationship between heart size and the size of a reentrant wave has been associated with the complexity and persistence of reentry (Garrey 1914; Winfree 1994; Kim et al. 1997). Obviously, there is a "critical mass" of tissue required to sustain reentrant VF and a single reentrant wave requires a minimal size of tissue. However, quantifying these quantities and their relationship to electrophysiological variables has been elusive. It has been suggested that λ plays a major role in the maintenance of reentrant arrhythmias. Mines (1913) first proposed that reentrant activity is only possible when λ is shorter than the available path length. Smeets et al. (1986) observed that the vulnerability of the rabbit atria to fibrillation increased when λ shortened due to changes in both APD and CV. Rensma et al. (1988), in an elegant study of the canine atria, established that λ was a more consistent electrophysiological index of atrial susceptibility to arrhythmias under a wide range of drug-induced conditions, compared to APD or CV individually. Girouard et al. (1996) found that, in guinea pigs, non-sustained VT transitioned to stable sustained VT when λ adapted to dimensions shorter than the path length but VT spontaneously terminated when λ failed to adapt to the dimensions of the circuit.

Similarly, myocardial thickness may affect the number, size and movement of reentrant waves. In thin (albeit small), 2-D preparations, reentrant waves have been shown to remain stable (Jalife et al. 1998; Kavanagh et al. 1992), while in whole hearts a single reentrant wave almost always degenerates into VF. According to Winfree (Winfree 1994), this 2-D stability is the result of a lack of sufficient thickness of the tissue. Two uniquely 3-D mechanisms could explain why reentry might be stable in 2-D but unstable in 3-D: (1) "negative filament tension"

(Biktashev and Holden 1994); and (2) high wavefront curvature resulting from the twisting of fiber rotation across the ventricular wall causing significant transmural gradients of transmembrane potential resulting in "twistons" that propagate along the filament (Fenton and Karma 1998). Both these instability mechanisms require a certain minimum thickness of tissue. Winfree proposed that this critical thickness for sustained VF (D_{crit}) is related to CV and the cycle length (CL):

$$D_{crit} = \frac{CV \times CL}{\pi}. \tag{13.1}$$

Equation (13.1) represents the hypothesis that the perimeter of the smallest possible reentrant circuit is $CV \times CL$ which is a different version of the wavelength (λ) described above; $CV \times CL > \lambda$ because $CL > APD$.

The optical mapping modality integrates fast and simultaneous data acquisition with the quantitative visualization of electrophysiological function combined with anatomy, and therefore, establishes it as an ideal and indispensable tool for understanding the mechanisms of VF dynamics. Due to the high spatiotemporal resolution afforded by optical mapping and the ability to record on the whole organ level, it is an essential complementary tool to the traditional electrophysiological recording methods, such as multi-electrode arrays, electrocardiographs, and patch-clamping. In optical mapping, cardiac tissue is stained with one (or more) dye which are excited via a light source and whose emission spectra (spectrum) are changed due to changes in the underlying variable, and subsequently recorded as emitted fluorescence as shown in Fig. 13.1. Figure 13.1 illustrates "dual camera" optical mapping in which one charged-coupled-device (CCD) camera records transmembrane potential (V_m) and a second camera records intracellular calcium (Ca_i) simultaneously from the same surface (Gray et al. 2006). Dual camera optical mapping can also be used to record from both the anterior and posterior epicardial surfaces of the isolated heart (Banville et al. 1999; Banville and Gray 2002) or epi- and endocardial surfaces of tissue slabs simultaneously (Evans and Gray 2004). The middle panel of Fig. 13.1 illustrates examples of V_m (red) and Ca_i (green) signals as a function of time from one location. The great advantage of video imaging is the very high-spatial resolution which is exemplified in the bottom panel which shows uninterpolated "isochrones maps" of the depolarization process for both V_m (left) and Ca_i (right) which illustrate, in one image, the sequence of activation. Typically, fluorescence signals in optical mapping provide measures of only relative (not absolute) changes of the variable of interest, and therefore are normalized. An example of dual camera optical mapping of V_m (left) and Ca_i during VF is presented in Fig. 13.2 (Gray et al. 2000). The top panel shows examples of V_m (red) and Ca_i (green) signals from one site on the epicardial surface during VF. The middlle panel shows the same data as in the top panel, but with time parameterized, i.e., $V_m(t)$ versus $Ca_i(t)$. The inset of the middle panel shows similar data for the pacing signals in Fig. 13.1 for comparison. The middle panel shows trajectories for one site in "true" state space since both V_m and Ca_i are measured variables (Gray et al. 2006). One can choose a "centroid", i.e., (V_m*, Ca_i*),

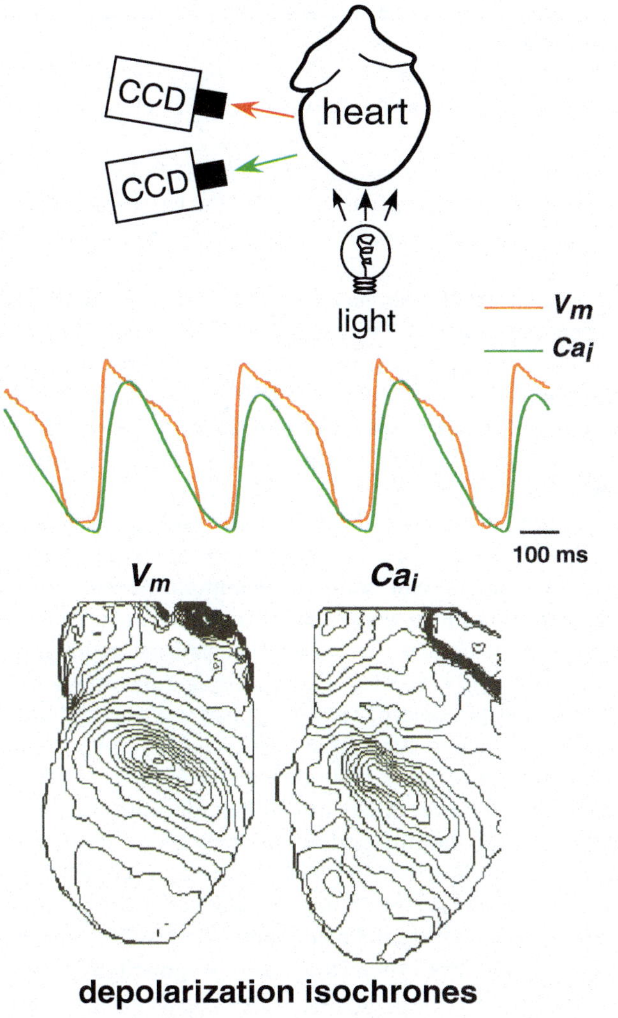

Fig. 13.1 Schematic of optical mapping methodology and example output during pacing. (*Top*) A dual camera setup to record both transmembrane potential (V_m) intracellular calcium concentration (Ca_i) simultaneous from the surface of an isolated Langendorff-perfused heart. A heart stained with two fluorescence dyes, RH237 for V_m (*red*), and Rhod2 for Ca_i (*green*). The dyes are excited by a light source (532 nm laser) and the emitted fluorescence signals are recorded via two charged-coupled-device (CCD) cameras. Bandpass emission filters were placed in front of the CCDs (centered at 645 nm for Vm and 590 nm for Cai). (*Middle*) Example of V_m (*red*) and Ca_i (*green*) signals from one location. (*Bottom*) Isochrone maps illustrating the position of the activation wavefront for both V_m (*left*) and Ca_i (*right*) every 2 ms resulting from pacing from the center of the anterior surface of a rabbit heart

in state space (indicated by white asterisks) and generate a new phase "state-space encoded phase variable" as

$$\theta(x, y, t) = \arctan\left[Ca_i(x, y, t) - Ca_i^*, V_m(x, y, t) - V_m^*\right]. \tag{13.2}$$

As described previously, this phase transformation allows for the identification of the location, number, and chirality of rotors on the heart surface as phase singularities, even if they are unstable or short-lived (see bottom panel of Fig. 13.2) (Jalife et al. 1998).

For many years, fibrillation has been described qualitatively as randomly meandering reentrant wavelets (Wiggers et al. 1930; Moe and Abildskov 1959). According to this "multiple wavelet" hypothesis, fibrillation is maintained by multiple reentrant waves distributed randomly throughout the heart; thus each region of the heart is equally capable of contributing to sustaining the arrhythmia (Wiggers et al. 1930; Moe and Abildskov 1959). An alternative "mother rotor" hypothesis, which states that fibrillation is maintained by a single stable reentrant wave with fibrillatory conduction away from the source, has been recently revived (Scherf et al. 1948; Wellner et al. 2002b; Tabereaux et al. 2009). The ability to image arrhytmia dynamics with high spatiotemporal resolution in various animal models, along with non-linear wave theory including reentrant spiral waves, has greatly expanded the understanding of the mechanisms underlying fibrillation. In 1992, Davidenko et al. (1992) optically imaged epicardial muscles of dog and sheep and demonstrated that the core of reentant spiral waves could either be anchored or drifting and contained an "excitable gap". Then in 1995, Gray et al. (1995a) used optical mapping to capture a single rapidly moving rotor in isolated rabbit hearts that resembles fibrillation electrocardiographically, thereby experimentally establishing the concept that the underlying activity of fibrillation can be due to the movement of a single rotor. In 1998, Jalife et al. (1998) introduced cardiac phase mapping and demonstrated an unprecedented amount of organization during VF, and provided a quantitative and mathematically justified means to identify both stable and unstable short-lived rotors via "phase singularities" (PSs). Subsequently, Choi et al. (2002) studied VF dynamics in time, space, and frequency domains in rabbit hearts to show that during fibrillation multiple wavelets are spatially unanchored, temporally unstable, and result in continuous wave breaks. Kay et al. (2006) explored the possibility of a mother rotor driving fibrillation in hearts comparable in size to human hearts. By optically imaging pig hearts, they observed that persistent stable rotors are absent and rather, short-lived rotors were common. They concluded that VF maintenance relies on the formation of new rotors. Additionally, Chen et al. (2000b) noted that in the isolated rabbit heart, some forms of VF are attributable to a single or small number of stable, high frequency periodic sources (a mother rotor) accompanied by fibrillatory conduction and short-lived wave breaks at some distance from this dominant stable rotor. Alternatively, Liu et al. (2004) proposed that both multiple wavelets and a mother rotor can co-exist in acutely ischemic rabbit ventricles. In short, it is likely that the tissue condition,

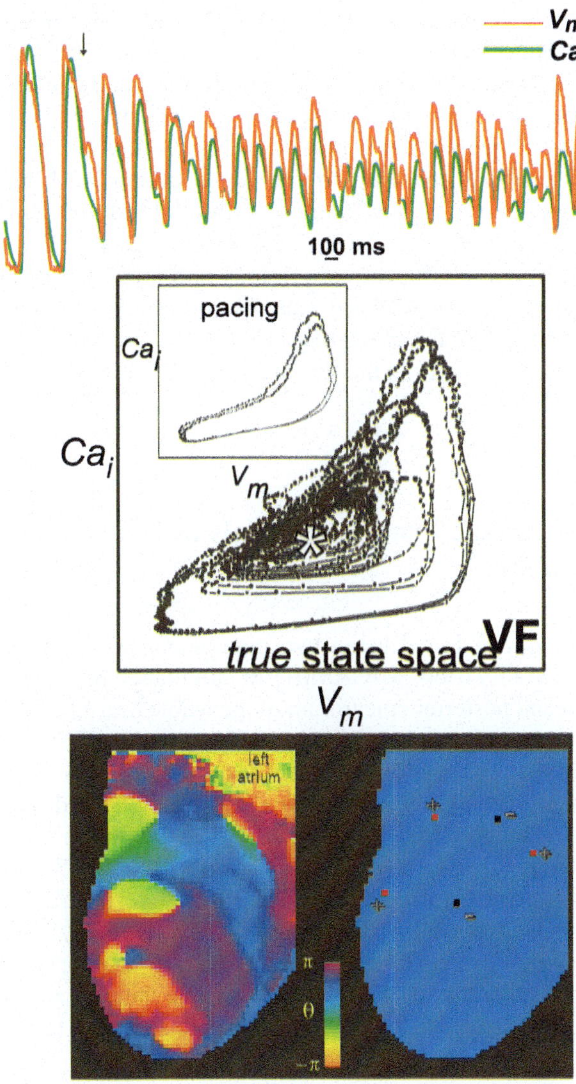

Fig. 13.2 Example optical mapping output during ventricular fibtrillation (VF) in the rabbit heart (using dual camera system shown in Fig. 13.1). (*Top*) Example of transmembrane potential V_m (*red*), and intracellular calcium concentration, Ca_i (*green*) signals from one location. (*Middle*) The data from the *Top panel* plotted in state space, i.e., Ca_i versus V_m. The *inset* shows the analogous plot during pacing (data from Fig. 13.1. *Middle panel*) in which trajectories tranverse counter-clockwise and small Ca_i alternans can been seen. During VF, the trajectories are still predominately counter-clockwise, but exhibit a large variety in "loop" shape exemplifying the variations of minimum and maximum values of V_m and Ca_i. (*Bottom*) Phase map (*left*) and phase singularity map (*right*) of one frame during VF; clockwise phase singularities are *red* and counter-clockwise phase singularities are *black*. At this instance there are three clockwise and two counterclockwise phase singularities on the epicardial anterior surface

electrophysiological variables, APD restitution, and the stage of VF all factor into the mechanism that underlies a particular fibrillatory activity and the ability to spatiotemporally monitor the VF dynamics using optical mapping will lead to better and improved therapies that target the specific mechanism.

In this chapter we quantitatively investigate the relationship between heart size D_{crit} by analyzing optical mapping data acquired during VF from four different animal models, small (rabbit) and large (pig) hearts perfused with two different electromechanical uncoupling drugs, cytochalasin D (CytoD), and diacetyl monoxime (DAM), that alter electrophysiology and affect VF dynamics. Specifically, we compare: (1) actual phase singularity density versus the "theoretical maximum" (assuming that the critical surface area for reentry is a region of tissue with area equal to D^2_{crit}) and (2) the "theoretical minimum rotor diameter" (D_{crit}) and heart wall thickness.

2 Signal and Image Processing Tools in Cardiac Optical Imaging of VF

Data were previously acquired using two high-resolution charge-coupled device (CCD) cameras (BrainVision, MICAM01) to simultaneously record fluorescence from the anterior and posterior epicardium of the whole heart. Data from previously published studies from four animal models of heart size and drug were analyzed: small hearts (rabbits) perfused with CytoD (SC) (Banville and Gray 2002); with DAM (SD) (Banville and Gray 2002; Banville et al. 1999) and large hearts (pigs) perfused with CytoD (LC) (Banville et al. 2004); and with DAM (LD) (Chattipakorn et al. 2001). Nineteen episodes of SC (dt = 3.5 ms), 14 episodes of SD, 27 episodes of LC (dt = 2.0 ms), and 33 episodes of LD (dt = 4 ms) were selected for analysis (dt is the frame rate for the respective study). The experimental protocols have been described elsewhere (Banville and Gray 2002; Banville et al. 1999; Chattipakorn et al. 2001). Images of the anterior and posterior views of a rabbit and pig heart recorded by the camera are shown in Fig. 13.3.

2.1 Signal Processing

We analyzed episodes of both VF initiation (all models) and sustained VF for LC and LD (neither SC nor SD exhibited sustained VF) (Banville and Gray 2002). The episodes of VF initiation were time aligned to the time of shock (t = 0 ms). All raw data were processed and filtered as described in the original publications (Banville and Gray 2002; Banville et al. 1999), with the exception of LD in which a 3 × 3 × 3 spatio-temporal boxcar filter was applied since no spatial filtering was done originally (Chattipakorn et al. 2001). For VF initiation episodes, the fluorescence signal

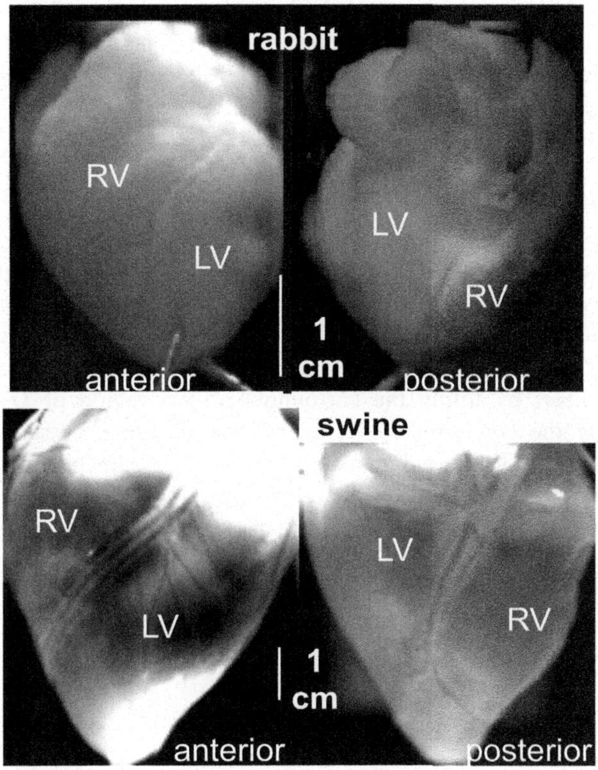

Fig. 13.3 Small rabbit and large swine heart images illustrating the high spatial resolution provided by video imaging. (*Top*) Images of the anterior and posterior surfaces of the rabbit (*small*) heart. (*Bottom*) Images of the anterior and posterior surfaces of the swine (*large*) hearts. The ventricular chambers are identified as LV (left ventricle) and RV (right ventricle). Scale bars indicate 1 cm

from each site was normalized using the paced action potential and was converted to phase (θ) by translating the signals to a polar coordinate system using "reconstructed" (as opposed to true, described above) state space (Jalife et al. 1998):

$$\theta(x, y, t) = \arctan\left[F(x, y, t + \tau) - F^*, F(x, y, t) - F^*\right], \tag{13.3}$$

where t is time in milliseconds, τ is the delay factor (10 ms) and F^* is the polar origin in reconstructed state space. We choose F^* as follows: $F^* = 67\%$ of the fluorescence level of the action potential amplitude of the last paced beat for VF initiation (Gray and Jalife 1998); or $F^* = 50\%$ of range during sustained VF. Depolarization time (DT) and repolarization time (RT) were computed as the time when $\theta = \pi/2$ and $\theta = -2\pi/3$, respectively (DTs and RTs were linearly interpolated between frames); this method for computing DT and RT has been validated over a wide range of pacing intervals in the rabbit heart (Banville and Gray 2002). A 5-point median spatial filter was applied to each DT map to reduce noise artifacts in the computation of CV.

2.2 Conduction Velocity

Conduction velocity (CV) was computed using a modified method of Bayly et al. (1998) as illustrated in Fig. 13.4. Briefly, DT movies were generated by filling every interval of DT_n to DT_{n+1} for each pixel with the corresponding DT_n value to generate a DT (isochrone) map at every time frame. CV was calculated at every pixel on the heart by fitting the surrounding 5×5 region of DTs to a plane, represented by the equation $ax + by + c = DT(x,y)$, using a least-squares fitting method of a surface in three dimensions. When the center pixel was at the edge of the mapped surface, at least four neighboring DTs within the surrounding 5×5 region were required to compute CV. The activated pixels were represented by a set of linear equations, $T(x,y) = A \cdot K$, to perform the planar fit:

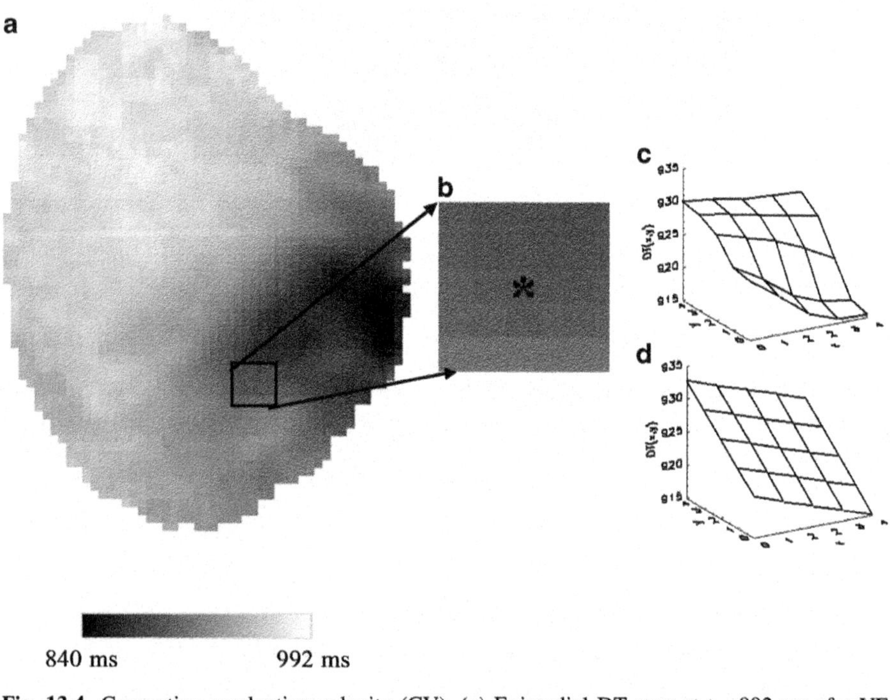

Fig. 13.4 Computing conduction velocity (CV). (**a**) Epicardial DT map at t = 992 ms of a VF episode of a small heart (rabbit) with cytochalasin D; the greyscale bar illustrates the range of values. (**b**) A 5×5 region to which the least-squares fitting of a plane was applied. An *asterisk* marks the pixel of interest at which CV (0.16 mm/ms) was calculated for this 5×5 region. (**c**) Surface plot of the original DTs of the 5×5 region. (**d**) Planar fit of the same 5×5 region

$$\begin{bmatrix} DT_1 \\ DT_2 \\ \vdots \\ \vdots \\ DT_M \end{bmatrix} = \begin{bmatrix} x_1 & y_1 & 1 \\ x_2 & y_2 & 1 \\ \vdots & \vdots & \vdots \\ \vdots & \vdots & \vdots \\ x_M & y_M & 1 \end{bmatrix} \cdot \begin{bmatrix} a \\ b \\ c \end{bmatrix}, \qquad (13.4)$$

where M is the total number of pixels in the region; A is an $M \times 3$ column matrix of the x- and y-coordinates in the first and second columns, respectively; T(x,y) is an $M \times 1$ column matrix of DT values; and K is a 3×1 column matrix of unknown coefficients a, b, and c. The matrix A^{-1} was found using singular value decomposition and multiplied by T(x,y) to get the unknowns of K. The fitted plane, $T_{fit}(x,y)$, was determined by finding the dot product of A and K. CV was computed by determining the gradient of $T_{fit}(x,y)$:

$$\nabla T_{fit} = [\partial T_{fit}/\partial x, \partial T_{fit}/\partial y]^T \qquad (13.5)$$

The x- and y-components of CV, dx/dT_{fit} and dy/dT_{fit}, were calculated using Eq. (13.5) where $T_{fit,x} = \partial T_{fit}/\partial x$ and $T_{fit,y} = \partial T_{fit}/\partial y$ as:

$$\begin{bmatrix} \dfrac{dx}{dT_{fit}} \\ \dfrac{dy}{dT_{fit}} \end{bmatrix} = \begin{bmatrix} \dfrac{T_{fit,x}}{T_{fit,x}^2 + T_{fit,y}^2} \\ \dfrac{T_{fit,y}}{T_{fit,x}^2 + T_{fit,y}^2} \end{bmatrix}, \qquad (13.6)$$

because CV cannot be obtained by simply inverting the elements of ∇T. Two differences exist between our method and Bayly et al.'s method. First, we did not use a time window when performing the least squares fit because we created DT "movies" as described above. Second, a planar fit was used instead of a polynomial fit due to the higher spatial resolution of our optical mapping data compared to the electrical mapping data used by Bayly et al. (1998).

2.3 Standard Electrophysiological Variables: CL, APD, and CV

Cycle length (CL) was calculated as the time interval between consecutive DTs, APD was calculated as the time interval between RT and the previous DT was calculated as the time interval between DT and the previous RT. CL and DT movies were generated by filling the DT-DT intervals of all pixels with the corresponding CL or DT value. Therefore, once a pixel was activated, it retained the CL or DT value until the next activation, at which time the pixel is set equal to the new CL or APD value. APD movies were similarly constructed, and updated every RT. This method overcomes problems associated with beat-to-beat analysis. CLs and APDs falling outside the range of [median CL/2, 2*median CL] and [median APD/2, 2*median APD], respectively, were considered to be outside a physiologically acceptable range and were not included in the analysis. Median CL and APD were determined from CLs and APDs from every pixel of each episode of each animal over the full duration of VF. The global spatial mean and the global spatial standard deviation (spatial dispersion) were computed from the movies of all electrophysiological factors for the four animal models. In addition, the mean spatial dispersion of APD (APD_{sd}) and of CV (CV_{sd}) were computed from the average of the spatial standard deviation. The electrophysiological variables are plotted versus time at 28 ms intervals (28 ms was the least common multiple of the recording rates of the four animal models).

2.4 *Heart Geometry*

The thicknesses and the rabbit right ventricle (RV), left ventricle (LV) and septum were taken from Table 2 of Vetter and McCulloch (1998) and are equal to 0.17, 0.49, and 0.49 cm respectively. The corresponding values for the pig were provided by Flavio Fenton (Georgia Tech University) from the model generated by Stevens et al. (2003) and are: 0.5 cm for RV; 0.81 cm for LV, and 1.19 cm for the septum.

2.5 *D_{crit} and Phase Singularity Density*

We assumed that the measured CV represents the geometrical mean of longitudinal (L) and transverse (T) velocities, i.e., $CV = \sqrt{CV_L \times CV_T}$. Since fibers are aligned perpendicular to the transmural direction, CV_T is used in Eq. (13.1) for the comparisons with the wall thickness values (using an anisotropy ratio of 3) (Schalij et al. 1992). Phase singularities (PS) were identified automatically by computing the line integral of the change in phase,

$$\oint \nabla\theta \cdot d\ell, \tag{13.7}$$

around each pixel (for each frame) using the 8 nearest neighbors. This integral is equal to zero everywhere except where a PS exists where it is equal to $\pm 2\pi$ where the sign indicates the spiral wave chirality on the epicardial surface. This line integral was computed at every pixel and *then* the resulting movies were "masked" to manually select rectangular regions. This was done to avoid the difficulties associated with identification of PS's such as edge effects, motion, and low signal-to-noise ratio.

3 Optical Imaging Study of the Critical Mass Hypothesis for VF

The tachyarrhythmias of the four animal models and control were analyzed for various electrophysiological variables, APD restitution, phase singularity, and heart size and thickness. The results suggest a relationship between electrophysiological variables and heart size in regards to VF dynamics.

3.1 Pacing

Isochrone maps illustrating the pattern of wave propagation during pacing are shown in Fig. 13.5. We computed all electrophysiological variables during pacing (last beat before VF initiation via a T-wave shock) to allow comparison to values recorded during early VF. The values are presented in Table 13.1. There were differences among the four animal models in APD, CV, and APD dispersion.

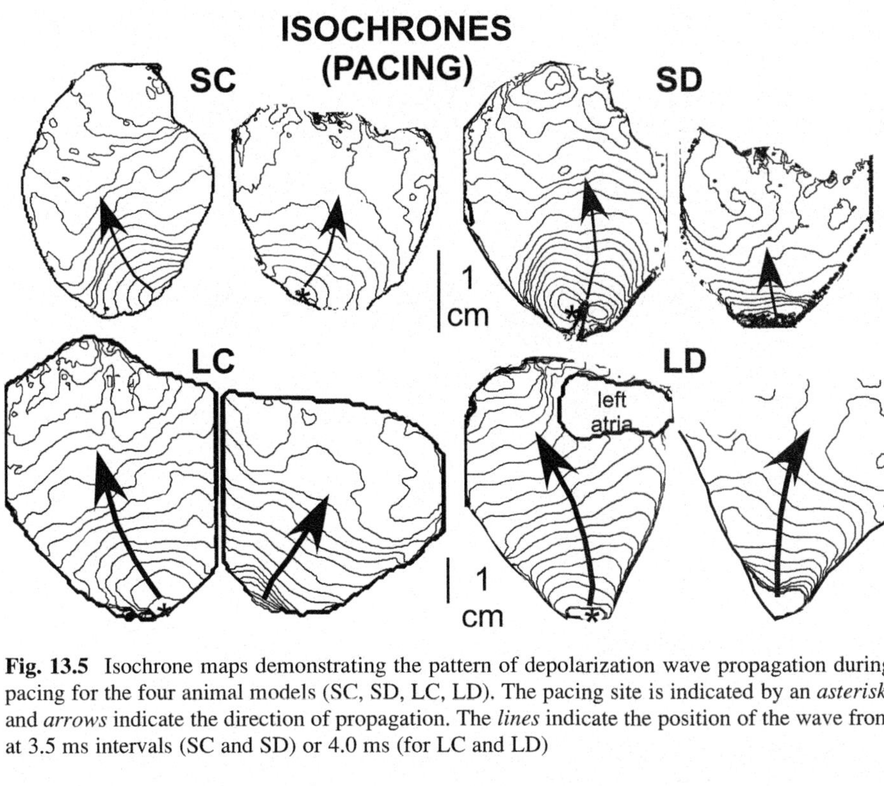

Fig. 13.5 Isochrone maps demonstrating the pattern of depolarization wave propagation during pacing for the four animal models (SC, SD, LC, LD). The pacing site is indicated by an *asterisk*, and *arrows* indicate the direction of propagation. The *lines* indicate the position of the wave front at 3.5 ms intervals (SC and SD) or 4.0 ms (for LC and LD)

Table 13.1 Average values of APD, CV, CL and of APD_{sd} and CV_{sd} before VF induction

Animal model	APD (ms)	CV (cm/s)	CL (ms)	APD_{sd} (ms)	CV_{sd} (cm/s)
SC	180 ± 12	41 ± 2	264 ± 22	9.9 ± 3.0	20 ± 2
SD	94 ± 16	52 ± 3	251 ± 1	14.9 ± 3.5	20 ± 1
LC	130 ± 5	38 ± 8	254 ± 16	18.1 ± 4.4	20 ± 4
LD	127 ± 17	49 ± 6	300 ± 11	17.1 ± 2.1	20 ± 2

Values presented as mean ± SEM

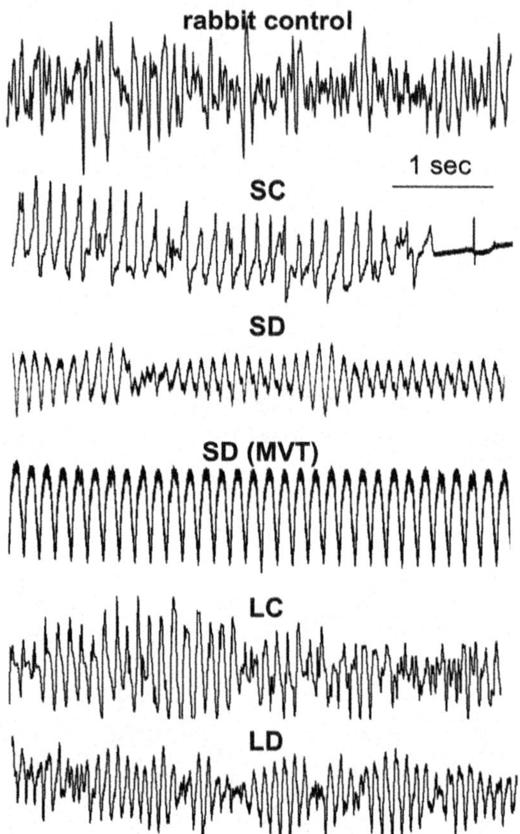

Fig. 13.6 Global electrocardiograms (ECGs) recorded from the bath for the four animal models (SC, SD, LC, LD) as well as for the rabbit without any uncoupling agents (*top*). Neither SD nor SC resulted in sustained VF, while both LC and LD episodes required electrical defibrillation

3.2 ECG Morphology

Representative ECGs of tachyarrhythmias exhibited by the four animal models (as well as from a control rabbit heart that exhibited sustained VF) are shown in Fig. 13.6. By visual determination of the fluorescence movies, episodes that appeared to be monomorphic VT (see SD MVT in Fig. 13.6) were excluded in the analysis. VF of SC always spontaneously terminated in contrast to the sustained arrhythmias for LC, LD and SD (SD episodes usually stabilized to MVT).

3.3 Electrophysiological Variables

Example traces of fluorescence (*F*) during VF initiation are shown for the four animal models in Fig. 13.7 (all episodes were aligned to the time of the shock, dashed vertical line). For SC the action potential maintained somewhat of a plateau and the take-off "potential" of *F* immediately increased but sometimes decreased toward resting levels, and only the minimum values of *F* changed from beat-to-beat. For SD

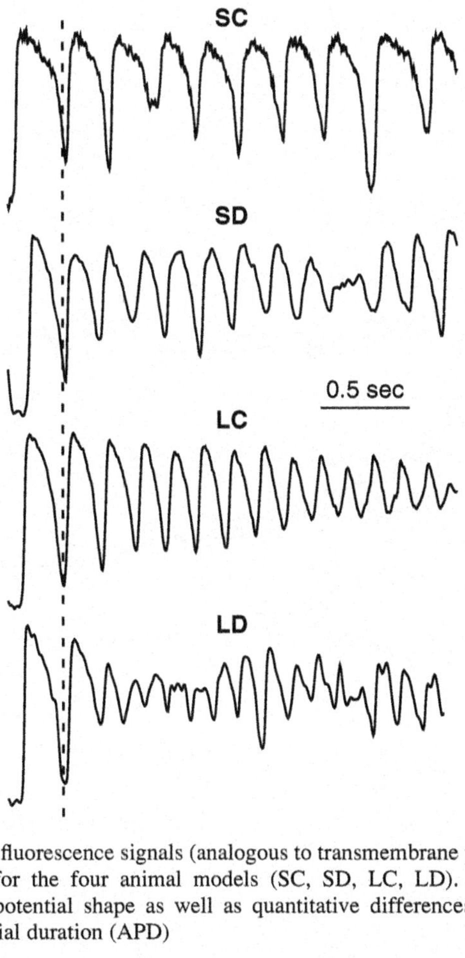

Fig. 13.7 Normalized fluorescence signals (analogous to transmembrane potential) from one site during VF initiation for the four animal models (SC, SD, LC, LD). Notice the qualitative differences in action potential shape as well as quantitative differences such as cycle length (CL) and action potential duration (APD)

LC, and LD, the take-off "potential" of F also immediately increased but remained elevated and both maximum and minimum values of F changed from beat-to-beat.

The spatial mean of cycle length (CL), APD, CV, and λ are shown for the four animal models during early VF in Fig. 13.8. The spatial means of all

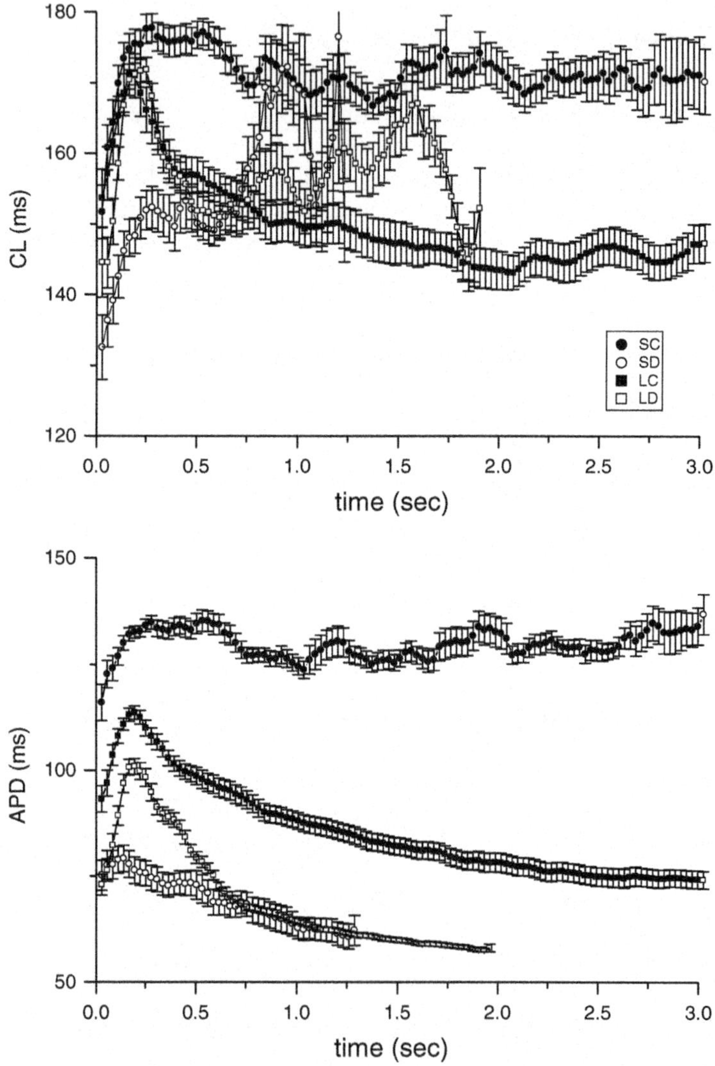

Fig. 13.8 Quantitative results of the change in electrophysiological variables during early VF for the four animal models (SC: *filled circle*; SD: *open circle*; LC: *filled square*; LD: *open square*). The different lengths of signals for the four models does not indicate the duration of the arrhythmia, only the shorter durations of recordings for SC and SD. The following electrophysiological variables are shown (from *top* to *bottom*): cycle length (CL); action potential duration (APD); conduction velocity magnitude (CV); and wavelength ($\lambda = \text{APD*CV}$). These values are smaller than the respective values computed during pacing as shown in Table 13.1

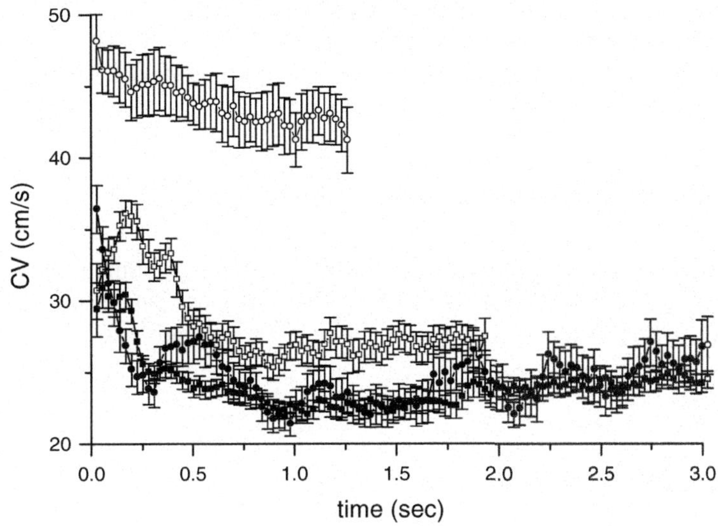

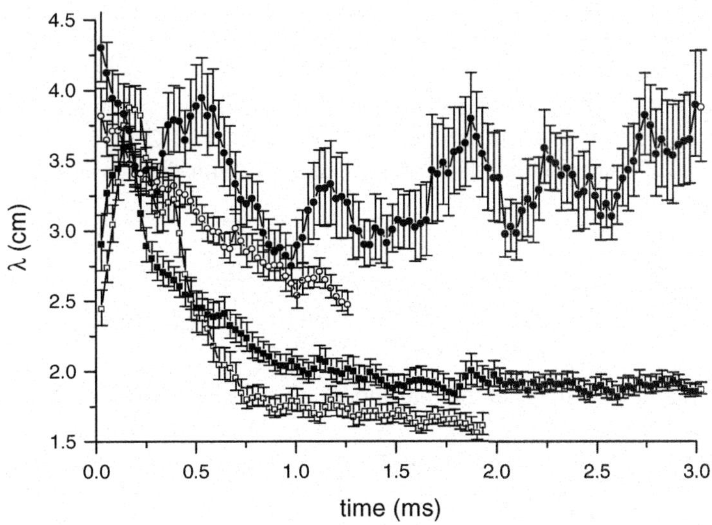

Fig. 13.8 (continued)

electrophysiological factors during VF were consistently below their respective pacing means in Table 13.1 ($p < 0.05$). The episode durations for the four models were not identical; variables for SC and LC could be computed for up to 3 s while SD (~1.2 s) and LD (~2 s) could not (this does not indicate that arrhythmias were not sustained, only that the recording durations were shorter). CL shortened immediately as a result of the shock, then increased and then: decreased for LC; remained constant for SC; and increased and oscillated for SD and LD. APD showed similar trends as CL except for SD and LD which decreased asymptotically after the initial increase. Conduction velocity was significantly larger (and remained elevated) for SD, while it decreased and leveled off for LD as well as for LC and SC (to similar values). λ first increased and then decreased and leveled out at 2 cm or less for SD and LD while it remained elevated at ~3.5 cm for SC and decreased exponentially for LC.

3.4 APD Restitution

APD, DI pairs were plotted to determine whether APD restitution might be playing a role during the first second of VF. The slopes of these data relationships were computed by performing linear regression and were negative for all four models (data not shown).

3.5 Spatial Dispersion

The possible role of APD_{sd} and CV_{sd} in VF maintenance was examined. Surprisingly, after an initial transition period (presumably due to shock-induced dispersion via virtual electrode polarization), APD_{sd} and CV_{sd} values during early VF (Fig. 13.9) were not consistently statistically different from the pacing spatial dispersions (see Table 13.1). Values for both APD_{sd} and CV_{sd} appeared to reach "steady state" values within 1–2 s for all models (with the possible exception of APD_{sd} for SD).

3.6 D_{crit} and Phase Singularity Density (Predictions)

D_{crit} was compared for each animal model as described above (using Eq. (13.1) with $CV = CV_T$) as shown in Fig. 13.10. For the rabbit, D_{crit} remained significantly above all wall thickness estimates for the duration of the VF initiation episodes for both SC and SD. However, for the pig, both LC and LD were always less than the septum thickness (1.19 cm) and decreased to values smaller than the LV thickness of 0.81 cm within 1 s. These data suggest that VF in the rabbit heart remains 2-D while in the pig heart it will be 2-D in the RV but always 3-D in the septum and at the border between 2-D and 3-D after 1 s in the LV.

The inverse of D^2_{crit} (using Eq. (13.1) with the computed CV) was calculated during early VF to provide an estimate of the "theoretical maximum phase singularity density". $\frac{1}{D^2_{crit}}$ for all four models is shown in Fig. 13.11 (error bars are not plotted because they are very large because these values are close to zero). $\frac{1}{D^2_{crit}}$ remains low and constant for SD while it increases to similar values for SC and LD and is greatest for LC. It is not clear if these values computed early in VF will be predictive, but they

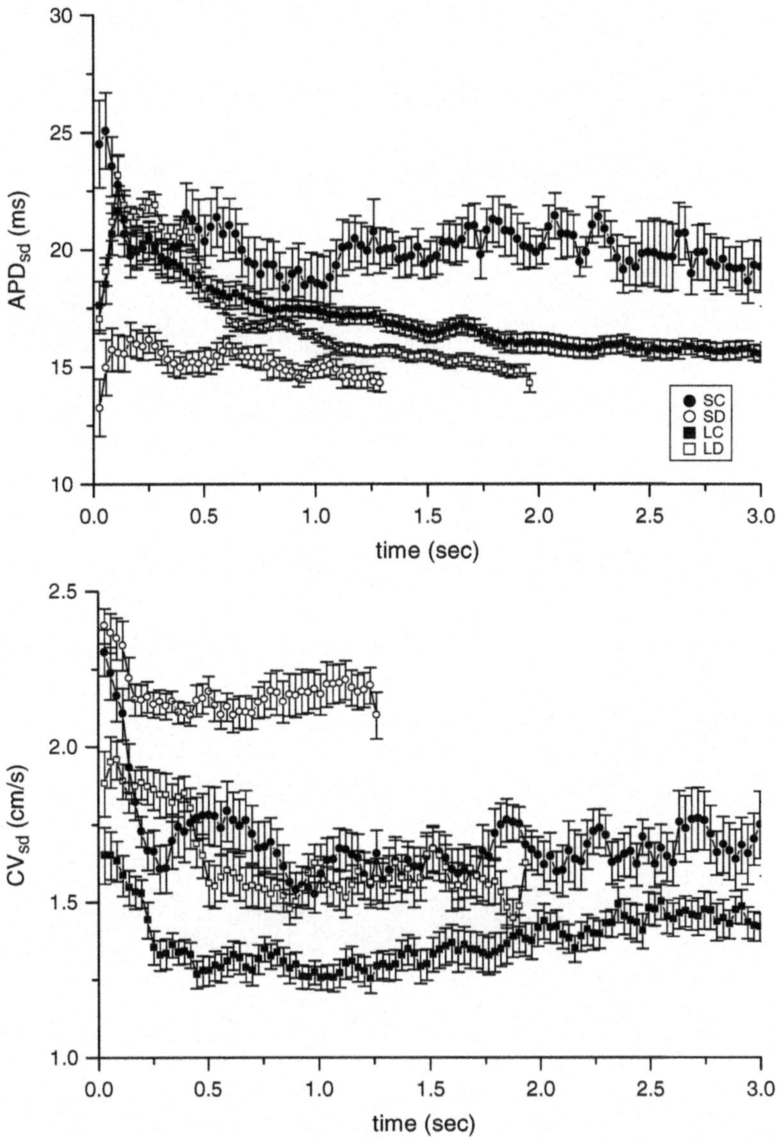

Fig. 13.9 Change in the spatial dispersion of action potential duration (APD) and conduction velocity (CV) during early VF for the four animal models (SC: *filled circle*; SD: *open circle*; LC: *filled square*; LD: *open square*) as quantified by the spatial standard deviation. These values are not different from the dispersion computed during pacing as shown in Table 13.1

suggest that SD will have the lowest PS density (and presumably, be the most organized) while LC will be the most disorganized on the epicardial surface. We consider the maximum value from this graph as the "theoretical maximum phase singularity density" which is compared to the actual values below.

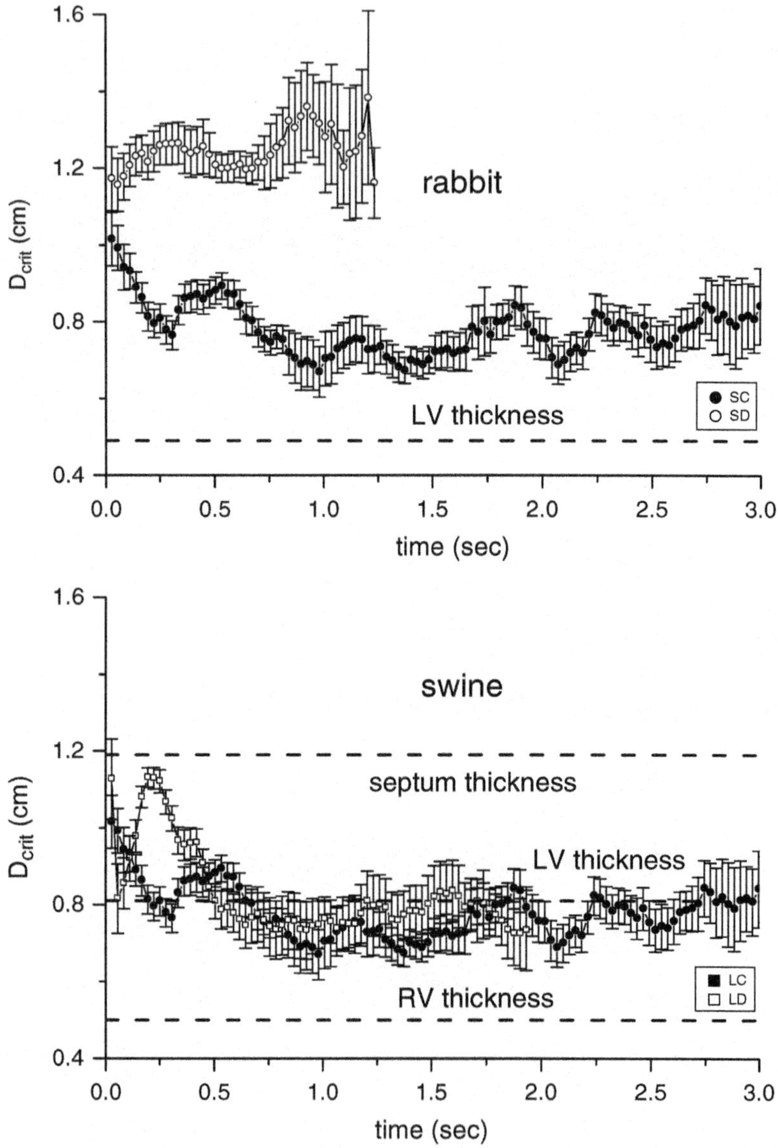

Fig. 13.10 Change of the computed critical thickness of a rotor (D_{crit}, see Eq. (13.2) and text) during early VF for the four animal models (SC: *filled circle*; SD: *open circle*; LC: *filled square*; LD: *open square*). Estimates of wall thicknesses for the rabbit and swine hearts are illustrated as *dashed horizontal lines*

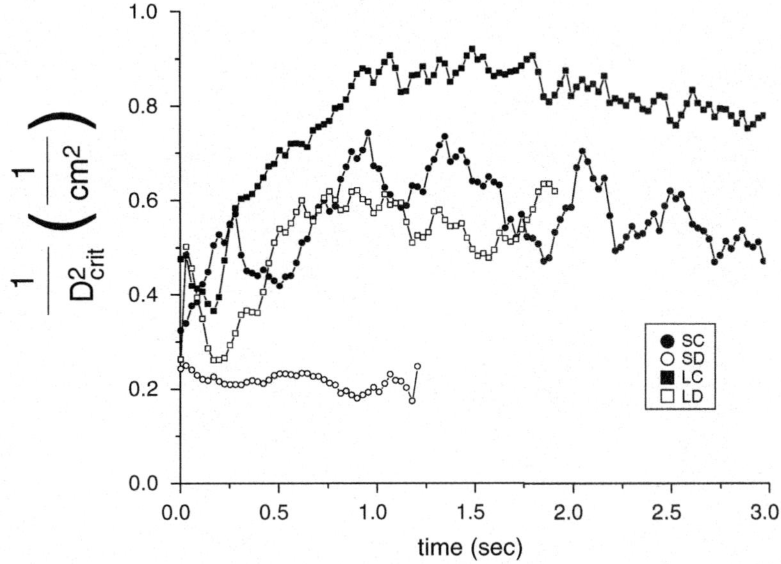

Fig. 13.11 Change of the computed theoretical phase singularity density $\left(\frac{1}{D^2_{crit}}, \text{see text}\right)$, see text) during early VF for the four animal models (SC: *filled circle*; SD: *open circle*; LC: *filled square*; LD: *open square*)

3.7 Phase Singularity Density (Actual)

Example phase maps for all models are shown in Fig. 13.12. Only one PS is observed on the anterior heart surface of the rabbit for both SC and SD. However, for the pig, multiple PSs can be seen on both anterior and posterior surfaces for both LC and LD. As mentioned above, SC and SD did not result in sustained VF; only the large heart models exhibited sustained VF. "Actual PS density" was computed for sustained VF as described above for LC and LD for a 2 s epoch as shown in Fig. 13.13. The average PS density for LC (0.7 ± 0.1 cm^{-2}) was significantly different than LD (0.5 ± 0.1 cm^{-2}) and these values were comparable to the maximum theoretical value (shown as dashed horizontal lines).

4 Discussion

Our results suggest that neither APD restitution nor spatial dispersion of APD or CV play a major role in early VF in these four normal, healthy heart models. These results are consistent with some (Salem et al. 2004), but not other (Han and Moe 1964; Kuo et al. 1983; Lammers et al. 1990), previous studies. Our findings are consistent with the hypothesis that a critical thickness is required for sustained VF (Winfree 1994) which occurred for our two large heart models but not the small

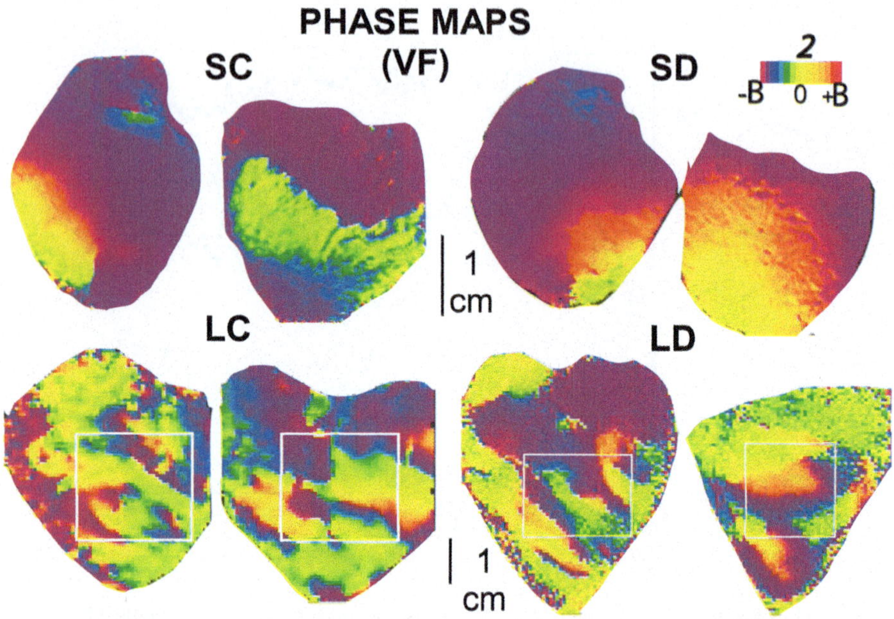

Fig. 13.12 Phase maps demonstrating a snapshot of the pattern of phase during VF for the four animal models (SC, SD, LC, LD). Phase singularities (PSs) are located at sites where phase values (*colors*) converge (see Eq. (13.3)). The *white boxes* illustrate the regions used to compute PS density for these episodes

heart models (Fig. 13.10), *but are also consistent* with the idea that our SC and SD models do not have sufficient surface area to sustain VF. Our results (see Fig. 13.10) are consistent with the findings of Rogers et al. (2000) that VF in swine is more organized in LV compared to the RV.

Our results provide support for the notion that the relationship between electrophysiological variables and heart size is quantitatively important in regards to VF dynamics. Despite identical heart sizes, we found significant differences between SC and SD as well as between LC and LD. This fact illustrates the wide range of VF dynamics, even within a species, that (presumably) result from the spatio-temporal dynamics of electrophysiological variables. For example, sustained fairly regular tachycardia was almost always the outcome for SC (CL = 157 ms, APD = 108 ms) and SD (CL = 139 ms, APD = 87 ms with 20 mM DAM) (Banville and Gray 2002), while VF was sustained (CL = 125 ms) for 10 mM DAM (Jalife et al. 1998) and in control (no uncoupling agents) with CL in the range of 64–80 ms (Kwaku and Dillon 1996; Chorro et al. 2000; Samie et al. 2001b). Estimating D_{crit} for VF in the normal Langendorff-perfused rabbit heart (CL = 72 ms, CV = 47 cm/s, and anisotropic ratio of 3) gives a value of 1.96 cm which is much *greater* than our estimates for SC and SD in early VF (see Fig. 13.10); therefore, we believe that rabbit VF is essentially 2-D. The diversity of the values of electrophysiological variables during VF are reviewed for a variety of species (mouse, guinea pig, rabbit, dog, pig,

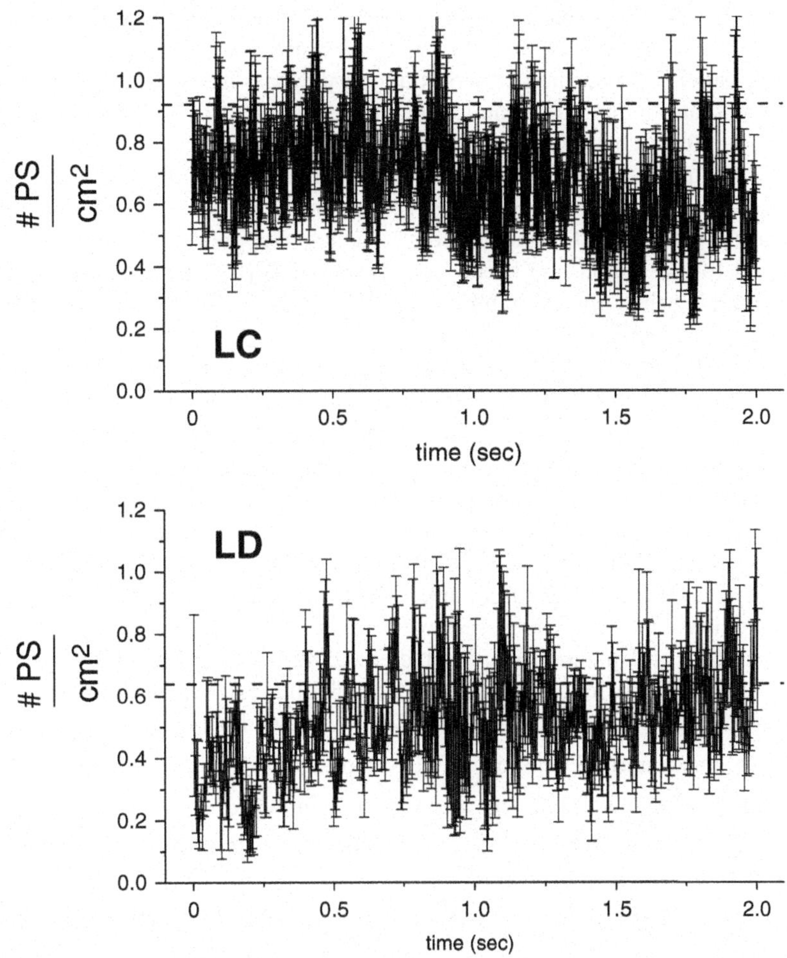

Fig. 13.13 Measured phase singularity density (PS/cm^2) during sustained VF for LC (*top*) and LD (*bottom*); the theoretical maximum density computed from Fig. 13.8 are shown as *horizontal dashed lines*

and human) in Gray (2009). Unlike our "abnormally" long VF CL in rabbit hearts, the CL during VF from our pig hearts (CL = 105 ms for LC and CL = 92 ms for LD) (Banville et al. 2004; Chattipakorn et al. 2001) are slightly shorter than CL without uncoupling agents (Salem et al. 2004; Hillsley et al. 1995; Huang et al. 1998.) Estimating D_{crit} for VF in the in-vivo pig heart (CL = 120 ms, CV = 45 cm/s) (Salem et al. 2004; Hillsley et al. 1995; Huang et al. 1998) gives a value of 1.72 cm which is much *greater* than our estimates for LC and LD (see Fig. 13.11) because we measured a much lower value of CV.

The concept of a "critical mass" required for VF has generally been interpreted that VF complexity increases with heart size (Garrey 1914; Winfree 1994;

Kim et al. 1997). This tends to be true when comparing animal models of VF among species of various sizes. However, recently recorded data of electrical mapping of VF from patients undergoing cardio-pulmonary bypass have exhibited much more organized spatio-temporal dynamics than expected. (Garrey 1914; Winfree 1994; Kim et al. 1997) Our view is similar to that of others, that human VF is *more* organized compared to VF in healthy dog and pig hearts. (Nash et al. 2006; Masse et al. 2007; Nair et al. 2011) For our large heart models (LC and LD) we computed PS density values of 0.7 and 0.5 cm^{-2} which are greater than the value we computed for the rabbit and sheep with 10 mM DAM (0.4 cm^{-2}) and appear to be much higher than observed in patients (Nash et al. 2006; Masse et al. 2007; ten Tusscher et al. 2009).

Panfilov (Panfilov 2006) has suggested that VF in the rabbit most resembles that of the human because they have the most similar "effective heart size" as quantified by the ratio of $\sqrt[3]{body \quad weight}$ and CL. Our result that reentry in rabbit hearts is primarily 2-D is consistent with this idea because Nanthakumar's group has demonstrated that reentry in humans tends to have transmural filaments (Masse et al. 2007; Nair et al. 2011). Following the ideas of Winfree (1995) and Panfilov (2006) we believe that both heart wall thickness and surface area and their relationship with rotor size are important variables in determining VF dynamics and "sustainability". In this chapter we have extended the critical mass hypothesis via these two ideas and quantitatively characterized multiple electrophysiological variables during VF for four animal models, including estimating critical thickness required to support intramural reentry (D_{crit}, computed using transmural CV) and theoretical maximum PS density $\left(\frac{1}{D_{crit}^2}\right)$. Assuming that VF is a result of reentry, we predict that another variable (the ratio of the number of PSs and the number of rotors, $\frac{\#PS}{\#rotors}$) will help predict VF organization. It has been shown previously that during VF in many cases, most PSs do not complete a full rotation (Jalife et al. 1998), while in other cases VF can be the result of a single rotor without any wavebreaks $\left(\frac{\#PS}{\#rotors} = 1\right)$ if it is moving fast enough (Gray et al. 1995a). Since only rotors (which are PSs that complete more than one rotation) are capable of *sustaining* VF (Gray 2012), we suggest that $\frac{\#PS}{\#rotors}$ will be related to VF complexity. We estimate that PS density for LC and LD are near their theoretical maximum values, indicating that reentrant waves in these models are inherently unstable and continuously form new wave breaks until the PS density reaches the maximum possible. Numerous computer simulations are consistent with the idea that VF in the whole heart is the result of numerous unstable reentrant waves (rotors) (Clayton et al. 2006; Clayton and Holden 2004; Cherry and Fenton 2004; Xie et al. 2004), however, these models must be validated and refined by a detailed comparison of electrophysiological variables and estimates relating these variables to physiological (*e.g.*, rotor size, #PS, and #rotors) and geometrical (*e.g.*, wall thickness and surface area) factors.

5 Conclusion

This work demonstrates the potential of analyzing the spatio-temporal patterns on the surface of the heart during VF. We quantified several electrophysiological variables (e.g., APD, CV as well as their spatial dispersion, wavelength, PS density) from four animal models and related these variables to various aspects of heart size. Our results suggest that the relationship between heart size and some electrophysiological variables play a role in VF dynamics. Surprisingly, our results suggest that APD restitution and spatial dispersion may not be essential causes of VF dynamics. Our analyses of VF in the four animal models are consistent with the idea that VF is essentially a two-dimensional phenomenon in small rabbit hearts whose size are near the boundary of the "critical mass" required to sustain VF, while VF in large pig hearts is three-dimensional and exhibits the maximal theoretical phase singularity density, thus will not terminate spontaneously. This study was possible only because of the advantages of high temporal and spatial resolution provided by optical imaging which allows for the visualization and quantification of tachyarrhythmias and the study of multiple interrelated variables (including heart geometry) that could affect VF dynamics. Optical imaging will continue to play an integral role in advancing the understanding of VF dynamics and the development of therapies for tachyarrhythmias.

Acknowledgements We would like to extend special thanks to Drs. Isabelle Banville and Nipon Chattipakorn who performed the experiments analyzed here. A portion of this work is contained in Dr. Park's Masters Thesis from the Department of Biomedical Engineering at the University of Alabama at Birmingham.

Glossary

λ	Wavelength=APD*CV (cm).
APD	Action potential duration (ms).
$\mathbf{APD_{sd}}$	Action potential duration dispersion (ms).
CL	Cycle length (ms).
CV	Conduction velocity (cm/s).
$\mathbf{CV_{sd}}$	Conduction velocity dispersion (cm/s).
D_{crit}	Critical rotor diameter (cm).
DI	Diastolic interval (ms).
LC	Large (pig) cytoD animal model.
LD	Large (pig) DAM animal model.
PS	Phase singularity (none).
SC	Small (rabbit) cytoD animal model.
SD	Small (rabbit) DAM animal model.

References

Banville I, Gray RA (2002) Effect of action potential duration and conduction velocity restitution and their spatial dispersion on alternans and the stability of arrhythmias. J Cardiovasc Electrophysiol 13:1141–1149

Banville I, Gray RA, Ideker RE, Smith WM (1999) Shock-induced figure-of-eight reentry in the isolated rabbit heart. Circ Res 85:742–752. http://www.ncbi.nlm.nih.gov/entrez/query.fcgi?cmd=Retrieve&db=PubMed&dopt=Citation&list_uids=10576949

Banville I, Chattipakorn N, Gray RA (2004) Restitution dynamics during pacing and arrhythmias in isolated pig hearts. J Cardiovasc Electrophysiol 15:455–463. http://www.ncbi.nlm.nih.gov/entrez/query.fcgi?cmd=Retrieve&db=PubMed&dopt=Citation&list_uids=15089996

Bayly PV, KenKnight BH, Rogers JM, Hillsley RE, Ideker RE, Smith WM (1998) Estimation of conduction velocity vector fields from epicardial mapping data. IEEE Trans Biomed Eng 45:563–571. http://www.ncbi.nlm.nih.gov/entrez/query.fcgi?cmd=Retrieve&db=PubMed&dopt=Citation&list_uids=9581054

Biktashev VN, Holden AV (1994) Design principles of a low voltage cardiac defibrillator based on the effect of feedback resonant drift. J Theor Biol 169:101–112. http://www.ncbi.nlm.nih.gov/entrez/query.fcgi?cmd=Retrieve&db=PubMed&dopt=Citation&list_uids=7934075

Cabo C, Pertsov AM, Baxter WT, Davidenko JM, Gray RA, Jalife J (1994) Wave-front curvature as a cause of slow conduction and block in isolated cardiac muscle. Circ Res 75:1014–1028. http://www.ncbi.nlm.nih.gov/entrez/query.fcgi?cmd=Retrieve&db=PubMed&dopt=Citation&list_uids=7525101

Cao JM, Qu Z, Kim YH, Wu TJ, Garfinkel A, Weiss JN, Karagueuzian HS, Chen PS (1999) Spatiotemporal heterogeneity in the induction of ventricular fibrillation by rapid pacing: importance of cardiac restitution properties. Circ Res 84:1318–1331. http://www.ncbi.nlm.nih.gov/entrez/query.fcgi?cmd=Retrieve&db=PubMed&dopt=Citation&list_uids=10364570

Chattipakorn N, Banville I, Gray RA, Ideker RE (2001) Mechanism of ventricular defibrillation for near-defibrillation threshold shocks: a whole-heart optical mapping study in swine. Circulation 104:1313–1319. http://www.ncbi.nlm.nih.gov/entrez/query.fcgi?cmd=Retrieve&db=PubMed&dopt=Citation&list_uids=11551885

Chen J, Mandapati R, Berenfeld O, Skanes AC, Gray RA, Jalife J (2000) Dynamics of wavelets and their role in atrial fibrillation in the isolated sheep heart. Cardiovasc Res 48:220–232. http://www.ncbi.nlm.nih.gov/entrez/query.fcgi?cmd=Retrieve&db=PubMed&dopt=Citation&list_uids=11054469

Chen J,Mandapati R, Berenfeld O, Skanes AC, Jalife J (2000) High-frequency periodic sources underlie ventricular fibrillation in the isolated rabbit heart. Circ Res 86:86–93. http://www.ncbi.nlm.nih.gov/entrez/query.fcgi?cmd=Retrieve&db=PubMed&dopt=Citation&list_uids=10625309

Cherry EM, Fenton FH (2004) Suppression of alternans and conduction blocks despite steep APD restitution: electrotonic, memory, and conduction velocity restitution effects. Am J Physiol Heart Circ Physiol 286:H2332–H2341. http://www.ncbi.nlm.nih.gov/entrez/query.fcgi?cmd=Retrieve&db=PubMed&dopt=Citation&list_uids=14751863

Choi BR,Nho W, Liu T, Salama G (2002) Life span of ventricular fibrillation frequencies. Circ Res 91:339–345. http://www.ncbi.nlm.nih.gov/entrez/query.fcgi?cmd=Retrieve&db=PubMed&dopt=Citation&list_uids=12193467

Choi BR, Liu T, Lavasani M, Salama G (2003) Fiber orientation and cell-cell coupling influence ventricular fibrillation dynamics. J Cardiovasc Electrophysiol 14:851–860. http://www.ncbi.nlm.nih.gov/entrez/query.fcgi?cmd=Retrieve&db=PubMed&dopt=Citation&list_uids=12890049

Chorro FJ, Canoves J, Guerrero J, Mainar L, Sanchis J, Such L, Lopez-Merino V (2000) Alteration of ventricular fibrillation by flecainide, verapamil, and sotalol: an experimental study. Circulation 101:1606–1615. http://www.ncbi.nlm.nih.gov/entrez/query.fcgi?cmd=Retrieve&db=PubMed&dopt=Citation&list_uids=10747356

Clayton RH, Holden AV (2004) Filament behavior in a computational model of ventricular fibrillation in the canine heart. IEEE Trans Biomed Eng 51:28–34. http://www.ncbi.nlm.nih.gov/entrez/query.fcgi?cmd=Retrieve&db=PubMed&dopt=Citation&list_uids=14723491
Clayton RH, Zhuchkova EA, Panfilov AV (2006) Phase singularities and filaments: simplifying complexity in computational models of ventricular fibrillation. Prog Biophys Mol Biol 90:378–398. http://www.ncbi.nlm.nih.gov/entrez/query.fcgi?cmd=Retrieve&db=PubMed&dopt=Citation&list_uids=16098568
Courtemanche M, Glass L, Keener JP (1993) Instabilities of a propagating pulse in a ring of excitable media. Phys Rev Lett 70:2182–2185. http://www.ncbi.nlm.nih.gov/entrez/query.fcgi?cmd=Retrieve&db=PubMed&dopt=Citation&list_uids=10053491
Cranefield PF, Wit AL, Hoffman BF (1972) Conduction of the cardiac impulse. 3. Characteristics of very slow conduction. J Gen Physiol 59:227–246. http://www.ncbi.nlm.nih.gov/entrez/query.fcgi?cmd=Retrieve&db=PubMed&dopt=Citation&list_uids=5058476
Cytrynbaum E, Keener JP (2002) Stability conditions for the traveling pulse: modifying the restitution hypothesis. Chaos 12:788–799. http://www.ncbi.nlm.nih.gov/entrez/query.fcgi?cmd=Retrieve&db=PubMed&dopt=Citation&list_uids=12779607
Davidenko JM, Pertsov AV, Salomonsz R, Baxter W, Jalife J (1992) Stationary and drifting spiral waves of excitation in isolated cardiac muscle. Nature 355:349–351. http://www.ncbi.nlm.nih.gov/entrez/query.fcgi?cmd=Retrieve&db=PubMed&dopt=Citation&list_uids=1731248
Evans FG, Gray RA (2004) Shock-induced epi- and endocardial virtual electrodes leading to ventricular fibrillation via reentry, graded responses, and transmural activation. J Cardiovasc Electrophysiol 51: 79–87. http://www.ncbi.nlm.nih.gov/entrez/query.fcgi?cmd=Retrieve&db=PubMed&dopt=Citation&list_uids=12475103
Fenton F, Karma A (1998) Vortex dynamics in three-dimensional continuous myocardium with fiber rotation: filament instability and fibrillation. Chaos 8:20–47. http://www.ncbi.nlm.nih.gov/entrez/query.fcgi?cmd=Retrieve&db=PubMed&dopt=Citation&list_uids=12779708
Fenton FH, Cherry EM, Hastings HM, Evans SJ (2002) Multiple mechanisms of spiral wave breakup in a model of cardiac electrical activity. Chaos 12:852–892
Garrey WE (1914) The nature of fibrillary contraction of the heart: its relation to tissue mass and form. AJP 33:397–414
Girouard SD, Pastore JM, Laurita KR, Gregory KW, Rosenbaum DS (1996) Optical mapping in a new guinea pig model of ventricular tachycardia reveals mechanisms for multiple wavelengths in a single reentrant circuit. Circulation 93:603–613. http://www.ncbi.nlm.nih.gov/entrez/query.fcgi?cmd=Retrieve&db=PubMed&dopt=Citation&list_uids=8565181
Gray RA (2009) Rotors and spiral waves in the heart. In: Zipes DP, Jalife J (eds) Cardiac electrophysiology: from cell to bedside. W.B. Saunders Co., Ltd., New York, pp 349–357
Gray RA (2012) Theory of rotors and arrhythmias. In: Zipes DP, Jalife J (eds) Cardiac electrophysiology: from cell to bedside. W.B. Saunders Co., Ltd., New York, pp 349–359
Gray RA, Jalife J (1998) Effects of atrial defibrillation shocks on the ventricles in isolated sheep hearts. Circulation 97:1613–1622. http://www.ncbi.nlm.nih.gov/entrez/query.fcgi?cmd=Retrieve&db=PubMed&dopt=Citation&list_uids=9593567
Gray RA, Jalife J, Panfilov AV, Baxter WT, Cabo C, Davidenko JM, Pertsov AM (1995a) Mechanisms of cardiac fibrillation. Science 270:1222–1223. http://www.ncbi.nlm.nih.gov/entrez/query.fcgi?cmd=Retrieve&db=PubMed&dopt=Citation&list_uids=7502055
Gray RA, Jalife J, Panfilov A, Baxter WT, Cabo C, Davidenko JM, Pertsov AM (1995b) Nonstationary vortexlike reentrant activity as a mechanism of polymorphic ventricular tachycardia in the isolated rabbit heart. Circulation 91:2454–2469. http://www.ncbi.nlm.nih.gov/entrez/query.fcgi?cmd=Retrieve&db=PubMed&dopt=Citation&list_uids=7729033
Gray RA, Wikswo JP, Lin SF, Baudenbacher F (2000) Phase mapping using both transmembrane potential and calcium. PACE 23:608
Gray RA, Iyer A, Bray MA, Wikswo JP (2006) Voltage-calcium state-space dynamics during initiation of reentry. Heart Rhythm 3:247–248. http://www.ncbi.nlm.nih.gov/entrez/query.fcgi?cmd=Retrieve&db=PubMed&dopt=Citation&list_uids=16443544

Han J, Moe GK (1964) Nonuniform recovery of excitability in ventricular muscle. Circ Res 14:44–60. http://www.ncbi.nlm.nih.gov/entrez/query.fcgi?cmd=Retrieve&db=PubMed&dopt=Citation&list_uids=14104163

Han J, Millet D, Chizzonitti B, Moe GK (1966) Temporal dispersion of recovery of excitability in atrium and ventricle as a function of heart rate. Am Heart J 71:481–487, PM:4951481

Hillsley RE, Bollacker KD, Simpson EV, Rollins DL, Yarger MD, Wolf PD, Smith WM, Ideker RE (1995) Alteration of ventricular fibrillation by propranolol and isoproterenol detected by epicardial mapping with 506 electrodes. J Cardiovasc Electrophysiol 6:471–485. http://www.ncbi.nlm.nih.gov/entrez/query.fcgi?cmd=Retrieve&db=PubMed&dopt=Citation&list_uids=7551316

Huang J, Rogers JM, KenKnight BH, Rollins DL, Smith WM, Ideker RE (1998) Evolution of the organization of epicardial activation patterns during ventricular fibrillation. J Cardiovasc Electrophysiol 9:1291–1304. http://www.ncbi.nlm.nih.gov/entrez/query.fcgi?cmd=Retrieve&db=PubMed&dopt=Citation&list_uids=9869529

Jalife J, Gray RA, Morley GE, Davidenko JM (1998) Self-organization and the dynamical nature of ventricular fibrillation. Chaos 8:79–93. http://www.ncbi.nlm.nih.gov/entrez/query.fcgi?cmd=Retrieve&db=PubMed&dopt=Citation&list_uids=12779712

Karma A (1994) Electrical alternans and spiral wave breakup in cardiac tissue. Chaos 4:461–472, PM:12780121

Kavanagh KM, Kabas JS, Rollins DL, Melnick SB, Smith WM, Ideker RE (1992) High-current stimuli to the spared epicardium of a large infarct induce ventricular tachycardia. Circulation 85:680–698. http://www.ncbi.nlm.nih.gov/entrez/query.fcgi?cmd=Retrieve&db=PubMed&dopt=Citation&list_uids=1735161

Kay MW, Walcott GP, Gladden JD, Melnick SB, Rogers JM (2006) Lifetimes of epicardial rotors in panoramic optical maps of fibrillating swine ventricles. Am J Physiol Heart Circ Physiol 291: H1935–H1941. http://www.ncbi.nlm.nih.gov/entrez/query.fcgi?cmd=Retrieve&db=PubMed&dopt=Citation&list_uids=16632545

Kim YH, Garfinkel A, Ikeda T, Wu TJ, Athill CA, Weiss JN, Karagueuzian HS, Chen PS (1997) Spatiotemporal complexity of ventricular fibrillation revealed by tissue mass reduction in isolated swine right ventricle. JCI 100:2486–2500

Kuo CS, Munakata K, Reddy CP, Surawicz B (1983) Characteristics and possible mechanism of ventricular arrhythmia dependent on the dispersion of action potential durations. Circulation 67:1356–1367. http://www.ncbi.nlm.nih.gov/entrez/query.fcgi?cmd=Retrieve&db=PubMed&dopt=Citation&list_uids=6851031

Kwaku KF, Dillon SM (1996) Shock-induced depolarization of refractory myocardium prevents wave-front propagation in defibrillation. Circ Res 79:957–973. http://www.ncbi.nlm.nih.gov/entrez/query.fcgi?cmd=Retrieve&db=PubMed&dopt=Citation&list_uids=8888688

Lammers WJ, Schalij MJ, Kirchhof CJ, Allessie MA (1990) Quantification of spatial inhomogeneity in conduction and initiation of reentrant atrial arrhythmias. Am J Physiol 259:H1254–H1263. http://www.ncbi.nlm.nih.gov/entrez/query.fcgi?cmd=Retrieve&db=PubMed&dopt=Citation&list_uids=1699438

Liu YB, Pak HN, Lamp ST, Okuyama Y, Hayashi H, Wu TJ, Weiss JN, Chen PS, Lin SF (2004) Coexistence of two types of ventricular fibrillation during acute regional ischemia in rabbit ventricle. J Cardiovasc Electrophysiol 15:1433–1440. http://www.ncbi.nlm.nih.gov/entrez/query.fcgi?cmd=Retrieve&db=PubMed&dopt=Citation&list_uids=15610292

Masse S, Downar E, Chauhan V, Sevaptsidis E, Nanthakumar K (2007) Ventricular fibrillation in myopathic human hearts: mechanistic insights from in vivo global endocardial and epicardial mapping. Am J Physiol Heart Circ Physiol 292:H2589–H2597. http://www.ncbi.nlm.nih.gov/entrez/query.fcgi?cmd=Retrieve&db=PubMed&dopt=Citation&list_uids=17259437

Mines GR (1913) On dynamic equilibrium in the heart. J Physiol 46:349–383, PM:16993210

Moe GK (1975) Evidence for reentry as a mechanism of cardiac arrhythmias. Rev Physiol Biochem Pharmacol 72:55–81. http://www.ncbi.nlm.nih.gov/entrez/query.fcgi?cmd=Retrieve&db=PubMed&dopt=Citation&list_uids=1096271

Moe GK, Abildskov JA (1959) Atrial fibrillation as a self-sustaining arrhythmia independent of focal discharge. Am Heart J 58:59–70. http://www.ncbi.nlm.nih.gov/entrez/query.fcgi?cmd=Retrieve&db=PubMed&dopt=Citation&list_uids=13661062

Nair K, Umapathy K, Farid T, Masse S, Mueller E, Sivanandan RV, Poku K, Rao V, Nair V, Butany J, Ideker RE, Nanthakumar K (2011) Intramural activation during early human ventricular fibrillation. Circ Arrhythm Electrophysiol 4:692–703, PM:21750274

Nanthakumar K, Walcott GP, Melnick S, Rogers JM, Kay MW, Smith WM, Ideker RE, Holman W (2004) Epicardial organization of human ventricular fibrillation. Heart Rhythm 1:14–23. http://www.ncbi.nlm.nih.gov/entrez/query.fcgi?cmd=Retrieve&db=PubMed&dopt=Citation&list_uids=15851110

Nash MP, Mourad A, Clayton RH, Sutton PM, Bradley CP, Hayward M, Paterson DJ, Taggart P (2006) Evidence for multiple mechanisms in human ventricular fibrillation. Circulation 114:536–542. http://www.ncbi.nlm.nih.gov/entrez/query.fcgi?cmd=Retrieve&db=PubMed&dopt=Citation&list_uids=16880326

Nolasco JB, Dahlen RW (1968) A graphic method for the study of alternation in cardiac action potentials. J Appl Physiol 25:191–196, PM:5666097

Panfilov AV (2006) Is heart size a factor in ventricular fibrillation? Or how close are rabbit and human hearts? Heart Rhythm 3:862–864. http://www.ncbi.nlm.nih.gov/entrez/query.fcgi?cmd=Retrieve&db=PubMed&dopt=Citation&list_uids=16818223

Qu Z, Weiss JN, Garfinkel A (1999) Cardiac electrical restitution properties and stability of reentrant spiral waves: a simulation study. Am J Physiol 276:H269–H283. http://www.ncbi.nlm.nih.gov/entrez/query.fcgi?cmd=Retrieve&db=PubMed&dopt=Citation&list_uids=9887041

Rensma PL, Allessie MA, Lammers WJ, Bonke FI, Schalij MJ (1988) Length of excitation wave and susceptibility to reentrant atrial arrhythmias in normal conscious dogs. Circ Res 62:395–410. http://www.ncbi.nlm.nih.gov/entrez/query.fcgi?cmd=Retrieve&db=PubMed&dopt=Citation&list_uids=3338122

Riccio ML, Koller ML,Gilmour Jr RF (1999) Electrical restitution and spatiotemporal organization during ventricular fibrillation. Circ Res 84:955–963. http://www.ncbi.nlm.nih.gov/entrez/query.fcgi?cmd=Retrieve&db=PubMed&dopt=Citation&list_uids=10222343

Rogers JM, Huang J, Pedoto RW, Walker RG, Smith WM, Ideker RE (2000) Fibrillation is more complex in the left ventricle than in the right ventricle. J Cardiovasc Electrophysiol 11:1364–1371. http://www.ncbi.nlm.nih.gov/entrez/query.fcgi?cmd=Retrieve&db=PubMed&dopt=Citation&list_uids=11196560

Salem JE, Cabrera ME, Chandler MP, McElfresh TA, Huang H, Sterk JP, Stanley WC (2004) Step and ramp induction of myocardial ischemia: comparison of in vivo and in silico results. J Physiol Pharmacol 55:519–536. http://www.ncbi.nlm.nih.gov/entrez/query.fcgi?cmd=Retrieve&db=PubMed&dopt=Citation&list_uids=15381824

Samie FH, Berenfeld O, Anumonwo J, Mironov SF, Udassi S, Beaumont J, Taffet S, Pertsov AM, Jalife J (2001) Rectification of the background potassium current: a determinant of rotor dynamics in ventricular fibrillation. Circ Res 89:1216–1223. http://www.ncbi.nlm.nih.gov/entrez/query.fcgi?cmd=Retrieve&db=PubMed&dopt=Citation&list_uids=11739288

Samie FH, Berenfeld O, Anumonwo J, Mironov SF, Udassi S, Beaumont J, Taffet S, Pertsov AM, Jalife J (2001b) Rectification of the background potassium current: a determinant of rotor dynamics in ventricular fibrillation. Circ Res 89:1216–1223, PM:11739288

Schalij MJ, Lammers WJ, Rensma PL, Allessie MA (1992) Anisotropic conduction and reentry in perfused epicardium of rabbit left ventricle. Am J Physiol 263:H1466–H1478. http://www.ncbi.nlm.nih.gov/entrez/query.fcgi?cmd=Retrieve&db=PubMed&dopt=Citation&list_uids=1279990

Scherf D, Romano FJ, Terranova R (1948) Experimental studies on auricular flutter and auricular fibrillation. Am Heart J 36:241–251, PM:18873261

Smeets JL, Allessie MA, Lammers WJ, Bonke FI, Hollen J (1986) The wavelength of the cardiac impulse and reentrant arrhythmias in isolated rabbit atrium. The role of heart rate, autonomic transmitters, temperature, and potassium. Circ Res 58:96–108. http://www.ncbi.nlm.nih.gov/entrez/query.fcgi?cmd=Retrieve&db=PubMed&dopt=Citation&list_uids=3943157

Spach MS, Dolber PC, Heidlage JF, Kootsey JM, Johnson EA (1987) Propagating depolarization in anisotropic human and canine cardiac muscle: apparent directional differences in membrane capacitance. A simplified model for selective directional effects of modifying the sodium conductance on Vmax, tau foot, and the propagation safety factor. Circ Res 60:206–219. http://www.ncbi.nlm.nih.gov/entrez/query.fcgi?cmd=Retrieve&db=PubMed&dopt=Citation&list_uids=2436826

Stevens C, Remme E, LeGrice I, Hunter P (2003) Ventricular mechanics in diastole: material parameter sensitivity. J Biomech 36:737–748, PM:12695004

Tabereaux PB, Dosdall DJ, Ideker RE (2009) Mechanisms of VF maintenance: wandering wavelets, mother rotors, or foci. Heart Rhythm 6:405–415, PM:19251220

ten Tusscher KH, Mourad A, Nash MP, Clayton RH, Bradley CP, Paterson DJ, Hren R, Hayward M, Panfilov AV, Taggart P (2009) Organization of ventricular fibrillation in the human heart: experiments and models. Exp Physiol 94:553–562, PM:19168541

Vetter FJ, McCulloch AD (1998) Three-dimensional analysis of regional cardiac function: a model of rabbit ventricular anatomy. Prog Biophys Mol Biol 69:157–183, PM:9785937

Viswanathan PC, Shaw RM, Rudy Y (1999) Effects of IKr and IKs heterogeneity on action potential duration and its rate dependence: a simulation study. Circulation 99:2466–2474. http://www.ncbi.nlm.nih.gov/entrez/query.fcgi?cmd=Retrieve&db=PubMed&dopt=Citation&list_uids=10318671

Wellner M, Berenfeld O, Jalife J, Pertsov AM (2002) Minimal principle for rotor filaments. Proc Natl Acad Sci U S A 99: 8015–8018. http://www.ncbi.nlm.nih.gov/entrez/query.fcgi?cmd=Retrieve&db=PubMed&dopt=Citation&list_uids=12048234

Wellner M, Berenfeld O, Jalife J, Pertsov AM (2002b) Minimal principle for rotor filaments. Proc Natl Acad Sci U S A 99:8015–8018, PM:12048234

Wiggers CJ, Bell JR, Paine M (1930) Studies of ventricular fibrillation caused by electric shock: II Cinematographic and electrocardiographic observations of the natural process in the dog's heart: its inhibition by potassium and the revival of coordinated beats by calcium. AHJ 5:351–365

Winfree AT (1994) Electrical turbulence in three-dimensional heart muscle. Science 266:1003–1006. http://www.ncbi.nlm.nih.gov/entrez/query.fcgi?cmd=Retrieve&db=PubMed&dopt=Citation&list_uids=7973648

Winfree AT (1995) Theory of spirals. In: Zipes DP, Jalife J (eds) Cardiac electrophysiology: from cell to bedside. W.B. Saunders Co., Ltd., New York, pp 379–389

Xie F, Qu Z, Yang J, Baher A, Weiss JN, Garfinkel A (2004) A simulation study of the effects of cardiac anatomy in ventricular fibrillation. J Clin Invest 113:686–693. http://www.ncbi.nlm.nih.gov/entrez/query.fcgi?cmd=Retrieve&db=PubMed&dopt=Citation&list_uids=14991066

Chapter 14
Imaging of Ventricular Fibrillation and Defibrillation: The Virtual Electrode Hypothesis

Bastiaan J. Boukens, Sarah R. Gutbrod, and Igor R. Efimov

Abstract Ventricular fibrillation is the major underlying cause of sudden cardiac death. Understanding the complex activation patterns that give rise to ventricular fibrillation requires high resolution mapping of localized activation. The use of multi-electrode mapping unraveled re-entrant activation patterns that underlie ventricular fibrillation. However, optical mapping contributed critically to understanding the mechanism of defibrillation, where multi-electrode recordings could not measure activation patterns during and immediately after a shock. In addition, optical mapping visualizes the virtual electrodes that are generated during stimulation and defibrillation pulses, which contributed to the formulation of the virtual electrode hypothesis. The generation of virtual electrode induced phase singularities during defibrillation is arrhythmogenic and may lead to the induction of fibrillation subsequent to defibrillation. Defibrillating with low energy may circumvent this problem. Therefore, the current challenge is to use the knowledge provided by optical mapping to develop a low energy approach of defibrillation, which may lead to more successful defibrillation.

Keywords Optical mapping • Fibrillation • Defibrillation • Arrhythmias • Imaging • Virtual electrodes • Stimulation • Heart • Re-entry • Shock

B.J. Boukens • S.R. Gutbrod
Department of Biomedical Engineering, Washington University, 1 Brookings Drive, St Louis, MO, USA

I.R. Efimov (✉)
Department of Biomedical Engineering, Washington University, 1 Brookings Drive, St Louis, MO, USA

L'Institut de rythmologie et modélisation cardiaque LIRYC, Université de Bordeaux, Bordeaux, France
e-mail: igor@wustl.edu

M. Canepari et al. (eds.), *Membrane Potential Imaging in the Nervous System and Heart*, Advances in Experimental Medicine and Biology 859,
DOI 10.1007/978-3-319-17641-3_14

1 Introduction

Ventricular tachyarrhythmias are responsible for 300,000 sudden cardiac deaths (SCD) a year in the United States of America (Kong et al. 2011). The mechanisms underlying these arrhythmias in the majority of cases are ventricular tachycardia and/or ventricular fibrillation (Huikuri et al. 2001). Extensive research over the past decades have resulted in the development of internal and external cardiac defibrillators, which have contributed to a significant reduction of the number of SCD (Jeffrey 2001). The development of defibrillators has been preceded by a vast gain in understanding of the molecular and organ-level mechanism underlying complex arrhythmias. In the early 1900s, re-entry was discovered and proposed as a simple mechanism underlying tachycardia in patients with structurally abnormal hearts and later as a mechanism underlying atrial and ventricular fibrillation in patients with structurally normal hearts (Janse and Rosen 2006). The ability to visualize these complex arrhythmias by electrical mapping and later optical mapping gave insight into the complicated re-entrant activation patterns that give rise to the abnormal body surface electrocardiograms indicative of ventricular fibrillation.

The life-saving effects of electric defibrillation have been known for more than two centuries (Gutbrod and Efimov 2013). The mechanisms of defibrillation, however, have been difficult to study with conventional electrical instrumentation because the shock potential obscures the local electrograms, which contain critical electrophysiological information. The development of computer bidomain models offered the opportunity to simulate potential fields during electrical stimulation and defibrillation and led to the hypotheses of "virtual" electrodes (Sepulveda et al. 1989; Sepulveda and Wikswo 1994). The optimization of the optical mapping technique enabled testing and direct visualization of these simulated potential fields during defibrillation and the "virtual electrode hypothesis" (Cheng et al. 1999a; Efimov et al. 1997; Wikswo et al. 1995).

In this chapter we briefly discuss the mechanism underlying re-entrant arrhythmias and how the development of optical imaging techniques have unraveled the mechanisms of defibrillation and the critical role of virtual electrodes. Furthermore, we will elaborate on how optical mapping influenced defibrillation in the clinical setting.

2 Mechanisms of Arrhythmias and Fibrillation

Cardiac arrhythmias arise from either abnormal impulse formation (focal activity) or abnormal (re-entrant) conduction (Hoffman and Rosen 1981). In this paragraph we focus on how an ectopic beat is caused by re-entrant excitation. This theory is relevant for understanding the mechanism of arrhythmogenesis caused by a defibrillating shock, especially for understanding how shocks fail to restore sinus rhythm. For a detailed classification of other mechanisms of arrhythmias we refer to the review by Hoffman and Rosen (1981) and preceding chapters in this book.

An activation sequence can be considered as re-entry when an activation front re-enters the same myocardial area repeatedly with some periodicity (Mines 1913). The first prerequisite for re-entry is that the wavefront needs to meet an area of unidirectional block, where the wavefront is able to propagate in one direction but not the other. The second requirement is the presence of a substrate, or re-entrant path, which enables the activation front to travel around the area of block and "re-enter" the previously activated myocardium causing a second activation. An area of block can be caused by the presence of an anatomical obstacle, like a scar formed after a myocardial infarction (de Bakker et al. 1988), or it can have a functional origin, like heterogeneity in refractoriness (Allessie et al. 1977). The time period where one area is already excitable but the adjacent area is still refractory is called the vulnerable window. When an ectopic beat arises in this time window then re-entry may occur. In contrast, during a re-entrant activation sequence the vulnerable window becomes the period where an extra stimulus may arrest re-entry (Mines 2010). This principle is the basis for anti-tachycardia pacing.

The critical requirement for re-entry to occur is that the available path length must be longer than the wavelength, which is the product of conduction velocity and the refractory period (Wiener and Rosenblueth 1946; Garrey 1914). Reentry cannot sustain when the wavelength is longer than the available path length within the tissue. The leading circle theory (Fig. 14.1a) states that when a wavefront travels around an area of functional block it will take the shortest pathway that allows the wavefront to re-circulate. This leaves no excitable gap, or vulnerable window.

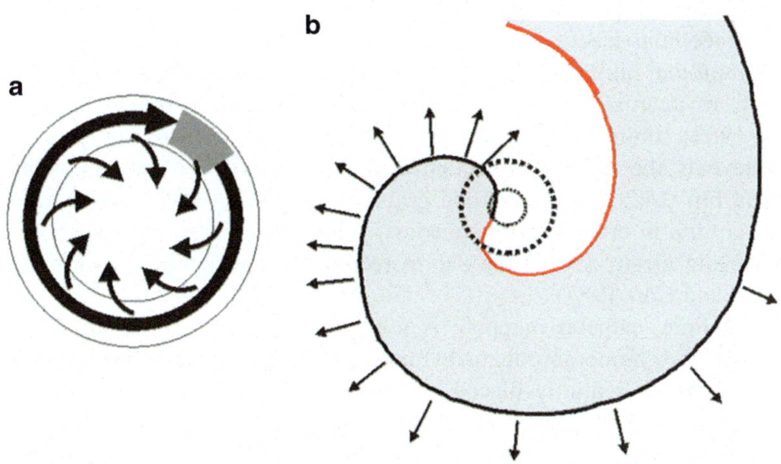

Fig. 14.1 (**a**) The leading circle concept: Activity establishes itself in the smallest pathway that can support reentry, shown as a *large black arrow*. Inside the leading circle, centripetal wavelets (*small arrows*) emanating from it constantly maintain the central core in a refractory state. (**b**) Spiral wave model: Schematic diagram of a spiral wave with the activation front shown in *black* and the repolarization front in *red*. The *point* at which the *red* and *black curves* meet has an undefined voltage state and is usually referred to as the phase singularity point. Modified from Comtois et al. (2005)

The core that the wavefront circles around or the region where a wavebreak occurs is often referred to as a phase singularity. The leading circle concept states that the activation front that circulates around a center will keep it refractory (Allessie et al. 1977). Another theory, the spiral wave theory, advocates that an activation front that travels around a center may reach a critical curvature, which prevents activation of the core of the rotating wavefront leaving an excitable gap in the middle (Fig. 14.1b) (Fast and Kleber 1997). These re-entrant waveforms are interchangeably called spiral waves, vortices, or rotors. Multiple rotors can be simultaneously present in the myocardium (Kleber and Rudy 2004) and generate the seemingly chaotic pattern on the electrocardiogram that is the hallmark of ventricular fibrillation.

3 Imaging of Fibrillation

3.1 Electrical Mapping

On the electrocardiogram, ventricular fibrillation presents as irregular and random waveforms with no clearly identifiable QRS complexes or P waves. Therefore, the electrocardiogram cannot be used for understanding the underlying mechanism of fibrillation. To study the re-entrant activation patterns, underlying local ventricular activation needs to be recorded. In the explanted perfused heart, multi-electrode grids enable the simultaneous recording of local extracellular electrograms at multiple locations. In a clean unipolar local electrogram the minimal dV/dt of the RS complex corresponds to the local time of myocardial activation under the electrode (Spach et al. 1972; Dower 1962). From these local times of activation a map can be generated, which reveals the activation sequence and the possible re-entrant pathway as shown in Fig. 14.2. Multi-electrode grids can be placed on the endocardium and the epicardium in order to simultaneously measure the activation pattern on both sides. Needle electrodes can be use to record the transmural activation patterns (Pogwizd and Corr 1987).

In the clinic, catheter mapping is used to record activation times (Shenassa et al. 2013). For monomorphic arrhythmias, a single roving catheter can be used, which will be sequentially placed on numerous sites (for instance the CARTO system (Biosense-Webster, Baldwin Park, CA, USA)). The position of the electrode is monitored using a heterogeneous magnetic field of which the strength can be detected by a sensor at the tip of the catheter. This enables the 3D reconstruction of the activation pattern. However, this technique is not suitable for mapping irregular arrhythmias because it is based on sequential recordings. Newer CARTO mapping systems use catheters with multiple electrodes and begin to circumvent this problem. Another often-used system is EnSite (St. Jude Medical, Inc., St. Paul, MN, USA), which is based on multi-electrode non-contact mapping. With this method

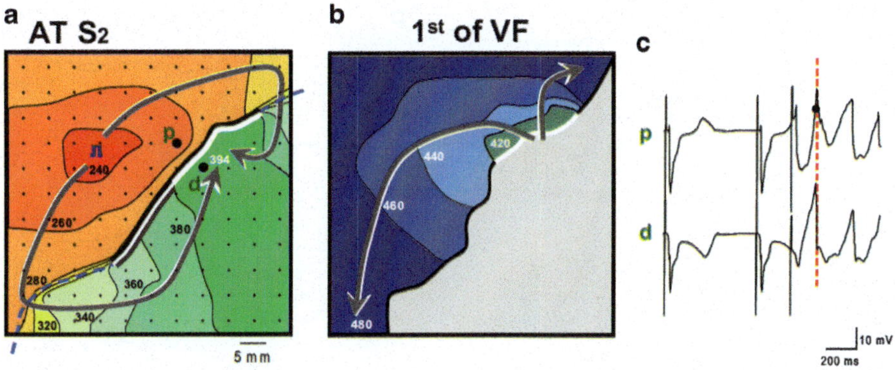

Fig. 14.2 Electrical mapping of ventricular fibrillation. At the *right*, an activation map is reconstructed of a premature beat (following S2). In the *middle*, the activation pattern is reconstructed VF is induced, because local distal activation coincided with local proximal repolarization. The *dashed line* indicates the epicardial interface between the perfusion beds. *Small dots* indicate electrode positions. *Bold dots* indicate positions from which electrograms (*right panels*) were recorded (p, proximal; d distal). Lines are 20 ms isochrones. *Red dotted vertical line* in the electrograms indicate the moment of distal activation. This figure is a courtesy of Dr. Ruben Coronel (Coronel et al. 2009)

the inverse solution is used to calculate the local moments of activation on the endocardium (Taccardi et al. 1987). The introduction of catheter non-contact and contact mapping has improved diagnosis and led to treatment of arrhythmias by ablation or antitachycardia pacing (Josephson et al. 1978; Durrer et al. 1967). However, electrical mapping techniques are not suitable for recording the tissue response directly after defibrillation because the overwhelming shock-induced artifacts mask the signal.

3.2 Optical Mapping

Electrical mapping has been a useful tool for studying the mechanisms of arrhythmia origin and perpetuation and even led to a variety of therapies. However, there are shortcomings of electrical mapping. The spatial resolution is limited and it is not possible to measure local electrical activity during or after a defibrillating shock (for reasons that will be discussed in detail later). Optical mapping provides an alternative method for measuring cardiac electrical activity and addresses the aforementioned shortcomings of electrical mapping (Boukens and Efimov 2014). Since optical mapping requires the use of fluorescent dyes the method is currently restricted to the experimental ex vivo settings.

Cardiac electrical activity can be transduced into fluorescent signals by using voltage sensitive dyes that bind to the membrane of cardiomyocytes. When a dye

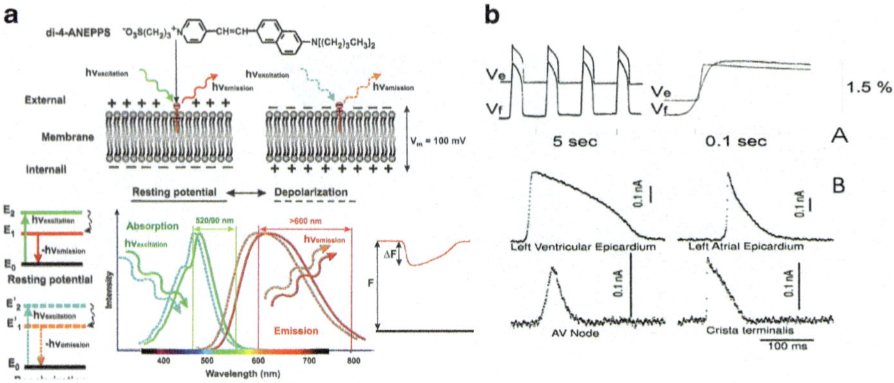

Fig. 14.3 Optical recording of cardiac action potentials using fluorescent dyes. (**a**) The *upper panel* illustrates the conformation change of di-4-ANEPPS upon a change in membrane potentials. The *lower panel* shows the shift in emission spectrum results in a change in the amount of recorded fluorescence. (**b**) The *upper panel*, Simultaneous fluorescent (Vf) and microelectrode (Ve) recordings of action potentials in the frog heart stained with Merocyanine-540. The *lower panel*, Optical recording of action potentials from different regions of the heart stained with di-4-ANEPPS: ventricular and atrial working myocardium, AV node, and Crista terminalis. Modified from Boukens et al. and Efimov et al. (Boukens and Efimov 2014; Efimov et al. 2004)

molecule binds to the membrane, emitted fluorescence can be recorded upon excitation of the dye. The wavelength of the emitted light depends on the membrane potential of the cardiomyocyte. When the membrane potential changes (i.e. during the action potential), the dynamic wavelength of the emitted light can be recorded as a loss in fluorescence with appropriate filter settings on the acquisition hardware. Filters are specific for particular dyes (e.g. excitation 520 ± 20 and emission >650 nm for di-4-ANEPPS). The change in emitted light is linearly related to the membrane potential, which enables the recording of "optical action potentials". This mechanism is illustrated in Fig. 14.3. The number of myocytes that give rise to an optical action potential signal depends on the type of the dye, the optical magnification, the focal plane and the pixel size of the sensor that is used for recording. Optical action potentials are not only derived from fluorescence collected from the surface of tissue but also of fluorescence from deeper layers (Bishop et al. 2007). Therefore, the optical action potential is an underestimation of the transmembrane potential on the surface and should be compared with transmembrane potential averaged over depth (Janks and Roth 2002). Commonly used sensors for recording optical action potentials are photodiode arrays, CCD and CMOS cameras (Efimov and Salama 2012).

Although, optical mapping is often the method of choice in experimental cardiology there are some disadvantages of this technique that need to be considered. Excitation of the fluorescent dye has been shown to produce toxic free radicals and may cause death of cardiomyoctyes. This side effect is most prominent when single myocytes or monolayers are measured. Also, the excitation of deeper located layers cause light scattering and distortion of the optical signal (Bishop et al. 2007; Ramshesh and Knisley 2003). Another disadvantage is that contractions of the heart lead to changes in emission intensity related to motion artifacts not the membrane potential. Therefore, in order to accurately measure the shape of optical action potentials, the excitation and contraction must be pharmacologically uncoupled. A common uncoupler agent is blebbistatin (Fedorov et al. 2007), but cytochalasin D and 2,3-butane-dion monoxime have also been used.

3.3 *Panoramic Imaging*

Although optical mapping and, to a lesser extent, conventional multi-electrode grids, increases spatial resolution, neither can track conduction when the activation front propagates outside the limited field of view. This hampers the reconstruction, and thereby interpretation, of the activation sequence, especially during complex re-entrant patterns. Panoramic imaging refers to optical imaging using multiple cameras or mirrors to measure from at least three different angles. This technique allows reconstruction of complete epicardial activation patterns (Kay et al. 2004) and has given insight into complex activation patterns during ventricular fibrillation. Figure 14.4 is an example of a re-entrant activation pattern of a rabbit heart using panoramic optical mapping (Ripplinger et al. 2009; Rogers et al. 2007).

Another problem with conventional optical mapping is that a curved epicardial surface is projected onto a 2-dimensional sensor. This means that distance in the "Z" direction is not taken into account. This is a limitation, when conduction velocity is calculated, because it leads to underestimation of conduction velocity, especially near the edges of the field of view. Panoramic imaging enables the projection of a 3-dimensional conduction pattern on the surface of the heart (Lou et al. 2012). Conduction velocity can be more precisely measured thereby improve our understanding of conduction during normal and re-entrant activation patterns.

3.4 *Optical Imaging of Ventricular Fibrillation*

As mentioned above the optics of a set-up can be fined tuned to specific applications, providing control over the spatial resolution and depth of the recorded signals. The fine spatial resolution of optical mapping makes it an ideal platform for tracking the dynamics of ventricular fibrillation, pinpointing focal discharges and the critical points of reentrant wavefronts as well as the critical points that are created when energy is applied during electrotherapy. Chapter 13 (Gray et al.) outlines how optical mapping can be used to study the mechanisms that maintain ventricular fibrillation including phase singularity tracking.

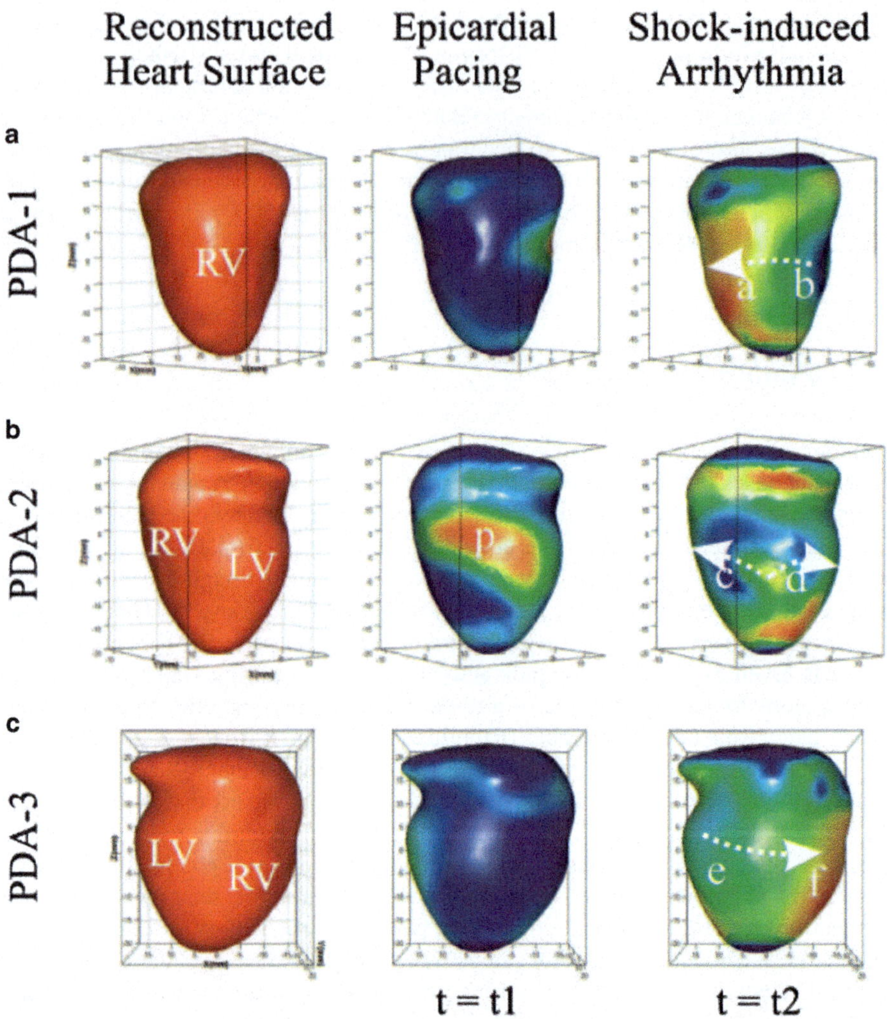

Fig. 14.4 Reconstructed heart surface and epicardial action potential texture mapping. *Left*: reconstructed heart surface visualized from 3 projections (**a**-**c**) of PDA arrays. *Middle*: epicardial action potential texture mapping during epicardial pacing (p in PDA-2 projection is the pacing site). *Right*: epicardial action potential texture mapping during shock-induced ventricular tachycardia. Modified from Qu et al. (Lou et al. 2008)

4 Bidomain Model of Cardiac Muscle

Activation of a myocyte causes passive flow of current into an adjacent quiescent myocyte through gap junctions. This leads to membrane depolarization of the adjacent myocyte and, when threshold is reached, activation of the sodium channels causing excitation. Since an electrical circuit needs to be closed in order to let

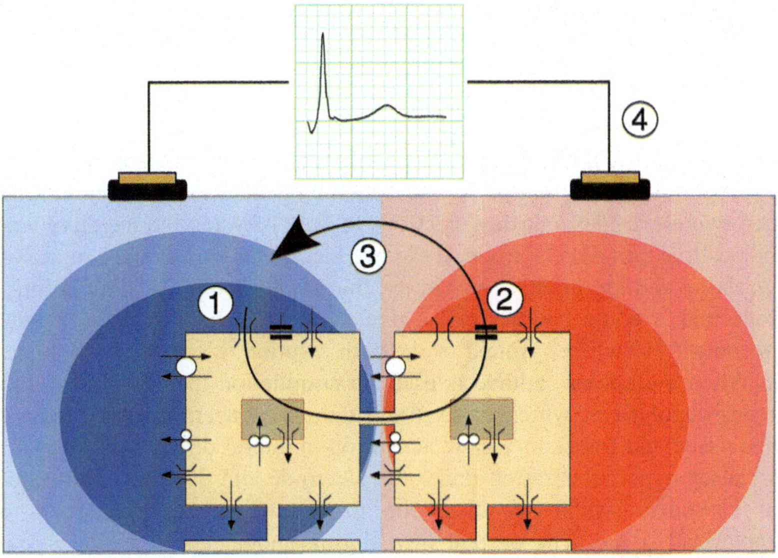

Fig. 14.5 Principle underlying membrane-based heart models. Two model cells are depicted schematically, with their ion channels, pumps, and exchangers. Gap junctions connect the cells. In depolarizing cells, a large inward sodium current flows along its electrochemical gradient (1). This current passes through gap junctions to neighboring cells. There, part of it serves to charge the cell membrane (2) represented by a capacitor symbol, and another part leaks through potassium channels that are still open (3). The two parts form an outward current in this cell, and the current loop is closed in the interstitium and outside the heart (4), where it generates a potential field, schematically indicated here in red for positive potentials and blue for negative potentials. This potential field is picked up as an ECG. This figure is a courtesy of Dr. Mark Potse (Potse 2012)

current flow the intracellular current that flows from cell 1 to cell 2 leads to extracellular flow of current from cell 2 to cell 1. A schematic representation of the relation between intra and extra cellular potentials is shown in Fig. 14.5. This mechanism underlies propagation of the activation front in the myocardium (Kleber and Rudy 2004). Thus, propagation can be sub-divided to an intracellular domain, representing activation of myocytes and passive conduction through gap junction, and an extracellular domain, composed of all other tissue components. The *Bidomain Model of Cardiac Muscle* is based on this concept and was first published in the late 1970s (Miller and Geselowitz 1978). The bidomain model describes the relation between intra and extracellular potentials thereby contributing to the understanding of unipolar electrograms and the simulation of the body surface electrocardiogram (Potse 2012). Current bidomain models use multiple membrane models to compute the intracellular domain and allow to simulate complete ventricular activation based on up to 100 million elements (Niederer et al. 2011). These so-called "Reaction-Diffusion" models make it possible to study the effect of a single ion current on local propagation and even on the body surface electrocardiogram. It is also possible to inject extracellular current and study the effect on intracellular electrophysiology. This approach was used by many groups to study the mechanism of defibrillation computationally (Trayanova et al. 2006).

5 Virtual Electrodes

5.1 *Electrical Stimulation*

Programmed electrical stimulation is essential for studying electrophysiology of the intact myocardium. Stimulation is established by generating a potential difference between two electrodes, causing electrons to flow between the negative (cathode) and the positive (anode) electrode. If the two electrodes both make contact with the cardiac tissue and are in close proximity, this is called bipolar stimulation. If one electrode makes contact with cardiac tissue and the other (indifferent) electrode makes contact elsewhere, ground or the bath solution, it is called unipolar stimulation. When a unipolar cathode is used for stimulation then the tissue under the electrodes depolarizes, which leads to excitation. Adjacent virtual anodes hyperpolarize tissue and transiently slow activation in lateral directions. Anodal stimulation causes hyperpolarization under the electrode and depolarization under two virtual cathodes, which stimulate the tissue at the lateral borders of the hyperpolarized tissue. The current flow during anodal and cathodal stimulation will be discussed in more detail in the Sect. 5.3.

The method of stimulation may influence stimulation threshold or refractoriness of the tissue. For instance, the repolarizing effect of anodal stimulation may shorten action potential duration and lead to short refractory periods or even completely de-excite the tissue restoring full excitability (Mehra et al. 1977). During bipolar stimulation refractory periods are shorter than during cathodal stimulation as well due to anodal stimulation at the proximal electrode (Mehra and Furman 1979). However, the incidence of inducing arrhythmias is not different between bipolar or unipolar stimulation, (Stevenson et al. 1986) unless anodal and cathodal shocks are considered, which do have significant differences in arrhythmogenic potential.

5.2 *Virtual Electrodes*

During cathodal stimulation the cardiac tissue right under the electrode depolarizes. However, the myocardium adjacent to the depolarized myocardium hyperpolarizes due to virtual anodes. Thus, the "real" cathode generates two "virtual" anodes. This phenomenon is referred to as "virtual electrodes" (Sepulveda et al. 1989; Furman et al. 1975). To avoid confusion in the nomenclature, often investigators refer to all areas of positive and negative polarizations as "virtual electrodes", regardless of their proximity to real stimulation electrodes. We will follow this convention. Figure 14.6a shows a computed (passive bidomain model) potential field during central anodal stimulation in a two-dimensional cardiac tissue sheet. The center is highly hyperpolarized (red) whereas the lateral adjacent sides are highly depolarized (blue). In this example the fiber direction is diagonal. The depolarized field expands up and down and shows a "dog-bone-like" morphology. The morphology of the

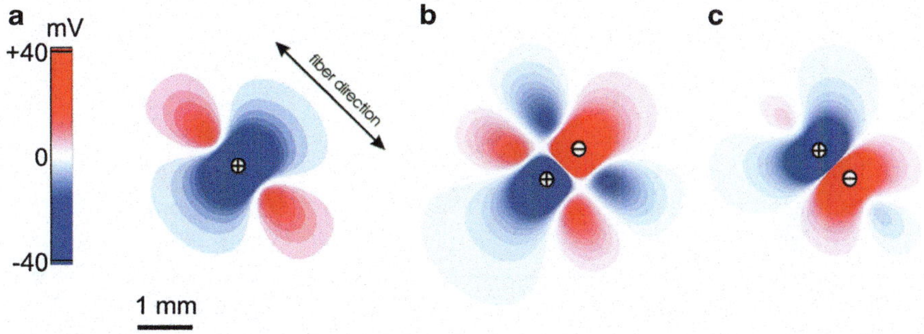

Fig. 14.6 Numerically simulated virtual electrode patterns for an interelectrode distance of 0.8 mm. Results were obtained from the passive two-dimensional bidomain model with unequal anisotropy ratios. (**a**) 40 mA/cm anodal current applied to the center of the 8 × 8-mm square sheet; (**b**) bipolar stimulation with the two electrodes located perpendicular to the direction of the fibers; (**c**) bipolar stimulation with the electrodes located along the direction of the fibers. Zero level corresponds to the resting transmembrane potential (*white*). This figure is modified from Nikolski et al. (2002)

virtual electrode depends on the stimulus strength, fiber direction, and both extra and intracellular resistance (Wikswo et al. 1991). A higher stimulus current causes an increase in the amplitude of positive and negative polarization of corresponding regions. Inversion of the current (i.e. cathodal stimulation) leads to the inverse potential field and virtual electrode pattern. Bipolar stimulation generates virtual electrodes as well (Nikolski and Efimov 2000). The number of virtual electrodes that is generated depends on the location and spacing of the bipole with respect to the fiber direction under the electrodes. When the anode and the cathode are positioned perpendicular to the fiber direction a total of six virtual electrodes will be generated (Fig. 14.6b). If the electrodes are positioned in parallel with the fiber direction a total of four virtual electrodes are observed (Fig. 14.6c) (Nikolski et al. 2002). For comparison, the unipolar stimulation produces a total of three virtual electrodes.

Optical imaging enables experimental testing of the virtual electrode hypothesis in *ex vivo* heart preparations. Figure 14.7 shows the transmembrane potential during stimulation of rabbit epicardium. The measured potential field closely resembles the expected morphology based on the computer simulation.

5.3 Virtual Electrode Make and Break Excitation

During electrical stimulation not only the "real" electrode can cause excitation but the simultaneously generated surrounding virtual electrodes can produce wavefronts as well. The wavefront will propagate from the virtual cathode to the virtual anode. The direction and lifetime of the induced wavefront is directly dependent of the specific VEP pattern. In this section we will only discuss unipolar cathodal and anodal stimulation, because it is directly relevant for understanding

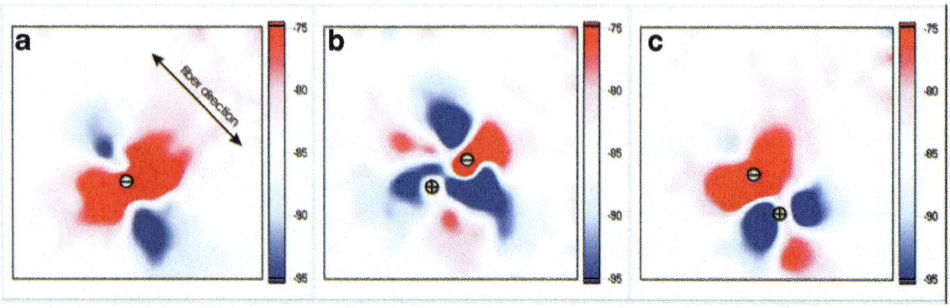

Fig. 14.7 Virtual electrode patterns optically recorded from a 4 × 4-mm area of the rabbit anterior epicardium during unipolar and bipolar stimulation. (**a**) Conventional "dogbone"-shaped virtual electrode polarizations (VEP) during unipolar cathodal stimulation. (**b**) VEP during bipolar stimulation with a pacing dipole placed perpendicular to myocardial fibers. (**c**) Results of bipolar stimulation with electrodes along the fibers. The interelectrode distance was 0.8 mm. Images were collected in the middle of a 2-ms diastolic stimulus. This figure is modified from Nikolski et al. (2002)

the mechanism of defibrillation, which will be reviewed below. There are four mechanisms of unipolar electrical stimulation: cathode make, anode make, cathode break and anode break (Dekker 1970). The "make" excitation occurs just after the beginning of the stimulus pulse, whereas "break" excitation happens after the end of the pulse.

Cathodal make stimulation resembles normal activation of the cardiac tissue during direct injection of depolarizing current via a microelectrode. The negatively charged real cathode causes depolarization of the myocardium right under the electrode. This depolarization causes the membrane to reach excitation threshold and initiates an activation front. Depending on the stimulus current the activation front will start directly under the electrode or from a point farther from the electrode. Nearby negative virtual anodes will slow the conduction but ultimately will be depolarized by the actively propagating wavefront.

The mechanism of excitation during anodal make stimulation resembles that of cathode make stimulation. However, the positively charged real anode causes hyperpolarization of the myocardium right under the electrode. This means that excitation starts at the lateral sides near the virtual cathode. The amplitude of depolarization near virtual cathode during anodal stimulation is lower than near real cathode during cathodal stimulation. Therefore, the threshold of excitation is lower during cathode make than for anode make stimulation.

As mentioned above, break excitation happens at the end or break of the stimulus. Cathode break stimulation occurs when the myocardial tissue directly under the electrode is refractory and unexcitable. However, the shock does still de-excite the tissue at the virtual anodes, rapidly repolarizing the membrane potential even if the tissue was refractory. At the end of the stimulus the current diffuses to the hyperpolarized tissue, exiting to the myocardium through the virtual anodes. This leads to excitation. The path that is created by the shock is parallel with the fiber direction, and the propagation direction into excitable tissue is different than cathode make excitation. The virtual anode is usually much weaker than the real cathode and therefore cathode make excitation will occur over cathode break excitation.

During anodal break excitation the hyper excitable path is perpendicular to the fiber direction. Thus, after the stimulus pulse, the tissue at the virtual cathode repolarizes slowly allowing current to flow into the tissue directly under the anode, which was rapidly de-excited during the shock. This will lead to excitation and a wave front that will propagate transversely before conducting into the normal myocardium.

6 Mechanisms of Defibrillation

As a therapy, defibrillation has a simple objective: restore a synchronized rhythm to the entire myocardium by terminating fibrillation and allowing restoration of normal sinus rhythm. A large electrical shock has been known to successfully achieve this objective for over a century with a long history of clinical use (Gutbrod and Efimov 2013). However, the precise physiological mechanism of the electrical requirements for success has proven to be a much more complex problem. Across decades the principle hypotheses have been shifting and sometimes even contradictory. Optical mapping has played, and continues to play, a critical role in the evolution of the mechanistic hypotheses for successful defibrillation. Before the development of this technique, investigators were hindered by limitations in available recording instrumentation. Electrical shocks of near defibrillation threshold strengths produce large artifacts in signals recorded with both extracellular and intracellular electrodes. These artifacts overwhelm all intrinsic electrical activity in the presence of the applied electric field and blind investigators to the immediate post-shock tissue interactions. The polarization of the electrodes delays the recovery of the recording integrity even after the shock is completely delivered. Figure 14.8 shows an example of this phenomenon from an ex vivo rabbit heart study. The optical and electrical signals are recorded simultaneously when the shock is delivered at the lightning bolt. Both the local bipolar electrogram in contact with the tissue and the far field bath electrode require >1,000 ms to recover with this instrumentation amplifier. Meanwhile the immediate shock-induced depolarization is visible in the optical signal as well as the delay until the arrhythmia resumes.

Some researchers developed creative attempts to minimize the time that the electrical signal was overloaded, including using a non-polarizable reference electrode (only useful if both the tissue and the field are uniform) or by switching amplifier gains or connectivity on the acquisition systems during shock delivery (Chattipakorn et al. 1998). These techniques decreased the recovery lapse in recording time down to 20 ms but it was still too long to observe the immediate shock response. Membrane bound potentiometric dyes translate the intrinsic electrical signal into an optical signal, which is isolated from the applied electric field but reflects the tissue's response. In so doing optical mapping allows for the uninterrupted exclusive visualization of the electrical activity of the tissue before,

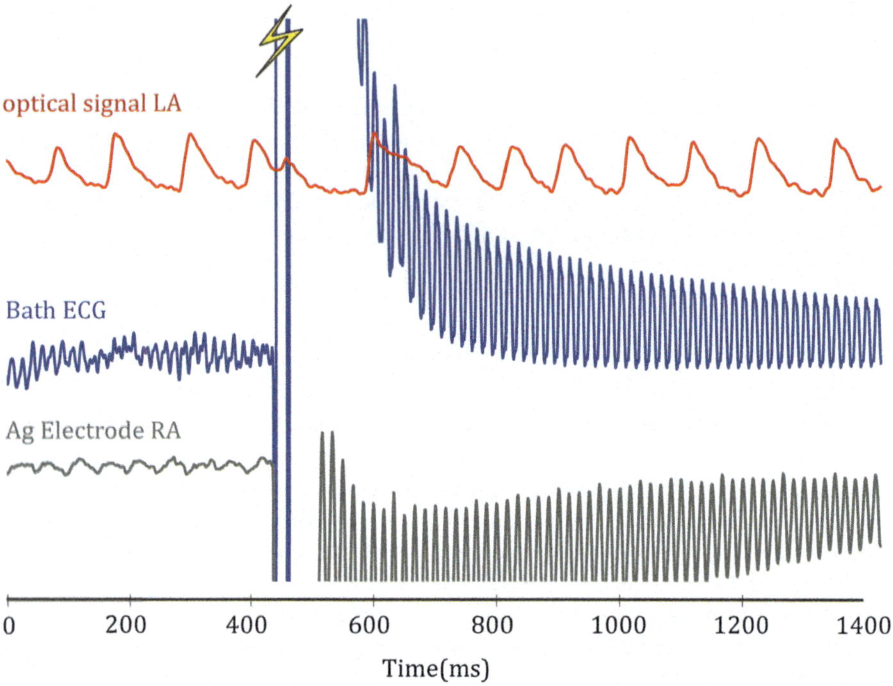

Fig. 14.8 Example of shock-induced artifact on simultaneously acquired electrical and optical signals. These artifacts can be reduced with changes to the instrumentation but cannot be eliminated

during, and immediately after the shock application. The path of current from an externally applied field flows through the extracellular and intracellular subdomains of the myocardium differently, creating voltage gradients that result in changes in transmembrane potentials. The transmembrane changes are the driving forces of the shock-induced restoration of sinus rhythm through initiated wavefronts. Membrane potential dyes measure the electrical activity with specificity for the signal that is mechanistically determining a successful defibrillation that cannot be achieved with extracellular electrodes.

6.1 Historical Mechanisms of Defibrillation

The lingering shock artifacts on electrical recordings limited the ability to predict the post-shock response to applied electrical shocks. However, there were still many theories postulated from the available data on what makes for a successful defibrillation attempt and these theories evolved with advancing technology. The mechanistic theories were also shaped be the changing hypotheses of the substrates that maintain fibrillatory patterns in the heart. Most mechanistic principles are founded in ventricular fibrillation research and are subsequently applied to atrial fibrillation. The main premise of the early electrophysiological theories was that the electric field produced by the shock needed to be strong enough to stimulate the entire myocardium, including regions in various states of repolarization and refractoriness. This theory of total extinction states that only a homogenous depolarization interrupted the fibrillatory patterns and reset the tissue (Hooker et al. 1933).

Some researchers amended this theory asserting that it was sufficient to capture enough tissue to halt all activation fronts, which did not necessarily include the entire excitable epicardium (Wiggers 1940). From this idea the critical mass (CM) hypothesis was formed. The CM hypothesis began as an investigation into the mechanism of maintaining continuous ventricular fibrillation but was quickly applied to achieving successful defibrillation as well. Since the early twentieth century, scientists proposed that there existed a minimum mass of tissue that was able to maintain a sustained fibrillatory pattern (Garrey 1914). If a mass below that limit was selectively isolated, a continuous arrhythmia could not be induced. Correspondingly, a shock could succeed if it interrupted enough activation fronts to isolate the patterns to small regions below the critical mass. Any remaining activation fronts would then self-terminate, leading to successful restoration of sinus rhythm (Zipes et al. 1975).

An alternate theory born of the same early concept is the extension of refractoriness. This idea suggested that an applied shock uniformly extended the refractory period of the captured tissue. Extending action potential durations across the tissue resulted in simultaneous recovery, which wipes away the "memory" of the fibrillatory waves maintaining the arrhythmia (Sweeney et al. 1990). A third theory began to gain ground in the late 1960s. Again, this theory was formed while considering VF induction. Experimentally, applying a shock to the tissue during the vulnerable period of repolarization can induce VF. It was observed that there was an upper limit in shock strength that would induce an arrhythmia (Fabiato et al. 1967). The value empirically corresponded well to the defibrillation threshold shock strength. This observation led researchers to consider whether they could predict the location of reentry caused by incrementally increasing shock strength until they reached the upper limit of vulnerability. From these studies, it was determined that the earliest activation after a shock delivery was located in regions of lowest potential gradient. Therefore in order to be successful a shock must halt activation fronts and achieve a sufficient potential gradient everywhere in the myocardium so as to not re-induce an arrhythmia.

6.2 *Optical Mapping Shapes Defibrillation Mechanism Theories*

Although these theories are not necessarily mutually exclusive, the introduction of optical mapping began to clarify the conflicting elements of existing hypotheses. Optical mapping experiments could target both root causes of the failure of defibrillation: (1) the persistence of pre-shock fibrillatory waves and (2) successful defibrillation followed by re-initiation of new shock-induced wavefronts. Dillon and colleagues conducted the first optical mapping study directly applied to defibrillation mechanisms in 1991. During this study defibrillation shocks were delivered during various phases of a paced rhythm between ring electrodes around the base and apex of isolated hearts. By analyzing the latency between upstroke and shock delivery, the authors could estimate the amount of tissue that is directly captured by various shock strengths compared to tissue that is excited by a shock induced wavefront originating elsewhere in the tissue. Additionally, the authors observed a coupling interval dependent increase in the optical action potential duration in the region of tissue they observed. A shock always elicited an increase in action potential duration in this study. However, the prolongation was not uniform across the ventricle because the tissue was spatially dispersed in states of refractoriness based on the coupling interval. They also did not clearly observe a successful defibrillation, followed by re-induction in the case of failure. Therefore a new theory was proposed called progressive depolarization, attempting to connect each of the existing theories with observed phenomena via optical mapping. This theory states, "Progressively stronger shocks depolarize, progressively more refractory myocardium, to progressively prevent postshock wavefronts, and prolong and synchronize postshock repolarization, in a progressively larger volume of the ventricle, to progressively decrease the probability of fibrillation after the shock" (Dillon and Kwaku 1998). Perhaps the most interesting concept out of this theory is the idea the defibrillation does not have to happen instantaneously, but rather in a continuous fashion as the response spreads spatially across the myocardium. This hypothesis failed to notice that electric shock is capable of both depolarizing and hyperpolarizing the tissue due to virtual cathode and virtual anode effects. If the hyperpolarization is ignored then the progressive depolarization hypothesis presents an incomplete description of defibrillation.

Over two decades of experiments conducted with optical mapping techniques have provided further insight into the mechanisms of defibrillation. This technique has been used to study the effects of electrostimulation across multiple scales, including single cell, cellular monolayers, tissue wedges, and intact hearts. As more questions arose based on how changes in the shape of the applied shocks altered the defibrillation mechanisms, optical mapping experiments rose to attempt to provide an answer. Although computer models were created to predict the cellular response to an applied electric field, voltage-sensitive dyes provided the experimental evidence to rigorously assess these predictions. By shocking isolated myocytes that were aligned with the applied electric field, researchers were able to observe that the transmembrane potential is distributed across the length of a cell when a shock is applied (Knisley et al. 1993). This phenomenon results in a hyperpolarized end near the anode and a depolarized end near the cathode. The strength, timing, and polarity of the shock determine the distribution of the transmembrane potential differences (Cheng et al. 1999b). However, an isolated myocyte cannot be used to study the response of intact tissue to electrical stimulation, therefore, many studies were conducted in cell culture and isolated hearts of various mammalian species.

The virtual electrode hypothesis of defibrillation is presently the most comprehensive explanation of defibrillation. It is based on solid experimental evidence and

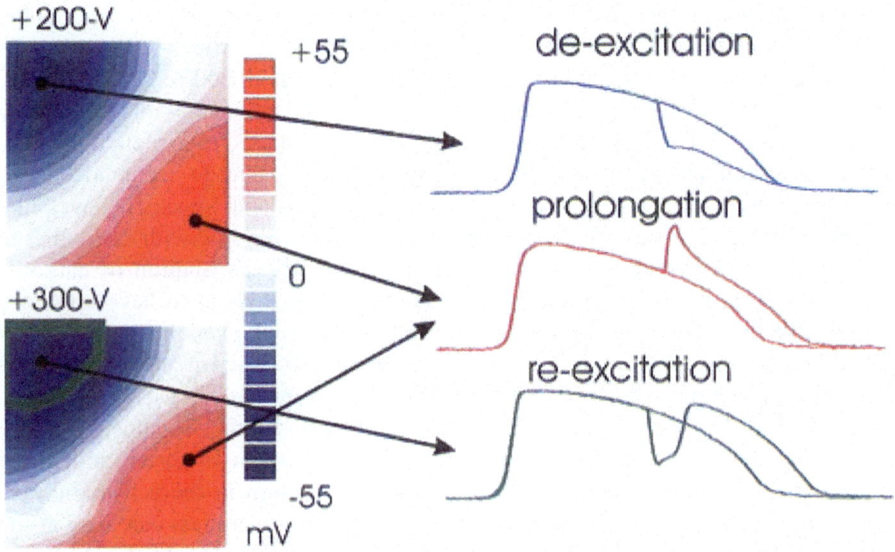

Fig. 14.9 Examples of three local effects to applied shocks: de-excitation, prolongation and re-excitation demonstrating heterogeneous effects on action potential duration. The shock induced responses are shown superimposed on top of control optical action potentials from the same point pre-shock. Modified from Efimov et al. (2000)

theoretical foundation (Ripplinger and Igor 2009). Numerous optical mapping studies presented evidence that in addition to depolarization, shock also was capable to hyper-polarize or de-excite other regions of the heart resulting in action potential shortening instead of the consistent prolongation observed in Dillon's first study. Figure 14.9 shows the results of one such study from an anterior rabbit ventricle with recorded optical action potentials displaying both shock-induced shortening (observed at the virtual anode) and prolongation (observed at the virtual cathode). This observation confirms the Virtual Electrode Hypothesis that was described in detail in the beginning of this chapter. The emergence of the virtual electrode hypothesis provided an alternative theory for defibrillation failure that was not tied to only potential field gradients. Subsequent generalized activation function theory was formulated by Sobie et al., which laid the theoretical foundation for the virtual electrode hypothesis (Sobie et al. 1997). The generalized activating function shows mathematically how a combination of field gradients, underlying heterogeneities in refractoriness and anatomical discontinuities in conduction properties could create neighboring regions of virtual anodes and cathodes, which hyperpolarize and depolarize tissue serving as secondary sources to stimulate the tissue or to induce new fibrillatory wavefronts (Efimov et al. 1997; Ripplinger and Igor 2009; Gillis et al. 2000; Fast et al. 1998).

Using cellular monolayers with patterned regions of discontinuity in the conduction properties, researchers were able to demonstrate how virtual electrode shapes and positions were predictably dependent on shock polarity (White et al. 1998). The spatial regions of hyperpolarization and depolarization mirrored each other with reversed polarity. Additional studies showed that the chirality of the VEP-induced reentrant path could be predicted from the postshock VEP pattern as the wavefront moves into the de-excited region (Cheng et al. 2000). Combining the research of polarity dependent spatial patterns of secondary sources and the reentry chirality dependence, optical mapping experiments were used to explain the mechanism for monophasic anodal shock superiority compared to cathodal shocks (Yamanouchi et al. 2001). The VEP distribution after anodal shocks behaves like a sink; with the induced pathways traveling inward and colliding with each other to self terminate any shock-induced wavefronts. The VEP distribution of cathodal shocks behaves more like a source, leaving more spatial room to develop into a sustainable reentrant wave. Figure 14.10a illustrates the sink phenomenon of anodal shocks with a schematic drawing showing a simple VEP pattern and the crowded inward propagation that is more difficult for a reentrant path to escape. Alternatively Fig. 14.10b shows the source behavior of cathodal shocks, illustrating the space that the virtual anodes provide for a reentrant path to thrive and direct propagation outward to the excitable regions. The pattern in the schematic also shows how phase singularities (represented by solid black circles) can arise at the transition between the virtual electrodes with opposing polarity. In this case, the cathodal shock induces quatrefoil reentry. The conduction velocity of the induced reentry is also dependent on the degree of polarization in the virtual anode; a more negative polarization increases conduction velocity. If the conduction is fast enough, the propagating wave will hit a line of conduction block before this tissue can recovery and an induced wavefront with self-terminate. Furthermore, optical mapping helped reveal the mechanism behind the empirical observation that biphasic shocks were superior compared to monophasic shocks. Biphasic shocks immediately reverse the polarity and distribution of the virtual electrodes, greatly decreasing the heterogeneity in the post-shock tissue response. The mirrored VEP pattern also removes the excitable de-excited tissue, reducing the risk of

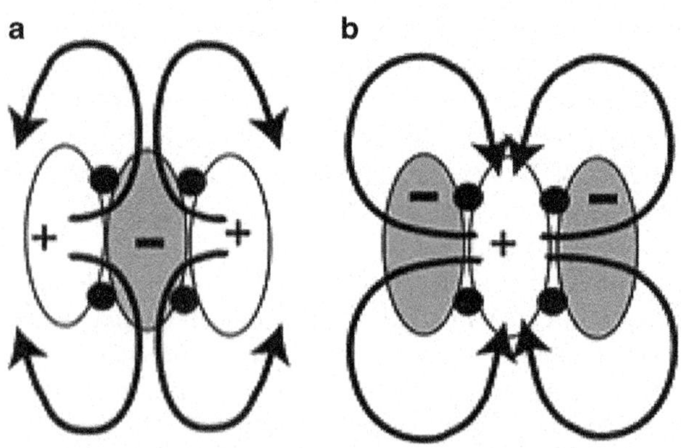

Fig. 14.10 Schematic of shock induced effects that can lead to re-entry. (**a**) Post-shock activation of an anodal shock. (**b**) Post-shock activation of a cathodal shock, which is more likely to sustain reentry. Modified from Yamanouchi et al. (2001)

VEP-induced phase singularities (Efimov et al. 1998). Another example of the application of optical mapping to explain defibrillation optimization is observed when comparing the VEP elicited by various waveform shapes. Multiple studies have shown that ascending waveforms are advantageous empirically, resulting in a lower defibrillation threshold. Mechanistically this is because the ascending VEP pattern achieves a greater degree of polarization, leading to a faster conduction velocity and an induced wavefront that is more susceptible to self-termination.

6.3 Virtual Electrode Hypothesis of Defibrillation in the Clinic

Optical mapping studies conducted in 1990s have helped to explain the mechanisms of electrical stimulation of the myocardium by point near-field stimulation and by far field shocks. Prior to the advent of voltage-sensitive dyes this was impossible. Subsequent optical mapping studies focusing on defibrillation explained numerous empirically known phenomena including: superiority of anodal versus cathodal monophasic shock defibrillation, superiority of cathodal versus anodal pacing, superiority of biphasic versus monophasic defibrillation, lower and upper limits of vulnerability, role of electroporation, etc. However, the defibrillation paradigm used presently is not without limitations. It requires significant shock strength to achieve successful defibrillation. Several studies have suggested that such strong electric shocks could be associated with myocardial damage and lead to increased mortality and progression of heart failure (Tereshchenko et al. 2009). Therefore, researchers are looking for new methods for reducing the energy required for defibrillation. One such method has emerged from the finding that local tissue heterogeneities lead to the development of virtual cathodes and anodes during far field stimulation (Fishler 1997). When a low energy shock is applied during critical phase of reentry, it can unpin and terminate a reentry-based arrhythmia (Ripplinger et al. 2006). Subsequent optical mapping studies have explained the mechanism of this low energy approach (Ripplinger et al. 2009) and can be used to guide the development of new defibrillation strategies, some of which are currently being tested clinically (Janardhan et al. 2012, 2014).

7 Conclusion

Optical mapping with voltage sensitive dyes has revolutionized cardiac electrophysiology, providing a platform for high-resolution investigation of fundamental mechanisms of normal and pathological electrophysiological processes. Cardiac conduction and repolarization, re-entry and focal activity, pacing and defibrillation could not be fully understood without this powerful technology. However, we would advocate that the studies of defibrillation and pacing are especially indebted to optical mapping, because alternative electrode-based techniques could not be used in studies involving application of strong electric fields. Optical mapping is the critically important methodology, which allowed elucidating the mechanisms of pacing and defibrillation.

References

Allessie MA, Bonke FI, Schopman FJ (1977) Circus movement in rabbit atrial muscle as a mechanism of tachycardia. III. The "leading circle" concept: a new model of circus movement in cardiac tissue without the involvement of an anatomical obstacle. Circ Res 41:9–18

Bishop MJ, Gavaghan DJ, Trayanova NA, Rodriguez B (2007) Photon scattering effects in optical mapping of propagation and arrhythmogenesis in the heart. J Electrocardiol 40:S75–S80

Boukens B, Efimov I (2014) A century of optocardiography. IEEE Rev Biomed Eng 7:115–125

Chattipakorn N, KenKnight BH, Rogers JM, Walker RG, Walcott GP, Rollins DL et al (1998) Locally propagated activation immediately after internal defibrillation. Circulation 97:1401–1410

Cheng Y, Mowrey KA, Van Wagoner DR, Tchou PJ, Efimov IR (1999a) Virtual electrode induced re-excitation: a basic mechanism of defibrillation. Circ Res 85:1056–1066

Cheng DK, Tung L, Sobie EA (1999b) Nonuniform responses of transmembrane potential during electric field stimulation of single cardiac cells. Am J Physiol 277:H351–H362

Cheng Y, Nikolski V, Efimov IR (2000) Reversal of repolarization gradient does not reverse the chirality of shock-induced reentry in the rabbit heart. J Cardiovasc Electrophysiol 11:998–1007

Comtois P, Kneller J, Nattel S (2005) Of circles and spirals: bridging the gap between the leading circle and spiral wave concepts of cardiac reentry. Europace 7(Suppl 2):10–20

Coronel R, Wilms-Schopman FJG, Opthof T, Janse MJ (2009) Dispersion of repolarization and arrhythmogenesis. Heart Rhythm 6:537–543

de Bakker JM, van Capelle FJ, Janse MJ, Wilde AA, Coronel R, Becker AE et al (1988) Reentry as a cause of ventricular tachycardia in patients with chronic ischemic heart disease: electrophysiologic and anatomic correlation. Circulation 77:589–606

Dekker E (1970) Direct current make and break thresholds for pacemaker electrodes on the canine ventricle. Circ Res 27:811–823

Dillon SM (1991) Optical recordings in the rabbit heart show that defibrillation strength shocks prolong the duration of depolarization and the refractory period. Circ Res 69:842–856

Dillon SM, Kwaku KF (1998) Progressive depolarization: a unified hypothesis for defibrillation and fibrillation induction by shocks. J Cardiovasc Electrophysiol 9:529–552

Dower GE (1962) In Defence of the Intrinsic Deflection. Br Heart J 24:55–60

Durrer D, Schoo L, Schuilenburg RM, Wellens HJ (1967) The role of premature beats in the initiation and the termination of supraventricular tachycardia in the Wolff-Parkinson-White syndrome. Circulation 36:644–662

Efimov I, Salama G (2012) The future of optical mapping is bright: RE: review on: "optical imaging of voltage and calcium in cardiac cells and tissues" by Herron, Lee, and Jalife. Circ Res 110:e70–e71

Efimov IR, Cheng YN, Biermann M, Van Wagoner DR, Mazgalev T, Tchou PJ (1997) Transmembrane voltage changes produced by real and virtual electrodes during monophasic defibrillation shock delivered by an implantable electrode. J Cardiovasc Electrophysiol 8:1031–1045

Efimov IR, Cheng Y, Van Wagoner DR, Mazgalev T, Tchou PJ (1998) Virtual electrode-induced phase singularity: a basic mechanism of defibrillation failure. Circ Res 82:918–925

Efimov IR, Gray RA, Roth BJ (2000) Virtual electrodes and deexcitation: new insights into fibrillation induction and defibrillation. J Cardiovasc Electrophysiol 11:339–353

Efimov IR, Nikolski VP, Salama G (2004) Optical imaging of the heart. Circ Res 95:21–33

Fabiato A, Coumel P, Gourgon R, Saumont R (1967) The threshold of synchronous response of the myocardial fibers. Application to the experimental comparison of the efficacy of different forms of electroshock defibrillation. Arch Mal Coeur Vaiss 60:527–544

Fast VG, Kleber AG (1997) Role of wavefront curvature in propagation of cardiac impulse. Cardiovasc Res 33:258–271

Fast VG, Rohr S, Gillis AM, Kleber AG (1998) Activation of cardiac tissue by extracellular electrical shocks: formation of 'secondary sources' at intercellular clefts in monolayers of cultured myocytes. Circ Res 82:375–385

Fedorov VV, Lozinsky IT, Sosunov EA, Anyukhovsky EP, Rosen MR, Balke CW et al (2007) Application of blebbistatin as an excitation-contraction uncoupler for electrophysiologic study of rat and rabbit hearts. Heart Rhythm 4:619–626

Fishler MG (1997) The transient far-field response of a discontinuous one-dimensional cardiac fiber to subthreshold stimuli. IEEE Trans Biomed Eng 44:10–18

Furman S, Hurzeler P, Parker B (1975) Clinical thresholds of endocardial cardiac stimulation: a long-term study. J Surg Res 19:149–155

Garrey W (1914) The nature of fibrillary contraction of the heart: its relation to tissue mass and form. Am J Physiol 33:17

Gillis AM, Fast VG, Rohr S, Kleber AG (2000) Mechanism of ventricular defibrillation. The role of tissue geometry in the changes in transmembrane potential in patterned myocyte cultures. Circulation 101:2438–2445

Gutbrod SR, Efimov IR (2013) Two centuries of resuscitation. J Am Coll Cardiol 62:2110–2111

Hoffman BF, Rosen MR (1981) Cellular mechanisms for cardiac arrhythmias. Circ Res 49:1–15

Hooker DR, Kouwenhoven WB, Langworthy OR (1933) The effect of alternating electrical currents on the heart. Am J Physiol 103:444–454

Huikuri HV, Castellanos A, Myerburg RJ (2001) Sudden death due to cardiac arrhythmias. N Engl J Med 345:1473–1482

Janardhan AH, Li W, Fedorov VV, Yeung M, Wallendorf MJ, Schuessler RB et al (2012) A novel low-energy electrotherapy that terminates ventricular tachycardia with lower energy than a biphasic shock when antitachycardia pacing fails. J Am Coll Cardiol 60:2393–2398

Janardhan AH, Gutbrod SR, Li W, Lang D, Schuessler RB, Efimov IR (2014) Multistage electrotherapy delivered through chronically-implanted leads terminates atrial fibrillation with lower energy than a single biphasic shock. J Am Coll Cardiol 63:40–48

Janks DL, Roth BJ (2002) Averaging over depth during optical mapping of unipolar stimulation. IEEE Trans Biomed Eng 49:1051–1054

Janse MJ, Rosen MR (2006) History of arrhythmias. Handb Exp Pharmacol (171): 1–39

Jeffrey K (2001) Machines in our hearts: the cardiac pacemaker, the implantable defibrillator, and american health care. Johns Hopkins University Press, Baltimore

Josephson ME, Horowitz LN, Farshidi A (1978) Continuous local electrical activity. A mechanism of recurrent ventricular tachycardia. Circulation 57:659–665

Kay MW, Amison PM, Rogers JM (2004) Three-dimensional surface reconstruction and panoramic optical mapping of large hearts. IEEE Trans Biomed Eng 51:1219–1229

Kleber AG, Rudy Y (2004) Basic mechanisms of cardiac impulse propagation and associated arrhythmias. Physiol Rev 84:431–488

Knisley SB, Blitchington TF, Hill BC, Grant AO, Smith WM, Pilkington TC et al (1993) Optical measurements of transmembrane potential changes during electric field stimulation of ventricular cells. Circ Res 72:255–270

Kong CY, Nattinger KJ, Hayeck TJ, Omer ZB, Wang YC, Spechler SJ et al (2011) The impact of obesity on the rise in esophageal adenocarcinoma incidence: estimates from a disease simulation model. Cancer Epidemiol Biomarkers Prev 20:2450–2456

Lou Q, Ripplinger CM, Bayly PV, Efimov IR (2008) Quantitative panoramic imaging of epicardial electrical activity. Ann Biomed Eng 36:1649–1658

Lou Q, Li W, Efimov IR (2012) The role of dynamic instability and wavelength in arrhythmia maintenance as revealed by panoramic imaging with blebbistatin vs. 2,3-butanedione monoxime. Am J Physiol Heart Circ Physiol 302:H262–H269

Mehra R, Furman S (1979) Comparison of cathodal, anodal, and bipolar strength-interval curves with temporary and permanent pacing electrodes. Br Heart J 41:468–476

Mehra R, Furman S, Crump JF (1977) Vulnerability of the mildly ischemic ventricle to cathodal, anodal, and bipolar stimulation. Circ Res 41:159–166

Miller WT, Geselowitz DB (1978) Simulation studies of the electrocardiogram. I. The normal heart. Circ Res 43:301–315

Mines GR (1913) On dynamic equilibrium of the heart. J Physiol 46:349–382

Mines GR (2010) On circulating excitations in heart muscle and their possible relation to tachycardia and fibrillation. Trans Roy Soc Can 8:43

Niederer S, Mitchell L, Smith N, Plank G (2011) Simulating human cardiac electrophysiology on clinical time-scales. Front Physiol 2:14

Nikolski V, Efimov IR (2000) Virtual electrode polarization of ventricular epicardium during bipolar stimulation. J Cardiovasc Electrophysiol 11:605

Nikolski VP, Sambelashvili AT, Efimov IR (2002) Mechanisms of make and break excitation revisited: paradoxical break excitation during diastolic stimulation. Am J Physiol Heart Circ Physiol 282:H565–H575

Pogwizd SM, Corr PB (1987) Reentrant and nonreentrant mechanisms contribute to arrhythmogenesis during early myocardial ischemia: results using three-dimensional mapping. Circ Res 61:352–371

Potse M (2012) Mathematical modeling and simulation of ventricular activation sequences: implications for cardiac resynchronization therapy. J Cardiovasc Transl Res 5:146–158

Ramshesh VK, Knisley SB (2003) Spatial localization of cardiac optical mapping with multiphoton excitation. J Biomed Opt 8:253–259

Ripplinger CME, Igor R (2009) The virtual electrode hypothesis of defibrillation. In: Efimov IR, Kroll MW, Tchou P (eds) Cardiac bioelectric therapy: mechanisms and practical implications. Springer Science + Business Media, New York

Ripplinger CM, Krinsky VI, Nikolski VP, Efimov IR (2006) Mechanisms of unpinning and termination of ventricular tachycardia. Am J Physiol Heart Circ Physiol 291:H184–H192

Ripplinger CM, Lou Q, Li W, Hadley J, Efimov IR (2009) Panoramic imaging reveals basic mechanisms of induction and termination of ventricular tachycardia in rabbit heart with chronic infarction: implications for low-voltage cardioversion. Heart Rhythm 6:87–97

Rogers JM, Walcott GP, Gladden JD, Melnick SB, Kay MW (2007) Panoramic optical mapping reveals continuations epicardial reentry during ventricular fibrillation in the isolated swine heart. Biophys J 92:1090–1095

Sepulveda NG, Wikswo JP (1994) Bipolar stimulation of cardiac tissue using an anisotropic bidomain model. J Cardiovasc Electrophysiol 5:258–267

Sepulveda NG, Roth BJ, Wikswo JP (1989) Current injection into a two-dimensional anisotropic bidomain. Biophys J 55:987–999

Shenassa M, Borggrefe M, Briethardt G (2013) Cardiac mapping, 4th edn. Elsmford Blackwell Publishing/Futura Division, Armonk, NY

Sobie EA, Susil RC, Tung L (1997) A generalized activating function for predicting virtual electrodes in cardiac tissue. Biophys J 73:1410–1423

Spach MS, Barr RC, Serwer GA, Kootsey JM, Johnson EA (1972) Extracellular potentials related to intracellular action potentials in the dog Purkinje system. Circ Res 30:505–519

Stevenson WG, Wiener I, Weiss JN (1986) Comparison of bipolar and unipolar programmed electrical stimulation for the initiation of ventricular arrhythmias: significance of anodal excitation during bipolar stimulation. Circulation 73:693–700

Sweeney RJ, Gill RM, Steinberg MI, Reid PR (1990) Ventricular refractory period extension caused by defibrillation shocks. Circulation 82:965–972

Taccardi B, Arisi G, Macchi E, Baruffi S, Spaggiari S (1987) A new intracavitary probe for detecting the site of origin of ectopic ventricular beats during one cardiac cycle. Circulation 75:272–281

Tereshchenko LG, Faddis MN, Fetics BJ, Zelik KE, Efimov IR, Berger RD (2009) Transient local injury current in right ventricular electrogram after implantable cardioverter-defibrillator shock predicts heart failure progression. J Am Coll Cardiol 54:822–828

Trayanova N, Plank G, Rodriguez B (2006) What have we learned from mathematical models of defibrillation and postshock arrhythmogenesis? Application of bidomain simulations. Heart Rhythm 3:1232–1235

White JB, Walcott GP, Pollard AE, Ideker RE (1998) Myocardial discontinuities: a substrate for producing virtual electrodes that directly excite the myocardium by shocks. Circulation 97:1738–1745

Wiener N, Rosenblueth A (1946) The mathematical formulation of the problem of conduction of impulses in a network of connected excitable elements, specifically in cardiac muscle. Arch Inst Cardiol Mex 16:205–265

Wiggers CJ (1940) The physiological basis for cardiac resuscitation from ventricular fibrillation-method for serial defibrillation. Am Heart J 20:413–422

Wikswo JP Jr, Wisialowski TA, Altemeier WA, Balser JR, Kopelman HA, Roden DM (1991) Virtual cathode effects during stimulation of cardiac muscle Two-dimensional in vivo experiments. Circ Res 68:513–530

Wikswo JP, Lin SF, Abbas RA (1995) Virtual electrodes in cardiac tissue: a common mechanism for anodal and cathodal stimulation. Biophys J 69:2195–2210

Yamanouchi Y, Cheng Y, Tchou PJ, Efimov IR (2001) The mechanisms of vulnerable window: the role of virtual electrodes and shock polarity. Can J Physiol Pharmacol 79:25–33

Zipes DP, Fischer J, King RM, Nicoll AD, Jolly WW (1975) Termination of ventricular fibrillation in dogs by depolarizing a critical amount of myocardium. Am J Cardiol 36:37–44

Chapter 15
Biophotonic Modelling of Cardiac Optical Imaging

Martin J. Bishop and Gernot Plank

Abstract Computational models have been recently applied to simulate and better understand the nature of fluorescent photon scattering and optical signal distortion during cardiac optical imaging. The goal of such models is both to provide a useful post-processing tool to facilitate a more accurate and faithful comparison between computational simulations of electrical activity and experiments, as well as providing essential insight into the mechanisms underlying this distortion, suggesting ways in which it may be controlled or indeed utilised to maximise the information derived from the recorded fluorescent signal. Here, we present different modelling methodologies developed and used in the field to simulate both the explicit processes involved in optical signal synthesis and the resulting consequences of the effects of photon scattering within the myocardium upon the optically-detected signal. We focus our attentions to two main types of modelling approaches used to simulate light transport in cardiac tissue, specifically continuous (reaction-diffusion) and discrete stochastic (Monte Carlo) methods. For each method, we provide both a summary of the necessary methodological details of such models, in addition to brief reviews of relevant application studies which have sought to apply these methods to elucidate important information regarding experimentally-recorded optical signals under different circumstances.

Keywords Optical mapping • Photon scattering • Fluorescent imaging • Cardiac electrophysiology • Bidomain modelling • Cardiac arrhythmias • Defibrillation • Cardiac action potential • Photon diffusion equation • Monte Carlo modelling

1 Introduction

Optical mapping is a widely used experimental technique capable of providing high-resolution spatiotemporal recordings of electrophysiological activity from the surface of the heart. The method is chiefly used in conjunction with specialized

M.J. Bishop (✉) • G. Plank
Department of Biomedical Engineering, King's College London,
St. Thomas' Hospital, London SE1 7EM, UK
e-mail: martin.bishop@kcl.ac.uk

M. Canepari et al. (eds.), *Membrane Potential Imaging in the Nervous System and Heart*, Advances in Experimental Medicine and Biology 859,
DOI 10.1007/978-3-319-17641-3_15

membrane-bound fluorescent dyes, which, upon illumination at the correct wavelength, transduce local changes in transmembrane potential as changes in fluorescent emission which are recorded by optical detectors. At the wavelengths commonly used in optical mapping, however, light is known to be highly scattering and relatively weakly absorbing in cardiac tissue. Illumination light incident onto the surface penetrates relatively deeply (a few millimeters) into the tissue depth, and fluorescence emitted from these depths may scatter and pass out of the surface to be detected. Consequently, ever since its inception, optical mapping signals recorded from intact three-dimensional tissue have been known not to provide a 100 % faithful representation of the electrical activity from the exact surface pixel imaged by the optical detector. Instead, the signals are thought to contain information regarding the electrophysiological state of tissue within a three-dimensional volume underneath the surface recording site from which the measured fluorescence originates. The resulting distortion in optical mapping signals from this averaging effect of electrical states within a subsurface volume compromises experimental data interpretation, and limits the use of optical recordings to validate computer simulations of electrical activity.

Knowledge of this artifact was originally suggested to primarily underlie the significant prolongation in the optical action potential upstroke duration, frequently recorded to be 5–10 ms in fluorescent signals compared to the 1–2 ms measured in microelectrode recordings (Gray 1999; Girouard et al. 1996). However, more recently, the distortion effects of signal collection from depth have been suggested to contain vital information, beyond just being a problematic artifact that prolongs upstrokes to be recognised and accounted for. Instead, this 'artifact' is now acknowledged to provide a useful tool in both measuring and understanding the *intramural* electrical activity occurring deep within the myocardial wall. For example, it has been suggested that collection of signatures of electrophysiological potential levels from depth may help to understand the surface and intramural response of cardiac tissue to strong electrical stimuli (Efimov et al. 2000; Rodriguez et al. 2005), the capturing of wholly intramural scroll-wave activity during reentrant arrhythmias (Efimov et al. 1999), as well as being able to deduce the specific orientation of subsurface propagating wavefronts during pacing (Hyatt et al. 2005).

Computational models of cardiac electrical activity have been used with much success to enrich our knowledge of a wide variety of physiological and pathological phenomena through integrating a fully three-dimensional understanding of electrophysiological behaviour from the sub-cellular to whole organ level. Despite such models often being constructed directly from experimental anatomical imaging data and containing parameters regarding tissue electrical conductivities and ionic membrane dynamics derived directly from experiment, a key component in fully utilising the predictions from such models lies in some form of experimental validation of the final overall results (Pathmanathan and Gray 2013). However, the presence of the distortion effects due to photon scattering in optical mapping—most often the measurement technique of choice due to its high spatial and temporal resolution and ability to measure activity during strong shocks—renders such direct comparison problematic. These issues therefore suggest the use of a combined electrophysiological and optical simulation model to synthesise the fluorescent signals recorded in optical mapping. Firstly, the use of such a combined model will provide an essential post-processing tool to be applied to the 'raw' electrophysiological data outputted from standard mono-/bidomain models to facilitate a much closer comparison with experimental recordings, improving the validation procedure of simulation results. Secondly, using a model to directly synthesise experimental signals facilitates a more complete understanding of the three-dimensional interaction between light scattering and the complex ventricular anatomy and wavefront propagation patterns that underlie specific characteristics of experimentally-recorded optical signals during paced rhythms as well as following strong shocks and during episodes of arrhythmia. Obtaining such detailed knowledge is imperative to fully utilise all the information contained within optical mapping signals regarding not only surface, but also intramural electrophysiological activity, and to even help guide the adaptation and development of novel optical imaging procedures to maximise this critical information.

In this Chapter, we review the methodological developments and applications of biophotonic models of cardiac optical imaging. We begin by discussing the experimental findings which highlighted the requirement for, and drove the further development of, such models. We then describe the original, more simplified models, which focussed purely on the averaging effects of signal 'pick-up' from depth along with the successful applications of such approaches. Finally, we move to discuss more detailed modelling approaches which aim to represent the full three-dimensional effects of photon scattering and subsequent signal collection from subsurface tissue volumes. We describe the methodological details underlying both continuum approaches based on the photon diffusion equation and the more powerful discrete stochastic (Monte Carlo) models which are able to simulate individual photon trajectories through tissue, along with a thorough review of the application studies which used these approaches to understand complex aspects of optically-recorded experimental signals.

2 Origin of Collected Fluorescent Signal

As it became generally accepted that the optical signal distortion arose due to signal pick-up from deeper layers of myocardium, controversy arose as to the exact depth of tissue which contributes to the recorded fluorescent signal. Assuming that light incident normal to the tissue surface decays in a monoexponential fashion, Girouard et al. (1996) derived decay constants from the transmission spectra of guinea-pig ventricular tissue, finding them to be 0.289 and 0.434 mm at the excitation and emission wavelengths of the voltage-sensitive dye DI-4-ANEPPS, respectively. Further, assuming that the total optical signal measured from the heart surface was the product of these two exponentials (one representing the transmission of light into the tissue, the other representing fluorescent light transmitted out of the tissue) the authors were able to estimate the total contribution of recorded light from varying depths. They suggested that over 95 % of the collected signal from the epicardium originated from a depth of < 500 μm with the first 100 μm contributing almost half of this signal. A study by Knisley (1995) also found fluorescent signals to originate from approximately 0.3 mm in recordings from rabbit ventricle preparations using DI-4-ANEPPS. These findings lead to the widely-held view that fluorescent recordings in optical mapping experiments originated from a thickness of just 0.3 mm beneath the recording surface.

However, later experimental investigations produced intriguing findings which challenged this widely-held view. In particular, in two separate studies by Efimov et al. (1999, 2000), the existence of strange 'dual-humped' action potentials were seen in optical recordings of sustained arrhythmias in isolated rabbit heart preparations. These dual-humped action potentials, whereby a second 'hump' is seen during the plateau phase of the optical trace, were attributed to transmural reentry with a wholly intramural filament lying close to the epicardial surface. It was suggested that the passage of an entirely intramural wavefront, associated with the filament of the reentrant circuit, with no signature itself on the epicardium, may produce a deflection in the surface recorded optical signal due to the effects of fluorescent signal averaging from depth. Due to the fact that, from activation wavelength considerations, wholly intramural reentry is not thought to be able to occur in just 0. 3 mm depth, these findings lead to the belief that optically-recorded signals can in fact originate from much larger depths beneath the tissue surface (Gray 1999).

Support for this interpretation of the source of optical mapping signals from greater than previously believed depths was also found in two other studies which used fluorescent recordings to investigate impulse propagation within the atrioventricular node region (Choi and Salama 1998; Efimov and Mazgalev 1998). The multiple upward deflections seen in the optically recorded action potentials from this region were attributed to signal 'pick-up' from different depths in this complex, multi-layered structure. The evidence provided by these studies required a re-evaluation of the depth from which optical signals were sampled (Gray 1999).

Experimental confirmation of the increased depth from which optical signals were thought to originate came later in two studies which sought to quantify the optical parameters associated with light penetration into cardiac tissue. Ding et al. (2001) used measurements of diffuse reflectance and transmittance from thin slices of rabbit ventricular tissue 0. 5 mm thick stained with DI-4-ANEPPS to determine optical coefficients for absorption and scattering. These measurements predicted exponential decay constants of 0. 58 mm at the excitation (488 nm) and 1. 84 mm at the emission (669 nm) wavelengths used. Baxter et al. (2001) performed measurements of light attenuation within thick (8 mm) slabs of sheep ventricular myocardium, again at the excitation (520 ± 30 nm) and emission (640 ± 50 nm wavelengths of DI-4-ANEPPS. Fitted decay constants showed the *penetration depth* to be significantly shorter at the excitation wavelength ($\delta_{ex} = 0.77$ mm) compared to the emission wavelength

($\delta_{em} = 1.36$ mm), highlighting that light is attenuated more strongly by the tissue during illumination. Both of these studies provided strong evidence that light at the wavelengths typically encountered in optical mapping experiments could penetrate deeply through the myocardial tissue. Thus, a significant fraction of the total fluorescent signal was shown to be collected from greater than 1 mm beneath the tissue surface which could have hugely important implications in the interpretation of the collected signal. This information was, in fact, also used later on in both of the above studies within mathematical models to attempt to better understand such implications of this light penetration.

2.1 Controversy Over Optical Properties of Myocardium

A potential issue closely related to the different collection depths and findings in the above studies could be related to the variability in optical properties of myocardial tissue, both between different animal species and within a population. A significant variation in optical penetration depths at both the excitation and emission wavelengths of DI-4-ANEPPS was witnessed between measurements described in guinea pig (Girouard et al. 1996), sheep (Baxter et al. 2001) and rabbit (Knisley 1995; Ding et al. 2001), ranging from approximately 0.3–0.8 mm during excitation and 0.4–1.8 mm during emission. Such a wide variation in optical properties of myocardial tissue between species is also highlighted in a thorough review of optical properties of biological tissue by Cheong et al. (1990). Although such differences could be partially explained by discrepancies between the specific optical measurement techniques employed in the studies, inter-species differences in histo-anatomical myocardial tissue structure could also be important.

In addition to large differences between species, there is also a large variability in optical properties between different tissue preparations from the same species. For example, in the study by Ding et al. (2001) the optical absorption coefficient measured from five similar rabbit ventricular preparations co-loaded with dyes Rh237 and Oregon Green at 488 nm was found to have a value of $0.15\,\text{mm}^{-1}$ with a standard deviation as large as $0.10\,\text{mm}^{-1}$ between the five measurements. This study also highlighted the large difference in optical properties of tissue dependent upon the specific dye with which it is stained. The authors measured the absorption coefficient of tissue stained with DI-4-ANEPPS to be $0.52 \pm 0.13\,\text{mm}^{-1}$, over threefold larger than the co-loaded tissue described above. Furthermore, as the wavelength was increased from 488 to 669 nm, the absorption coefficient was seen to decrease significantly to just 0.10 ± 0.04^{-1}. It is important to consider these studies which highlight the variability in optical properties and measured penetration depths between preparations when interpreting fluorescent signals collected in optical mapping studies.

3 Modelling of Depth-Weighted Averaging of Recorded Fluorescent Signal

As it became widely accepted that light in optical mapping experiments penetrated significant depths into the myocardium, so there developed the need to quantitatively understand the specific origin of the recorded fluorescent signals and to help interpret the *effects* this had on the optical measurements. Computational modelling provides an ideal approach to achieve these goals, facilitating a detailed understanding of the exact mechanisms by which signal collection from depth may cause some of the unusual experimentally-documented characteristics of the optical signal.

3.1 Mathematical Representation of Signal Collection from Depth

The Baxter et al. (2001) study, introduced above, explicitly determined the fluence profiles of light attenuation at both the excitation and emission wavelengths of DI-4-ANEPPS, shown in Fig. 15.1a. Both profiles showed distinct subsurface peaks due to the combined effects of escape of photons from the tissue surface along with back-scattering from the surface interface. For tissue depths > 1 mm, the profiles demonstrated a mono-exponential decay. The study then used this information to mathematically represent the excitation profile, accounting for this subsurface peak, by the following expression (Gardner and Jacques 1996)

$$\phi(z) = C_1 \exp(-k_1 z/\delta_{ex}) - C_2 \exp(-k_2 z/\delta_{ex}), \tag{15.1}$$

where $\phi(z)$ is the excitation fluence as a function of depth (z). The emission profile was represented by a simple mono-exponential function, as the subsurface peak is not relevant for fluorescent photons originating within the tissue, only their rate of attenuation, given by

$$G(z) = C_3 \exp(-k_3 z/\delta_{em}), \tag{15.2}$$

where $G(z)$ is the emission fluence as a function of depth z. The free parameters k_i, C_i in Eqs. (15.1) and (15.2) were fitted from the data to have values $k_1, k_3, C_3 = 1$, $k_2 = 1.8$, $C_1 = 927$, $C_2 = 702$ (Gardner and Jacques 1996). Figure 15.1b shows experimentally-derived profiles at the excitation and emission wavelengths along with the fitted mathematical functions used to describe them.

Importantly, Baxter et al. (2001) then used the expressions for $\phi(z)$ and $G(z)$ to derive a *weighting function* $w(z)$, defined as the convolution of the excitation and emission profiles (Girouard et al. 1996), normalised by the total fluorescent light emitted from the entire tissue

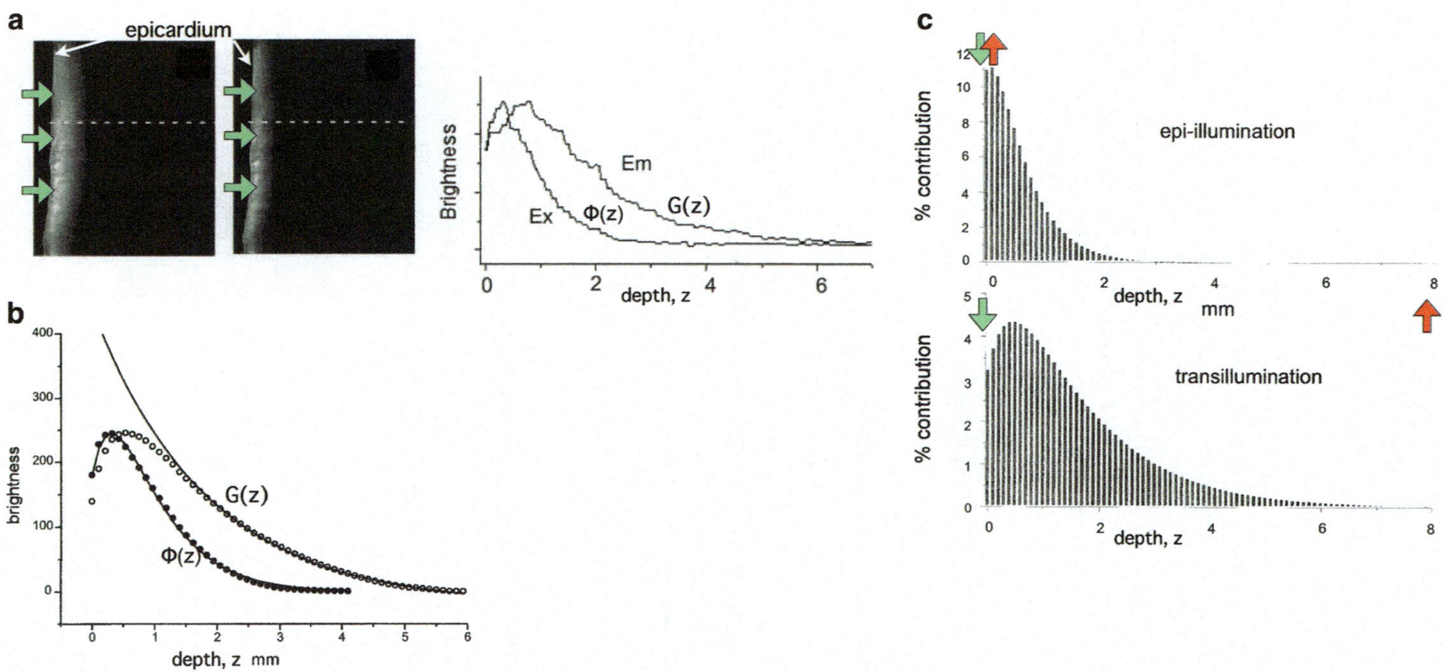

Fig. 15.1 Model of fluence profiles of excitation and emission light derived from experimental measurements. (**a**) Experimental transmural cross-section of ventricular slab showing decay of light incident of epicardium (shown by *green arrows*) at excitation (*left*) and emission (*centre*) wavelengths. (*Right*) individual profiles obtained from dotted-lines in experimental images plotted with depth. (**b**) Average excitation (*filled circles*) and emission (*open circles*) profiles, including *solid lines* showing fit of mathematical relations described above. (**c**) Percentage contributions to total collected signal during epi-illumination and transillumination modalities with illumination and detection directions shown by *green* and *red arrows*, respectively. Used with permission from Baxter et al. (2001)

$$w(z) = \frac{\phi(z)G(z)}{\int_0^T \phi(z)G(z)dz}, \tag{15.3}$$

where $z = 0$ is the illuminated surface and T the tissue thickness. Using this function, the authors were able to directly calculate the relative contribution to the total collected signal from the tissue depth, showing that 82 % of collected fluorescent originates from within the first 1 mm of the epicardium. A similar analysis was also performed for transillumination in which the optical detector is located on the opposite side of the tissue to the illuminated surface. Figure 15.1c shows the percentage contributions to the recorded signal from both setups.

3.2 *Simulation of Depth-Weighted Averaging of Fluorescent Signals*

3.2.1 Combing with Micro-Electrode Recordings

In the same study in which the weighting function in Eq. (15.3) was mathematically defined, Baxter et al. (2001) went on to use this expression to simulate optically-recorded action potentials using a digitised action potential from a microelectrode recording as input. The spatial profile of the microelectrode action potential was combined with the weighting function over successive time-points. The results clearly showed the expected prolongation of the simulated optical action potential upstroke, qualitatively similar to the corresponding experimental optical measurements.

3.2.2 Combing with Simulated Transmembrane Potential Dynamics

Soon after, studies by Janks and Roth (2002) and Bray and Wikswo (2003) applied the weighting function proposed by Baxter et al. (2001) to assist in the interpretation of a number of important electrophysiological findings, recorded by optical mapping, but whose physiological explanation was as yet not fully understood.

Both studies combined the weighting function in Eq. (15.3) along with the transmembrane potential levels $V_m(x, y, z)$ calculated explicitly at each point in space using the bidomain (Janks and Roth 2002) or monodomain (Bray and Wikswo 2003) equations to derive an expression for the weighted average of the transmembrane potential recorded at the tissue surface $\bar{V}_m(x, y)$ given by

$$\bar{V}_m(x, y) = \frac{\int_0^\infty V_m(x, y, z)w(z)dz}{\int_0^\infty w(z)dz}, \tag{15.4}$$

where z is the depth beneath the surface.

The study by Janks and Roth (2002) suggested that the averaging of fluorescent signals over depth may account for the under-estimation of the optically-recorded polarisation levels of cardiac tissue close to unipolar electrodes following strong stimulation, relative to the much stronger epicardial polarisation values predicted by computational bidomain simulations. Simulating unipolar electrode stimulus within a simple slab geometry, in combination with the modelling approach of Eq. (15.4), the authors demonstrated an approximate threefold reduction in the amplitude of the simulated optical potential close to the electrode, relative to the predicted electrical potential at a corresponding location on the epicardial surface, significantly improving the match with corresponding experimental measurements. Although strongly polarised on the tissue surface, polarisation levels are known to decay intramurally with decay constant equal to the cardiac length constant. Thus, such a reduction in $\bar{V}_m$ was suggested by the authors to result from the additional contributions of these more weakly polarised mid-myocardial layers of tissue, introduced through the use of the weighting function, reducing the apparent optically-recorded epicardial polarisation level.

Bray and Wikswo (2003) used a similar method to examine the depth-averaging effects upon simulated epicardially-recorded fluorescent signals during episodes of sustained reentry. The study demonstrated that transduction of information from depth regarding entirely intramural wavefronts propagating beneath an epicardial recording site during reentry could be responsible for the existence of the so-called dual-humped action potentials. These dual-humped action potentials had morphologies which closely matched those witnessed previously in experimental measurements (Efimov et al. 1999, 2000), and thus provided strong evidence that averaging over tissue depth of the collected optical signal was the mechanism behind their formation. In addition to confirming the prolongation of the optically-recorded action potential upstroke, the authors also showed that, following spiral-wave break-up, the averaging of transmembrane potential signals from depth resulted in a dependence of the local optical action potential amplitude upon the immediate subepicardial activity. This, combined with a spatial blurring of the optical signal, compromised the accuracy of surface phase singularity localisation (the organising centre of the reentrant wave), including both false positive and false negative findings.

4 Simulation of 3D Effects of Photon Scattering Using the Diffusion Equation

The studies described above all used depth-dependent weighting functions to simulate the effects of sampling transmembrane potential levels from internal layers of tissue beneath the surface recording sites. Consequently, they considered only the effects of fluorescent signal collection from *linear* depths beneath the recording site—they had not, as yet, accounted for the effects of *lateral photon scattering*.

As light passes through biological tissue it is both absorbed and scattered, meaning the trajectories of photons may significantly change on their journey. The effects of lateral photon scattering in planes parallel to the recording tissue surface, where both excitation light intensity and fluorescent emission are strongest, are thus thought highly likely to contribute to signal distortion. In fact, at a similar time to the above studies, Ding et al. (2001) used a stochastic modelling approach (discussed later in Sect. 5) to simulate the trajectories of individual photons (or photon 'packets'). They demonstrated, for the first time in this context, that photon scattering was an inherently three-dimensional phenomena, and that optical signals from a given pixel are collected from an *interrogation volume*, extending a similar, and in some cases greater, distance in the plane of the recording surface as from depth.

The landmark study by Hyatt et al. (2003) was the first of its kind to simulate, and analyse, the *effects* of three-dimensional photon scattering in cardiac tissue on fluorescent signal distortion during optical mapping recordings. The authors, for the first time, developed the application of the *photon diffusion equation* for the simulation of optical mapping signals; a general method which would be followed by many other important subsequent studies.

4.1 The Photon Diffusion Equation

At the wavelengths commonly used in fluorescent imaging, biological tissue is highly scattering and relatively weakly absorbing. In such instances, the Boltzmann transport equation—which describes the movement of electromagnetic waves in scattering media—can be approximated to a continuum representation of photon transport, termed the *photon diffusion equation* (Haskell et al. 1994; Hielscher et al. 1995). For application to optical mapping, the steady-state form of the photon diffusion equation is appropriate as the time-scales which govern changes in electrical activation (10^{-6} s) are many orders of magnitude larger than the time-scales involved in photons reaching equilibrium within the tissue (10^{-15} s). The steady-state photon diffusion equation provides a computationally-efficient method of simulating photon transport in biological tissue which has been shown to agree well with more accurate discrete stochastic approaches such as the Monte Carlo method (see Sect. 5).

The steady-state photon diffusion equation for highly-scattering media is given by

$$D\nabla^2\Phi(\mathbf{r}) - \mu_a\Phi(\mathbf{r}) = -\omega(\mathbf{r}), \tag{15.5}$$

where Φ is the photon density (photons/mm^3) at any point in the tissue with position vector $\mathbf{r}$ (mm), D (mm) and μ_a (mm^{-1}) are the optical *diffusivity* and *absorption coefficients*, respectively, and ω describes the source of photons within the tissue at position $\mathbf{r}$. It should be noted that, as we are dealing with the steady-state form of the photon diffusion equation, the time-derivative term $\frac{1}{c}\frac{\partial \Phi}{\partial t}$ which appears in the full form of the photon diffusion equation (Hielscher et al. 1995), has been dropped, and thus a common factor of the speed of light c is missing from Eq. (15.5). Hence, we adopt the standard procedure of giving D and μ_a the unorthodox units above.

The diffusion coefficient D can also be defined in terms of the explicit optical properties of the tissue by Jacques (1998)

$$D = \frac{1}{3(\mu_a + \mu_s^{'})}, \tag{15.6}$$

where $\mu_s^{'}$ is the *reduced scattering coefficient*, related to the *scattering coefficient* μ_s by $\mu_s^{'} = \mu_s(1 - g)$, where g is the anisotropy of scattering and represents the average cosine $\langle \cos\theta \rangle$ of each scattering deflection angle θ (Jacques 1998). Values of μ_a, μ_s and g can be obtained experimentally from measurements of diffuse reflectance and transmittance, as performed in Ding et al. (2001). It should be noted that the use of D as a scalar quantity implies an assumption of isotropic scattering properties of the tissue.

4.2 Application of the Photon Diffusion Equation to Simulation Optical Mapping Signals

The photon diffusion equation can be applied directly to simulate the distribution of light during optical mapping recordings using the method originally proposed by Hyatt et al. (2003). In the original study by Hyatt et al. (2003), the distribution of illuminating light was modelled by simply using a mono-exponential decay profile, made possible due to the regular cuboid domain over which the simulations were performed, with the photon diffusion equation used to simulate the distribution of fluorescent light during the process of emission. In later works (Bishop et al. 2006c, 2007b), however, the photon diffusion equation was also used to model the excitation process.

4.2.1 Simulating Illumination

As in Bishop et al. (2006c), the photon diffusion equation can be used to calculate the distribution of photon density during the process of illumination. In this case, a modified version of Eq. (15.5) is used

$$D^{ex}\nabla^2\Phi_{ex} - \mu_a^{ex}\Phi_{ex} = 0, \tag{15.7}$$

where the source term w is set to zero, as there are no sources of photons within the myocardium during illumination. Instead, we define boundary conditions on

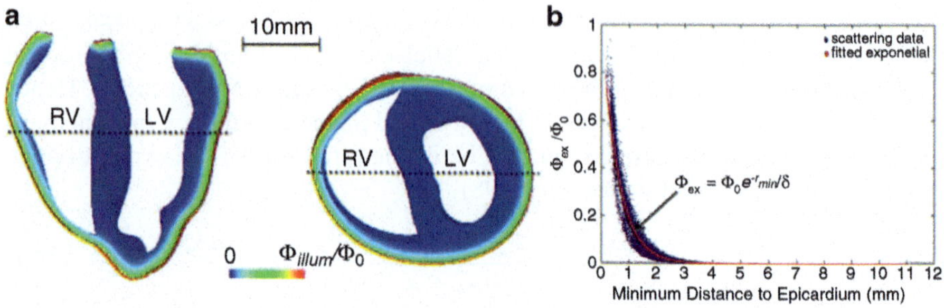

Fig. 15.2 (**a**) Distribution of photon density due to uniform epicardial illumination within an anatomically-realistic rabbit ventricular model derived from numerical solution to Eq. (15.7). (**b**) Decay of simulated photon density with depth into tissue, approximating monoexponential decay. Used with permission from Bishop et al. (2006c)

Eq. (15.7) to replicate uniform illumination of the tissue surface, setting $\Phi_{ex} = \Phi_0$, an arbitrary constant which just defines the magnitude of the illuminating source. The constants D^{ex} and $\mu_a{}^{ex}$ are taken from measurements specifically at the illuminating wavelength.

Figure 15.2 shows the solution to Eq. (15.7) over an anatomically-realistic rabbit ventricular model, demonstrating the distribution of illumination photon density throughout the ventricular volume with optical parameter values taken from Ding et al. (2001). In this particular case, a numerical finite element solution method has been used (see Sect. 4.3) to solve Eq. (15.7). Boundary conditions (Dirichlet) of $\Phi_{ex} = 1$ have been applied all over the epicardial surface to replicate uniform panoramic illumination. The Figure clearly shows the attenuation of the illuminating light intensity into the tissue depth, such that very little light in fact reaches the endocardial surface. Figure 15.2b shows that within the tissue the decay of the photon density approximates mono-exponential with penetration depth equal to $\sqrt{D/\mu_a}$. The distribution of Φ_{ex} throughout the tissue is then used as input to the simulation of fluorescent emission.

4.2.2 Simulating Voltage-Sensitive Fluorescent Emission

The second process, of fluorescent emission, is simulated as in Hyatt et al. (2003) (and later studies) by solving the photon diffusion equation of the form

$$D^{em}\nabla^2\Phi_{em} - \mu_a^{em}\Phi_{em} = -k\Phi_{ex}V_m. \tag{15.8}$$

In this specific case, the right-hand side source term ω has been represented as being proportional to the product of the photon density at a given point in the tissue and also the specific level of transmembrane voltage (V_m), and where k is a constant.

This approach successfully models the principles by which voltage-sensitive fluorescent dyes work: the total fluorescence given-off by a region of tissue varies in proportion to the total amount of illuminating light it receives, and also the value of transmembrane potential at that point. Analogous to the case of illumination, the optical coefficients in Eq. (15.8) are obtained at the emission wavelength of the fluorescent light.

The distribution of V_m within the tissue is usually derived in an entirely separate simulation procedure, whereby the bidomain (or monodomain) model of cardiac tissue electrophysiology is used, in a similar manner to the simple depth-weighted averaging approaches of Eq. (15.4). Thus, it is often convenient to think of this method of optical signal simulation as a useful post-processing tool which can be subsequently applied to V_m simulations in order to better compare with experiments. It should be noted that, practically, it is necessary to add an off-set to the V_m distribution, (such as + 1 V) to ensure that the source term ω is never negative. This offset replicates the fact that the optical mapping technique records small changes in intensity upon a large background of fluorescence, with a typical 10 % change in fluorescence occurring following a 100 mV change in membrane voltage.

4.2.3 Derivation of the Optical Signal

V_m distributions at each required time-step (usually 1 ms) are sequentially used as input into Eq. (15.8) and solved to give the corresponding distribution of Φ_{em} throughout the tissue. Φ_{em} is then used to directly derive the value of the recorded optical signal V_{opt} at the tissue surface from which the signal is recorded. In optical mapping experiments, the optical detector records the photons exiting the epicardial surface which are subsequently incident on the detector. Here, this flux is calculated by applying Fick's Law at the recording tissue surface in accordance with Hyatt et al. (2003). Fick's Law is a commonly used formula to relate the flux of some diffusive variable exiting a boundary to the spatial gradient of that variable across the boundary. Applying this, we can write-down

$$V_{opt} = -D^{em} \nabla \Phi_{em} \cdot \mathbf{n}, \tag{15.9}$$

where **n** is the normal to the recording surface. Uniform detection is assumed, which approximates the conditions in panoramic optical mapping experimental set-ups, such that we assume that all photon flux exiting the epicardium is detected.

In optical mapping experiments, spatial variations across the tissue in dye loading/uptake, photo-bleaching, illuminating light intensity amongst other factors, means that absolute changes in recorded fluorescent intensity from a specific region are hard to directly relate to absolute levels of local membrane potential. Consequently, fluorescent signals from each detector element are normalised with respect to the amplitude of an action potential, recorded during pacing. A similar process is followed in simulating such signals.

4.3 Numerical (Finite Element) Solution to the Photon Diffusion Equation

The first approaches which applied the photon diffusion equation to simulate optical mapping signals, by the Pertsov laboratory, used explicit analytic solution methods to solve for the fluorescent photon density throughout the tissue (Hyatt et al. 2003, 2005; Bernus et al. 2004). These methods used a convolution of the Green's function over the tissue volume with the distribution of the voltage-sensitive fluorescence source term ω. Although mathematically elegant and computationally efficient, these methods were restricted to solving the problem over simple, regular domains (primarily a 3D infinite slab or cuboid geometry). Numerical approaches, such as the finite element method, were suggested as an alternative approach to simulate optical mapping signals in the first studies by our Group (Bishop et al. 2006c, 2007b). The key advantage of using a finite element solution method to solve the photon diffusion equation is that it can be applied to any geometrical domain, allowing signals to be simulated over anatomically-realistic ventricular models (being used more frequently in computational modelling), and facilitating a more direct comparison with experimental investigations.

4.3.1 Boundary Conditions

The optical domain Ω is bounded by Γ

$$\Gamma := \partial\Omega = \bar{\Gamma}_R \cup \bar{\Gamma}_D \tag{15.10}$$

where Γ_R is a mixed Robin-type boundary and Γ_D is of Dirichlet type. In the case of illumination, a given surface of the tissue, Γ_{ex}, is illuminated. This is modeled by applying a Dirichlet boundary condition along Γ_{ex}, that is

$$\Phi_\xi = \Phi_{ex} \quad \text{on} \quad \Gamma_D := \Gamma_{ex}. \tag{15.11}$$

Here $\Phi_{ex} = 1$ is a typical choice. A schematic representation of the boundary conditions used during illumination and emission is shown in Fig. 15.3.

The choice of boundary conditions during illumination on other tissue surfaces, and the choice on *all* surfaces during the process of fluorescent emission is less apparent, however. Here, all other boundaries are assumed to be of mixed Robin-type boundaries where the *partial current boundary condition* (Haskell et al. 1994; Hielscher et al. 1995) holds. This type of boundary condition is noted for its accurate representation of the interface between biological tissue and its surrounding media which accounts for the refractive index mismatches across this interface. The partial current boundary condition is derived by considering conservation of flux at the boundary and accounts for the total internal reflection of

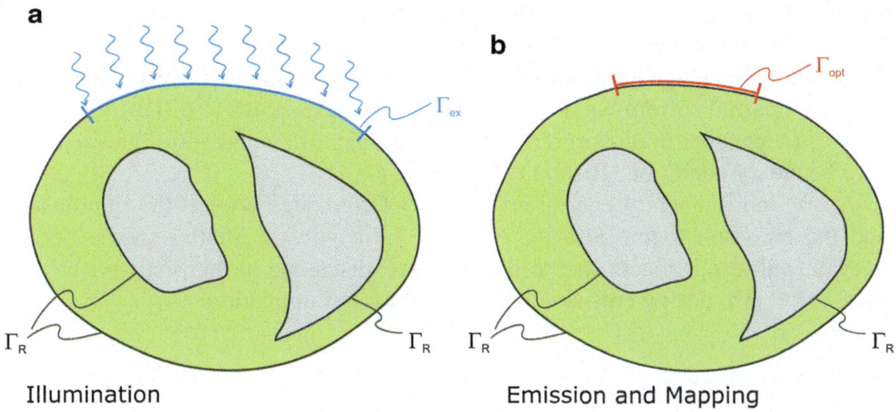

Fig. 15.3 Boundary conditions for photon diffusion equations during (**a**) illumination and (**b**) emission. Note that Γ_{opt} is not a boundary condition used for solving the emission equation, rather, over Γ_{opt} optical traces are derived from Φ_{em} to compute optical maps

photons incident on the boundary. Increased photon reflection at a boundary will increase scattering in this region, creating a greater distortion in the optical signal. In particular, the partial current boundary condition relates the value of photon density on the boundary to the value of the gradient of the photon density at the same point on the boundary by

$$\Phi_\xi - L_s \nabla \Phi_\xi \cdot \mathbf{n} = 0 \quad \text{on} \quad \Gamma_R, \tag{15.12}$$

where $\mathbf{n}$ is the outward pointing normal vector at the boundary Γ_R. L_s is defined by

$$L_s = \frac{1 + R_{eff}}{1 - R_{eff}} 2D, \tag{15.13}$$

where R_{eff} is the effective reflection coefficient, and is a function of the refractive index mismatch between the tissue and the surrounding media (Haskell et al. 1994).

Many previous studies which have used the photon diffusion equation to simulate photon transport in biological tissue have favoured the use of a zero boundary condition

$$\Phi_\xi = 0 \quad \text{on} \quad \Gamma_R, \tag{15.14}$$

on the bounding external tissue surfaces (for example in Arridge 1993). Such a choice has also been used in the other studies which have used the photon diffusion equation to simulate photon movement during optical mapping experiments (Hyatt et al. 2003, 2005; Bernus et al. 2004). The appeal of this condition lies in the simplicity of its implementation, although in terms of physical representation, this approach is less accurate than the partial current boundary condition. The details of these differences are discussed in many studies that have assessed the use of different boundary conditions on the photon diffusion equation (Haskell et al. 1994; Hielscher et al. 1995). However, in relation to specific metrics of simulated optical signals (such as optical action potential upstroke duration), the use of the partial current boundary condition has been shown to differ by less than 5–10 % (dependent upon specific interface simulated) compared to the use of the zero boundary condition (Bishop et al. 2006c).

Apart from boundary conditions the only difference between the illumination and the emission scenario is the presence of a volume photon source density over the entire domain Ω due to fluorescence, discussed in the sections above to be $\omega = k\Phi_{ex}V_m$ during emission and $\omega = 0$ during illumination.

4.4 Finite Element Discretisation

In terms of numerical treatment there is no difference between illumination and emission scenario. Thus, the general form of Eq. (15.5) is used. A weak form of Eq. (15.5) is found by multiplying with a suitable test function, ψ, which fulfills any prescribed boundary conditions, and integration over the entire domain Ω

$$\int_{\Omega} \psi \nabla \cdot (D^{\xi} \nabla \Phi_{\xi}) d\Omega - \int_{\Omega} \psi \mu_a^{\xi} \Phi_{\xi} d\Omega = \int_{\Omega} \psi \omega d\Omega, \tag{15.15}$$

where here ξ represents either *ex* or *em*, representing either excitation illumination or emission, respectively. Using the identity

$$\int_{\Omega} \psi \nabla \cdot \mathbf{j} d\Omega = \int_{\Omega} \nabla \cdot (\psi \mathbf{j}) d\Omega - \int_{\Omega} \nabla \psi \cdot \mathbf{j} d\Omega = \int_{\Gamma} \psi \mathbf{j} d\Gamma - \int_{\Omega} \nabla \psi \cdot \mathbf{j} d\Omega \tag{15.16}$$

with $\mathbf{j} = D^{\xi} \nabla \Phi_{\xi}$ and rearranging terms yields

$$-\int_{\Omega} \nabla \psi \cdot D^{\xi} \nabla \Phi_{\xi} d\Omega + \int_{\partial\Omega} \psi D^{\xi} \nabla \Phi_{\xi} \cdot \mathbf{n} d\Gamma - \int_{\Omega} \psi \mu_a^{\xi} \Phi_{\xi} d\Omega = \int_{\Omega} \psi \omega_{\xi} d\Omega. \tag{15.17}$$

Prescribing the partial current boundary condition given in Eq. (15.12) we rewrite

$$\int_{\partial\Gamma_N} \psi D^{\xi} \nabla \Phi_{\xi} \cdot \mathbf{n} d\Gamma = \int_{\partial\Gamma_N} \psi D^{\xi} L_s^{-1} \Phi_{\xi} d\Gamma. \tag{15.18}$$

For notational convenience we use

$$a_1(\psi, \Phi_{\xi}) = \int_{\Omega} \nabla \psi \cdot D^{\xi} \nabla \Phi_{\xi} d\Omega \tag{15.19}$$

$$a_2(\psi, \Phi_\xi) = \int_{\partial\Omega} \psi D^\xi L_s^{-1} \Phi_\xi d\Gamma \tag{15.20}$$

$$a_3(\psi, \Phi_\xi) = \int_{\Omega} \psi \mu_a^\xi \Phi_\xi d\Omega \tag{15.21}$$

$$a_4(\psi, \omega_\xi) = \int_{\Omega} \psi \omega_\xi d\Omega \tag{15.22}$$

hence the weak form is given as

$$-a_1(\psi, \Phi_\xi) + a_2(\psi, \Phi_\xi) - a_3(\psi, \Phi_\xi) = a_4(\psi, \omega_\xi) \tag{15.23}$$

which is satisfied for all choices ψ within a suitable function space.

To apply the finite element method, the unknown field Φ_ξ is approximated as

$$\Phi_\xi = \sum_{i=1}^{N} \alpha_i \Phi_{\xi,i} \tag{15.24}$$

where $\Phi_{\xi,i}$ are time-dependent coefficients and α_i are appropriate (spatial) basis functions. By inserting Eq. (15.24) into Eq. (15.18) and using the basis functions α to construct the test function ψ one obtains a linear system of equations given as

$$\mathbf{K}\Phi_\xi + \mathbf{M}_\Gamma \Phi_\xi - \mu_a^\xi \mathbf{M} \Phi_\xi = \mathbf{M}\omega_\xi \tag{15.25}$$

where

$$\mathbf{K}(i,j) = -\int_{\Omega} \nabla \alpha_j \cdot D^\xi \nabla \alpha_i \, d\Omega \tag{15.26}$$

$$\mathbf{M}(i,j) = \int_{\Omega} \alpha_j \alpha_i \, d\Omega \tag{15.27}$$

$$\mathbf{M}_\Gamma(i,j) = \int_{\Gamma_N} \alpha_j D^\xi L_s^{-1} \alpha_i \, d\Gamma. \tag{15.28}$$

The resulting matrices are referred to as stiffness matrix **K**, mass matrix **M**, and boundary mass matrix $\mathbf{M}_\Gamma$. Note that the boundary mass matrix $\mathbf{M}_\Gamma$ is only used if proper partial current boundary conditions are enforced. Otherwise the term $\mathbf{M}_\Gamma \Phi_\xi$ is dropped and homogeneous Dirichlet boundary conditions are enforced over all surfaces $\Gamma \backslash \Gamma_{ex}$. The resulting system of equations is solved iteratively.

4.5 Application Studies Using the Diffusion Approach

4.5.1 Simulation of Optical Action Potential Upstroke Distortion

As mentioned previously, the prolongation of the optical action potential upstroke has long been suggested to be an important quantitative measure of the degree of distortion of optical mapping signals, relative to upstroke recorded from micro-electrodes. Hyatt et al. (2003) represented the first Group to use the photon diffusion equation to quantify the effects of three-dimensional photon scattering upon signal distortion during optical recordings from cardiac tissue during pacing. The authors used an analytic solution to Eq. (15.8) over a simplified (cuboid) slab of ventricular tissue, with uniform illumination being modelled by a simple mono-exponential decay of excitation light intensity with depth. They showed that the upstroke of the optically-simulated action potential associated with a planar paced wavefront passing beneath the surface recording site was prolonged and distorted with respect to the 'true' epicardial action potential upstroke derived from electrophysiological simulations (which acted as input to the optical simulation model). Importantly, though, the authors also showed that the morphology of distortion which characterised the optical upstroke was specific to the intramural direction of wavefront propagation beneath the epicardium: propagation parallel to the epicardial recording surface gave rise to a symmetrically distorted upstroke, whilst propagation towards or away from the epicardium resulted in an asymmetrical distortion. More specifically, they suggested that the potential level corresponding to the maximal rate of rise of the optical upstroke (termed Vf^*) depends upon the angle θ of the wavefront direction with respect to the imaged tissue surface. These results were later validated experimentally (Hyatt et al. 2005).

The simulation studies by Hyatt et al. (2003, 2005) were restricted to simulating optical signals from regular geometries, due to the analytic solution method used to solve the photon diffusion problem. The work in this field by our Group has sought to use a numerical finite element solution approach, discussed above in Sect. 4.3, to allow the synthesis of voltage-sensitive fluorescent signals over anatomically-realistic ventricular models. Such an approach has allowed us to quantify and understand the three-dimensional interaction between photon scattering and the complex ventricular anatomy and wavefront morphologies facilitating a close comparison with experimental measurements. The first study by Bishop et al. (2006c) simulated optical mapping signals over a detailed rabbit ventricular model. The authors demonstrated similar distortions in the optically-simulated upstroke morphologies for global wavefronts travelling parallel or perpendicular to the imaged surface to those seen in the Hyatt et al. studies (Hyatt et al. 2003, 2005), although they did not witness such a close explicit relationship between Vf^* and the global angle of wavefront approach (Bishop et al. 2006b). Furthermore, the study by Bishop et al. (2006c) also demonstrated that the specific distortion of the optical action potential upstroke depends sensitively upon the specific values of optical properties (μ_a, μ_s, g) used as input to the optical simulation model, highlighting the importance of accurate measurement and use of these quantities when quantitatively comparing simulation with experiment. However, a low sensitivity was found in respect of the optical upstroke distortion relative to the specific external optical medium (glass, air, water) surrounding the tissue, represented through the use of the more physically-realistic partial current boundary condition, and thus the similar overall results seen compared to using the zero boundary condition.

The study by Bernus et al. (2004) used an analytic diffusion approach to simulate and assess the utility of different optical imaging modes in effectively inferring three-dimensional electrophysiological activity. Here, coaxial scanning and broad-field illumination methods were simulated, which were combined with models of both trans- and epi-illumination, in conjunction with the use of fluorescent and absorptive dyes. The study examined differences in spatial resolutions of the different optical setups and the resulting differences in recorded optical signals, including simulated action potential upstrokes. Of the methods examined, fluorescent coaxial scanning was seen to provide the best resolution and provide the most information regarding intramural electrophysiological activity.

4.5.2 Interpretation of Optical Recordings During Strong Electrical Shocks

The early study by Janks and Roth (2002) had demonstrated that averaging of fluorescent signals purely from depth could explain the reduction in the optically-recorded surface polarisation levels following unipolar stimulation, relative to predictions by the bidomain model. However, in-light of the above works demonstrating the importance of the averaging of fluorescent signals from a three-dimensional tissue *volume*, the corresponding interaction with complex ventricular anatomy and subsurface voltage distributions, and the important information this transduces into the surface optical signal, how these results may be affected when considering 3D photon scattering effects in more realistic ventricular models was not known.

A series of works by the Group of Trayanova had demonstrated the highly complex distribution of transmembrane potential levels throughout the three-dimensional tissue volume during the application of strong electrical defibrillation strength shocks applied via external plate electrodes to anatomically-realistic rabbit ventricular models (Rodriguez et al. 2005; Trayanova et al. 2006). Importantly, although very strong de- and hyperpolarisation was seen on the epicardial and endocardial surfaces, highly heterogeneous distributions of intermediate polarisation levels were witnessed

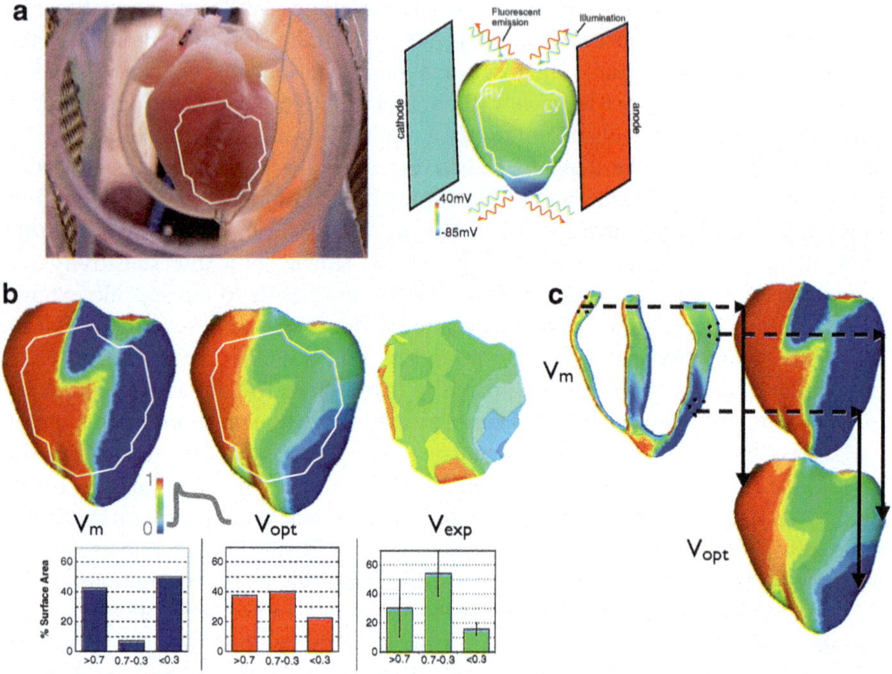

Fig. 15.4 (**a**) Experimental setup for application of strong external shocks (*left*), along with schematic of computational model of optical signal synthesis (*right*). (**b**) Epicardial shock-end polarisation distributions comparing direct bidomain simulations (V_m), simulated fluorescent optical signal (V_{opt}) and experimental recording (V_{exp}), normalised with respect to paced action potential amplitude. Below shows quantification of the improved match between V_{opt} and V_{exp}, relative to V_m, showing the percentage of the epicardial surface within the recording window (circumscribed by a *white line* in (**a**)) with different polarisation levels. *Vertical black lines* in the case of the V_{exp} plots show one standard deviation. (**c**) Demonstration of how averaging of signals from scattering volume beneath surface recording site (shown schematically as *dotted hemispheres*) acts to significantly modulate simulated fluorescent signal. Used with permission from Bishop et al. (2007b)

intramurally. Although a reasonably close qualitative agreement was seen between the epicardial potential distribution predicted by bidomain simulations and corresponding optical mapping experiments, quantitatively the simulations predicted much stronger surface polarisations levels, with a less distinct transition between depolarisation and hyperpolarisation regions across the ventricles (Rodriguez et al. 2005).

The studies by Bishop et al. (2006a,c) applied the optical simulation model described above to simulate the affects of three-dimensional photon scattering upon these bidomain predictions. Application of the model was shown to significantly improve the match between bidomain simulations and experiment, as highlighted in Fig. 15.4, in respect of both the quantitative magnitude and the qualitative pattern of surface polarisation levels. As shown in Fig. 15.4c, despite the bidomain simulations predicting very strongly polarised epicardial surface V_m levels, the effects of three-dimensional averaging within a subsurface volume (shown schematically as dotted hemi-spheres) included information regarding the much more weakly-polarised mid-myocardial V_m levels, significantly reducing the magnitude of the epicardial surface polarisation levels predicted in the simulated V_{opt} signal. Importantly, the lateral effects of averaging within the plane of the tissue surface were also shown to be important, reducing the polarisation gradient across the tissue surface, in line with experiments. These simulations provided an important means of both validating predictions made by the bidomain model and a more complete understanding of the effects of strong shocks on the ventricles. They also demonstrated how such knowledge can be used to help better interpret experimental shock-end recordings to elucidate information regarding intramural polarisation levels from surface recorded signals, for example providing essential information regarding the presence and location of intramural excitable gaps which govern post-shock propagation.

4.5.3 Elucidation of Intramural Activity During Arrhythmias

During pacing and sinus rhythm, optically-recorded fluorescent signals show a distortion primarily only during the action potential upstroke, representing the only time transmembrane potential levels differ significantly within the scattering volume associated with a surface recording site. Transmembrane potential levels differ significantly as the front of the wave passes throughout the scattering volume associated with a surface recording site, but are relatively homogeneous before it passes through (all the tissue is rest), during the plateau and during repolarisation due to the relatively larger width of the wave-back with respect to the scattering volume dimensions. However, the complex wavefront dynamics that occurs during arrhythmia results in highly heterogeneous distributions of transmembrane potentials within a particular scattering volume, particularly in the vicinity of the core of the reentrant wave (the filament (3D) or phase singularity (2D)).

Studies by Bishop et al. (2007a,b) used the same methods as in their previous work (Bishop et al. 2006c) to simulate such complex episodes of arrhythmias within an anatomically-realistic ventricular model. Their aim was to understand how three-dimensional photon scattering conveys information regarding these heterogeneous transmembrane potential distributions, and the mechanisms by which this information is manifested within the surface optically-recorded signal. Results showed that, despite a blurring in the spatial extent of the wavefront itself (as also documented during previous pacing studies (Bishop et al. 2006c; Hyatt et al. 2003)), overall patterns of simulated fluorescent signal (V_{opt}) matched closely to those of the V_m patterns directly from the bidomain simulation. However, the highly heterogeneous distribution of V_m close to the core of the reentrant wave, combined with the distortion effects of scattering, were seen to lead to an error in the calculated spatial location of the epicardial phase singularities, shown to be >4 mm in cases, as shown in Fig. 15.5. These noted shifts in epicardial phase singularities locations were significantly greater than those seen previously by

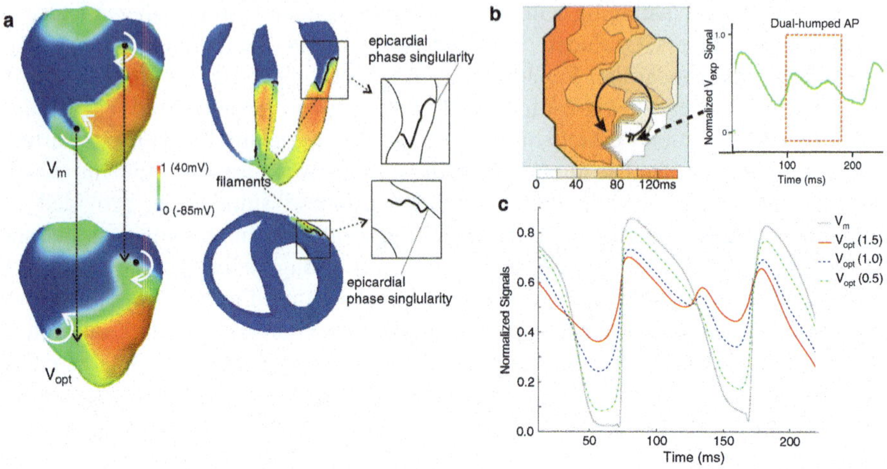

Fig. 15.5 (**a**) (*Left*) V_m and V_{opt} distributions during an episode of shock-induced arrhythmogenesis, with *black dotted arrows* showing the shift in measured phase singularity location on the epicardial surface. *White curved arrows* show direction of rotation on reentrant circuits. (*Right*) associated filaments showing complex morphologies throughout the myocardial wall. [Note that different angled clipping planes are used to visualise intramural filaments in different panels.] (**b**) Experimental optical mapping measurement of an episode of sustained arrhythmia, showing detection of a dual-humped action potential close to core of reentry. (**c**) Comparison between simulated V_{opt} (*blue*) and V_m (*grey*) signals from the same recording site, highlighted the existence of dual-humped morphology in the simulated optical signal. As the optical penetration depths (at both illumination and emission wavelengths) are concurrently increased by a factor of 1.5 (*red*), both relative dual-humped amplitude and resting potential level correspondingly increase; conversely, as penetration depths are decreased by a factor of 0.5, these metrics are seen to decrease. Used with permission from Bishop et al. (2007b)

Bray and Wikswo (2003) (approximately $0.5\,\text{mm}$) who used a more simple 1D depth-averaging model. Such larger shifts were attributed to the representation of full three-dimensional photon scattering effects, in addition to the more complex filament orientations associated with the more realistic representation of ventricular anatomy, emphasised in Fig. 15.5.

Analysis of the optically-recorded fluorescent signals during episodes of arrhythmias in the same study, also confirmed the existence of ‘dual-humped’ action potentials due to the passage of a wholly intramural wavefront close to the core of the reentry, also shown in the previous study by Bray and Wikswo (2003). However, the representation of three-dimension photon scattering effects combined with the realistic ventricular anatomy allowed the discovery of two other novel mechanisms by the authors of dual-humped action potential formation: (1) wavefront collision on the epicardium combining with scattering in the direction parallel to the recording surface plane; and, (2) delayed propagation into the septum combining with signal pick-up through the thin RV cavity close to the apex, agreeing with a similar phenomena previously recorded experimentally (Al-Khadra et al. 2000). An example of an experimentally-recorded dual-humped action potential is shown in Fig. 15.5b, along with an example of a corresponding simulated trace in Fig. 15.5c both taken from Bishop et al. (2007b). All dual-humped morphologies of simulated traces were shown to vary in amplitude as corresponding variations in optical properties used in the optical simulation model were made, emphasising the mechanism of photon scattering effects responsible for their formation (shown in Fig. 15.5c). Finally, the study also demonstrated that the experimentally noted elevation of resting potential in optical signals recorded close to reentrant cores, and the reduction in the observed action potential amplitude were both explained by the effects of three-dimensional photon scattering. Specifically, the model was used to explain how the complexity of the wavefront dynamics in this region gives rise to tissue within the scattering volume being activated at different times, and so is not all either at rest or at maximum amplitude at the same time.

Finally, a study by Bernus et al. (2007) used a diffusion-based model, but this time adapted to simulate trans-illumination imaging in conjunction with near infra-red voltage-sensitive absorptive dyes to detect intramural filaments. The longer penetration depth of the near infra-red dye allowed for reliable detection of intramural scroll waves throughout the whole thickness (10 mm) of the ventricular wall simulated, demonstrating the utility of such a novel imaging approach to measure complex electrical dynamics.

4.5.4 Dual Voltage-Calcium Imaging

Although optical mapping studies primarily use potentiometric voltage-sensitive fluorescent dyes to measure the dynamics of transmembrane potential, a number of different fluorescent dyes also exist which allow for the measurement of intracellular calcium dynamics. If a voltage- and calcium-sensitive dye are chosen with good separation in their fluorescent emission spectra, but which can be excited at the same wavelength, simultaneous imaging of V_m and Ca_i signals is possible. The first study to consider the effects of photon scattering on optically-recorded fluorescent calcium signals was by Walton and Bernus (2009). The authors created a novel diffusion-based model to simulate fluorescent signals during dual-voltage calcium optical mapping recordings, allowing representation of two different calcium dyes (Rhod-2 and Rhod-FF). In calcium imaging, the amount of fluorescence emitted depends upon the current concentration of calcium bound to the fluorescent dye, CaD. Here, the authors applied a model of chemical equilibrium to determine the relationship between the current calcium concentration Ca_i and CaD

$$CaD = \frac{Ca_i D_T}{K_d + Ca_i}, \tag{15.29}$$

where D_T is the total dye concentration and K_d is the dissociation constant of the dye. Once CaD had been derived from Ca_i, which itself was outputted directly from electrophysiological simulations of ionic membrane dynamics, it was used in a

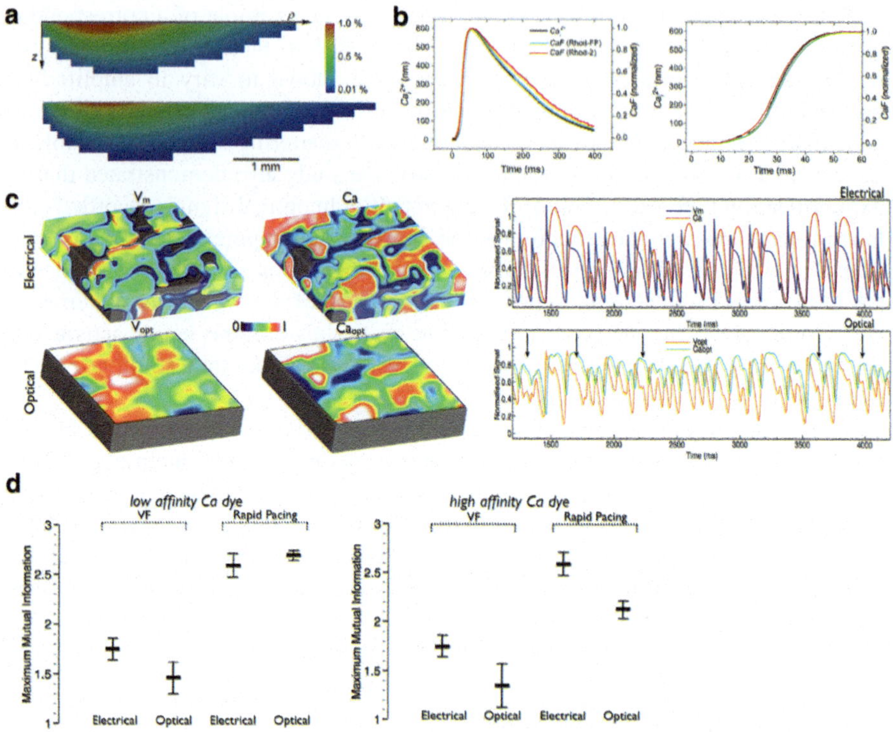

Fig. 15.6 (**a**) Optical interrogation volume for the calcium sensitive fluorescent dyes Rhod-2 and Rhod-FF (*top*) along with corresponding volume for voltage-sensitive dye RH-237 (*bottom*) recorded from a single surface 'pixel'. Colours encode the overall relative contribution. (**b**) Optically-recorded fluorescent calcium transients (*left*) and rise-phases (*right*) for Rhod-2 (*red*) and Rhod-FF (*green*), compared to intracellular calcium signals directly from monodomain simulations (*black*). (**c**) Spatial distributions of simulated electrical and optical voltage and calcium signals during an episode of VF (*left*) along with temporal traces of signals taken from a central surface recording site. *Black arrows* highlight additional inflections in fluorescent V_{opt} signal not witnessed in Ca_{opt} or V_m traces. (**d**) Maximum MI scores during VF and rapid pacing between V_m/Ca_i and V_{opt}/Ca_{opt} signals using a low (*left*) and high (*right*) affinity calcium-sensitive fluorescent dye. Used with permission from Bishop et al. (2011)

similar manner to V_m to scale the source of fluorescent emission at each point in the tissue. Results using the model showed that surface-recorded calcium transients originate from depths up to 0.8 mm beneath the recording surface, leading to significant differences in optically-recorded calcium transient upstrokes and activation times relative to epicardial calcium concentrations, highlighted in Fig. 15.6a,b.

A later study by Bishop et al. (2011) applied a similar approach with an aim to simulate, quantify and understand the recent dissociation witnessed experimentally (Omichi 2004; Wu et al. 2005) between optically-recorded voltage and calcium signals during complex episodes of fibrillation. Mutual information (MI) scores were used to quantify the dissociation in simulated fluorescent voltage and calcium signals during episodes of fibrillation and rapid pacing protocols over simple slab geometries. Figure 15.6c shows snap-shots of the spatial distribution of simulated V_m, V_{opt}, Ca_i and Ca_{opt} signals during an episode of VF, along with temporal traces from a central surface measurement site. The study showed that the effects of photon scattering decreased the MI of V_{opt}/Ca_{opt} signals relative to the raw electrical V_m/Ca_i signals (obtained directly from monodomain simulations) during VF (Fig. 15.6d), relative to the higher MI scores witnessed during rapid pacing for both electrical and optical signals.

This dissociation was shown to be driven by differences in fluorescent wavelengths between the two dyes leading to signal averaging from scattering volumes of different dimensions between voltage and calcium signals (as demonstrated previously in Fig. 15.6a). Thus, during complex fibrillatory activity, the larger scattering volume of the voltage fluorescent signal may pick-up information regarding intramural wavefronts (appearing as 'humps' in the V_{opt} signal), not seen within the smaller scattering volume of the fluorescent calcium signal, giving rise to dissociation between fluorescent voltage and calcium signals. Such an effect is highlighted by the black arrows in the traces of Fig. 15.6c. Increasing the wavelength separation between fluorescent voltage/calcium signals increased this dissociation, emphasising this mechanism. In contrast, during rapid pacing, fluorescent voltage and calcium signals were seen to be largely closely associated shown in Fig. 15.6d, left. However, high calcium dye affinity (representing a measure of the nonlinearity of calcium dyes) was seen to decrease MI by attenuating alternans in Ca_{opt} during pacing, but not in V_{opt} signals, although dye affinity had less effect on MI scores during VF (Fig. 15.6d, right).

5 Stochastic Modelling of Photon Scattering and Signal Collection

The above approaches used a continuum approximation to simulate the movement of photons within biological tissue as a diffusive process. Although such approaches provide a useful and relatively fast method to simulate photon scattering in tissue (Arridge 1993), the potential information derived from each simulation is limited. Although photon densities and fluxes exiting tissue surfaces may be derived, its major disadvantage is that there is no means of deriving individual photon histories. Furthermore, diffusion methods are restricted to simulating the distribution of photon densities within *entirely diffusive media* i.e. the domain must consist entirely of highly scattering biological tissue, in which the diffusion approximation is valid. It thus cannot be used to accurately represent voids or cavities containing media in which photons are not strongly scattered. Alternatively, discrete stochastic methods can be used to simulate the movement of individual photons (or, more specifically, groups or 'packets' of photons) as they propagate through the tissue. Such approaches use a Monte Carlo (MC) method to model the individual scattering, absorption, reflection and refraction processes that occur during the movement of these packets as they propagate through the tissue, and thus can provide greater insight into the effects of photon scattering in biological tissue. Of key importance is that, as individual photon paths are simulated, the explicit *origin* of each photon recorded by the optical detector is known, providing information regarding the distribution of tissue contributing to the recorded signal which can be progressively built-up.

MC methods have been used extensively, and with much success, in a wide variety of applications to biomedical optics phenomena, as they may be used to faithfully simulate light propagation within any medium, regardless of its optical properties. They are often referred to as the 'Gold Standard' to which other methods (such as the photon diffusion method) are compared against due to the accuracy of their predictions, combined with the wealth of information which goes into, and which it is possible to extract from, each simulation (Okada et al. 1996; Wang et al. 1995; Jacques 1998). As each photon interaction event occurs stochastically, each photon trajectory is unique and thus large numbers of photons need to be simulated to obtain good statistics. MC methods are therefore highly computationally expensive (Arridge 1993; Wang et al. 1995).

5.1 The Monte Carlo Algorithm

The algorithm below describes the MC simulation procedure presented in Wang et al. (1995). In the MC method, photons are assumed to move in large numbers, called photon packets. The spatial tissue domain through which photons propagate is discretised into equal cubic elements (forming a grid) into which physical quantities (such as photon density deposited within the tissue due to absorption) can be stored. At any time, the photon packet's position is described by Cartesian coordinates x, y, z. Its current direction of movement is defined by two angles, the deflection angle θ and the azimuthal angle ψ, from which the *directional cosines* μ_x, μ_y and μ_z may be defined

$$\mu_x = \mathbf{r} \cdot \hat{x}, \qquad \mu_y = \mathbf{r} \cdot \hat{y}, \qquad \mu_z = \mathbf{r} \cdot \hat{z} \tag{15.30}$$

where $\mathbf{r}$ represents the current direction of propagation and $\hat{x}$, $\hat{y}$ and $\hat{z}$ are cartesian unit vectors.

Photons begin their journey (or are 'launched') with an initial packet weight of $W = 1.0$ and in a given direction. The photons move in successive steps along their current direction. The magnitude of each step-size s ($0 \leq s < \infty$) is determined by sampling from a probability distribution

$$s = -\ln(\xi)/\mu_t \tag{15.31}$$

where ξ is a random uniformly distributed variable $0 \leq \xi \leq 1$ and μ_t is the optical interaction coefficient, defined as $\mu_t = \mu_a + \mu_s$. The photon packet is then advanced forward by s (so long as interaction with an external boundary or an interface between media of different optical properties does not occur) into its new position (x', y', z')

$$x' = x + \mu_x s \qquad y' = y + \mu_y s \qquad z' = z + \mu_z s. \tag{15.32}$$

Following each step, if the packet has not hit a surface, photon absorption is modelled by assuming that each packet deposits a proportion of its weight $dW = W\mu_a/\mu_t$ in the current optical element (i) in which it resides after moving its step s. This quantity can be logged in an array (A), which is successively added to throughout the simulation, and overall represents the photon density distribution within the tissue. It is important to note that determining the optical element in which a photon packet will reside after moving to its new location (x', y', z'), although relatively trivial in regular geometries (cuboid/slab, etc), is a significantly greater challenge within irregular, unstructured domains. Such a method must be highly computationally-efficient as it must be evaluated numerous times for each individual photon packet trajectory for each of the many millions of packets launched.

Following absorption, the remaining packet weight is then scattered into a new direction. The scattering angles are determined by the Henyey-Greenstein function (Wang et al. 1995), depending upon the packets current direction and the anisotropy of scattering coefficient, g. Note that a value of $g = 0$ represents isotropic scattering, whereas $g = 1$ or $g = -1$ represents forward and back scattering, respectively. The scattering function also includes additional stochasticity from the use of additional random numbers. The photon packet then takes an additional step in its new direction. The procedure of scattering and depositing weight through absorption continues until the packet's weight falls below a critical threshold.

If at any stage of the photon packet's journey a given step takes it across a boundary (be it an external boundary or an interface between two regions of differing optical properties), the packet may be either internally reflected or else transmitted through the interface according to Snell's Law. The packet is first stepped-back onto the boundary to the point at which it intersected the boundary. Then, the *reflection coefficient* R is calculated (Wang et al. 1995), based on the angle of incidence to the surface α_i and the angle of transmission α_t derived from Snell's Law. Assuming α_i is greater than the critical angle, another random number $(0 \leq \xi \leq 1)$ is then used to decide whether in fact the packet is reflected $(\xi \leq R)$ or transmitted $(\xi > R)$. The boundary may be either an external boundary or internal interface between two regions with different optical properties. For external boundaries, upon transmission the packet may be tracked through the surrounding media until it interacts with the detection device (Bishop et al. 2009).

5.2 *Adaptation of Stochastic Methods to Simulate Optical Mapping Signals*

Fundamentally, to use a MC approach to simulate optical mapping signals, the separate processes of excitation illumination and subsequent fluorescent emission are simulated using the main photon packet propagation algorithm, described above, using optical properties (μ_a, μ_s, g) obtained at the wavelengths specific to illumination and emission (Ding et al. 2001).

5.2.1 Simulating Illumination

To simulate uniform illumination of the tissue surface by an external source, packets are incident upon the external tissue boundary at an illumination angle θ_{ex}. Usually, θ_{ex} is taken at 0 to replicate illuminated light normal to the tissue surface (Ding et al. 2001; Bishop et al. 2009). Dependant upon the refractive index mis-match between the tissue (n_t) and external media (n_m), the initial packet weight is reduced by specular reflectance R_{sp} to $W - R_{sp}$ where

$$R_{sp} = \frac{(n_t - n_m)^2}{(n_t + n_m)^2}. \tag{15.33}$$

For $\theta_{ex} \neq 0$, the initial direction of propagation within the tissue is found from Snell's Law. As the photon packets propagate through the tissue they deposit weight within the tissue which is logged in the spatial array A which typically has a spatial discretisation of 0.05 mm (or the order of the mean step-size $< s >$ of a photon packet). Therefore A represents that photon density (photons per volume) due to illumination throughout the tissue. For simulating illumination from a single point source (infinitely narrow beam), typically 10^5 photon packets are required to obtain a smooth distribution within the tissue. For uniform surface illumination, it has been shown that > 500 photon packets are required to be incident at each surface element to ensure good statistics (Bishop et al. 2009). Note that other illumination strategies may also be performed whereby the distribution, number and initial illumination angles of excitation photon packets on the illuminated surface are changed to replicate strategies such as two-photon excitation (Ramshesh and Knisley 2003). Photon packets are terminated either when their weight falls below the critical threshold (subject to energy conservation restrictions (Ding et al. 2001; Wang et al. 1995; Jacques 1998)), or else they pass out of an external tissue boundary in which case they are deemed lost.

5.2.2 Simulating Fluorescent Emission

In the case of fluorescent emission, the fluorescent photons originate from dye molecules within the tissue itself. In an optical mapping experiment, the more photons received by a region of tissue during illumination, the more fluorescent dye molecules get excited and fluoresce. Thus, to simulate this process, the total number of fluorescent photon packets (N^{em}) emitted from each optical element (i) within the tissue is directly proportional to the corresponding excitation photon density at that same element A_i, each starting its journey with a full photon packet weight $W_{initial} = 1.0$ i.e. $N^{em} \propto A_i$. Note that alternatively, a fixed number of fluorescent photon packets could be emitted from every element within the tissue, where the initial starting weight at each point is instead scaled in proportion to the excitation illumination photon density (i.e. $W_{initial} \propto A_i$). However, this method is significantly more computationally expensive than the former case, as because A_i gets very small for regions of tissue that do not receive much illuminating light, no photon packets are launched at all.

In contrast to the process of illumination, fluorescent photon packets are emitted at randomly distributed angles (isotropic emission) from their point source using additional random numbers (one each for θ and ψ) within the Henyey-Greenstein scattering functions. As fluorescent emission occurs from a 3D volume of tissue (not a 2D surface), typically fewer photon packets are required to be simulated from each tissue element, such that $N^{em} > 100$ is sufficient. Photon packets then undergo absorption and scattering events in a similar manner to illumination. However, when a fluorescent photon packet exits the tissue surface it may simply be lost, or alternatively, it may constitute the collected fluorescent signal.

5.2.3 Simulating Optical Detection

If a photon packet exits the tissue surface within an area from which signals are being recorded, it deposits its total weight at the time of exit as the recorded fluorescent signal from that particular region, or 'pixel', of imaged tissue. Successive photon packets exiting the same region build-up the total signal recorded from this pixel, with the total recorded intensity thus representing the final accumulated packet weight. Furthermore, each and every fluorescent photon packet which exits from this specific region on the tissue surface contains information regarding its point of origin within the tissue (practically whilst performing a simulation, this is the current optical element location within the tissue from which fluorescence is being emitted). With this information, along with the weight of the exiting packet, it is possible to build-up a distribution which shows the relative fraction of recorded fluorescence that originates from a given volume of tissue. Practically, each detection pixel has an associated 3D array D detailing the relative contribution from each optical tissue element in the domain to the signal recorded from that particular pixel. The dimensions of such *interrogation* (Ding et al. 2001) or *scattering* (Bishop et al. 2006c, 2009) volumes of tissue thus provide essential information regarding the origin of optical mapping signals under different circumstances, which is not possible to obtain with other simulation methods.

The above method represents the most simple form of optical detection, whereby the assumption is that all photons exiting the imaging tissue region successfully reach the optical detector (Ding et al. 2001). Alternatively, more elaborate detection strategies may be modelled, which continue to track the photon packet trajectory upon exiting the tissue surface, including its passage through lens systems (Bishop et al. 2009). Such models allow the effects of focal plane location and numerical aperture upon the relative location of the recorded signal to be investigated.

5.2.4 Simulating Voltage-Sensitive Fluorescent Emission

Many of the original studies which applied MC methods to simulate optical mapping signals assumed a uniform distribution of transmembrane potential (Ding et al. 2001; Ramshesh and Knisley 2003), assuming that the amount of fluorescent light given-off by a region of tissue depended only on the photon density due to excitation illumination. This allowed these studies to quantify the region of tissue contributing to the recorded optical signal, independent of the level of transmembrane potential or its distribution being recorded. Later studies (Bishop et al. 2009; Hyatt et al. 2008), however, used MC methods to simulate voltage-sensitive fluorescent emission.

In this case, one possible approach involves scaling the relative number of emitted fluorescent photon packets N^{em} with respect to both the excitation illumination photon intensity at each optical element (A_i) *and* the level of transmembrane potential (V_m) at the same location (derived from previous electrophysiological monodomain/bidomain simulations), in a similar manner to continuum approaches. However, such a method is immensely computationally expensive as a full MC simulation of photon scattering during fluorescent emission must be repeated for each and every V_m distribution (at each outputted V_m time-step, usually 1 ms).

Instead, the approach described in Sect. 5.2.3 provides quantification of the relative fraction of recorded fluorescence that originates from different regions of tissue beneath the detection pixel. This information, recorded in the array A, can then be convoluted with the calculated distribution of V_m at corresponding points throughout the tissue such that the total signal intensity recorded from that pixel, V_{opt}, is given by

$$V_{opt} = \sum_i A_i V_{m,i}, \tag{15.34}$$

where A_i is the relative signal intensity detected from optical tissue element i and $V_{m,i}$ is the V_m level of the tissue at the same location. Note that $V_{m,i}$ can be normalised and scaled such that the total recorded fluorescence faithfully replicates the experimental scenario of an approximate 10 % change upon a background of fluorescence.

Although useful in simulating signals from individual recording detector pixels (Bishop et al. 2009), using the above approach to simulate numerous pixels (i.e. a full detector array) is more challenging due to the significant memory demands required to store the numerous data arrays. As each individual detector array (A) has a size equal to the number of optical elements within the tissue, and an optical element typically has dimensions of $1.25 \times 10^{-4}\,\mathrm{mm}^3$, each detector array often represents a significant memory footprint.

5.3 Application Studies Using Stochastic Methods

Despite its accuracy in modelling light scattering in tissue, the stochastic methods have sparsely been applied to the simulation of the scattering effects encountered in optical mapping experiments. The first major study to attempt this by Ding et al. (2001) utilised the fact that MC simulations can provide information regarding the individual histories of detected photons to examine the spatial localisation of the recorded fluorescent signal. The authors used a combination of MC simulations (based on the algorithm outlined in Wang et al. 1995) and optical mapping experiments to quantify the size of the interrogation region associated with a particular epicardial recording site. Results showed that the majority (approximately 80 %) of the content of the total fluorescent signal recorded at a given epicardial site originated from an extended three-dimensional volume of tissue beneath the recording site itself, extending to approximately 1–2 mm both radially and in depth. The specific dimensions of this volume were also seen to depend upon the optical scattering and absorption properties of the tissue, explicitly measured in the study. The study focussed on the comparison between specific different illumination strategies (broad-field and laser-scanning excitation), and concluded that tissue regions larger than the laser beam or collecting area of the detector array element contribute to the recorded fluorescent signal, compromising spatial resolution of the optical signal. A series of follow-up studies by the same Group examined how this interrogation region could be limited either through the use of controlled optical illumination strategies and multi-photon excitation (Ramshesh and Knisley 2003), or through co-loading with a highly absorbing optical dye to limit light scattering (Ramshesh and Knisley 2006).

The study by Bishop et al. (2009) used a MC approach to simulate voltage-sensitive fluorescent signals, using the methods described above in Sect. 5.2. Here, the authors were able to explicitly quantify how the spatial origin of the fluorescent signal recorded from individual pixels on the epicardial surface varied with the specific pixel dimensions, and the implications this had on the optically-recorded action potential upstroke. Results highlighted the importance of lateral scattering, demonstrating that the scattering volume (defined as the tissue volume from which 80 % of the total detected fluorescence originates) extends over three times further in the surface plane than in depth. This spatial origin of the recorded fluorescent signal, along with the total percentage contributions from certain tissue depths and

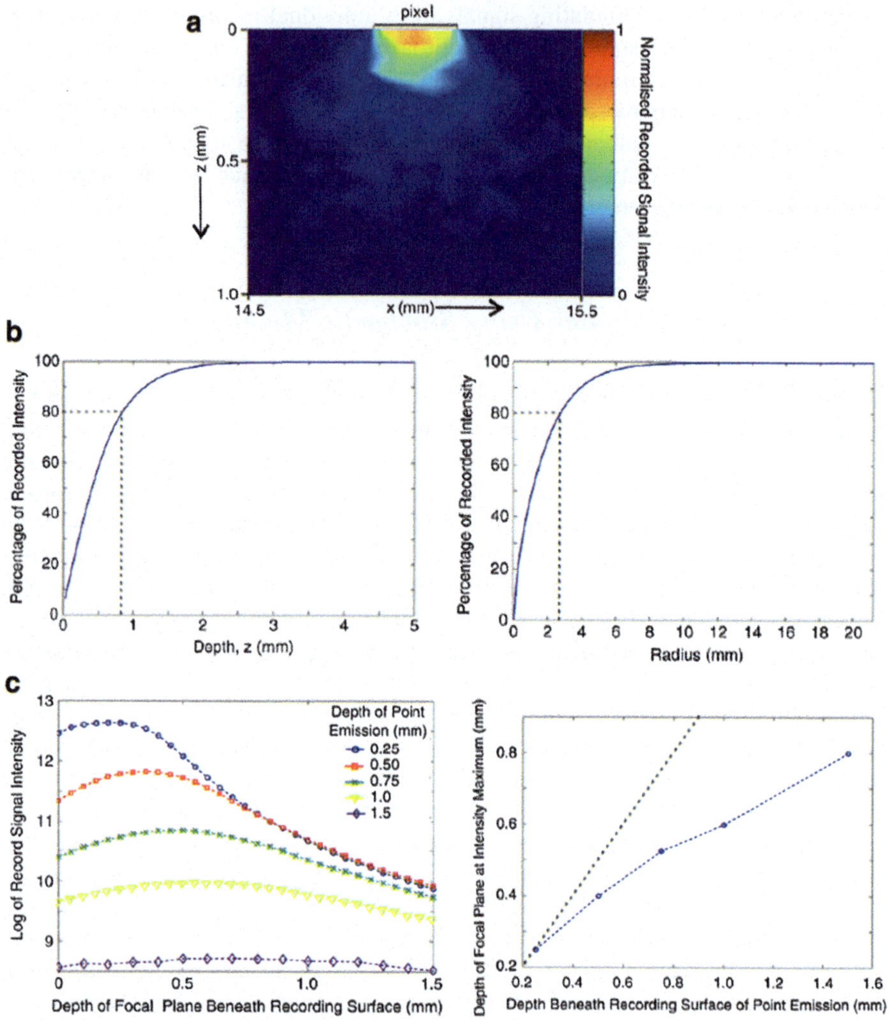

Fig. 15.7 (**a**) Spatial distribution of scattering volume for a 0.25×0.25 mm pixel, showing cross-section of the normalised signal intensity recorded by the pixel (shown as a *black square*) in depth. (**b**) Percentage of total signal intensity recorded by detector originating from different depths (*left*) and from different radii (*right*) from the centre of the pixel. *Dashed lines* show location of 80 % cut-off for scattering volume definition. (**c**) (*Left*) logarithm of recorded signal intensity as depth of focal plane into the tissue is varied for point emission of photons from different depths beneath the tissue surface. (*Right*) focal plane depth at which signal in panel (**a**) is maximum for each depth of point emission, including line of unit slope for comparison (*dashed-line*). Used with permission from Bishop et al. (2009)

radii from the pixel, are shown in Fig. 15.7a. Consequently, only a small fraction of the total signal (11 % for a 0.25×0.25 mm pixel) originates from tissue directly beneath the recording site. Although increasing pixel size was shown to increase this fraction, this was at the cost of spatial resolution, causing a corresponding increase in overall scattering volume and signal distortion, as optical action potential upstroke duration also increased. The study also went on to advance the standard MC simulation method by explicitly simulating a realistic model of a tandem-lens optical detection apparatus with a view to understanding how changes in the detector characteristics may alter the specific tissue regions from which fluorescent signals are obtained. Interestingly, the authors demonstrated that although including realistic optical detection through a lens set-up excludes photons exiting the surface at shallow angles from being detected, these effects appeared negligible, with the dimensions of the scattering volume being very similar to the case of ideal detection. Furthermore, changes in the location of the focal depth position into the tissue and increases in the numerical aperture of the system also were correspondingly shown to have very little impact on the scattering volume, suggested to be due to the fact that fluorescence close to the tissue surface completely dominates the recorded signal for uniform illumination, even when the lens system is tuned to record light from deeper into the tissue. Finally, simulations also showed that scattering effects significantly limit the ability of optical detection system tuning in accurately locating individual fluorescent emission events (similar to two-photon emission) within the tissue depth, consistently providing an underestimation of the emission depth (Fig. 15.7c).

Another early study to explicitly combine a MC model of photon scattering with an electrophysiological model of electrical activity to simulate voltage-sensitive fluorescent emission was the study by Hyatt et al. (2008). Here, the authors compared simulated optical signals from a diffusion based model with those from a MC based model, using a similar method to that described above, over a simple slab geometry. As found in Bishop et al. (2009) study, a noticeable difference was seen in the profiles of simulated photon density with depth with the MC model predicting a distinct subsurface peak (as noted in the experimental work of Baxter et al. (2001)) compared to the mono-exponential decay given by the diffusion model. The existence of such a subsurface peak was seen to maintain the excitation intensity at a relatively higher level with depth, shown by the intensity profiles with depth in Fig. 15.8a. Consequently, the authors found that the integrating volume contributing to the surface-recorded fluorescent signal extended deeper into the tissue using a MC model than the continuum approach resulting in the MC model predicting a longer optical action potential upstroke. Such a longer upstroke duration matched more closely with experimental measurements in guinea pig tissue also performed in the study (shown in Fig. 15.8b). The effect of the tissue thickness, however, was noted to have only a modest impact on these differences.

A similar MC model by the same Group was also used to demonstrate the utility of longer-wavelength dyes (DI-4-ANBDQBS) at measuring important electrophysiological features throughout the depth of the myocardial wall in the rat ventricle. Simulations using the MC model showed that the deeper penetration of the near infra-red dye produced approximately 30 % longer optical action potential upstrokes than the conventional DI-4-ANEPPS signals. The model was further used to confirm the experimental observations that such dyes allow for the faithful measurement of transmural heterogeneity in action potential duration and restitution as epicardially-recorded signals were shown to contain a significant endocardial component.

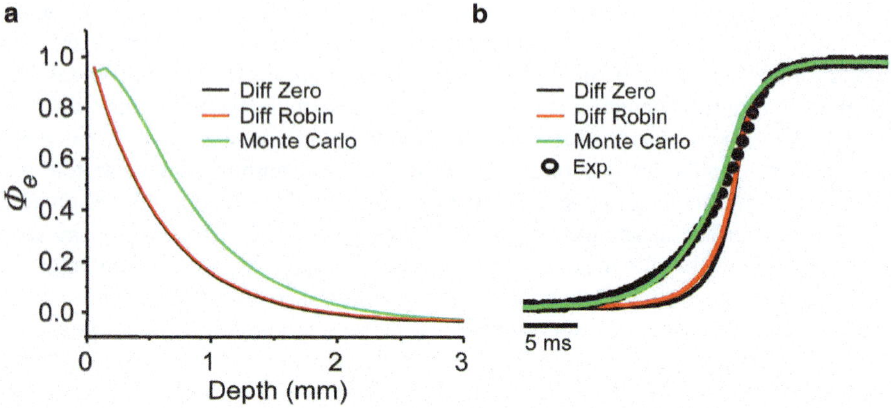

Fig. 15.8 (**a**) Distribution of excitation photon density (Φ_e) as a function of depth beneath the tissue surface comparing predictions from MC simulations (*green*) versus diffusion-based solutions using the zero boundary condition (*black*) and Robin (or partial current) boundary condition (*red*). (**b**) Simulated action potential upstrokes showing the closer agreement of the Monte Carlo (*green*) with experiment data (*open circles*) relative to the diffusion approach (*red, black*). Used with permission from Hyatt et al. (2008)

6 Future Directions

6.1 *Modelling Approaches for Use in Conjunction with High-Resolution Imaged-Derived Models*

An important future direction in the development of biophotonic models of optical mapping is currently being driven by the significant recent advance in the level of anatomical detail computational models are capable of representing due to corresponding advances in anatomical imaging (MR, histology) techniques. Previous 'anatomically-realistic' geometric computational models (such as those used in the studies discussed above, and in Figs. 15.2, 15.4, and 15.5) represented the myocardial wall as a solid structure (typically 2–6 mm in the rabbit ventricular models used) containing only myocardial tissue throughout. High-resolution imaging datasets ($< 25\,\mu m$) have recently been used to directly construct computational finite element models containing faithful representations of fine-scale anatomical structures including the coronary vasculature and endocardial features such as trabeculations and papillary muscles. Figure 15.9 shows an example of such a high-resolution model of a rabbit ventricular wedge (Bishop et al. 2010), derived directly from high-resolution MR data (Bishop et al. 2010).

However, an important limitation in the use of these models in conjunction with optical simulation methods, such as those described above, lies in the presence of intramural cavities (vessels, extracellular clefts), highlighted in Fig. 15.9. In optical mapping experiments, in which the heart is perfused with saline solution, these

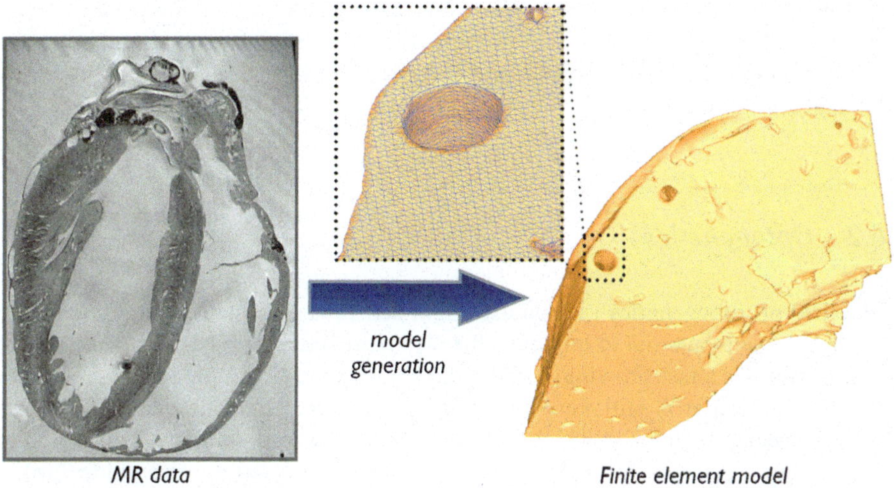

Fig. 15.9 Example of a high-resolution MR-derived finite element computational model representing a left-ventricular wedge of the rabbit and highlighting the presence of cavities such as blood vessels within the myocardial wall (Bishop et al. 2010). Both models are derived from MR data published in Bishop et al. (2010)

structures represent regions of almost transparent fluid, very weakly scattering to photons at the excitation/emission wavelengths used. In such regions, the photon diffusion equation is not valid and cannot be used. It is possible, however, to use stochastic methods, which can represent photon propagation within any media, regardless of its optical properties. Nonetheless, up until now, MC methods for optical mapping signal synthesis have exclusively been used in conjunction with highly simplified geometric models, usually representing regular slabs of cardiac tissue. In such regular geometries (consisting of regular, structured meshes), it becomes trivial to compute the optical element a given photon packet resides after taking a step along its path between interaction events. In unstructured meshes, such as those necessarily used to construct the latest anatomically detailed models, tracking the packets' position, and in addition, checking for potential boundary interactions, in a highly-optimised computational manner becomes a significant challenge. However, recently MC models of photon propagation within unstructured tetrahedral meshes have been proposed, where the interaction between photon packet and triangle (a face of a tetrahedral element) is computed recursively and rapidly (Shen and Wang 2011). This method has even been applied to model light propagation within a whole body mouse mesh, including organs. Application of similar techniques, suitably adapted for specific optical mapping setups, will allow the simulation of optical signals over the latest computational anatomical models, intramural cavities and all. An interesting first line of application of such models will be during defibrillation shocks due to the recent suggestion of the importance of 'virtual-electrode' formation around large subepicardial vessels (Bishop et al. 2012; Luther et al. 2012). It is thought that the interaction of photon scattering with the fine-scale anatomical features present in these models, particularly large sub-epicardial blood vessels, will provide a further set of interesting mechanisms regarding optical signal distortion and facilitate an even closer match with experimental recordings.

6.2 Optogenetics

The field of ‘optogenetics’, which uses light-sensitive ion channels introduced into cardiac tissue to attempt to influence the electrical activity of the heart with optical illumination, is a fast-growing, exciting area of cardiac electrophysiology research. As this technique is still very much in its infancy, computational models are currently being developed which aim to provide a generic framework in which to simulate optogenetics in order to better understand its utility and guide further developments (Boyle et al. 2013). The wavelengths involved in optogenetics are within the visible spectrum, and thus the process of modelling photon propagation within the tissue is similar to that undertaken in optical mapping. Thus far, only simple models of light attenuation into the tissue have be applied, using mono-exponential decay functions. However, as the modelling frameworks become more advanced and aim to replicate the tissue excitation within more anatomically-complex models combined with more intricate optical illumination strategies, so will the need to apply more complex methodologies to simulate photon propagation. Nonetheless, it is believed that optogenetics will represent an important application of biophotonic modelling applied to cardiac electrophysiology in the coming years.

References

Al-Khadra A, Nikolski V, Efimov IR (2000) The role of electroporation in defibrillation. Circ Res 87(9):797–804

Arridge SR (1993) A finite element approach for modeling photon transport in tissue. Med Phys 20 (2):299–309

Baxter WT, Mironov SF, Zaitsev AV, Jalife J (2001) Biophysical journal – visualizing excitation waves inside cardiac muscle using transillumination. Biophys J 80(1):516–530

Bernus O, Wellner M, Mironov SF, Pertsov AM (2004) Simulation of voltage-sensitive optical signals in three-dimensional slabs of cardiac tissue: application to transillumination and coaxial imaging methods. Phys Med Biol 50(2):215–229

Bernus O, Mukund KS, Pertsov AM (2007) Detection of intramyocardial scroll waves using absorptive transillumination imaging. J Biomed Opt 12(1):014035

Bishop MJ, Rodriguez B, Trayanova N, Gavaghan DJ (2006a) Modulation of shock-end virtual electrode polarisation as a direct result of 3D fluorescent photon scattering. In: 28th Annual international conference of the IEEE engineering in medicine and biology society, 2006. EMBS’06, pp 1556–1559

Bishop MJ, Rodriguez B, Trayanova N, Gavaghan DJ (2006b) Inference of intramural wavefront orientation from optical recordings in realistic whole-heart models. Biophys J 91(10):3957–3958
Bishop MJ, Rodriguez J, Eason B, Whiteley JP, Trayanova N, Gavaghan DJ (2006c) Synthesis of voltage-sensitive optical signals: application to panoramic optical mapping. Biophys J 90(8):2938–2945
Bishop MJ, Gavaghan DJ, Trayanova NA, Rodriguez B (2007a) Photon scattering effects in optical mapping of propagation and arrhythmogenesis in the heart. J Electrocardiol 40 (6 Suppl):S75–S80
Bishop MJ, Rodriguez B, Qu F, Efimov IR, Gavaghan DJ, Trayanova NA (2007b) The role of photon scattering in optical signal distortion during Arrhythmia and defibrillation. Biophys J 93 (10):3714–3726
Bishop MJ, Bub G, Garny A, Gavaghan DJ, Rodriguez B (2009) An investigation into the role of the optical detection set-up in the recording of cardiac optical mapping signals: a Monte Carlo simulation study. Physica D 238(11–12):1008–1018
Bishop MJ, Boyle PM, Plank G, Welsh DG, Vigmond EJ (2010a) Modeling the role of the coronary vasculature during external field stimulation. IEEE Trans Biomed Eng 57(10):2335–2345
Bishop MJ, Plank G, Burton RAB, Schneider JE, Gavaghan DJ, Grau V, Kohl P (2010b) Development of an anatomically detailed MRI-derived rabbit ventricular model and assessment of its impact on simulations of electrophysiological function. Am J Physiol Heart Circ Physiol 298(2):H699–H718
Bishop MJ, Rowley A, Rodriguez B, Plank G, Gavaghan DJ, Bub G (2011) The role of photon scattering in voltage-calcium fluorescent recordings of ventricular fibrillation. Biophys J 101(2):307–318
Bishop MJ, Plank G, Vigmond E (2012) Investigating the role of the coronary vasculature in the mechanisms of defibrillation. Circ Arrhyth Electrophysiol 5(1):210–219
Boyle PM, Williams JC, Ambrosi CM, Entcheva E, Trayanova NA (2013) A comprehensive multiscale framework for simulating optogenetics in the heart. Nat Commun 4:2370
Bray M, Wikswo JP (2003) Examination of optical depth effects on fluorescence imaging of cardiac propagation. Biophys J 85(6):4134–4145
Cheong WF, Prahl SA, Welch AJ (1990) A review of the optical properties of biological tissues. IEEE Quantum Electron 26(12):2166–2185
Choi BR, Salama G (1998) Optical mapping of atrioventricular node reveals a conduction barrier between atrial and nodal cells. Am J Physiol Heart Circ Physiol 274(3):H829–H845
Ding L, Splinter R, Knisley SB (2001) Quantifying spatial localization of optical mapping using Monte Carlo simulations. IEEE Trans Biomed Eng 48(10):1098–1107
Efimov IR, Mazgalev TN (1998) High-resolution, three-dimensional fluorescent imaging reveals multilayer conduction pattern in the atrioventricular node. Circulation 98(1):54–57
Efimov IR, Sidorov V, Cheng Y, Wollenzier B (1999) Evidence of three dimensional scroll waves with ribbon shaped filament as a mechanism of ventricular tachycardia in the isolated rabbit heart. J Cardiovasc Electrophysiol 10(11):1452–1462
Efimov IR, Aguel F, Cheng Y, Wollenzier B, Trayanova N (2000) Virtual electrode polarization in the far field: implications for external defibrillation. Am J Physiol Heart Circ Physiol 48(3):H1055
Gardner CM, Jacques SL (1996) Light transport in tissue: accurate expressions for one dimensional fluence rate and escape function based upon Monte Carlo simulation. Lasers Surg Med 18:129–138
Girouard SD, Laurita KR, Rosenbaum DS (1996) Unique properties of cardiac action potentials recorded with voltage-sensitive dyes. J Cardiovasc Electrophysiol 7(11):1024–1038
Gray RA (1999) What exactly are optically recorded action potentials? J Cardiovasc Electrophysiol 10(11):1463–1466
Haskell RC, Svaasand LO, Tsay T, Feng T, McAdams MS, Tromberg BJ (1994) Boundary conditions for the diffusion equation in radiative transfer. J Opt Soc Am A 11(10):2727

Hielscher AH, Jacques SL, Wang L, Tittel FK (1995) The influence of boundary conditions on the accuracy of diffusion theory in time-resolved reflectance spectroscopy of biological tissues. Phys Med Biol 40(11):1957–1975

Hyatt CJ, Mironov SF, Wellner M, Berenfeld O, Popp AK, Weitz DA, Jalife J, Pertsov AM (2003) Synthesis of voltage-sensitive fluorescence signals from three-dimensional myocardial activation patterns. Biophys J 85(4):2673–2683

Hyatt CJ, Mironov SF, Vetter FJ, Zemlin CW Pertsov AM (2005) Optical action potential upstroke morphology reveals near-surface transmural propagation direction. Cir Res 97(3):277–284

Hyatt CJ, Zemlin CW, Smith RM, Matiukas A, Pertsov AM, Bernus O (2008) Reconstructing subsurface electrical wave orientation from cardiac epi-fluorescence recordings: Monte Carlo versus diffusion approximation. Opt Express 16(18):13758–15772

Jacques SL (1998) Light distributions from point, line and plane sources for photochemical reactions and fluorescence in turbid biological tissues. Photochem Photobiol 67(1):23–32

Janks DL, Roth BJ (2002) Averaging over depth during optical mapping of unipolar stimulation. IEEE Trans Biomed Eng 49(9):1051–1054

Knisley SB (1995) Transmembrane voltage changes during unipolar stimulation of rabbit ventricle. Circ Res 77(6):1229–1239

Luther S, Fenton FH, Kornreich BG, Squires A, Bittihn P, Hornung D, Zabel M, Flanders J, Gladuli A, Campoy L, Cherry EM, Luther G, Hasenfuss G, Krinsky VI, Pumir A, Gilmour RF, Bodenschatz E (2012) Low-energy control of electrical turbulence in the heart. Nature 475(7355):235–239

Okada E, Schweiger M, Arridge SR, Firbank M, Delpy DT (1996) Experimental validation of Monte Carlo and finite-element methods for the estimation of the optical path length in inhomogeneous tissue. App Opt 35(19):3362

Omichi C (2004) Intracellular Ca dynamics in ventricular fibrillation. Am J Physiol Heart Circ Physiol 286(5):H1836–H1844

Pathmanathan P, Gray RA (2013) Ensuring reliability of safety-critical clinical applications of computational cardiac models. Front Physiol 4:358

Ramshesh VK, Knisley SB (2003) Spatial localization of cardiac optical mapping with multiphoton excitation. J Biomed Opt 8(2):253–259

Ramshesh VK, Knisley SB (2006) Use of light absorbers to alter optical interrogation with epi-illumination and transillumination in three-dimensional cardiac models. J Biomed Opt 11(2):024019

Rodriguez B, Li L, Eason JC, Efimov IR, Trayanova NA (2005) Differences between left and right ventricular chamber geometry affect cardiac vulnerability to electric shocks. Circ Res 97(2):168–175

Shen H, Wang G (2011) A study on tetrahedron-based inhomogeneous Monte Carlo optical simulation. Biomed Opt Express 2(1):44–57

Trayanova N, Plank G, Rodriguez B (2006) What have we learned from mathematical models of defibrillation and postshock arrhythmogenesis? Application of bidomain simulations. Heart Rhythm 3(10):1232–1235

Walton RD, Bernus O (2009) Computational modeling of cardiac dual calcium-voltage optical mapping. In: Annual international conference on IEEE engineering in medicine biology society, pp 2827–2830

Wang L, Jacques SL, Zheng L (1995) MCML—Monte Carlo modeling of light transport in multi-layered tissues. Comput Methods Programs Biomed 47(2):131–146

Wu S, Weiss JN, Chou C-C, Attin M, Hayashi H, Lin S-F (2005) Dissociation of membrane potential and intracellular calcium during ventricular fibrillation. J Cardiovasc Electrophysiol 16(2):186–192

Chapter 16
Towards Depth-Resolved Optical Imaging of Cardiac Electrical Activity

Richard D. Walton and Olivier Bernus

Abstract The spatiotemporal dynamics of arrhythmias are likely to be complex three-dimensional phenomena. Yet, the lack of high-resolution three-dimensional imaging techniques, both in the clinic and the experimental lab, limits our ability to better understand the mechanisms of such arrhythmias. Optical mapping using voltage-sensitive dyes is a widely used tool in experimental electrophysiology. It has been known for decades that even in its most basic application, epi-fluorescence, the optical signal contains information from within a certain intramural volume. Understanding of this fundamental property of optical signals has paved the way towards novel three-dimensional optical imaging techniques. Here, we review our current understanding of the three-dimensional nature of optical signals; how penetration depths of cardiac optical imaging can be improved by using novel imaging modalities and finally, we highlight new techniques inspired from optical tomography and aiming at full depth-resolved optical mapping of cardiac electrical activity.

Keywords Depth-resolved optical imaging • Transillumination • Near-infrared voltage sensitive dyes • Optrodes • Optical tomography

1 Introduction

In Europe and the USA, over 250,000 sudden cardiac deaths (SCDs) occurring each year are due to ventricular fibrillation (VF). Currently, the majority of individuals dying suddenly cannot be recognized pre-emptively. Furthermore, the mechanisms of human VF remain poorly understood mainly due to the urgency to defibrillate the patient. The spatiotemporal dynamics of VF are likely to be a complex three-dimensional (3D) phenomenon, with contribution from and interplay between various triggers and substrates (both electrophysiological and structural) throughout the thickness of the ventricular wall. Yet, the lack of high-resolution three-

R.D. Walton, Ph.D. (✉) • O. Bernus
L'Institut de rythmologie et modélisation cardiaque LIRYC, Centre de Recherche Cardiothoracique de Bordeaux, Université de Bordeaux, Inserm U1045, Pessac, France
e-mail: Richard.walton@ihu-liryc.fr; Olivier.bernus@u-bordeaux.fr

M. Canepari et al. (eds.), *Membrane Potential Imaging in the Nervous System and Heart*, Advances in Experimental Medicine and Biology 859,
DOI 10.1007/978-3-319-17641-3_16

dimensional (3D) imaging, both in the clinic and the experimental lab, limits our ability to better understand the mechanisms of VF and potentially develop patient-specific targeted therapies to this life-threatening arrhythmia.

Current clinical measurements of electrical activity in the heart are limited to the endo- or epicardium using sequential localized measurements or low-resolution multi-electrode catheters. These measurements are inadequate to fully capture the electrophysiological and structural heterogeneities in the myocardial wall that may play a role in arrhythmogenesis. Thus intramural triggers and substrate of VF often remain elusive. In the experimental lab, wealth of techniques have been developed with the endeavor of characterizing intramural electrical activity. However, due to limitations of spatiotemporal resolution, range in depth of penetration into the myocardium, signal-to-noise ratios, or the invasiveness of procedures, characterization of 3D organization of electrical activity remains challenging.

Optical mapping using voltage-sensitive dyes is a widely used tool in experimental electrophysiology (see Chap. 12). The capacity of high spatial and temporal resolution of signals proportionate to the transmembrane potential afforded by optical mapping has led to significant interest in optical approaches. The most common application of optical mapping is epi-fluorescence, where a tissue surface is illuminated at the excitation wavelength of the dye and fluorescence is recorded from the same surface. It is important to note, that even in epi-fluorescence, optical signals do not originate solely from the tissue surface but represent a weighted sum of single-cell transmembrane potentials within a certain volume (Bishop et al. 2007; Girouard et al. 1996; Hyatt et al. 2005; Janks and Roth 2002). This volume is determined by the scattering and absorptive properties of the tissue as discussed in Chap. 15. The size and depth of this integration volume greatly depends on the wavelengths used and imaging modality. Understanding of this fundamental property of optical signals has paved the way towards novel 3D optical imaging techniques.

In this chapter we report on the current understanding of 3D electrical activity in the heart recorded through optical approaches and focus firstly, on interpretation of the depth contributions to optical signals; secondly, how optical signals and their depth contributions are modulated by the different imaging modalities currently available and finally, we highlight current techniques that are paving the way towards full depth-resolved optical mapping of electrical activity.

2 The Optical *vs.* Electrical Action Potential Upstroke

The upstroke of the action potential is the phase that appears to be the most affected by optical averaging effects. Single cell electrical action potentials depolarize rapidly over durations of 1–2 ms. However, the optical upstroke recorded from tissue is typically between 5 and 15 ms in duration (Hyatt et al. 2005; Walton et al. 2012). The main reason is that fluorescence continues to increase while the electrical wave front is propagating across the optical integration volume, only reaching a maximum signal amplitude when all, or the largest proportion with the greatest contribution, of the integration volume has depolarized. This phenomenon significantly prolongs and blurs the optical upstroke compared to the cellular electrical upstroke. This effect can be observed to an extreme in certain cases whereby two distinct layers of poorly coupled tissue reside within the optical integration volume. In such cases, if conductivity between the two layers was sufficiently low so that the time of activation of each region were significantly delayed from one another, the optical upstroke reveals a two-phase depolarization profile and very long upstroke durations. Such signal morphologies are observed at the sino-atrial node whereby early activation of these specialized cells precedes activation of the surrounding myocardium. The occurrence of biphasic upstroke morphologies is indicative of boundaries of conduction discontinuities residing within the optical integration volume. Their relative contributions to the total amplitude of the optical signal are proportional to their dominance within the optical integration volume and spatial gradients of their contributions infer changes in proximity to the imaged surface.

2.1 Optical Activation Time

The significantly longer duration of optical upstrokes compared to electrical upstrokes raises a question on the most accurate determination of local activation time. Two techniques for measuring activation times from optical recordings have been proposed and used by various groups: the time of maximal slope, in analogy to electrical measurements, and the time to reach a fixed fluorescence threshold (typically 50 % of the maximal amplitude). The former approach was thought to be more accurate, whereas the latter more robust in noisy signals. However, it was only recently that the accuracy of each technique was assessed by combined optical and electrical measurements. In a first study, Knisley and Neuman (2003) found that the time of maximal slope of the optical upstroke correlated better with the time of maximal derivative from simultaneously acquired unipolar electrograms. More recently, Walton et al. (2012) performed a systematic comparison between optical upstrokes and true transmembrane electrical upstrokes obtained by microelectrode recordings. They showed that the time of the maximal slope of the optical upstroke (tF*) corresponded well to the electrical activation time (tE) at the tissue surface with differences as small as 0.7 ms. However, using a fixed threshold was found to be a poor predictor of the surface electrical activation time, leading to significant errors on estimation of conduction velocities (Fig. 16.1).

2.2 Upstroke Morphology and Sub-surface Wave Front Orientation

The optical blurring of the upstroke can be seen as a limitation on the effective spatial resolution that can be achieved independent of the optical setup. However, it

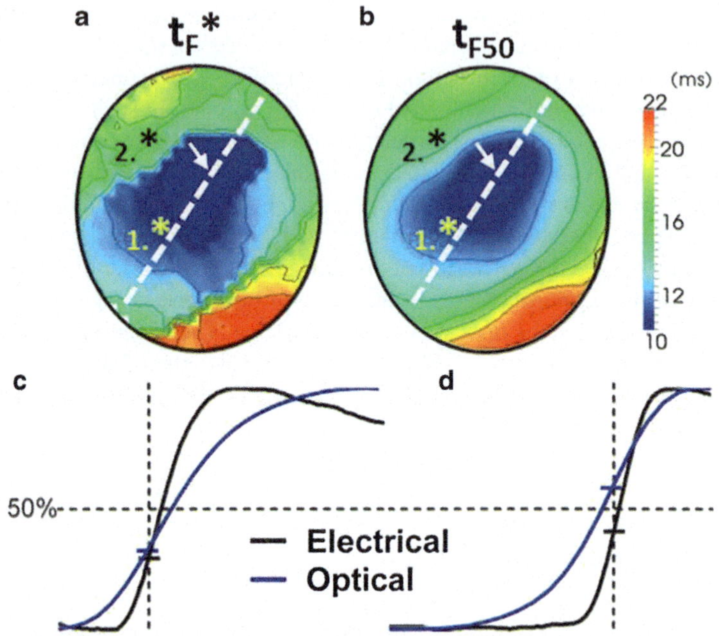

Fig. 16.1 Comparison of optical and electrical activation times in isolated rat hearts. (**a**) Activation map based on time of maximal upstroke velocity (tF*) following stimulation in the center of the field-of-view. (**b**) Activation map based on a 50 % fluorescence threshold (tF50) for the same activation sequence as in Panel **a**. The *white dashed lines* indicate the directions of fastest propagation. (**c**) Comparison between electrical (*black*) and optical (*blue*) upstrokes measured in point 1 in Panels **a** and **b**. (**d**) Comparison between electrical (*black*) and optical (*blue*) upstrokes measured in point 2 in Panels **a** and **b**. The *crosses* indicated the point of maximal upstroke velocity both in the optical (*blue*) and electrical (*black*) traces. The *vertical dashed lines* indicate the electrical activation time, while the *horizontal line* indicates the 50 % fluorescence level. Modified from Walton et al. (2012)

has been shown that the optical upstroke also contains useful 3D information. Indeed, the slope of the optical upstroke is steepest when myocardium depolarizes in layers that contribute the most to the optical signal. In the case of epi-fluorescence optical mapping, where the imaged surface is also the illuminated surface, the near-surface layers of the optical integration volume contribute more than the deeper layers. The timing of the steepest portion of the optical upstroke is therefore determined by the orientation of the wave front (Hyatt et al. 2005, 2008; Zemlin et al. 2008). In the case of a wave front originating and propagating from the imaged surface toward deeper layers, the maximum positive slope of the fluorescent signal occurs early during the rise of the optical upstroke (Fig. 16.2a). Conversely, a wave front propagating from the deeper myocardium, toward the imaged surface produces the maximal slope later in the optical upstroke (Fig. 16.2b). Therefore, the orientation of the electrical wave front, with respect to the imaged surface, can be extrapolated from the relative level at which the maximal slope occurs during the

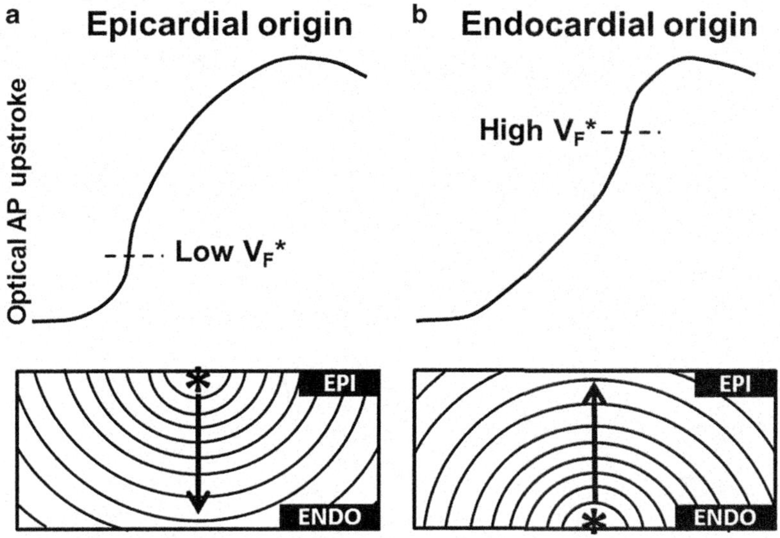

Fig. 16.2 Optical upstroke morphology *vs* sub-surface wave front orientation. (**a**) Representative optical upstroke (*top*) measured from the epicardium for a wave front originating at the epicardium and propagating intramurally towards the endocardium (*bottom*), with a typical low relative fluorescence level at which the steepest portion of the upstroke occurs (V_F*). (**b**) Representative optical upstroke (*top*) measured from the epicardium for a wave front originating at the endocardium and propagating towards the epicardium (*bottom*), with a typical high relative fluorescence level at which the steepest portion of the upstroke occurs (V_F*)

optical upstroke, denoted as V_F*. Furthermore, there appears to be a good correlation between V_F* and the angle of the sub-surface wave front with respect to the imaged surface. This important observation has been verified experimentally in various conditions (Hyatt et al. 2008; Zemlin et al. 2008).

3 Probing Deeper in to the Myocardium

3.1 Transillumination

As discussed above, the most common application of optical mapping is the epi-fluorescence mode. In this configuration measured electrical activity is confined predominantly to a thin layer close to the tissue surface (<2 mm) constituting only a small portion of the ventricular wall. Several studies have attempted to measure electrical activity from a greater range of depths of the myocardium in the transmural plane, including multiple plunge needle electrodes and measuring directly from the cut surface in wedge preparations (Di Diego et al. 2013; Glukhov et al. 2010). The former approach has relatively limited spatial resolution and can only yield electrograms (no action potentials), whereas the latter only provides

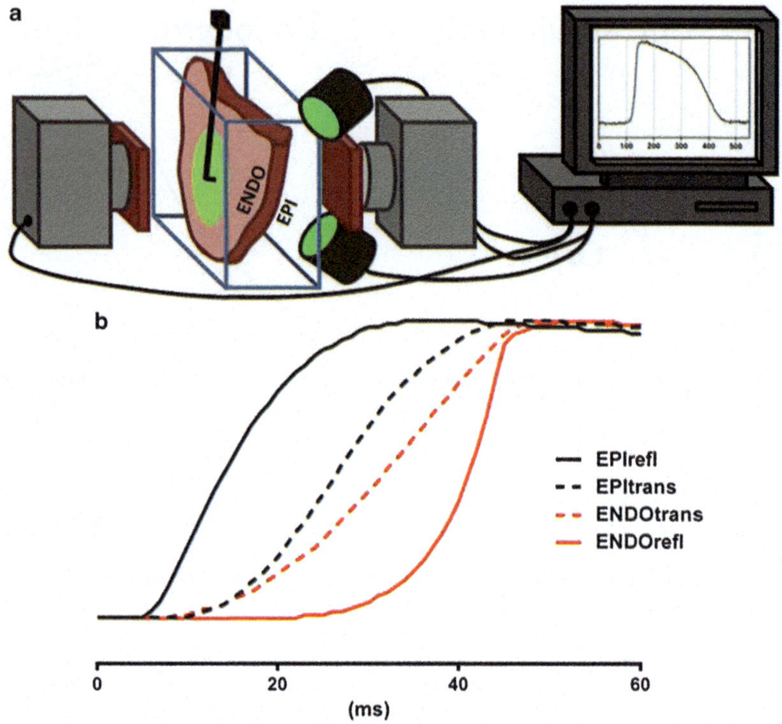

Fig. 16.3 (**a**) Schematic representation of a typical dual camera transillumination setup where one surface (epicardium) is illuminated and fluorescence is acquired both in epi-fluorescence and transillumination. (**b**) Optical action potential traces from different imaging modalities acquired from both epi- and endocardium illustrating a typical epi- to endocardium activation sequence. Modified from Baxter et al. (2001)

information from a specific transmural plane. In order to circumvent the limitations associated with cut surfaces and extend the optical penetration depth of optical mapping, a new technique using a *transillumination* approach was proposed by Baxter et al. (2001). Transillumination consists of aiming the excitation light on to the opposite surface of tissue to the photodetector, as shown schematically for a ventricular wedge preparation in Fig. 16.3a. Baxter et al. (2001) examined transillumination signals in right ventricular wedge preparations dissected from sheep hearts. A CCD camera was aimed at one tissue surface and the contralateral surface illuminated for acquisition in transillumination mode. Surface pacing was used to drive the excitation wave front in a regular and controlled manner so that wave fronts originated from the imaged surface and propagated transmurally, activating lastly the opposing illuminated surface. Under such activation sequences, there is a latency in the earliest AT identified during transillumination imaging relative to reflected imaging whereby illumination was switched to the imaged surface (Fig. 16.3b). The AT latency is due to the orientation of the propagating wave front, which originated at the imaged surface and travelled toward the opposing surface therefore confirming that transillumination signals originate from deeper layers of the myocardium than reflected signals. A biophotonic model showing that the depth contribution to the optical signal was significantly increased in transillumination vs. epi-fluorescence corroborated these various observations.

Re-entrant activation patterns in cardiac tissue are believed to underlie ventricular tachycardia and ventricular fibrillation. In thick ventricular myocardium, functional re-entrant activity is organized as a scroll wave rotating around a singularity line called a filament (Medvinsky et al. 1984; Winfree 1994). Where filaments terminate at the tissue surface, activation patterns reveal distinguishable rotating activation wave fronts around the filament. However, filaments are dynamic and non-linear and their trajectories are not strictly transmural. Therefore, the main body of the filament, which is intramural, commonly runs obliquely or parallel to the tissue surface (Gray et al. 1996). Often a center of rotation is not observed at the surface and activation patterns generally occur as breakthroughs and/or planar waves emerging from the edges of the mapped region (Chen et al. 1988; El-Sherif et al. 1996; Frazier et al. 1989; Gray et al. 1998; Janse et al. 1980; Pogwizd and Corr 1990; Witkowski et al. 1998), characteristic of intramural scroll waves with filaments located deeper in the myocardium. For a more detailed description of scroll waves and arrhythmias, see Chaps. 13 and 14.

Scroll wave filaments are thought to be the drivers of lethal ventricular arrhythmias. Visualizing and locating these filaments is therefore key to a better understanding and treatment of SCD. The filaments have various shapes and orientations and can also drift within the myocardium. Identifying and locating an intramural scroll wave filament is often challenging, particularly when limited to mapping the tissue surface. Transillumination imaging has been shown to allow visualizing filaments located intramurally that would otherwise go unnoticed with epi-fluorescence (Baxter et al. 2001). More recently, new image processing techniques were proposed to directly visualize intramural scroll wave filaments by Bernus et al. (2007) who proposed three approaches based on computer simulations of intramural scroll waves in ventricular slabs: (1) amplitude maps; (2) time-space plots and (3) power-of-dominant-frequency maps (Fig. 16.4).

Amplitude maps provide the maximal amplitude of the optical signal throughout a single activation cycle. It was shown that filaments can be identified as the minimum in the amplitude maps due to the unexcited core around which a scroll wave rotates (in other words no fluorescence is produced in the core or filament of a scroll wave). Amplitude maps were successful in determining the in-plane location of filaments positioned at a range of 1.8–4.6 mm depth from the imaged surface. Time-space plots show the spatiotemporal dynamics of electrical activity along a profile. A profile perpendicular to the filament plotted against time reveals the rotating excitation gives rise to a zigzag pattern with two mutually phase-shifted branches either side of the filament, which is a signature of rotating waves. The intersection of the phase-shifted portions of the plots indicates the location of the filament. Finally, power-of-dominant-frequency analysis revealed reduced power of the dominant frequency component along the filament compared to distant regions of the tissue. A filament depth-dependent second harmonic at twice the dominant frequency was also observed

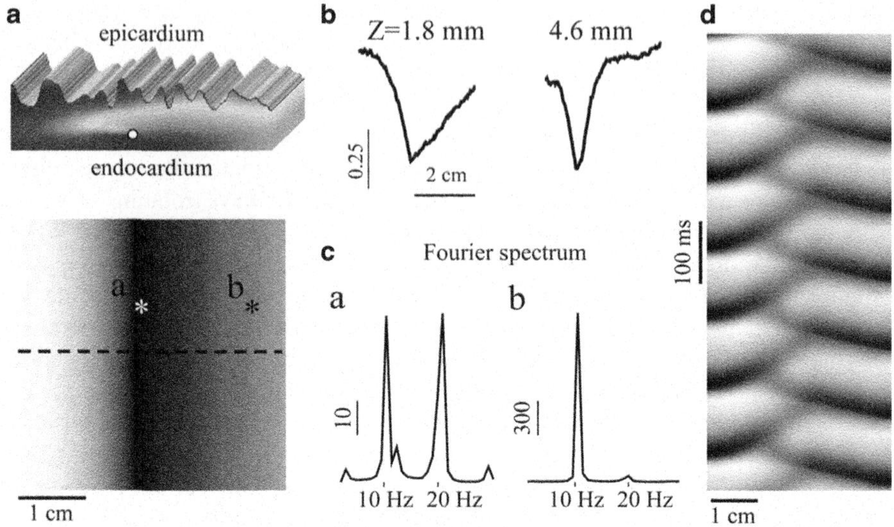

Fig. 16.4 Transillumination imaging of intramural scroll waves. (**a**) Epicardial amplitude map (*bottom*) showing a region of low amplitude directly above the intramural scroll wave's location (*top*—filament position indicated by the *white disk*). (**b**) Amplitude profiles along the *dashed line* in panel **a** for two different depths Z of the intramural filament. (C) Power spectra obtained directly above (point *a* in panel **a**) and away from the filament (point *b*). (**d**) Time-space plot along the *dashed line* in panel **a**. Modified from Bernus et al. (2007)

exclusively along filaments. A second harmonic was attributed to dual humped action potentials recorded above the filament due to the optical integration volume being excited twice, superficial and deep to the filament, during each activation cycle. This property had originally been observed and described in epi-fluorescence recordings of VT in rabbit hearts by Efimov et al. (1999).

While the approaches proposed by Bernus et al. (2007) do not strictly reconstruct the depth of the filament, they infer their locations within the imaging plane. Amplitude mapping requires however that the filament remains stable for at least one cycle and the power-of-dominant-frequency approach must analyse frequency over multiple activation cycles and therefore requires highly stable filaments over longer periods of time. Conversely, time-space plotting enables tracking of the spatiotemporal dynamics of meandering filaments. Interestingly, it was found that absorptive transillumination, rather than fluorescence transillumination, would yield the highest depth penetration. Transillumination enables identification and localization of intramural filaments, whereas surface mapping would only reveal surface breakthrough activity.

3.2 *Near-Infrared Optical Mapping*

Conventional voltage-sensitive dyes such as the styryl dye, Di-4-ANEPPS, have been used successfully for optical mapping in cardiac cells and near-surface electrical activity in the myocardium. However, their utility for probing at deeper layers in the myocardium has been limited due to light scattering and absorption by endogenous chromophores at blue-green excitation wavelengths. An emergence of novel voltage-sensitive dyes with absorptive properties in the near infrared (NIR) range has pushed depth-resolved optical mapping to new bounds. A range of such dyes were developed (Matiukas et al. 2006, 2007), the most effective being DI-4-ANBDQBS, which has low susceptibility to photobleaching, internalization or phototoxicity and relatively strong signal-to-background ratios (10–15 %). These NIR dyes were developed in order to overcome enhanced scattering and absorption conditions in blood-perfused tissue, due to the presence of hemoglobin, which strongly absorbs green light so that even conventional surface recordings are difficult to obtain (Huizar et al. 2007). The absorption spectra of NIR dyes are shifted towards the red beyond the absorption maxima of hemoglobin and therefore improve significantly both transillumination and reflected fluorescent signals in blood-perfused tissue (Matiukas et al. 2007). In addition, NIR dyes allow for better transillumination signals, which are orders of magnitude smaller than epi-fluorescence signals in the blue-green excitation range (Baxter et al. 2001). The absorption spectra of NIR dyes are shifted towards the red beyond the absorption maxima of hemoglobin and therefore improve significantly both transillumination and reflected fluorescent signals in blood-perfused tissue (Matiukas et al. 2007). In addition to imaging blood-perfused preparations, NIR dyes have applications in saline-perfused myocardium. Shifting excitation light from green to red increases transmittance through cardiac tissue by more than an order of magnitude (Van Gemert et al. 1990). Photon penetration of tissue upon uniform illumination of the epicardium with 530 and 660 nm excitation light has been estimated using Monte Carlo photon transport models (Walton et al. 2010). Figure 16.5 shows normalized fluence of photons, i.e. the total number of photons intersecting a unit area in a specific time, as a function of depth for green and red illumination. The peak fluence of green light is observed at the epicardial surface and decays rapidly with depth. Red light however peaks intramurally and shows reduced rates of decay. A lower absorption in the red part of the spectra is accompanied with decreased light scattering. Both factors contribute to improved fluorescent signal-to-background ratios. Attenuation of absorption and light scattering by shifting to the red

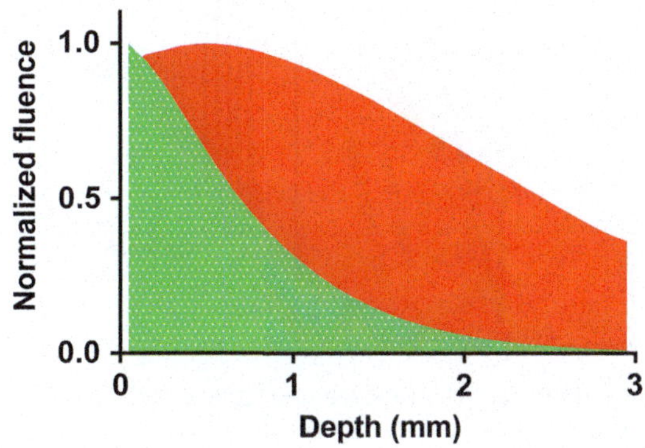

Fig. 16.5 Normalized fluence of excitation photons as a function of depth in tissue for *green* (532 nm) and *red* (660 nm) light

coincidentally increases the optical integration volume and therefore tissue penetration. Up to 90 % of the total contribution to the optical signal has been estimated to originate from depths of just 0.75 mm using green excitation light, but is extended twofold to 1.55 mm for NIR.

4 Towards 3D Optical Imaging of Cardiac Electrical Activity

Thus far in this chapter we have discussed the impact of scattering and absorption on optical recordings, and techniques to increase the depth penetration of optical imaging through the development of novel dyes or application of new imaging modalities such as transillumination. Recently, new techniques have emerged that specifically seek to probe transmural electrophysiological properties by optical means with a view in obtaining 3D information. These methods are described in detail in this section (Fig. 16.6).

4.1 Optrodes

Enhanced optical integration volumes by imaging using transillumination mode probes electrical activity over deeper layers of the myocardium compared to the

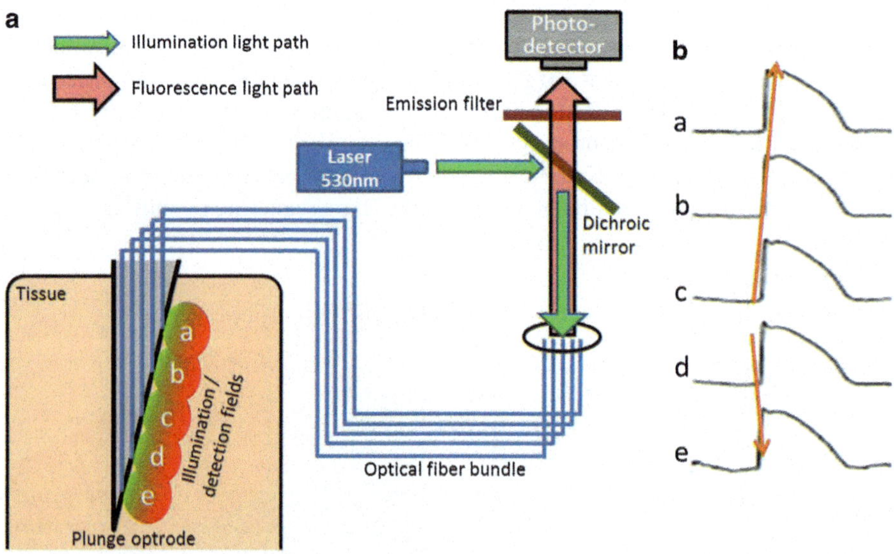

Fig. 16.6 (**a**) Schematic representation of a multi-channel optrode imaging systems for intramural optical recordings. (**b**) Optical recordings using a 5-channel optrode showing spread of action potentials from an intramural location (*red arrows*)

epi-fluorescence mode. However, the transmural plane continues to be represented only by a single value. Furthermore, transillumination imaging is not readily applicable to intact hearts. Alternative optical approaches applied to better resolve electrical activity in the transmural plane includes the use of optrodes. These are optical fibers that permit excitation and collection of emission light at multiple localized sites along an insertion needle that may be injected intramurally (Fig. 16.6). Optical recordings may be made at discrete sites along the optrode. Unlike epi-fluorescence or transillumination modalities, optrodes are not limited by the depth of light penetration as excitation light is delivered via the optical fibers directly to the optrode tips to illuminate the myocardium from the same site as fluorescence is detected. Caldwell et al. (2005) assessed the usability of optrodes for measuring optical electrical activity in the ventricles. Due to recording from such discrete localized sites, voltage-sensitive signal-to-noise ratio was typically superior to epi-fluorescent imaging. Moreover, optical AT and repolarization gradients across the ventricular wall reliably reproduced those measured from unipolar electrical recordings at adjacent sites in the myocardium. Optrodes have since been further employed to characterize complex electrical activity transmurally, such as ventricular fibrillation (Kong et al. 2009). However, the drawbacks of optrode recordings echo those of electrical plunge electrode recordings. The spatial resolution in the surface plane is limited to the sites and number of optrodes inserted in to the myocardium and the insertion of needles is in itself a destructive approach.

4.2 Dual Excitation Wavelength Imaging

Transmural gradients of electrophysiological properties have been well characterized in mammalian species at a molecular and cellular level. In the left ventricles of human and also small mammalian species such as the rat heart, there appears to be a transmural gradient of action potential duration (APD); increasing from epicardium to endocardium (Nerbonne and Kass 2005). However, APD and repolarization gradients have been less well characterized in intact tissues. Due to the relatively small thickness of the left ventricular free wall of the rat heart (2–3 mm), epi-fluorescent NIR imaging probes almost the entire depth of the free wall, whereas depth contributions using blue-green light remains relatively superficial. Comparing signals obtained with NIR vs blue-green excitation should therefore allow to qualitatively probe transmural heterogeneity in these hearts.

Walton et al. (2010) combined NIR imaging with conventional green excitation light to investigate transmural heterogeneities in intact hearts. This was achievable due to a broad excitation spectrum of DI-4-ANBDQBS dye, which enabled excitation from two wavelengths with a separation in excess of 100 nm. Two wavelengths of excitation light were rapidly alternated synchronously with the acquisition rate of the photodetector. Therefore, alternate frames from the complete recording contained fluorescent signals from both excitation sources. As this configuration acquires both signals pseudo-simultaneously, recordings are not limited to stable

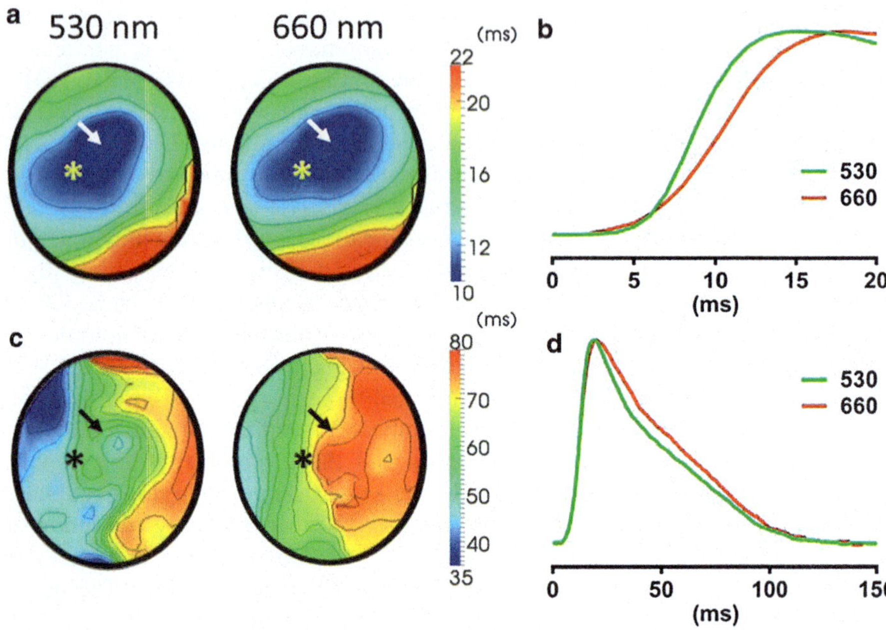

Fig. 16.7 Dual excitation wavelength imaging in isolated rat hearts. (**a**) Activation maps obtained with 530 nm (*left*) and 660 nm (*right*) excitation. The *white arrow* indicates the stimulation site. (**b**) Optical upstrokes obtained with 530 nm (*green*) and 660 nm (*red*) excitation from the point indicated by an *asterisk* in panel **a**. (**c**) Action potential duration maps for 530 nm (*left*) and 660 nm (*right*) excitation. (**d**) Action potentials obtained with 530 nm (*green*) and 660 nm (*red*) excitation from the point indicated by an *asterisk* in panel **c**. Modified from Walton et al. (2010)

rhythms such as regular pacing protocols or require gating acquisitions. Separation of alternate frames and reconstruction in to two independent recordings for each excitation wavelength is performed during the post-processing steps.

Dual excitation wavelength imaging alternating between green (530 nm) and red (660 nm) illumination wavelengths of left ventricles of rat hearts paced at the epicardial surface revealed differences in activation sequences and patterns of action potential duration due to the different penetration depths of each imaging modality (Fig. 16.7) (Walton et al. 2010). Excitation with red light modulated optical signals, relative to green light, in the following ways: the delay of the earliest AT following stimulation were greater; elliptical activation sequences, due to preferential propagation along muscle fibers than across adjacent fibers, propagating away from the site of stimulation displayed a consistent rotation of the long axes of the elliptical pattern. This phenomenon is associated with the rotation of muscle fibers in the transmural plane; optical upstroke durations were longer due to the enhanced optical integration volume and action potential durations were significantly greater as a result of greater contributions from sub-endocardial layers. Finally, rate-dependent changes of action potential duration more closely matched electrical recordings acquired directly from the endocardial surface than the epicardium. Dual wavelength epi-fluorescence was the first non-destructive approach that enabled detection of transmural gradients of electrophysiological properties in the intact heart.

4.3 Depth-Resolved Localization of Intramural Focal Sources by Alternating Transillumination

Ectopic focal sources of electrical activity can be major contributors to arrhythmogenicity and sudden cardiac death. Precise localization of these sources may play an important role in our understanding and treatment of life-threatening arrhythmias. Khait et al. (2006) proposed a theoretical optical approach to determine the depth and spatial co-ordinates of ectopic foci using a combination of reflection (epi-fluorescence) and transillumination imaging modalities. This method requires simultaneous imaging of the epicardial and endocardial surfaces and is thus applicable to coronary-perfused wedge preparations. Photodetectors, such as CCD or CMOS cameras, image opposing surfaces and are aligned along their central axis so that each pixel images juxtaposed regions of the preparation with the same field of view. The method was described for application in ventricular preparations of large mammalian species, equivalent in ventricular wall thickness to humans. As the approach relies on acquisition in transillumination mode, the excitation and emission wavelengths suitable for near infrared voltage sensitive dyes, due to increased light penetration at such wavelengths, were preferable.

The proposed technique relies on the theoretical observation that the ratio of spatially integrated reflected J^{refl} and transilluminated J^{trans} signals only depends on the depth of the fluorescent source, for a given medium of known optical properties and in the approximation of a rectangular slab geometry. It follows from diffusion theory (see Chap. 15) that the depth Z of a fluorescent point source can be estimated by solving the following equation:

$$\frac{J^{refl}}{J^{trans}} = \frac{sinh[(L - Z + d)/\delta]}{sinh[(Z + d)/\delta]} \tag{16.1}$$

Where: L is the thickness of the preparation, d is the extrapolation distance for emission light and δ is the emission light attenuation length.

During the early phase of excitation, while the optical source is small, retaining an ellipsoidal form and prior to surface breakthrough, Z describes the depth of the elliptical centroid, therefore the site of origin of the excitation wave. Theoretically, Z solved for illumination of either surface should identify the site of origin (Khait et al. 2006), however in practice, when the source is located too deep from the illuminated surface, the signal amplitude is not sufficiently isolated from the noise to accurately determine the depth of the source (Mitrea et al. 2011). Therefore, to overcome this limitation, Khait et al. (2006) further proposed to couple epicardial and endocardial illumination images: each site of illumination would produce a pair of reflected and transilluminated images that could be used independently to estimate the depth of the source according to Eq. (16.1).

The use of this alternating illumination, combining depth estimates from epicardial and endocardial illumination, significantly reduces the depth dependence of the localization accuracy of ectopic foci in the presence of noise. Using a straight forward algorithm, a selection criterion was developed for deciding which reflection and transillumination pair of images was chosen for the depth estimate. This was based on the signal to noise ratio in each image pair, whereby the image pair with the greatest signal-to-noise ratio was selected. If ratios were similar between pairs, then an average of independent depth estimates is most accurate.

In order to aptly capture spontaneous ectopic activity, it is important to acquire reflection and transillumination measurements simultaneously to ensure good temporal alignment of acquisitions and reliably reproduce the activation sequence. Practical implementation of this approach can be achieved by alternating illumination between surfaces rapidly and in synchrony with the acquisition rate of the photodetector (Mitrea et al. 2011). Alternating illumination however reduces the effective frame rate for each imaged surface which poses some limitations and practical challenges. Namely, the dynamic nature of cardiac excitation requires millisecond time resolution but the exposure time of the photodetector to fluorescence must be adequate to separate signal from background fluorescence. Mitrea et al. (2011) have developed such a system based on laser illumination and a beam chopper and provided the first experimental validation of the technique proposed by Khait et al. (2006). They were able to resolve the depth of foci (induced by intramural stimulation using plunge needle electrodes) with an accuracy of about 1 mm in pig coronary-perfused ventricular wedge preparations.

Khait et al. showed that the depth estimation technique could not only be applied to retrieve the intramural location of a point source, but could also yield information on the transmural propagation (or expansion rate) of the electrical wave. Indeed, as the wave expands, the independent depth estimates for epi- and endocardial illumination tended away from the centroid towards the surface of illumination. However, in order to obtain the correct depth, or intramural distance from the illuminated surface, of the wave front a so-called form factor needs to be introduced, which takes into account the distributed nature of the source (as opposed to a single point source). This form factor depends only on the transmural axis of an ellipsoid and therefore systematic correction to the original depth estimates apply only to the portion of the wave front closest to each imaged surface. The resulting corrected depth estimate provides a means to track wave fronts in the transmural plane and determine transmural propagation from which transmural conduction velocities can readily be estimated.

4.4 Towards Depth Resolved Optical Tomography

The previous sections have introduced techniques that allow gaining information on the depth of cardiac electrical sources and electrophysiological properties. However, none of these techniques allows for a fully time- and depth-resolved 3D reconstruction of cardiac electrical activity. This would, in principle, be feasible using Diffuse Optical Tomography (DOT), a technique, or rather a collection of techniques, solving the optical inverse problem for a given optical imaging modality when the diffusion regime is applicable. The most common application for DOT is imaging of anatomical features or inclusions (Arridge 1993; Gibson et al. 2005; Markel and Schotland 2004) and functional DOT algorithms have been developed for example for exposed-cortex functional brain imaging via hemoglobin absorption and cortical voltage-sensitive dye fluorescence (Hebden et al. 2004; Hillman et al. 2006, 2007; Siegel et al. 2003). However, cardiac dynamics impose severe constraints on the acquisition rate (of the order of milliseconds) therefore greatly limiting the applicability of most DOT techniques to cardiac electrophysiology, or restricting their use to stable and rhythmic phenomena. Nonetheless, several approaches have been proposed and applied to cardiac imaging in the last decade, which are reviewed below (Fig. 16.7).

Wellner et al. (2006) proposed a theoretical biaxial illumination and imaging setup. This is comparable to coaxial scanning where by the illumination focus and detector are aligned along the same axis but on opposite sides of the tissue preparation. Biaxial scanning differs due to a fixed lateral offset of the illumination and detector. This lateral shift creates a displacement of the fluorescence intensity maximum on the 2D imaged surface. The relative displacement is dependent on the intramural depth of the source. For example a source located close to the illumination surface would be further displaced on the opposing surface than a source located nearer to the imaged surface. The fluorescent profile of the biaxial image is sharp and tall for sources located near to either the illuminated or imaged surface but wide and short for sources located in the middle of the preparation. The method was termed multiplicative optical tomography due to the offset of the illumination and detector being carried out with equivalent magnitudes but opposite directions along one axis, i.e. $\pm$x, and in directions perpendicular ($\pm$y). Therefore, four biaxial images were acquired. The algorithm employed simply determines the shift required in each image to overlay points of maximum intensity of the biaxial images. Each image is then multiplied in a point-by-point manner to remove non-overlapping signals that would otherwise produce spurious images. Source depth was correlated directly to the shift imposed for the realignment of each biaxial image. This approach was tested on simulated data and shown to yield accurate 3D reconstructions of various source configurations (Fig. 16.8).

Hillman et al. investigated another approach based on the previously developed technique of Laminar Optical Tomography (LOT) (Hillman et al. 2006, 2007). Experiments were performed on Langendorff perfused rat hearts. A square region of the right ventricular epicardium was scanned with an excitation laser beam and fluorescence recorded from selected points at predetermined distances from the

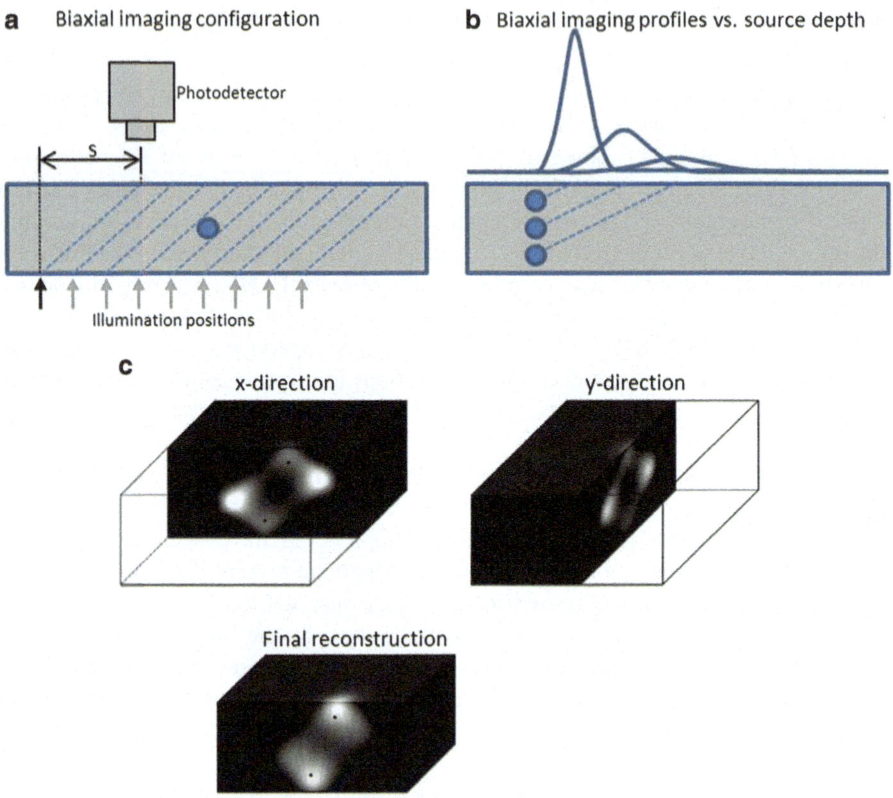

Fig. 16.8 Multiplicative optical tomography. (**a**) Schematic representation of a biaxial imaging system with an offset *s* between the excitation spot on one tissue surface and the acquisition site on the opposite tissue surface. (**b**) Schematic representation of biaxial fluorescence profiles across the imaged surface for point sources at three different depths. (**c**) Three-dimensional reconstruction of a pair of intramural point sources (*bottom*) by combination (multiplication) of biaxial scanning images in two independent x- and y-directions (*top*). The *black dots* indicate the true source positions. Modified from Wellner et al. (2006)

excitation beam. The underlying idea is that the further away fluorescence is collected from the laser beam, the deeper the signal originated in the myocardium. Solutions for the inverse problem using Tikhonov regularization can then provide a fully resolved 3D reconstruction. Here, acquisitions were achieved at 667 Hz, equating to 1.5 ms per line. Up to 800 lines were scanned across a scanning field continuously over a period of 1.2 s (7–9 beats of a rat heart) following a stimulus trigger. The scanning field was shifted by a tenth of the total scanning field before repeating the scan upon the next stimulus trigger. This was repeated for ten steps. 3D reconstructions of action potential propagation was achieved for the first time and validated against simulation (Fig. 16.9) (Hillman et al. 2006).

Although the reconstructions remained limited in depth-penetration due to the use of conventional blue-green dyes and the relatively slow scanning and in

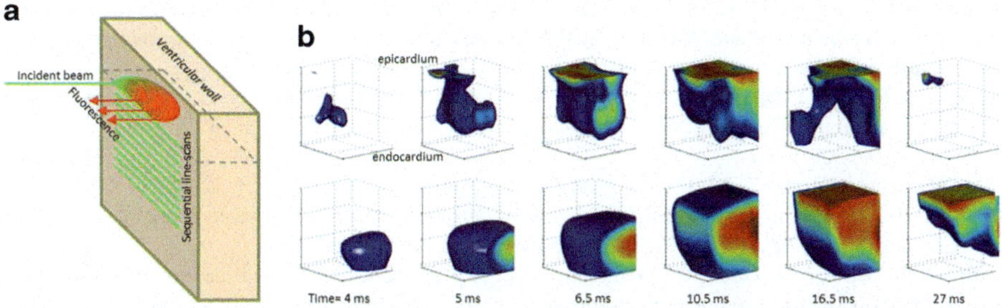

Fig. 16.9 Laminar optical tomography (LOT). (**a**) Schematic representation of the LOT acquisition with up to seven fluorescence measurements at varying distances from excitation laser beam. (**b**) Three-dimensional reconstruction of cardiac electrical activity in a 3.6 × 3.6 × 2 mm tissue volume at different time points for a wave initiated by endocardial stimulation. Comparison between experimental (*top*) and simulated (*bottom*) data. Modified from Hillman et al. (2006)

acquisition rate requiring gating over several beats preventing application of this technique to more complex activity such as observed during arrhythmias, the study by Hillman et al. (2007) represents a major breakthrough in depth-resolved optical imaging of cardiac electrical activity.

5 Conclusions

Several novel approaches have been developed in the last decade to infer intramural cardiac electrical properties using optical imaging allowing for unprecedented measurements. Although depth-resolved optical imaging is still in its early days, the development of new near-infrared dyes and novel rapid acquisition modalities and hardware should allow for unrestricted and full 3D optical imaging of cardiac electrical activity in the near future.

References

Arridge SR (1993) Forward and inverse problems in time-resolved infrared imaging. Medical optical tomography: functional imaging and monitoring. SPIE Optical Engineering Press, Bellingham, WA, pp 35–64

Baxter WT, Mironov SF, Zaitsev AV, Jalife J, Pertsov AM (2001) Visualizing excitation waves inside cardiac muscle using transillumination. Biophys J 80(1):516–530, PubMed Pubmed Central PMCID: 1301253

Bernus O, Mukund KS, Pertsov AM (2007) Detection of intramyocardial scroll waves using absorptive transillumination imaging. J Biomed Opt 12(1):014035

Bishop MJ, Rodriguez B, Qu F, Efimov IR, Gavaghan DJ, Trayanova NA (2007) The role of photon scattering in optical signal distortion during arrhythmia and defibrillation. Biophys J 93(10):3714–3726, PMCID: 2072057

Caldwell BJ, Legrice IJ, Hooks DA, Tai DC, Pullan AJ, Smaill BH (2005) Intramural measurement of transmembrane potential in the isolated pig heart: validation of a novel technique. J Cardiovasc Electrophysiol 16(9):1001–1010

Chen PS, Wolf PD, Dixon EG, Danieley ND, Frazier DW, Smith WM et al (1988) Mechanism of ventricular vulnerability to single premature stimuli in open-chest dogs. Circ Res 62 (6):1191–1209

Di Diego JM, Sicouri S, Myles RC, Burton FL, Smith GL, Antzelevitch C (2013) Optical and electrical recordings from isolated coronary-perfused ventricular wedge preparations. J Mol Cell Cardiol 54:53–64, PubMed Pubmed Central PMCID: 3535682

Efimov IR, Sidorov V, Cheng Y, Wollenzier B (1999) Evidence of three-dimensional scroll waves with ribbon-shaped filament as a mechanism of ventricular tachycardia in the isolated rabbit heart. J Cardiovasc Electrophysiol 10(11):1452–1462

El-Sherif N, Caref EB, Yin H, Restivo M (1996) The electrophysiological mechanism of ventricular arrhythmias in the long QT syndrome. Tridimensional mapping of activation and recovery patterns. Circ Res 79(3):474–492

Frazier DW, Wolf PD, Wharton JM, Tang AS, Smith WM, Ideker RE (1989) Stimulus-induced critical point Mechanism for electrical initiation of reentry in normal canine myocardium. J Clin Invest 83(3):1039–1052, PubMed Pubmed Central PMCID: 303781

Gibson AP, Hebden JC, Arridge SR (2005) Recent advances in diffuse optical imaging. Phys Med Biol 50(4):R1–R43

Girouard SD, Laurita KR, Rosenbaum DS (1996) Unique properties of cardiac action potentials recorded with voltage-sensitive dyes. J Cardiovasc Electrophysiol 7(11):1024–1038

Glukhov AV, Fedorov VV, Lou Q, Ravikumar VK, Kalish PW, Schuessler RB et al (2010) Transmural dispersion of repolarization in failing and nonfailing human ventricle. Circ Res 106(5):981–991, PubMed Pubmed Central PMCID: 2842469

Gray RA, Jalife J, Panfilov AV, Baxter WT, Cabo C, Davidenko JM et al (1996) Mechanism of cardiac fibrillation. Science 270:1222–1225

Gray RA, Pertsov AM, Jalife J (1998) Spatial and temporal organization during cardiac fibrillation. Nature 392(6671):75–78

Hebden JC, Gibson A, Austin T, Yusof RM, Everdell N, Delpy DT et al (2004) Imaging changes in blood volume and oxygenation in the newborn infant brain using three-dimensional optical tomography. Phys Med Biol 49(7):1117–1130

Hillman EMC, Devor A, Dunn AK, Boas DA (2006) Laminar optical tomography: high-resolution 3D functional imaging of superficial tissues. Proc SPIE 6143:1M

Hillman EM, Bernus O, Pease E, Bouchard MB, Pertsov A (2007) Depth-resolved optical imaging of transmural electrical propagation in perfused heart. Opt Express 15(26):17827–17841, PubMed Pubmed Central PMCID: 2441893

Huizar JF, Warren MD, Shvedko AG, Kalifa J, Moreno J, Mironov S et al (2007) Three distinct phases of VF during global ischemia in the isolated blood-perfused pig heart. Am J Physiol Heart Circ Physiol 293(3):H1617–H1628

Hyatt CJ, Mironov SF, Vetter FJ, Zemlin CW, Pertsov AM (2005) Optical action potential upstroke morphology reveals near-surface transmural propagation direction. Circ Res 97(3):277–284

Hyatt CJ, Zemlin CW, Smith RM, Matiukas A, Pertsov AM, Bernus O (2008) Reconstructing subsurface electrical wave orientation from cardiac epi-fluorescence recordings: Monte Carlo versus diffusion approximation. Opt Express 16(18):13758–13772

Janks DL, Roth BJ (2002) Averaging over depth during optical mapping of unipolar stimulation. IEEE Trans Biomed Eng 49(9):1051–1054

Janse MJ, van Capelle FJ, Morsink H, Kleber AG, Wilms-Schopman F, Cardinal R et al (1980) Flow of "injury" current and patterns of excitation during early ventricular arrhythmias in acute regional myocardial ischemia in isolated porcine and canine hearts. Evidence for two different arrhythmogenic mechanisms. Circ Res 47(2):151–165

Khait VD, Bernus O, Mironov SF, Pertsov AM (2006) Method for the three-dimensional localization of intramyocardial excitation centers using optical imaging. J Biomed Opt 11(3):34007
Knisley SB, Neuman MR (2003) Simultaneous electrical and optical mapping in rabbit hearts. Ann Biomed Eng 31(1):32–41
Kong W, Ideker RE, Fast VG (2009) Transmural optical measurements of Vm dynamics during long-duration ventricular fibrillation in canine hearts. Heart Rhythm 6(6):796–802
Markel VA, Schotland JC (2004) Dual-projection optical diffusion tomography. Opt Lett 29 (17):2019–2021
Matiukas A, Mitrea BG, Pertsov AM, Wuskell JP, Wei MD, Watras J et al (2006) New near-infrared optical probes of cardiac electrical activity. Am J Physiol Heart Circ Physiol 290(6): H2633–H2643
Matiukas A, Mitrea BG, Qin M, Pertsov AM, Shvedko AG, Warren MD et al (2007) Near-infrared voltage-sensitive fluorescent dyes optimized for optical mapping in blood-perfused myocardium. Heart Rhythm 4(11):1441–1451
Medvinsky AB, Panfilov AV, Pertsov AM (1984) Properties of rotating waves in three dimensions. Scroll rings in myocard. Self-organization: autowaves and structures far from equilibrium. Springer, Berlin, pp 195–199
Mitrea BG, Caldwell BJ, Pertsov AM (2011) Imaging electrical excitation inside the myocardial wall. Biomed Opt Express 2(3):620–633, PubMed Pubmed Central PMCID: 3047367
Nerbonne JM, Kass RS (2005) Molecular physiology of cardiac repolarization. Physiol Rev 85 (4):1205–1253
Pogwizd SM, Corr PB (1990) Mechanisms underlying the development of ventricular fibrillation during early myocardial ischemia. Circ Res 66(3):672–695
Siegel AM, Culver JP, Mandeville JB, Boas DA (2003) Temporal comparison of functional brain imaging with diffuse optical tomography and fMRI during rat forepaw stimulation. Phys Med Biol 48(10):1391–1403
Van Gemert MJC, Welch AJ, Jacques SL, Cheong QF, Star WM (1990) Light distribution, optical properties, and cardiovascular tissues. In: Abela GS (ed) Lasers in cardiovascular medicine and surgery: fundementals and techniques. Kluwer Academic Publishers, Boston
Walton RD, Benoist D, Hyatt CJ, Gilbert SH, White E, Bernus O (2010) Dual excitation wavelength epifluorescence imaging of transmural electrophysiological properties in intact hearts. Heart Rhythm 7(12):1843–1849
Walton RD, Smith RM, Mitrea BG, White E, Bernus O, Pertsov AM (2012) Extracting surface activation time from the optically recorded action potential in three-dimensional myocardium. Biophys J 102(1):30–38
Wellner M, Bernus O, Mironov SF, Pertsov AM (2006) Multiplicative optical tomography of cardiac electrical activity. Phys Med Biol 51(18):4429–4446
Winfree AT (1994) Electrical turbulence in three-dimensional heart muscle. Science 266 (5187):1003–1006
Witkowski FX, Leon LJ, Penkoske PA, Giles WR, Spano ML, Ditto WL et al (1998) Spatiotemporal evolution of ventricular fibrillation. Nature 392(6671):78–82
Zemlin CW, Bernus O, Matiukas A, Hyatt CJ, Pertsov AM (2008) Extracting intramural wavefront orientation from optical upstroke shapes in whole hearts. Biophys J 95(2):942–950

Part V
New Approaches: Potentials and Limitations

Chapter 17
Two-Photon Excitation of Fluorescent Voltage-Sensitive Dyes: Monitoring Membrane Potential in the Infrared

Jonathan A.N. Fisher and Brian M. Salzberg

Abstract Functional imaging microscopy based on voltage-sensitive dyes (VSDs) has proven effective for revealing spatio-temporal patterns of activity *in vivo* and *in vitro*. Microscopy based on two-photon excitation of fluorescent VSDs offers the possibility of recording sub-millisecond membrane potential changes on micron length scales in cells that lie upwards of one millimeter below the brain's surface. Here we describe progress in monitoring membrane voltage using two-photon excitation (TPE) of VSD fluorescence, and detail an application of this emerging technology in which action potentials were recorded in single trials from individual mammalian nerve terminals *in situ*. Prospects for, and limitations of this method are reviewed.

Keywords Two photon microscopy • Voltage-sensitive dye • Fluorescence • Nerve terminals • Action potential • Cross section

1 Introduction

Voltage-sensitive dyes, or potentiometric probes, are molecular voltmeters that insert into, but do not cross cell membranes, where they intercalate among the phospholipids that compose either leaflet of the bilayer. There, by mechanisms that are not entirely understood, they sense a representative portion of the membrane electric field and alter their light absorption and/or emission in response to changes in that field, thereby providing the physical basis for optical measurement of membrane potential (Salzberg et al. 1973; Cohen and Salzberg 1978). From its

J.A.N. Fisher
New York Medical College, Valhalla, NY, USA

B.M. Salzberg (✉)
University of Pennsylvania School of Medicine, Philadelphia, PA, USA

Department of Neuroscience, Perelman School of Medicine, University of Pennsylvania, 418 Guardian Drive, Philadelphia, PA 19104-6074, USA
e-mail: bmsalzbe@mail.med.upenn.edu

M. Canepari et al. (eds.), *Membrane Potential Imaging in the Nervous System and Heart*, Advances in Experimental Medicine and Biology 859,
DOI 10.1007/978-3-319-17641-3_17

inception in the early 1970s, optical measurement of electrical activity promised advantages under circumstances where electrodes were difficult or impossible to use for reasons of size, complexity, or membrane topology. But it was the prospect of voltage-sensitive dye functional imaging of nervous systems, whether invertebrate ganglia (Salzberg et al. 1977) or mammalian brains (Grinvald 1985; Orbach et al. 1985; Obaid et al. 1992) that really impelled the field of optical recording of electrical activity. Indeed, beginning with some of the earliest reports of molecular probes of membrane potential, ambitious ideas for the application of these nanometer scale voltmeters to the study of the central nervous system were never far from consciousness. The tantalizing metaphor of Sherrington's (1951) "enchanted loom where millions of flashing shuttles weave a dissolving pattern, always a meaningful pattern though never an abiding one; a shifting harmony of sub patterns. . ." inspired scientists to try to find new ways of exploiting potentiometric probes to learn more about the brain. Since the functional architecture of the mammalian brain is emphatically three-dimensional, it is the goal of potentiometric neuroimaging to accurately resolve neuronal electrical activity in all dimensions with high spatial- and temporal-resolution. Further, the work of Llinás (1988) and Llinás et al. (1998) has clearly demonstrated that mesoscale aspects of brain function can emerge from the electrical properties of individual neurons, emphasizing the need for cellular and sub-cellular scale imaging in three dimensions. Only recently, however, has the promise of functional imaging of electrical activity in three dimensions using voltage-sensitive dyes seemed close to being fully realized.

The use of voltage-sensitive (potentiometric) dyes (VSDs) as molecular voltmeters (Salzberg et al. 1973; Cohen and Salzberg 1978; Salzberg 1983) is presently the only optical technique enabling direct measurement of sub-millisecond changes in neuronal membrane potential, and optical detection of transmembrane electrical events offers numerous advantages over more conventional measurement techniques. Spatial resolution is usually limited only by microscope optics and noise considerations; it is now possible to monitor changes in membrane potential from regions of a cell having linear dimensions smaller than 1 μm. Temporal resolution is limited, not by the response time constant of the probe (k_{on} and k_{off}) as in measurements of $[Ca^{2+}]_i$, nor by macromolecular kinetics, in the case of fluorescent proteins, but rather by the bandwidth imposed upon the recording by noise considerations. Indeed, optical response times faster than any known membrane time constant are readily achieved (Salzberg et al. 1993). Finally, since no electrodes are employed, and the measurement is actually made at a distance from the preparation, in a convenient image plane, it is possible to monitor changes in membrane voltage from a very large number of discrete sites, virtually simultaneously. Spatial resolution, in this case, is limited primarily by the resolution of the camera, or other image dissector, the statistics of photon detection, and the efficiency with which photons are delivered to and detected from the membranes of interest.

Voltage-sensitive dyes, probed *via* one-photon excitation or absorption, have proven effective for measuring electrical activity from neurons *in vitro* (e.g., Salzberg et al. 1973, 1977; Grinvald et al. 1983; Parsons et al. 1989, 1991;

Obaid et al. 1992; Yuste et al. 1997; Contreras and Llinás 2001) and *in vivo* (Orbach and Cohen 1983; Grinvald et al. 1994; Petersen et al. 2003). To date, several studies have demonstrated voltage-sensitive dye responses in three-dimensions using one-photon fluorescence *in vivo* (Kleinfeld and Delaney 1996; Petersen et al. 2003; Fisher et al. 2004). However, in all of these studies, optical sectioning was fundamentally limited by the optical properties intrinsic to one-photon fluorescence microscopy; viz., the images contain contributions from outside the focus of the objective, since image formation does not explicitly reject out-of-focus fluorescence.

Two-photon laser scanning microscopy (Denk et al. 1990) enables true three-dimensional imaging because of its intrinsic optical sectioning properties. It depends upon the nonlinear phenomenon of quasi-simultaneous two-photon absorption by a molecular receptor, first predicted in her 1931 doctoral thesis by Maria Goeppert-Mayer (Goeppert-Mayer 1931). The absorption cross-sections that Goeppert-Mayer predicted from a quantum mechanical analysis remained only theoretical possibilities for three decades, until the advent of mode-locked lasers gave flesh to the calculations, and the physical phenomenon of two-photon laser excitation lay dormant for a further three decades before Denk and colleagues (1990) devised a practicable two-photon laser scanning fluorescence microscope. What Denk and colleagues achieved was based upon the realization that the two-photon molecular excitation predicted by Goeppert-Mayer (1931) was made possible by the very high local instantaneous intensity provided by the tight focusing in a laser scanning microscope coupled with the temporal concentration intrinsic to a femtosecond pulsed mode-locked laser. They emphasized the capabilities of the system for three-dimensional, spatially resolved photochemistry, "particularly photolytic release of caged effecter molecules" (Denk et al. 1990).

In this chapter, we will describe the present state of imaging *membrane voltage* using two-photon excitation of voltage-sensitive dye fluorescence. We will restrict our purview to neuronal systems and we will emphasize two-photon imaging of electrical activity from dense, cortex-like, neuronal tissues. We will begin with a requisite discussion of the two-photon excitation cross-sections of VSDs, including a general theoretical overview of nonlinear optical phenomena as well as an analysis of the two-photon action spectrum of a representative group of VSDs, the aminonaphthylstyrylpyridinium chimeras, di-n-ANEPPDHQ (Obaid et al. 2004). A brief summary of the optical considerations that dominate efforts to measure voltage-dependent changes in fluorescence from thick specimens consisting of highly scattering nervous tissue will follow. The bulk of the chapter will consist of a discussion, in some detail, of an application of these principles in which two-photon optical recording of action potentials from individual mammalian nerve terminals was achieved *in situ*. Finally, we will consider the outlook for two-photon fluorescence excitation of VSDs as a tool for three-dimensional functional imaging, capable of single-trial detection of fast discrete electrical events with single cell resolution, deep in cortex. Anticipated advances to be discussed include novel fast-scanning techniques as well as design of voltage-sensitive probes with extended spectral ranges and larger two-photon excitation cross-sections.

2 Two-Photon Cross-Sections of VSDs

2.1 Background Theory

While a detailed description of second- and third-order nonlinear optical processes is beyond the scope of this chapter, a qualitative picture of nonlinear optical processes, including general frequency mixing properties, can be drawn by considering an intuitive classical model. Light interacts with molecule's bound charges, and causes them to oscillate. The lowest-order contribution to the charge motion is simple harmonic oscillation at the frequency of the incident beam (assuming it is monochromatic). But the restoring forces on the bound charges are never purely linear (i.e. they are not bound by a purely quadratic potential), and are also, typically, asymmetric. Thus, the bound charges oscillate (and, hence, radiate electromagnetic fields) with frequency components not necessarily present in the incident light. These charges constitute oscillating dipoles, and the dipole moment per unit volume, the polarization, will be some function of the applied electric field, i.e. $P = f(E)$. The polarization can then be approximated as a Taylor series, i.e.

$$P = \chi^{(1)}E + \chi^{(2)}E^2 + \chi^{(3)}E^3 + \cdots,$$

where the expansion coefficients are abbreviated as $\chi^{(n)}$, defined as the nth-order electric susceptibilities. The higher order (nonlinear) terms, are typically noticed only when the strength of the applied electric field approaches the electric field strength binding these charges. In fact, it was not until the introduction of the laser that the first nonlinear optical phenomena were measured, beginning with second-harmonic generation reported by Franken and colleagues in 1961 (Franken et al. 1961).

This power-series expansion of the nonlinear polarization says nothing about the underlying physical processes, but it accurately predicts the general classes of nonlinear optical phenomena. This is because for sinusoidal time-varying electric fields, multiplication of the fields in the higher order terms of the polarization expansion accurately predicts the array of induced polarization frequencies. For example, for a sinusoidal time-varying electric field containing two distinct frequency components, ω_1 and ω_2, the term E^2 leads to second-order polarization terms at frequencies 0, $\omega_1 + \omega_2$, $\omega_1 - \omega_2$, $2\omega_1$, and $2\omega_2$. Macroscopically, these polarization components at different frequencies are responsible for the nonlinear optical phenomena observed experimentally. Second-order effects (due to $\chi^{(2)}$) include second harmonic generation (SHG), sum- and difference-frequency generation, and optical rectification ($\omega = 0$, a static electric field). Third-order effects (due to $\chi^{(3)}$) include third harmonic generation, multi-photon absorption, and stimulated Raman scattering.

Although two-photon excitation of fluorescent molecules and SHG both lead to the production of photons with shorter wavelength (higher energy) relative to the incident radiation field, it is important to note some key differences. Most importantly, in SHG (described in detail in Chap. 19), photon energy is conserved (Fig. 17.1, right), whereas in two-photon absorption (TPA), photon energy is

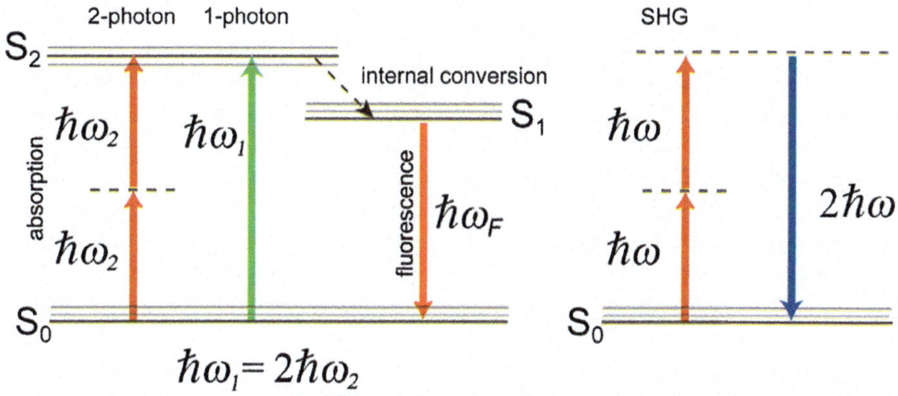

Fig. 17.1 Jablonsky energy level diagram illustrating one- and two-photon absorption processes followed by fluorescence (*left*) as well as second harmonic generation (*right*). S_0 is the ground state, S_1 and S_2 are excited states. The multiple lines at each state represent different vibrational energy levels. ω_1 is the angular frequency of a photon at the peak of a one-photon absorption resonance, $\omega_2 = \frac{1}{2}\omega_1$, and $\hbar = \frac{h}{2\pi} \approx 1.054 \times 10^{-34}$ J s. The *dotted lines* allude to virtual states not associated with an absorption resonance *per se*

transferred to the molecule (Fig. 17.1, left). The dotted lines in Fig. 17.1 represent "virtual states," or excited electronic states in which an electron can reside for a duration $\Delta t < \hbar/2\Delta E$, by the time-energy uncertainty principle; here ΔE is the energy difference between the virtual state and the nearest molecular eigenstate. Physically, therefore, SHG is a scattering event (hyper-Rayleigh scattering) and the resultant signal propagates largely in the direction of the excitation beam. TPA, on the other hand, involves absorption of photon energy followed by fluorescence emitted roughly isotropically. Furthermore, SHG requires a non-centrosymmetric material geometry for production, which may be realized, in practice, by the insertion of a SHG contrast agent (dye) into only one leaflet of the cell membrane bilayer (Moreaux et al. 2003; Pons et al. 2003; Dombeck et al. 2005; Nuriya et al. 2006; Teisseyre et al. 2007). Finally, the emission spectrum for TPA-induced fluorescence is typically identical to the one-photon emission spectrum, whereas the SHG signal is always exactly half of the excitation wavelength (twice the frequency) (Fig. 17.1, *right*).

2.2 The Two-Photon Absorption Cross-Section

The third-order susceptibility, $\chi^{(3)}$, allows for nonlinear polarization contributions at the same frequency as the incident light, thereby leading to an intensity-dependent index of refraction. The rate of absorption, R, which in the linear case is proportional to I via the one-photon absorption cross-section, σ_1, i.e. $R \propto \sigma_1 I$, is now proportional to an intensity-dependent cross-section, $\sigma = \Sigma I$, and $R \propto \Sigma I^2$, where Σ is the two-photon absorption (TPA) cross-section.

The informal unit for Σ is the Göppert-Mayer (GM), where 1 GM = 10^{-50} cm^4 s $photon^{-1}$. Σ is described as a "cross-section" in order to establish a two-photon analogy to the linear (one-photon) absorption cross-section, which has true units of area. The extra factor of cm^2 is due to the extra factor of I in the nonlinear index of refraction. The two-photon excitation (TPE) cross-section, σ_2 is the product of Σ and the two-photon quantum efficiency, q_2, defined as the number of emitted photons by the molecule per *pair* of absorbed photons, i.e. $\sigma_2 = q_2\Sigma$. In two-photon microscopy the measured signal is typically fluorescence emission, and therefore σ_2 is generally a more useful quantity than Σ.

2.3 Two-Photon Cross-Section Measurements

Xu and Webb (1996) modeled the TPA-induced fluorescence excited by focused, ultrafast mode-locked laser sources, and used this model to extract two-photon cross-sections for a variety of compounds. Because this method for extracting the cross-sections requires accurate characterization of both the spatial and temporal profiles of the excitation volume as well as the detection efficiency, Albota and colleagues (1998) introduced a ratiometric measurement technique, wherein many of these error-prone quantities cancel out. Using this method, σ_2 of an unknown sample is deduced from identical measurements made on a reference sample with known values of σ_2. Using this approach the two-photon excitation cross-section, σ_{2S}, of an unknown sample is

$$\sigma_{2S}(\lambda) = q_{2S}\Sigma_S(\lambda) = q_{2R}\Sigma_R(\lambda)\frac{\varphi_R C_R}{\varphi_S C_S}\frac{\langle P_R(t)\rangle^2}{\langle P_S(t)\rangle^2}\frac{\langle F(t)\rangle_S \eta_R}{\langle F(t)\rangle_R \eta_S},$$

where the subscripts S and R indicate "sample" and "reference," respectively, φ is the emission collection efficiency of the spectroscopic system, C is the concentration (cm^{-3}), $\langle P(t)\rangle$ (photons s^{-1}) is the time-averaged illumination power, $\langle F(t)\rangle$ (photons s^{-1}) is the time-averaged fluorescence signal, and η is the sample index of refraction. This equation is valid only when there is no simultaneous one-photon fluorescence, and care should be taken to avoid misinterpreting mixed one- and two-photon fluorescence signals (cf. Fisher et al. 2009).

2.4 Two-Photon Cross-Sections of Voltage-Sensitive Dyes

Xu, Albota, Webb, Zipfel and their colleagues performed one of the earliest surveys of two-photon excitation cross-sections for a variety of biologically relevant molecular fluorophores (Xu et al. 1996; Albota et al. 1998; Xu 2000). Their survey included some potentiometric dyes, e.g., DiOC5 (3,3′-dipentyloxacarbocyanine iodide) (Cohen et al. 1974). Those probes were molecules that provide "slow," redistribution-based optical measures of membrane voltage (Cohen et al. 1974; Cohen and Salzberg 1978). Fisher and colleagues (Fisher et al. 2005) produced the first survey of near-infrared TPE cross-sections for a series of nine "fast," membrane bound VSDs (Cohen and Salzberg 1978) (structures shown in Fig. 17.2). TPE cross-section spectra from this work are reproduced in Fig. 17.3. In general, two-photon absorption is strongest at wavelengths approaching twice the peak of the one-photon absorption wavelength. RH 795, RH 414, RH 421, di-8-ANEPP dyes, and Merocyanine 540, all blue-green-absorbing dyes, exhibit increases of more than an order of magnitude as the excitation wavelength varies from 790 to 960 nm. Interestingly, TPE cross-sections for di-8-ANEPPDHQ, a chimera

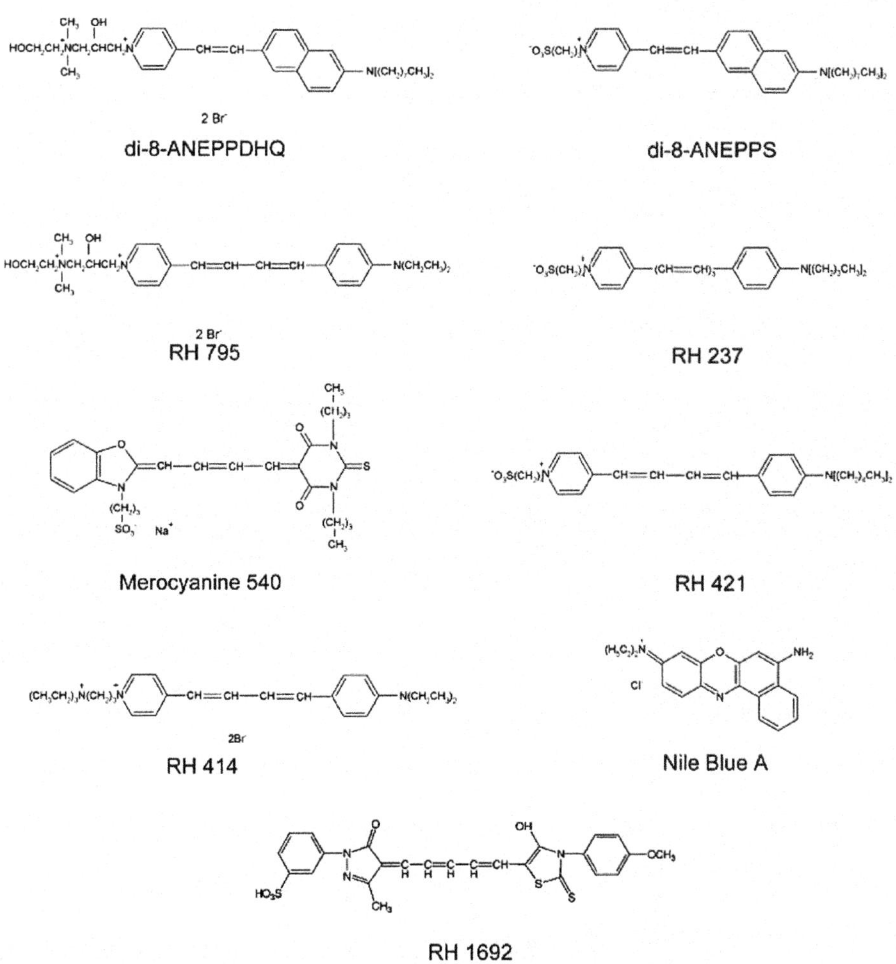

Fig. 17.2 Chemical structures for the voltage-sensitive dyes discussed in this chapter

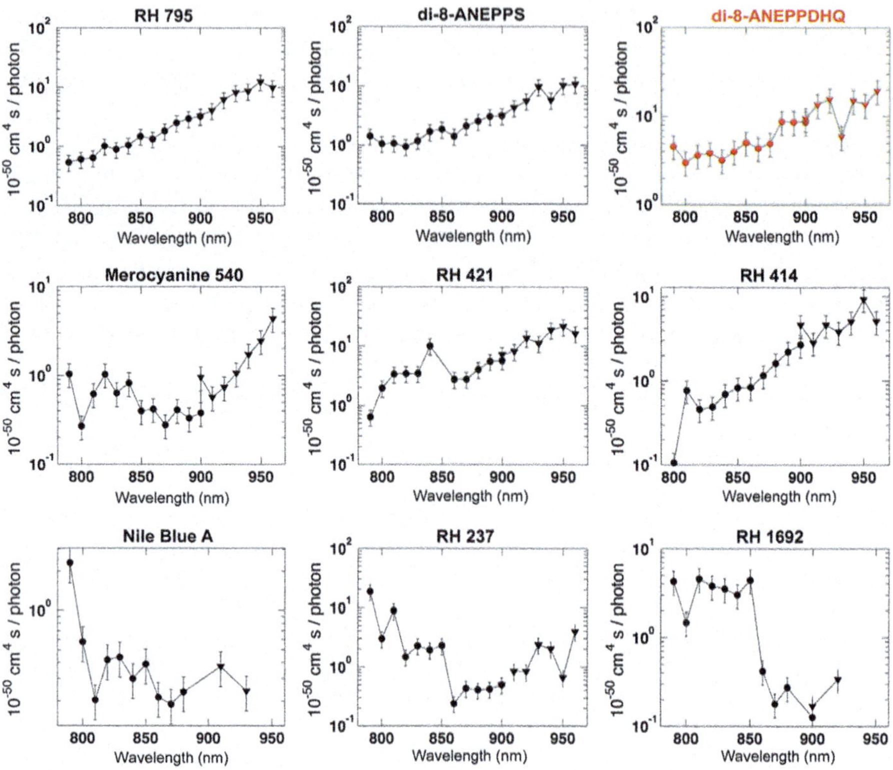

Fig. 17.3 Log-scale plots of two-photon excitation (TPE) cross-sections of voltage-sensitive dyes. Values of σ_2 are plotted in units of 10^{-50} cm^4 s photon^{-1} (GM) as a function of excitation wavelength (nm). The TPE spectrum for di-3-ANEPPDHQ, discussed in detail in this chapter, is highlighted in *red*. The cross-section measurements utilized a Mira Basic mode locked Ti: Sapphire laser (Coherent Inc, Santa Clara, CA, USA) which required two optics sets to span the wavelength range investigated. *Circular markers*: mid-wavelength optics set (790–900 nm); *triangular markers*: long-wavelength optics set (900–960 nm). Systematic error for σ_2 is ±30 % for each point

constructed from the potentiometric dyes RH795 and di-8-ANEPPS (Obaid et al. 2004), are nearly double those of its constitutive components over the entire tuning range (Fig. 17.3).

3 Optical Considerations for Measurements in Thick Specimens

3.1 TPE Excitation Volume

The lateral and axial dimensions of the focal volume can be calculated from high numerical-aperture formulations of the illumination point spread function (IPSF)

(Richards and Wolf 1959) which is the three-dimensional illumination intensity distribution. Since two-photon excitation is proportional to the square of the excitation intensity, measures of the lateral and axial widths of the two-photon illumination point spread function ($IPSF^2$) such as FWHMs and $1/e$ widths of Gaussian fits to the $IPSF^2$ function are effective measures of lateral and axial resolution (Zipfel et al. 2003). For the case of an over-filled 0.95 NA objective at 850 nm excitation, assuming the refractive index of the sample is ~1.4, the lateral and axial $1/e$ diameters (which are greater than the FWHM) are ~0.4 and ~1.8 μm, respectively.

The optical sectioning ability of two-photon microscopy relies primarily on the improbability of two-photon absorption. While the quadratic excitation intensity dependence described above facilitates significantly enhanced imaging resolution, these gains can be diminished if the TPE rate is high enough to cause fluorophore saturation within the focal volume. As a third-order nonlinear optical process, relatively high excitation intensity is typically required to generate TPE. Ultrafast pulsed (i.e. femtosecond-scale pulse width) lasers, however, achieve peak illumination power several orders of magnitude higher than average output. Even with an unfocused beam, current commercial Ti:Sapphire lasers (peak pulse power ~100 kW) can generate significant TPE in materials with appreciable (e.g. >10 GM) TPE cross-sections. In fact, Hell and colleagues (1998) have reported TPE fluorescence imaging of some biological samples utilizing purely CW laser excitation, albeit at relatively high excitation intensity.

Functional TPE imaging of small optical signals (e.g., from VSDs), in particular, can easily lead to near-saturation conditions. This is because small fractional fluorescence changes ($\Delta F/F$) necessitate high baseline fluorescence in order to overcome shot noise. Besides possibly expanding the focal volume, high levels of fluorescence near, but outside, the focal plane can effect a reduction in functional recording signal-to-noise by adding shot noise-contributing background fluorescence without (necessarily) constructively adding function-related $\Delta F/F$ signal.

The time-averaged two-photon fluorescence signal $\langle F(t)\rangle_{2-photon}$ detected by an imaging system with collection efficiency φ and utilizing mode-locked laser sources with time-averaged illumination power $\langle P(t)\rangle$ is

$$\langle F(t)\rangle_{2-photon} = \frac{1}{2}\varphi q_2 \Sigma_2 C \frac{g_p}{f\tau} \frac{8n\langle P(t)\rangle^2}{\pi\lambda}.$$

Here C is the concentration of fluorophores in the sample (cm^{-3}), f is the repetition rate of the pulsed laser, τ is the laser pulse width, g_p is a numerical constant of order unity, λ is the center wavelength of the laser pulse, and n is the real part of the linear index of refraction of the sample (Xu and Webb 1996). Thus, in addition to illumination focus geometry, the TPE fluorescence rate, on which the technique's high resolution depends, is also a function of excitation intensity, fluorophore concentration and TPE cross-section. In practice, high concentrations of bulk-loaded exogenous dyes (such as VSDs) can lead to a reduction of functional signal-to-noise ratio by adding background fluorescence. Further, the consequent high rate of fluorescence can lead to an "apparent" expansion of the focal volume via saturation of the detection device. To avoid these pitfalls, the dye concentration and excitation laser intensity should, therefore, be kept low enough (typically <100 μM for compounds having TPE >5 GM, and <30 mW, respectively), given the application-specific integration time.

3.2 Penetration Depth and Resolution in Scattering Media

Biological tissues are relatively transparent to near-infrared light. While some small amount of absorption by tissue and exogenous compounds does still occur, scattering is typically the dominant interaction. Each scattering event will alter the direction of propagation slightly; after multiple scattering events, however, photons will be observed to diffuse randomly through these media. The degree to which a material scatters light can be quantified by the scattering coefficient, μ_s (cm^{-1}), which is a measure of the scattering cross-section (cm^2) per unit volume (cm^3) of the medium. A related, and more intuitive quantity is the reduced scattering coefficient, $\mu_s^{'} = \mu_s(1 - g)$, where g is the average cosine of the scattered angle. The "mean free path," equivalent to the random walk step size, is $1/\mu_s^{'}$ (cm). Bulk values of $1/\mu_s^{'}$ for human cortex *in vivo* have been measured experimentally, and are roughly 1.1–1.2 mm at 956 nm (Bevilacqua et al. 1999).

TPE microscopy requires both efficient two-photon excitation and efficient detection of the emitted fluorescence signal; scattering and, to a lesser extent, absorption, are detrimental to both of these requirements. A nominally focused beam of near-infrared light will exhibit degraded focus at increasing tissue penetration depth. This is because scattering reduces the proportion of ballistic (i.e. unscattered) photons exponentially as a function of focal depth (Dunn et al. 2000), reducing the peak excitation intensity at the focal point and increasing the relative proportion of TPE due to scattered light (Theer et al. 2003). This re-shaping of the excitation beam focal profile imposes a slight numerical aperture (NA) dependence on the TPE efficiency (a dependence which also varies with imaging depth) (Dunn et al. 2000; Oheim et al. 2001).

When using mode-locked lasers, one strategy to overcome this loss of power is to increase the peak pulse power (at the expense of repetition rate) using a regenerative amplifier (Oheim et al. 2001; Theer et al. 2003). While higher pulse power increases both the TPE due to ballistic *and* non-ballistic photons, neither models nor experiments have shown large changes in imaging resolution as a function of imaging depth, even at depths beyond the mean-free-path length (Dunn et al. 2000; Theer et al. 2003). For functional two-photon imaging, a practical definition of "maximum imaging depth" therefore depends not only on the optical parameters of the microscope and tissue, but on the functional intensity changes expected.

4 VSD Two-Photon Action Spectra

The physical mechanisms underlying optical changes exhibited by fast potentiometric dyes are not yet completely understood. Very generally, the two main categories of proposed physical mechanisms involve (1) a voltage-dependent reorientation of the dye molecules within their phospholipid-membrane environment resulting in dimer formation (Dragsten and Webb 1978), and (2) a voltage-dependent shift of the dyes' electrical resonances (molecular Stark effect; electrochromism) (Waggoner 1979). "Slow" dyes exhibit Nernstian accumulation, followed by aggregation or binding (Sims et al. 1974). While some spectroscopic studies have described hemicyanine dyes that appear to display predominantly Stark-effect related voltage sensitivity (Kuhn and Fromherz 2003; Fromherz et al. 2008), a combination of multiple physical mechanisms seems likely for some other dyes (Loew et al. 1985; Fromherz and Muller 1993; Clarke et al. 1995).

The voltage-dependent optical changes displayed by VSDs include both shifts in spectra as well as changes in quantum yield. Thus, in an ideal experiment one would continuously monitor the absorption and emission spectra of each image pixel. In practice, however, this is generally unrealistic because it requires a highly versatile illumination system as well as a superbly efficient and fast spectral detector. The most common compromise is to excite fluorescence and detect emission at carefully selected, fixed wavelengths. Using fixed-wavelength excitation and detection, the choice of these wavelengths is, theoretically, based on an understanding of the particular dye's voltage-dependent spectral shifts and changes in quantum yield. Unfortunately, as of the time of writing, these properties have only been measured for a select few dyes (e.g. Dragsten and Webb 1978; Clarke et al. 1995; Foley and Muschol 2008). In light of this incomplete understanding, a popular strategy is to detect fluorescence from a wide wavelength region of high emissivity, while exciting at a relatively narrow wavelength band centered at a point of steep incline on the absorption spectrum. Monochromatic laser illumination, then, required for TPE microscopy, is the ideal manifestation of this excitation scheme.

Based upon the TPE cross-section measurements (Fig. 17.3), some of the most promising candidates for TPE fluorescence measurement of changes in membrane voltage are currently the naphthylstyrylpyridinium dyes of the di-n-ANEPPDHQ type (Obaid et al. 2004). Figure 17.4 shows one- and two-photon action spectra for this dye (Fig. 17.4b), which plot $\Delta F/F$ responses to identical electrical stimuli, applied to the intact neurohypophysis, as a function of excitation wavelength with the emission detection wavelength range held constant. Comparing the excitation and action spectra (Fig. 17.4a, b, respectively) for both one- and two-photon modes, two trends can be noted. First, the biphasic nature of the one-photon action spectrum derives from a stereotyped shifting of the excitation peak. In response to depolarization, the absorption spectrum of di-3-ANEPPDHQ is blue-shifted. The zero-crossing point on the one-photon action spectrum, then, represents an excitation wavelength which, after the spectral shift, corresponds to approximately voltage-independent excitation efficiency. Second, while the data points have larger errors, the two-photon action spectrum's general downward trend and, more

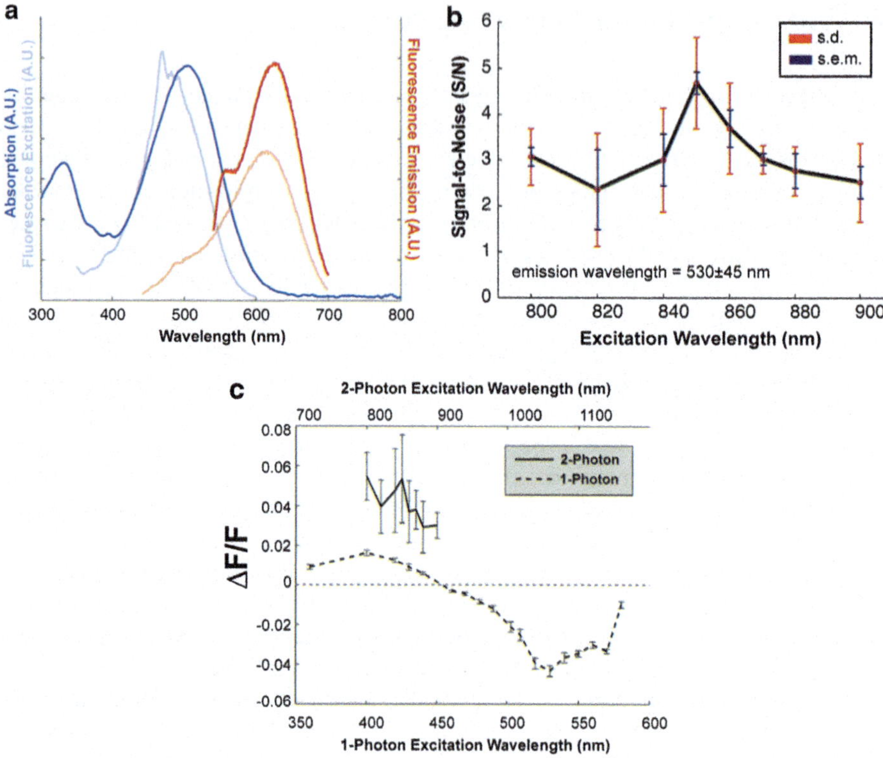

Fig. 17.4 Excitation, emission, and action spectra for di-3-ANEPPDHQ. (**a**) Absorption/excitation spectra (*dark blue/light blue*) plotted alongside fluorescence emission excited at two wavelengths (*dark red*: 530-nm excitation; *light red*: 425-nm excitation). The fluorescence excitation curve (*light blue*) measures emission at 610 nm as a function of excitation wavelength (arbitrary units). (**b**) S/N of 2-photon excited fluorescence signals as a function of excitation wavelength in single-trial optical recordings of action potentials (n = 10). The *red error bars* represent standard deviation; *blue bars* represent standard error. S/N has been defined here as the amplitude of the evoked signal (peak-to-peak) divided by the RMS noise, which is calculated from data recorded before the stimuli were delivered (i.e., 50 ms before). (**c**) Comparison of 1- and 2-photon action spectra. 1- and 2-photon action spectra for di-3-ANEPPDHQ are expressed as $\Delta F/F$. Note that the 2-photon action spectrum is shown on a different scale (*top axis*) for wavelength comparison

clearly, the polarity of the $\Delta F/F$ response are in agreement. This suggests that, at least for the di-n-ANEPPDHQ compounds, one-photon action spectra may be used to predict general features of their two-photon counterparts.

Compared with the one-photon action spectrum, the two-photon data points have relatively large errors (Fig. 17.4c). This emphasizes a general difference between the two modes of excitation: for synchronous electrical activity in membrane areas larger than the two-photon excitation volume (or, for global changes occurring in small features), the relative signal amplitude in two-photon recording ($\Delta F/F$) is compromised by the reduced spatial signal averaging. For the one-photon action spectra data points here, the smaller errors reflect the fact that those optical responses were measured from the entire neurohypophysis and, as such, represent the virtually synchronous firing of hundreds of thousands, if not millions of terminals.

For TPE microscopy, then, signal-to-noise ratio (S/N) is generally a more relevant experimental parameter than $\Delta F/F$. To this end, it should be noted that the S/N is proportional to $\Delta F/F$ and the square root of the two-photon excitation cross-section, σ_2, i.e. $S/N \propto \frac{\Delta F}{F}\sqrt{\sigma_2}$. A plot of S/N versus excitation wavelength for di-3-ANEPPDHQ is shown in Fig. 17.4b.

5 Illustrative Application: Two-Photon Recording of Action Potentials in Mammalian Nerve Terminals

5.1 *Two-Photon Microscope*

Two-photon imaging and recording was performed with the custom-built microscope illustrated in Fig. 17.5. The excitation source was a mode-locked Titanium:

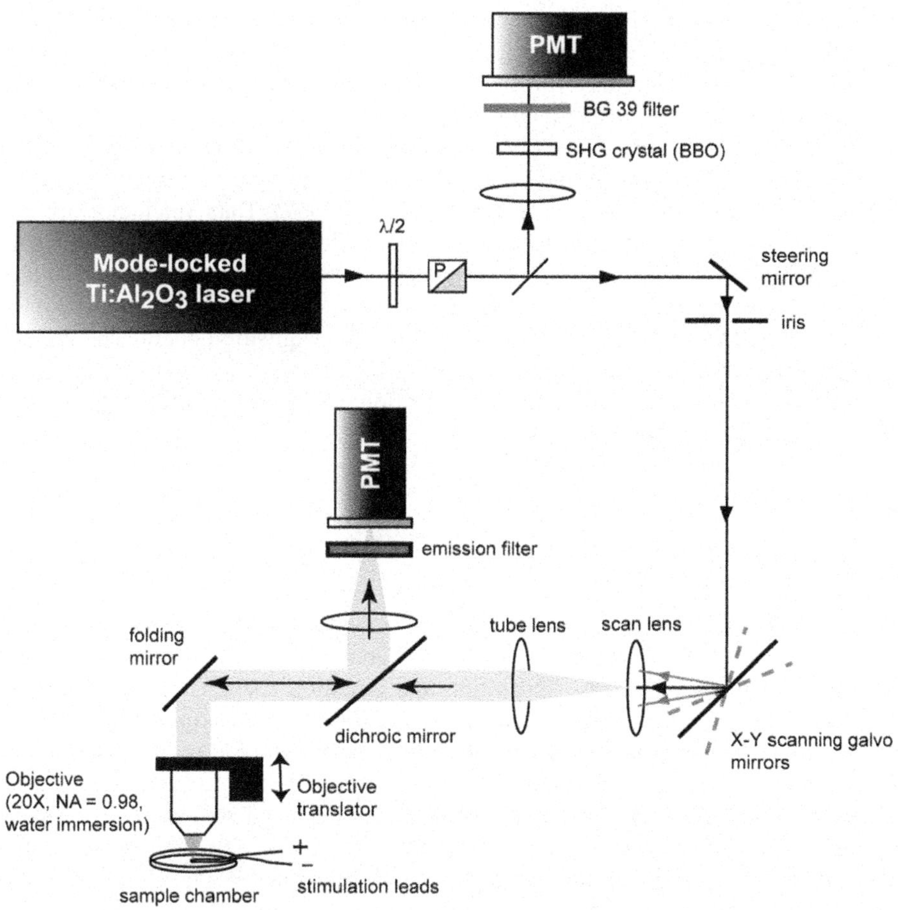

Fig. 17.5 Optical apparatus for 2-photon imaging and functional recording. Reproduced from Fisher et al. J. Neurophysiol 2008 fig. 1 with permission

Sapphire (Ti:Al_2O_3) laser (Chameleon, Coherent Inc., Santa Clara, CA, USA). Two-photon excited fluorescence was collected in the epifluorescence path by the objective, and redirected to a photomultiplier tube (PMT) using a long-pass dichroic mirror (reflect <700 nm; 90 % reflectance at 700 nm) followed by an interference bandpass emission filter. The dichroic mirror and emission interference filters were from Chroma Technology Corporation (Rockingham, VT, USA). The PMT was an R943 (Hamamatsu Photonics, Hamamatsu City, Japan) operating in analog mode, as the photon flux was typically too high for reliable photon counting. The average excitation power at the sample was adjusted to ~10–20 mW by rotating a $\lambda/2$ waveplate placed in front of a Glan–Thompson polarizer (P). The excitation beam was scanned in the sample plane using galvanometer scanning mirrors (Cambridge Technology Inc., Lexington, MA, USA), and scanned along the vertical (illumination) axis using a piezoelectric-driven objective translator (P721, PhysikInstrumente GmbH & Co. KG, Karlsruhe/Palmbach, Germany). The combination of scan lens and tube lens expanded the excitation laser beam and backfilled the microscope objective (20×/0.95 NA XLUMPlanFl, Olympus, Melville, NY). To monitor and correct for fluctuations in the square of the laser power, $\langle P(t)\rangle^2$, a small fraction of the beam was picked off with a microscope coverslip and focused onto a beta-barium borate (β-BaB_2O_4, or "BBO") second harmonic generation (SHG) crystal (Inrad, Northvale, NJ). This SHG signal was then filtered with a BG39 Schott glass absorption filter and detected by a second, auxiliary, PMT (model R955, Hamamatsu Photonics, Hamamatsu City, Japan) also operating in analog mode. Device operations were controlled using custom-written LabVIEW (National Instruments, Austin, TX) code (Fisher 2007). Data (including image) processing was performed using custom Matlab (Mathworks, Natick MA) code. The analog signals from both PMTs were amplified with a wide bandwidth transimpedance amplifier (13AMP005, MellesGriot, Carlsbad, CA) before digitization by the PCI-6052E National Instruments data acquisition board. The voltage outputs from the transimpedance amplifiers were oversampled at 150 kHz, and a number of points corresponding to the integration time were collected and averaged for each sampled point.

5.2 *Absorption and Emission Spectra for di-3-ANEPPDHQ*

Di-3-ANEPPDHQ is a more hydrophilic member of the di-n-ANEPPDHQ sequence than di-8-ANEPPDHQ, its di-octyl analogue, and is thus more suitable for staining neurons *in situ*. Absorption and emission spectra for this dye were measured in a cuvette, using octanol to approximate the environment of the dye when bound to membrane (Sims et al. 1974). Final sample concentrations were $\leq$100 μM. Absorption measurements were performed with a spectrophotometer (Ocean Optics, Dunedin, FL), and excitation and emission spectra were recorded using a spectrofluorometer (DeltaRAM, Photon Technology International, Birmingham, NJ). These one-photon spectra are all illustrated in Fig. 17.4a.

5.3 Biological Preparation

The neurohypophysis (*pars nervosa*), which together with the *pars intermedia* compose the neurointermediate lobe of the mammalian pituitary, is an extremely favorable preparation for the demonstration of a wide variety of optical phenomena. It is a neuroendocrine gland that secretes the peptide hormones arginine vasopressin and oxytocin together with their associated proteins, the neurophysins. From the hypothalamus, magnocellular neurons project their axons as bundles of fibers through the median eminence and infundibular stalk, to arborize extensively and terminate in the neurohypophysis. Here, the peptides are released into the circulation by a Ca^{2+}-dependent mechanism (Douglas 1963; Douglas and Poisner 1964). In the rat, some 20,000 neurons give rise to approximately 40,000,000 neurosecretory swellings and terminals, whose membranes make up nearly 99 % of the excitable membrane in this gland (Nordmann 1977). The very large surface-to-volume ratios found in vertebrate neurohypophyses explains the unusually robust extrinsic optical signals that can be recorded from this tissue with impermeant voltage-sensitive dyes (Salzberg et al. 1983; Obaid et al. 1985, 1989; Obaid and Salzberg 1996; Muschol et al. 2003), and the synchronicity with which the action potential (AP) invades hundreds of thousands, if not millions, of terminals explains the large intrinsic optical signals, including light scattering changes (Salzberg et al. 1985; Obaid et al. 1989) and changes in autofluorescence (Kosterin et al. 2005), that are observed in mammals. In the mouse, these neurosecretory terminals range in size from 1 to 10 μm, and their density closely resembles that of cortical tissue.

For the TPE fluorescence measurements of membrane voltage described in this chapter, we removed the neurointermediate lobe from 30- to 60-day-old female CD-1 mice that had been anesthetized by CO_2 inhalation and decapitated. The mouse head was pinned to the bottom of a Sylgard-lined dish and, following removal of the skin, the skull was opened along the dorsal midline and the bone removed bilaterally. The optic and olfactory nerves were cut and the brain was deflected caudally, severing the infundibular stalk. The infundibular stump, together with the entire pituitary gland, remained attached to the base of the skull. The gland was then gently removed using iridectomy scissors and fine forceps, and transferred to a dish containing mouse Ringer's solution (in mM: 154 NaCl, 5.6 KCl, 1.0 $MgCl_2$, 2.2 $CaCl_2$, 10 glucose, and 20 HEPES, adjusted to pH 7.4 with NaOH) where the anterior pituitary (*pars anterior*) was separated from the neurointermediate lobe and discarded. The *pars intermedia*, which consists of a delicate lacework of cells cradling the neurohypophysis, provided a convenient support through which the specimen could be pinned to the Sylgard-lined bottom of a shallow chamber, while preserving the integrity of the neurohypophysis (see Fig. 17.6). The tissue was stained by bathing it for 30–60 min in a Ringer's solution containing ~50 μg/ml (~78 μM) of the VSD di-3-ANEPPDHQ, together with 0.5 % EtOH (used as stock solvent). After staining, the neurohypophysis was washed in mouse Ringer's solution, which was replaced every 20–30 min to preserve tissue viability. The mouse neurohypophysis preparation is illustrated in Fig. 17.6.

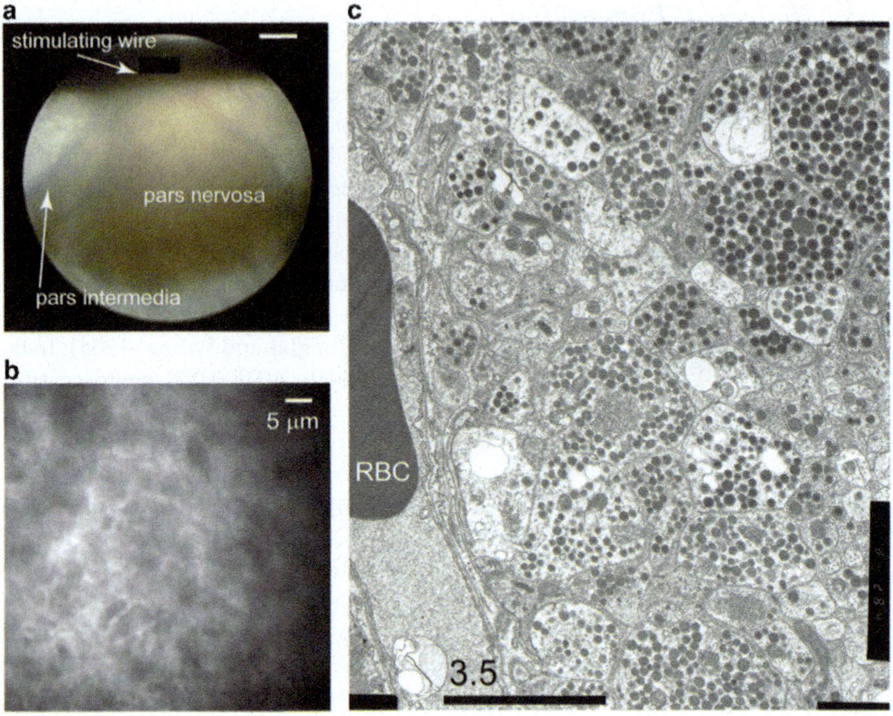

Fig. 17.6 Anatomy of the neurohypophysis. (**a**) Dissection microscope image of an isolated neurohypophysis with surrounding *pars intermedia* and infundibular stalk connecting to stimulation leads (scale bar = 200 μm). (**b**) 2-photon background image of neurohypophysis stained with the voltage-sensitive potentiometric probe di-3-ANEPPDHQ. Excitation wavelength = 850 nm. (**c**) Transmission electron microscope image of mouse neurohypophysis. Individual axon terminals contain peptide hormone-filled vesicles. A red blood cell (RBC) provides scale (diameter ~5.5 μm). Reproduced from Fisher et al. J. Neurophysiol 2008 fig. 2 with permission

Prior to TPE fluorescence functional imaging, two-photon scanning images such as that illustrated in Fig. 17.7a were recorded. Stimulation of the myriad terminals was achieved using a pair of Teflon-coated Pt-Ir (90–10 %) electrodes clasping the infundibular stump and consisted of brief (100–500 μs) shocks delivered through a stimulus isolator. The imaging depth was maintained at ~150 μm below the tissue surface. The focusing routine consisted of advancing the micrometer drive-adjustable stage upward until two-photon fluorescence was detected on the surface of the preparation (either visually, or by monitoring the photomultiplier voltage). Repeated experiments using stained preparations established that locating the neurohypophysis surface by eye was consistently accurate to within ~50 μm. After recording a surface image of the tissue, the objective was translated downward ~150 μm using the piezoelectric objective driver which has sub-micron precision and repeatability. Although TPE voltage-sensitive fluorescence signals could be recorded from depths up to 300 μm below the surface, a

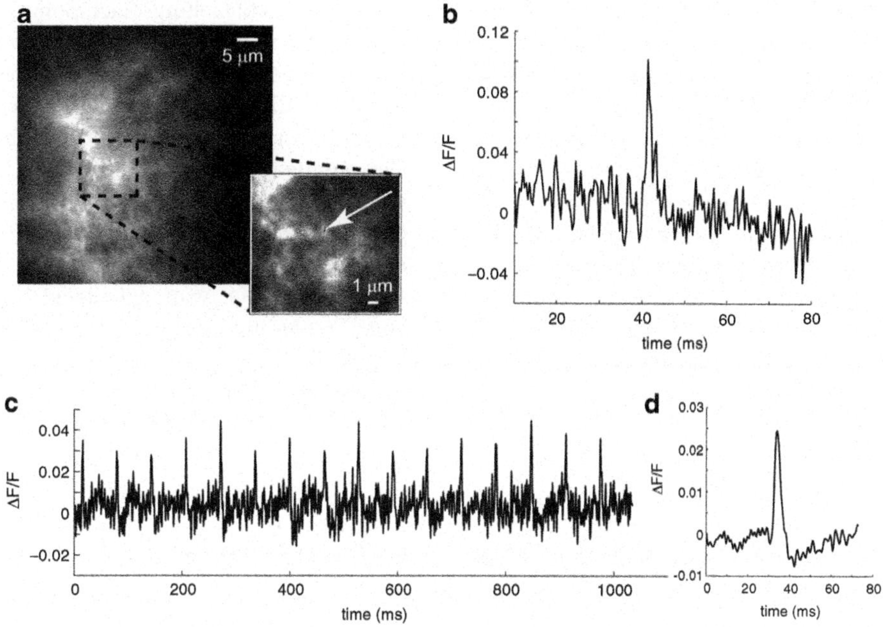

Fig. 17.7 Functional recording of action potentials from individual nerve terminals *in situ*. (**a**) Background 2-photon image of the region of the neurohypophysis used for functional optical recording (scale ~5 µm). *Inset*: magnified view (scale bar ~1 µm) of imaged region; the *short red line* indicates the actual location of the line-scan excitation. (**b**) A single-trial response to an electrical stimulus exhibiting both high S/N and large $\Delta F/F$ (nearly 10 %). (**c**) Single-trial data showing a train of 16 action potentials in response to repeated stimuli delivered at 16 Hz for 1 s. The optical data were band-pass filtered (ideal filter in the frequency domain) between 10 Hz and 0.7 kHz and smoothed using a 3-point sliding boxcar average. (**d**) An average of the 16 action potentials from the single trial shown in (**c**). For all data, the excitation wavelength was 850 nm and the emission wavelength was 530 ± 45 nm

depth of 150 µm was chosen as a standard imaging depth to remove one source of variation while manipulating the experimental parameters of excitation and emission wavelengths. This depth was selected because it offered reliable tissue viability, avoiding potentially damaged surface tissue that typically contributes non-specific two-photon fluorescence background, while exhibiting relatively little scattering-induced blurring.

5.4 Two-Photon Functional Recording Protocol

TPE fluorescence recording of action potentials consisted of short line-scans, typically ~3 points along a line ~0.5 µm long, as in Fig. 17.7a (inset), integrating for ~100 µs at each point. Line scanning was employed instead of static single-point "beam-parking" to reduce the effects of photodamage and bleaching. Each point in the raw data traces represents the average of the ~3 points along the short red line segment. Electrical stimuli (~1–10 mA) were delivered within ~50 ms of the beginning of the line-scan in order to further minimize photobleaching. When multiple stimuli were delivered, the stimulation frequency was 16 Hz. Otherwise, all trials were single-stimulus experiments. The records were 100 ms long with 5 s wait time between trials for single stimuli, and 500 ms long with 1 min wait time between trials for trains of stimuli. Averaging, when needed, was performed offline utilizing responses from within a single multi-stimulus trial. For each point scanned in functional recordings, fractional fluorescence ($\Delta F/F$) was calculated by subtracting the baseline, F, from each point in the recording, and then dividing by this same background. For our experiments, we took F to be the average of the first 10 data points in the recording. Functional recording data were filtered as indicated in Fig. 17.7 legend.

5.5 Two-Photon Optical Recording of Action Potentials

The results shown in Fig. 17.7 demonstrate single-trial optical recording of electrically-evoked action potentials from individual nerve terminals *in situ,* using a voltage sensitive dye. Figure 17.7a shows a two-photon background image of a mouse neurohypophysis stained with 78 μM di-3-ANEPPDHQ. Functional recordings, shown in Fig. 17.7b, c, were obtained by recording at the same location highlighted in the inset of Fig. 17.7a. Figure 17.7c shows a train of action potentials (16 stimuli delivered at 16 Hz) recorded optically, in a single trial, from the location indicated in Fig. 17.7a. Figure 17.7b illustrates an unfiltered single-trial recording of an action potential from an individual nerve terminal exemplifying the high S/N and large $\Delta F/F$ (~10 %/100 mV) achievable with this technology. Figure 17.7d illustrates a spike-triggered average of the 16 optical responses shown in Fig. 17.7c. The characteristic shape of the neurohypophysial action potential (Gainer et al. 1986; Muschol et al. 2003), including its after-hyperpolarization, is evident. Single-trial responses were 2–3 ms wide at half maximum (FWHM), and averaged responses were often 4–5 ms wide. The broadening in the averaged response is likely due to physiological jitter in the timing of the response, since the recordings represent the activity of individual nerve terminals. It is interesting to note that in one-photon fluorescence optical recordings from large regions of the neurohypophysis, there is an apparent after-depolarization that actually reflects the depolarization of nearby glial cells (Muschol et al. 2003); here, the TPE fluorescence recording is only from a single terminal and is not contaminated by the pituicyte response.

6 Conclusions and Perspective

It is now possible to monitor membrane voltage (action potentials) from micron-scale structures deep in cortex-like tissue using two-photon excitation of fluorescent voltage-sensitive dyes. In this chapter, for example, we have illustrated the recording of action potentials in single trials from ~1 μm patches of membrane on individual nerve terminals. The advantages of this modality are critical and numerous. Two-photon excitation (TPE) permits greater penetration into highly scattering tissue, as well as reduced likelihood of photodynamic damage (Denk and Svoboda 1997; Zipfel et al. 2003). It also provides exceptional lateral and axial spatial resolution, with a two-photon excitation focal volume defined by widths of Gaussian fits to the *square* of the illumination point spread function. Also, most salient is the fact that, unlike SHG, two-photon excitation of fluorescent VSDs can utilize epifluorescence detection of the optical signals (i.e., detection in the *backward direction*), a measurement modality that is optimal for non-invasive recording of electrical activity from intact brain.

It also is of interest that the application of two-photon fluorescence excitation of voltage-sensitive dyes reviewed here yielded fractional fluorescence changes, $\Delta F/F$, ~5 times larger than one-photon excited signals at "corresponding" ($\lambda_{2\text{-photon}}/2$) excitation wavelengths. This enhancement of $\Delta F/F$ using two-photon excitation, compared with one-photon excitation is consistent with measurements in cultured neurons stained with a hemicyanine potentiometric probe (Kuhn et al. 2004). While this agreement may be fortuitous (the measurements were made at opposite sides of the absorption spectra), it is significant, since the relatively poor S/N of one-photon measurements is a reflection of the small $\Delta F/F$ observed in this mode. There is good reason to believe that this circumstance is, at least in part, the result of the broad-band spectral excitation that is commonly employed for one-photon fluorescence studies. Evidence suggests that judiciously chosen, extremely narrow-band, excitation may yield much larger $\Delta F/F$, and, using quiet light sources of sufficient intensity, this will substantially increase the S/N for optical signals derived from one-photon excitation of VSD fluorescence. At the same time, our expectation is that precision selection of two-photon excitation wavelength, optimally matched to the individual potentiometric probe and its particular membrane environment (neuron class) will also contribute to greater $\Delta F/F$ and S/N.

The Ti:Al_2O_3 laser that we used for two-photon fluorescence excitation of di-3-ANEPPDHQ had a long-wavelength limit of 960 nm and, thus, did not permit us to exploit the red peak of the dye's one-photon action spectrum (~530 nm). However, we would argue that infrared two-photon fluorescence excitation for functional imaging of membrane potential offers exciting prospects that need to be explored. In this wavelength domain, some of the advantages of two-photon microscopy are even greater, particularly penetration into highly scattering tissue, and the diminished likelihood of photodynamic damage. Evidence clearly suggests (Fig. 17.3) that several voltage-sensitive dyes have the peak of their two-photon absorption cross-sections beyond 960 nm. Since the one-photon action spectra of some of these dyes indicate that larger signals can be expected for two-photon excitation between 1.0 and 1.15 μm, measurement of σ_2 at these longer wavelengths using newly available modelocked $Ti:Al_2O_3$ lasers and fiber-lasers (Teisseyre et al. 2007) should permit optimization of the excitation parameters, thereby increasing even further the sensitivity of the technique. It is also significant that while H_2O absorption in tissue rises dramatically between 920 and 970 nm, it *does* fall off with wavelength by 1.025 μm (Collins 1925). Furthermore, there is a local minimum in H_2O absorption at ~1.06 μm, exactly twice the "red" peak in the one-photon action spectrum of di-3-ANEPPDHQ.

The emergence of laser light sources with extended spectral range and greater amplitude stability, together with new objectives optimized for the infrared, calls for the design and synthesis of new dyes that are optimized for longer wavelengths, with large two-photon absorption cross-sections, and that produce sensitive, faithful responses to changes in membrane potential. The laboratory of Leslie Loew developed a series of dyes with varying chromophores and side chains that securely anchor them in cell membranes (Teisseyre et al. 2007); several of these dyes exhibit rapid and sensitive responses to membrane potential. Yan and colleagues (2012) recently described a series of novel fluorinated hemicyanine VSDs; by adjusting the position of fluorine substituents, the optical absorption and emission properties of these fluorescent molecules can be optimized for a given application. This flexibility affords additional latitude in designing experiments that combine TPE with other imaging modalities or optical manipulations such as photolysis of "caged" compounds. One of the recently reported fluorinated compounds in particular, di-2-AN(F)EPPTEA, displays fractional fluorescence voltage sensitivity greater than 16 % per 100 mV when intracellularly loaded in single cells in brain slices. This compound additionally features increased photostability, and has been used to detect back-propagating action potentials in single dendritic spines (Acker et al. 2011) as well as spikes in single axonal boutons (Rowan et al. 2014).

It is interesting that while some dyes showed optimal activity for two-photon fluorescence excitation, others were better suited for contrast-enhanced SHG. This systematic approach to dye synthesis and screening has now revealed differing patterns of sensitivity and kinetics for these two nonlinear optical modalities and suggests that the voltage sensitivity of these two dye-dependent phenomena originates from differing mechanisms.

While incremental improvements in the sensitivity and photostability of VSDs, as well as reduced phototoxicity of some organic probes can certainly be anticipated (Salama et al. 2005; Matiukas et al. 2006, 2007; Zhou et al. 2007; Yan et al. 2012), fluorescent *protein* voltage sensors are increasingly employed as the voltage-transducers of choice for *in vivo* applications. These proteins can be genetically targeted to the plasma membrane of the cells of interest, and their expression limited to particular classes of neurons (Guerrero and Isacoff 2001; Sakai et al. 2001; Ataka and Pieribone 2002; Guerrero et al. 2002; Knöpfel et al. 2003). While voltage-sensitive proteins are slow compared with organic dyes, with response time constants measured in milliseconds, versus *microsecond* response time of organic fast VSDs, they can be used with TPE to monitor subthreshold electrical events in higher nervous systems with high specificity and sub-cellular resolution (Jin et al. 2012; Cao et al. 2013).

Ultimately, the study of neuronal functional dynamics requires *multiple-site* optical recording of transmembrane voltage in the brain. In the example illustrated here, high spatial and temporal resolution was achieved by employing a very short line scan (~3 points over 0.5 μm) sampled at 150 kHz, and sacrificing actual "imaging". For high-resolution multiphoton functional *imaging*, two different approaches to increasing image acquisition speed have been proposed or implemented. One approach is to excite fluorescence at multiple sites simultaneously by splitting the excitation beam into several excitation beamlets. There have been three main techniques for multiplexing a single excitation laser beam: (1) passing an expanded laser beam through a rotating microlens-containing disk (based on a Nipkow disk, and similar to the Yokogawa microlens disk) which scans a sample in two-dimensions as the spiral pattern of lenses rotates (Bewersdorf et al. 1998), (2) passing an expanded laser beam through a stationary microlens array and scanning the excitation pattern over a sample via galvonometric mirrors (Kim et al. 2005), and (3) multiplexing a single unexpanded excitation laser beam by progressively dividing the beam through a system of beamsplitters (Nielsen et al. 2001; Kurtz et al. 2006). Such "multifocal" techniques, however, typically require that the detected fluorescence be re-focused onto a multiple-pixel detector in order to spatially separate the fluorescence signals attributed to the different excitation foci. This can lead to cross-talk between the foci, due mostly to scattering within the specimen. This approach is particularly well-suited for functional imaging of neuronal networks, where only sparsely distributed features within the larger field-of-view are of interest. Kurtz and colleagues (2006) have demonstrated how a multiline strategy could be applied to reconcile high spatial- with relatively high temporal-resolution in functional imaging of neuronal activity. By combining a central 50 %-beamsplitter mirror, to double incoming beams, with high reflectivity mirrors to either side of it sending the beams repeatedly back to the central beamsplitter, they could generate an array of up to 64 laser beams arranged in a row. Then, they used image detection in non-descanned mode with an electron-multiplying back-illuminated CCD with high quantum efficiency to obtain high sensitivity (Kurtz et al. 2006), and they achieved a "multiple line scan" at 89.3 Hz "frame" rate for imaging calcium dynamics in visual motion-sensitive neurons in fly eyes. (In frame scan mode, the rate was 8.2 Hz.)

A second approach for fast imaging employs high-speed random access scanning of a single laser beam (see Chap. 18), using either inertia-free acousto-optic deflectors (AODs) with appropriate dispersion compensation (Salome et al. 2006; Otsu et al. 2008), or fast galvanometer-driven mirrors (Lillis et al. 2008). Random access scanning of two-photon excitation laser beams for functional imaging has been used by several groups recently (see e.g., Vucinic and Sejnowski 2007), also for monitoring calcium dynamics in neurons. Random-access with two axis AOD scanning was first proposed for single photon laser scanning microscopy (Bullen et al. 1997). The basic idea has now been extended to multi-photon laser scanning microscopy by several laboratories. A random-access two-photon scanning microscope (RA-TPSM) (Salome et al. 2006), and a random-access multiphoton microscope (RAMP) (Iyer et al. 2006; Otsu et al. 2008) have both been described, using crossed AODs to steer the beam. Salome and colleagues (2006) reported an all-digital two-photon scanning microscope that included an acousto-optic modulator (AOM) placed at 45° to the AODs to pre-compensate for the large spatial distortions of femtosecond laser pulses that occur in AODs. Their microscope permitted recording from selected points of interest at high speed (1 kHz), demonstrated with calcium imaging of spontaneous action potentials in cultured hippocampal neurons, and is a significant step toward achieving multiunit optical recordings with millisecond resolution in neuronal tissues. Otsu et al. (2008) took a slightly different approach to resolving the chromatic angular dispersion and temporal chirp of the AODs in their optimized RAMP microscope. Essentially, they used longer laser pulses, which are less distorted because of their narrower spectral bandwidth (Iyer et al. 2003) to mitigate the AOD-induced distortions, and they were able to follow calcium transients in the spines and shafts throughout the dendritic tree of cerebellar Purkinje cells, at frame rates of 0.5–1.5 kHz (Otsu et al. 2008), and map dendritic heterogeneity at high spatial *and* temporal resolution. A related modality, called targeted-path scanning (TPS) (Lillis et al. 2008), is also a random-access strategy for two-photon fluorescence imaging, but, instead of AODs, it employs a two-axis galvanometer-mirror based scanner together with novel computer algorithms that generate highly efficient scan paths. Jerome Mertz's group was able to achieve 100 Hz scan rates over millimeter spatial domains when measuring calcium transients (Lillis et al. 2008).

The measurements of two-photon excited VSD fluorescence described in this chapter represent the first single-trial optical recordings of electrical activity at sub-micron resolution from an intact preparation from mammalian brain using epifluorescence detection. Although TPE permits imaging deep into tissue, functional recording using exogenous VSDs is often limited by background fluorescence owing to a lack of labeling specificity. Kuhn and colleagues (2008) performed the first recordings of neural activity *in vivo* using two-photon imaging of VSD fluorescence in the mouse primary somatosensory cortex. The optical signals they measured were modest, on the order of 1 % or less, due in large part to an abundance of background fluorescence. The development of VSDs that are more sensitive, but still bear high TPE cross-sections, can alleviate these limitations.

Acknowledgments Supported by USPHS grants NS40966 and NS16824 (B.M.S.) and by a Fellowship from Oak Ridge Institute for Science and Education (J.A.N.F.).

References

Acker CD, Yan P, Loew LM (2011) Single-voxel recording of voltage transients in dendritic spines. Biophys J 101:L11–L13

Albota MA, Xu C, Webb WW (1998) Two-photon fluorescence excitation cross sections of biomolecular probes from 690 to 960 nm. Appl Opt 37:7352–7356

Ataka K, Pieribone VA (2002) A genetically targetable fluorescent probe of channel gating with rapid kinetics. Biophys J 82:509–516

Bevilacqua F, Piguet D et al (1999) In vivo local determination of tissue optical properties: applications to human brain. Appl Opt 38:4939–4950

Bewersdorf J, Pick R, Hell SW (1998) Multifocal multiphoton microscopy. Opt Lett 23:655–657

Bullen A, Patel SS, Saggau P (1997) High-speed, random-access fluorescence microscopy: I. High-resolution optical recording with voltage-sensitive dyes and ion indicators. Biophys J 73:477–491

Cao G, Platisa J, Pieribone VA, Raccuglia D, Kunst M, Nitabach MN (2013) Genetically targeted optical electrophysiology in intact neural circuits. Cell 154(4):904–913

Clarke RJ, Zouni A, Holzwarth JF (1995) Voltage sensitivity of the fluorescent probe RH421 in a model membrane system. Biophys J 68:1406–1415

Cohen LB, Salzberg BM (1978) Optical measurement of membrane potential. Rev Physiol Biochem Pharmacol 83:35–88

Cohen LB, Salzberg BM et al (1974) Changes in axon fluorescence during activity: molecular probes of membrane potential. J Membr Biol 19:1–36

Collins JR (1925) Change in the infra-red absorption spectrum of water with temperature. Phys Rev 26:771–779

Contreras D, Llinás R (2001) Voltage-sensitive dye imaging of neocortical spatiotemporal dynamics to afferent activation frequency. J Neurosci 21:9403–9413

Denk W, Strickler JH, Webb WW (1990) Two-photon laser scanning fluorescence microscopy. Science 248:73–76

Denk W, Svoboda K (1997) Photon upmanship: why multiphoton imaging is more than a gimmick. Neuron 18:351–357

Dombeck DA, Sacconi L, Blanchard-Desce M, Webb WW (2005) Optical recording of fast neuronal membrane potential transients in acute mammalian brain slices by second-harmonic generation microscopy. J Neurophysiol 94:3628–3636

Douglas WW (1963) A possible mechanism of neurosecretion release of vasopressin by depolarization and its dependence on calcium. Nature 197:81–82

Douglas WW, Poisner AM (1964) Stimulus-secretion coupling in a neurosecretory organ: the role of calcium in the release of vasopressin from the neurohypophysis. J Physiol 172:1–18

Dragsten PR, Webb WW (1978) Mechanism of the membrane potential sensitivity of the fluorescent membrane probe merocyanine 540. Biochemistry 17:5228–5240

Dunn AK, Wallace VP, Coleno M, Berns MW, Tromberg BJ (2000) Influence of optical properties on two-photon fluorescence imaging in turbid samples. Appl Opt 39:1194–1201

Fisher JAN (2007) Linear and non-linear fluorescence imaging of neuronal activity (Ph.D.: 212). Department of Physics and Astronomy, University of Pennsylvania, Philadelphia, PA

Fisher JA, Civillico EF, Contreras D, Yodh AG (2004) In vivo fluorescence microscopy of neuronal activity in three dimensions by use of voltage-sensitive dyes. Opt Lett 29:71–73

Fisher JA, Salzberg BM, Yodh AG (2005) Near infrared two-photon excitation cross-sections of voltage-sensitive dyes. J Neurosci Meth 148:94–102

Fisher JA, Barchi JR, Welle CG, Kim GH, Kosterin P, Obaid AL, Yodh AG, Contreras D, Salzberg BM (2008) Two-photon excitation of potentiometric probes enables optical recording of action potentials from mammalian nerve terminals in situ. J Neurophysiol 99:1545–1553

Fisher JAN, Susumu K, Therien MJ, Yodh AG (2009) One- and two-photon absorption of highly conjugated multiporphyrin systems in the two-photon Soret transition region. J Chem Phys 130:134506 (2009)

Foley J, Muschol M (2008) Action spectra of electrochromic voltage-sensitive dyes in an intact excitable tissue. J Biomed Opt 13:064015
Franken PA, Hill AE, Peters CW, Weinreich G (1961) Generation of optical harmonics. Phys Rev Lett 7:118–119
Fromherz P, Muller CO (1993) Voltage-sensitive fluorescence of amphiphilichemicyanine dyes in neuron membrane. Biochim Biophys Acta 1150:111–122
Fromherz P, Hübener G, Kuhn B, Hinner MJ (2008) ANNINE-6plus, a voltage-sensitive dye with good solubility, strong membrane binding and high sensitivity. Eur Biophys J 37:509–514
Gainer H, Wolfe SA Jr, Obaid AL, Salzberg BM (1986) Action potentials and frequency-dependent secretion in the mouse neurohypophysis. Neuroendocrinology 43:557–563
Goeppert-Mayer M (1931) Ueber Elementarakte mit zwei Quantenspruengen. Ann Phys (Paris) 273
Grinvald A (1985) Real-time optical mapping of neuronal activity: from single growth cones to the intact mammalian brain. Annu Rev Neurosci 8:263–305
Grinvald A, Fine A, Farber IC, Hildesheim R (1983) Fluorescence monitoring of electrical responses from small neurons and their processes. Biophys J 42:195–198
Grinvald A, Lieke EE, Frostig RD, Hildesheim R (1994) Cortical point-spread function and long-range lateral interactions revealed by real-time optical imaging of macaque monkey primary visual cortex. J Neurosci 14:2545–2568
Guerrero G, Isacoff EY (2001) Genetically encoded optical sensors of neuronal activity and cellular function. Curr Opin Neurobiol 11:601–607
Guerrero G, Siegel MS, Roska B, Loots E, Isacoff EY (2002) Tuning FlaSh: redesign of the dynamics, voltage range, and color of the genetically encoded optical sensor of membrane potential. Biophys J 83:3607–3618
Hell SW, Booth M et al (1998) Two-photon near- and far-field fluorescence microscopy with continuous-wave excitation. Opt Lett 23:1238–1240
Iyer V, Losavio BE, Saggau P (2003) Compensation of spatial and temporal dispersion for acousto-optic multiphoton laser-scanning microscopy. J Biomed Opt 8:460–471
Iyer V, Hoogland TM, Saggau P (2006) Fast functional imaging of single neurons using random-access multiphoton (RAMP) microscopy. J Neurophysiol 95:535–545
Jin L, Han Z, Platisa J, Wooltorton JR, Cohen LB, Pieribone VA (2012) Single action potentials and subthreshold electrical events imaged in neurons with a fluorescent protein voltage probe. Neuron 75(5):779–785
Kim DH, Kim KH, Yazdanfar S, Peter TC (2005) Optical biopsy in high-speed handheld miniaturized multifocal microscopy. Proc SPIE 14:5700
Kleinfeld D, Delaney KR (1996) Distributed representation of vibrissa movement in the upper layers of somatosensory cortex revealed with voltage-sensitive dyes. J Comp Neurol 375:89–108
Knöpfel T, Tomita K, Shimazaki R, Sakai R (2003) Optical recordings of membrane potential using genetically targeted voltage-sensitive fluorescent proteins. Methods 30:42–48
Kosterin P, Kim GH, Muschol M, Obaid AL, Salzberg BM (2005) Changes in FAD and NADH fluorescence in neurosecretory terminals are triggered by calcium entry and by ADP production. J Membr Biol 208:113–124
Kuhn B, Fromherz P (2003) Anellatedhemicyanine dyes in a neuron membrane: molecular Stark effect and optical voltage recording. J Phys Chem B 107:7903–7913
Kuhn B, Fromherz P, Denk W (2004) High sensitivity of Stark-shift voltage-sensing dyes by one- or two-photon excitation near the red spectral edge. Biophys J 87:631–639
Kuhn B, Denk W, Bruno RM (2008) In vivo two-photon voltage-sensitive dye imaging reveals top-down control of cortical layers 1 and 2 during wakefulness.". Proc Natl Acad Sci U S A A105:7588–7593
Kurtz R, Fricke M, Kalb J, Tinnefeld P, Sauer M (2006) Application of multiline two-photon microscopy to functional in vivo imaging. J Neurosci Meth 151:276–286

Lillis KP, Eng A, White JA, Mertz J (2008) Two-photon imaging of spatially extended neuronal network dynamics with high temporal resolution. J Neurosci Meth 172:178–184

Llinás RR (1988) The intrinsic electrophysiological properties of mammalian neurons: insights into central nervous system function. Science 242:1654–1664

Llinás R, Ribary U, Contreras D, Pedroarena C (1998) The neuronal basis for consciousness. Philos Trans R Soc Lond B Biol Sci 353:1841–1849

Loew LM, Cohen LB, Salzberg BM, Obaid AL, Bezanilla F (1985) Charge-shift probes of membrane potential. Characterization of aminostyrylpyridinium dyes on the squid giant axon. Biophys J 47:71–77

Matiukas A, Mitrea BG et al (2006) New near-infrared optical probes of cardiac electrical activity. Am J Physiol Heart Circ Physiol 290:H2633–H2643

Matiukas A, Mitrea BG et al (2007) Near-infrared voltage-sensitive fluorescent dyes optimized for optical mapping in blood-perfused myocardium. Heart Rhythm 4:1441–1451

Moreaux L, Pons T, Dambrin V, Blanchard-Desce M, Mertz J (2003) Electro-optic response of second-harmonic generation membrane potential sensors. Opt Lett 28:625–627

Muschol M, Kosterin P, Ichikawa M, Salzberg BM (2003) Activity-dependent depression of excitability and calcium transients in the neurohypophysis suggests a model of "stuttering conduction". J Neurosci 23:11352–11362

Nielsen T, Fricke M, Hellweg D, Andresen P (2001) High efficiency beam splitter for multifocal multiphoton microscopy. J Microsc 201:368–376

Nordmann JJ (1977) Ultrastructuralmorphometry of the rat neurohypophysis. J Anat 123:213–218

Nuriya M, Jiang J, Nemet B, Eisenthal KB, Yuste R (2006) Imaging membrane potential in dendritic spines. Proc Natl Acad Sci U S A 103:786–790

Obaid AL, Salzberg BM (1996) Micromolar 4-aminopyridine enhances invasion of a vertebrate neurosecretory terminal arborization: optical recording of action potential propagation using an ultrafast photodiode-MOSFET camera and a photodiode array. J Gen Physiol 107:353–368

Obaid AL, Orkand RK, Gainer H, Salzberg BM (1985) Active calcium responses recorded optically from nerve terminals of the frog neurohypophysis. J Gen Physiol 85:481–489

Obaid AL, Flores R, Salzberg BM (1989) Calcium channels that are required for secretion from intact nerve terminals of vertebrates are sensitive to omega-conotoxin and relatively insensitive to dihydropyridines. Optical studies with and without voltage-sensitive dyes. J Gen Physiol 93:715–729

Obaid AL, Zou D-J, Rohr S, Salzberg BM (1992) Optical recording with single cell resolution from a simple mammalian nervous system: electrical activity in ganglia from the submucousplexus of the guinea-pig ileum. Biol Bull 183:344–346

Obaid AL, Loew LM, Wuskell JP, Salzberg BM (2004) Novel naphthylstyryl-pyridinium potentiometric dyes offer advantages for neural network analysis. J Neurosci Meth 134:179–190

Oheim M, Beaurepaire E, Chaigneau E, Mertz J, Charpak S (2001) Two-photon microscopy in brain tissue: parameters influencing the imaging depth. J Neurosci Methods 111:29–37

Orbach HS, Cohen LB (1983) Optical monitoring of activity from many areas of the in vitro and in vivo salamander olfactory bulb: a new method for studying functional organization in the vertebrate central nervous system. J Neurosci 3:2251–2262

Orbach HS, Cohen LB, Grinvald A (1985) Optical mapping of electrical activity in rat somatosensory and visual cortex. J Neurosci 5:1886–1895

Otsu Y, Bormuth V et al (2008) Optical monitoring of neuronal activity at high frame rate with a digital random-accessmultiphoton (RAMP) microscope. J Neurosci Methods 173:259–270

Parsons TD, Kleinfeld D, Raccuia-Behling F, Salzberg BM (1989) Optical recording of the electrical activity of synaptically interacting aplysia neurons in culture using potentiometric probes. Biophys J 56:213–221

Parsons TD, Salzberg BM, Obaid AL, Raccuia-Behling F, Kleinfeld D (1991) Long-term optical recording of patterns of electrical activity in ensembles of cultured aplysia neurons. J Neurophysiol 66:316–333

Petersen CC, Grinvald A, Sakmann B (2003) Spatiotemporal dynamics of sensory responses in layer 2/3 of rat barrel cortex measured in vivo by voltage-sensitive dye imaging combined with whole-cell voltage recordings and neuron reconstructions. J Neurosci 23:1298–1309

Pons T, Moreaux L, Mongin O, Blanchard-Desce M, Mertz J (2003) Mechanisms of membrane potential sensing with second-harmonic generation microscopy. J Biomed Opt 8:428–431

Richards B, Wolf E (1959) Electromagnetic diffraction in optical systems. II. Structure of the image field in an aplanatic system. Proc R Soc Lond A 253:358–379

Rowan MJM, Tranquil E, Christie JM (2014) Distinct K_v channel subtypes contribute to differences in spike signaling properties in the axon initial segment and presynaptic boutons of cerebellar interneurons. J Neurosci 34:6611–6623

Sakai R, Repunte-Canonigo V, Raj CD, Knöpfel T (2001) Design and characterization of a DNA-encoded, voltage-sensitive fluorescent protein. Eur J Neurosci 13:2314–2318

Salama G, Choi BR et al (2005) Properties of new, long-wavelength, voltage-sensitive dyes in the heart. J Membr Biol 208:125–140

Salome R, Kremer Y et al (2006) Ultrafast random-access scanning in two-photon microscopy using acousto-optic deflectors. J Neurosci Meth 154:161–174

Salzberg BM (1983) Optical recording of electrical activity in neurons using molecular probes. In: Barker J, McKelvy J (eds) Current methods in cellular neurobiology. Wiley, New York, NY

Salzberg BM, Davila HV, Cohen LB (1973) Optical recording of impulses in individual neurones of an invertebrate central nervous system. Nature 246:508–509

Salzberg BM, Grinvald A, Cohen LB, Davila HV, Ross WN (1977) Optical recording of neuronal activity in an invertebrate central nervous system: simultaneous monitoring of several neurons. J Neurophysiol 40:1281–1291

Salzberg BM, Obaid AL, Senseman DM, Gainer H (1983) Optical recording of action potentials from vertebrate nerve terminals using potentiometric probes provides evidence for sodium and calcium components. Nature 306:36–40

Salzberg BM, Obaid AL, Gainer H (1985) Large and rapid changes in light scattering accompany secretion by nerve terminals in the mammalian neurohypophysis. J Gen Physiol 86:395–411

Salzberg BM, Obaid AL, Bezanilla F (1993) Microsecond response of a voltage-sensitive merocyanine dye: fast voltage-clamp measurements on squid giant axon. Jpn J Physiol 43 (Suppl 1):S37–S41

Sherrington C (1951) Man on his nature. Cambridge University Press, Cambridge

Sims PJ, Waggoner AS, Wang CH, Hoffman JF (1974) Studies on the mechanism by which cyanine dyes measure membrane potential in red blood cells and phosphatidylcholine vesicles. Biochemistry 13:3315–3330

Svoboda DW (1997) Photon upmanship: why multiphoton imaging is more than a gimmick. Neuron 18:351–357

Teisseyre TZ, Millard AC et al (2007) Nonlinear optical potentiometric dyes optimized for imaging with 1064-nm light. J Biomed Opt 12:044001

Theer P, Hasan MT, Denk W (2003) Two-photon imaging to a depth of 1000 microns in living brains by use of a Ti:Al2O3 regenerative amplifier. Opt Lett 28:1022–1024

Vucinic D, Sejnowski TJ (2007) A compact multiphoton 3D imaging system for recording fast neuronal activity. PLoS One 2:e699

Waggoner AS (1979) Dye indicators of membrane potential. Annu Rev Biophys Bioeng 8:47–68

Xu C (2000) Two-photon cross-sections of indicators. In: Yuste R, Lanni F, Konnerth A (eds) Imaging neurons: a laboratory manual. Cold Spring Harbor Laboratory Press, Cold Spring Harbor, NY

Xu C, Webb WW (1996) Measurement of two-photon excitation cross sections of molecular fluorophores with data from 690 to 1050 nm. J Opt Soc Am 13:481–491

Xu C, Zipfel W, Shear JB, Williams RM, Webb WW (1996) Multiphoton fluorescence excitation: new spectral windows for biological nonlinear microscopy. Proc Natl Acad Sci U S A 93:10763–10768

Yan P, Acker CD, Zhou WL, Lee P, Bollensdorff C, Negrean A, Lotti J, Sacconi L, Srdjan A, Kohl P, Mansvelder HD, Pavone FS, Loew LM (2012) Palette of fluorinated voltage-sensitive hemicyanine dyes. Proc Natl Acad Sci U S A 109:20443–20448

Yuste R, Tank DW, Kleinfeld D (1997) Functional study of the rat cortical microcircuitry with voltage-sensitive dye imaging of neocortical slices. Cereb Cortex 7:546–558

Zhou W-L, Yan P, Wuskell JP, Loew LM, Antic SD (2007) Intracellular long-wavelength voltage-sensitive dyes for studying the dynamics of action potentials in axons and thin dendrites. J Neurosci Meth 164:225–239

Zipfel WR, Williams RM, Webb WW (2003) Nonlinear magic: multiphoton microscopy in the biosciences. Nat Biotechnol 21:1369–1377

Chapter 18
Random-Access Multiphoton Microscopy for Fast Three-Dimensional Imaging

Gaddum Duemani Reddy, Ronald J. Cotton, Andreas S. Tolias, and Peter Saggau

Abstract Studies in several important areas of neuroscience, including analysis of single neurons as well as neural networks, continue to be limited by currently available experimental tools. By combining molecular probes of cellular function, such as voltage-sensitive or calcium-sensitive dyes, with advanced microscopy techniques such as multiphoton microscopy, experimental neurophysiologists have been able to partially reduce this limitation. These approaches usually provide the needed spatial resolution along with convenient optical sectioning capabilities for isolating regions of interest. However, they often fall short in providing the necessary temporal resolution, primarily due to their restrained laser scanning mechanisms. In this regard, we review a method of laser scanning for multiphoton microscopy that overcomes the temporal limitations of pervious approaches and allows for what is known as 3D Random Access Multiphoton (3D RAMP) microscopy, an imaging technique that supports full three dimensional recording of many sites of interest on physiologically relevant time scales.

Keywords Multiphoton • Microscopy • Three-dimensional • Random-access scanning • Acousto-optic deflector • Spatial dispersion • Temporal dispersion • Motion tracking • di 8 ANEPPS

1 Introduction

In microscopy, random-access scanning refers to a form of laser beam scanning in which the beam focus is selectively positioned at individual, typically user-selected, sites in either a two-dimensional (2D) plane or a three-dimensional (3D) space.

G.D. Reddy • R.J. Cotton • A.S. Tolias
Department of Neuroscience, Baylor College of Medicine, Houston, TX 77030, USA

P. Saggau (✉)
Allen Institute for Brain Science, 551N 34th Street, Seattle, WA 98103, USA

Department of Neuroscience, Baylor College of Medicine, Houston, TX 77030, USA
e-mail: peters@alleninstitute.org

M. Canepari et al. (eds.), *Membrane Potential Imaging in the Nervous System and Heart*, Advances in Experimental Medicine and Biology 859,
DOI 10.1007/978-3-319-17641-3_18

The positioning is considered 'random' since there are no restrictions on the relationship between selected sites. This is in contrast to other, more conventional methods of beam scanning which physically constrain the scan to a predetermined, usually linear pattern.

In imaging modalities that utilize some form of beam scanning, such as multiphoton microscopy or confocal laser scanning microscopy, random-access scanning offers three distinct advantages. First, it increases the overall speed, or temporal resolution, since non-useful or void areas of space are selectively removed from the scan. Second, it improves the signal-to-noise ratio (S/N), since the dwell time in signal-producing areas is increased, enhancing the amount of signal acquired. Lastly, by allowing the user to determine not only the locations but also the total number of sites to visit, random-access scanning offers flexibility in adjusting the relative spatial resolution (as measured by number of points per scan) and temporal resolution (as measured by the number of scans per unit time). This is exemplified by the two extremes available in a random-access scan: maximizing temporal resolution at the expense of spatial resolution by fixing the laser focus at a single location, and maximizing spatial resolution at the expense of temporal resolution by scanning all sites of interest, e.g. raster scanning.

Given these advantages, random-access imaging modalities are very useful in monitoring electrophysiological parameters through optical indicators such as voltage-sensitive or calcium-sensitive dyes, where temporal resolutions and S/N are particularly important. Indeed the advantages obtained by combining fluorescent optical indicators with some form of random-access scanning has been demonstrated in several different imaging schemes over the past two decades, including simple scanning microscopes (Bullen et al. 1997; Bullen and Saggau 1998), confocal microscopes (Bansal et al. 2006) and multiphoton microscopes (Lechleiter et al. 2002; Iyer et al. 2006; Salome et al. 2006; Lv et al. 2006). The last group, random-access multiphoton (RAMP) microscopes, is the latest development of this scanning scheme. RAMP microscopes are unique in that they represent the only imaging technique developed so far that has been able to implement full 3D random-access scanning at speeds high enough to support multi-kilohertz sampling rates as they are necessary for voltage imaging in particular and for spike reconstruction techniques in general. Combined with the intrinsic spatial resolution of multiphoton excitation, this feature gives results in the unprecedented ability to monitor electrophysiological signals from selected sites in a volume of neural tissue, such as multiple spines on a dendrite or a group of neurons in a population.

2 Random-Access Scanning Principles

While it is possible, in principle, to achieve a random-access scan by selectively positioning a laser focus with the galvanometer driven pivoting mirrors and stepper motors found in most commercial scanning microscopes, this approach is of limited usefulness for fast imaging requirements. This is because these beam positioning mechanisms have a significant inertial barrier to overcome. This drastically reduces the speed at which they operate, particularly when frequently stopped and started for repositioning the beam as would be needed in a random-access scan.

In order to reduce the extent to which a scan is inertia-limited, different methods have been developed, such as driving galvanometers at their resonant frequency (Pawley 1995) or following a smooth trajectory through a number of selected sites (Gobel et al. 2007). While these techniques do significantly increase the temporal resolution, they still spend significant time transitioning between points of interest. The optimal solution would be a completely inertia-free scanning mechanism. In this regard, acousto-optic deflectors (AODs) have become one of the most used methods for generating 2D and 3D random-access scans.

As shown in Fig. 18.1a, AODs are devices in which acoustic waves propagate in an optically and acoustically transparent medium, usually a crystal, to create a diffraction grating that is utilized to deflect incoming light by an angle, θ, determined by Eq. (18.1),

$$\theta = \frac{\lambda f}{v}, \tag{18.1}$$

where λ is the wavelength of the incident light and f and v are the frequencies and velocities of the propagating acoustic waves, respectively. By utilizing a pair of

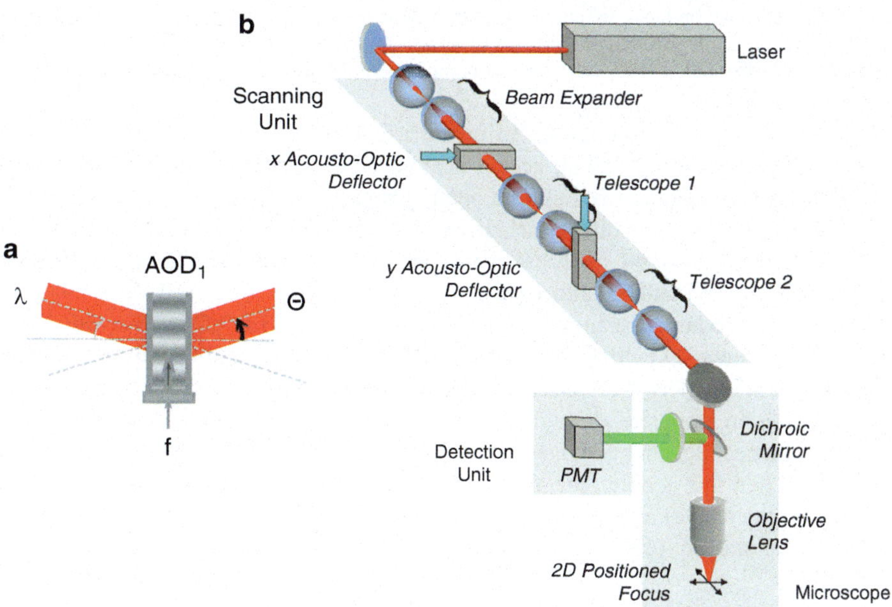

Fig. 18.1 2D AOD Scanning Microscope. (**a**) Angular deflection (Θ) of a collimated laser beam by a single AOD with a fixed frequency (f) acoustic wave. (**b**) Fast scanning microscope employing two orthogonal AODs for two-dimensional random-access scanning

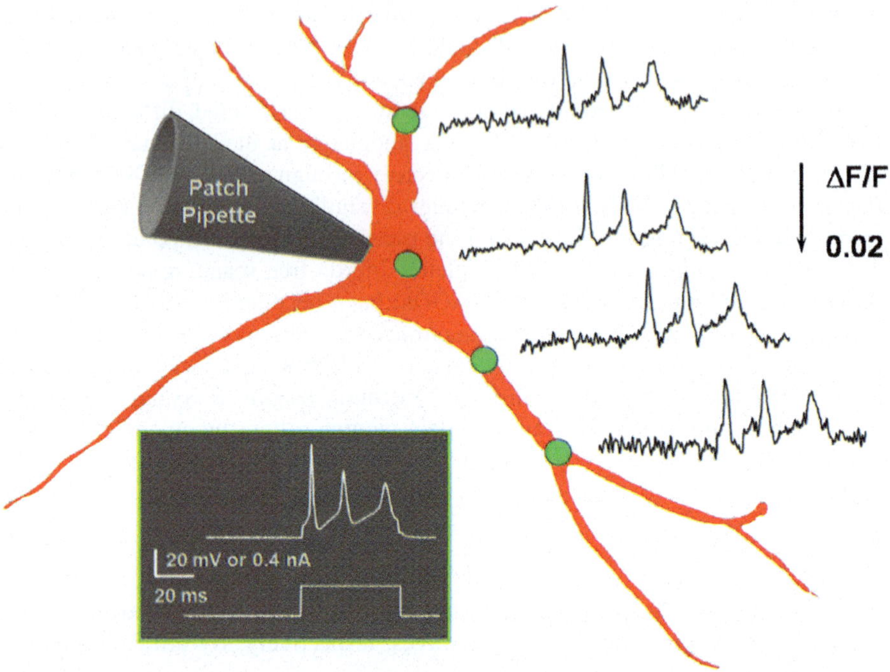

Fig. 18.2 Fast voltage imaging with AOD scanning microscope. Concurrent multi-site recording of action potentials, induced by current injection into soma, propagating into dendrites. Hippocampal neuron in dissociated low-density culture, stained with 100 μM VSD di-8-ANEPPS. Single-photon excitation at 514 nm, no averaging applied. Modified from Saggau et al. (1998)

orthogonal AODs, it is possible to create a random-access scan in a 2D plane (Fig. 18.1b). Indeed, a random-access beam scanner consisting of two orthogonal AODs was incorporated in a simple beam scanning microscope that used single-photon excitation of voltage sensitive dyes to monitor action potentials at non-contiguous sites on neurons in dissociated cell culture (Fig. 18.2) (Bullen et al. 1997; Bullen and Saggau 1998).

In addition to lateral scanning, it is also possible to utilize AODs to create either an axial or even a 3D random-access scan (Reddy and Saggau 2005). In this scheme (see Fig. 18.3a, b) two AODs are used per dimension, with their sound waves propagating in opposite directions. The frequencies of these sound waves are equally chirped, i.e. changed over time, resulting in a linearly modulated diffraction grating. When this is done, the angle of deflection $\theta(x)$ is dependent upon position (x) within the crystal and can be described by Eq. (18.2) for a purely axial random-access scan and Eq. (18.3) for a complete 3D random-access scan.

$$\theta\ (x) = -\left(2\alpha\frac{\lambda}{v^2}\right)x \tag{18.2}$$

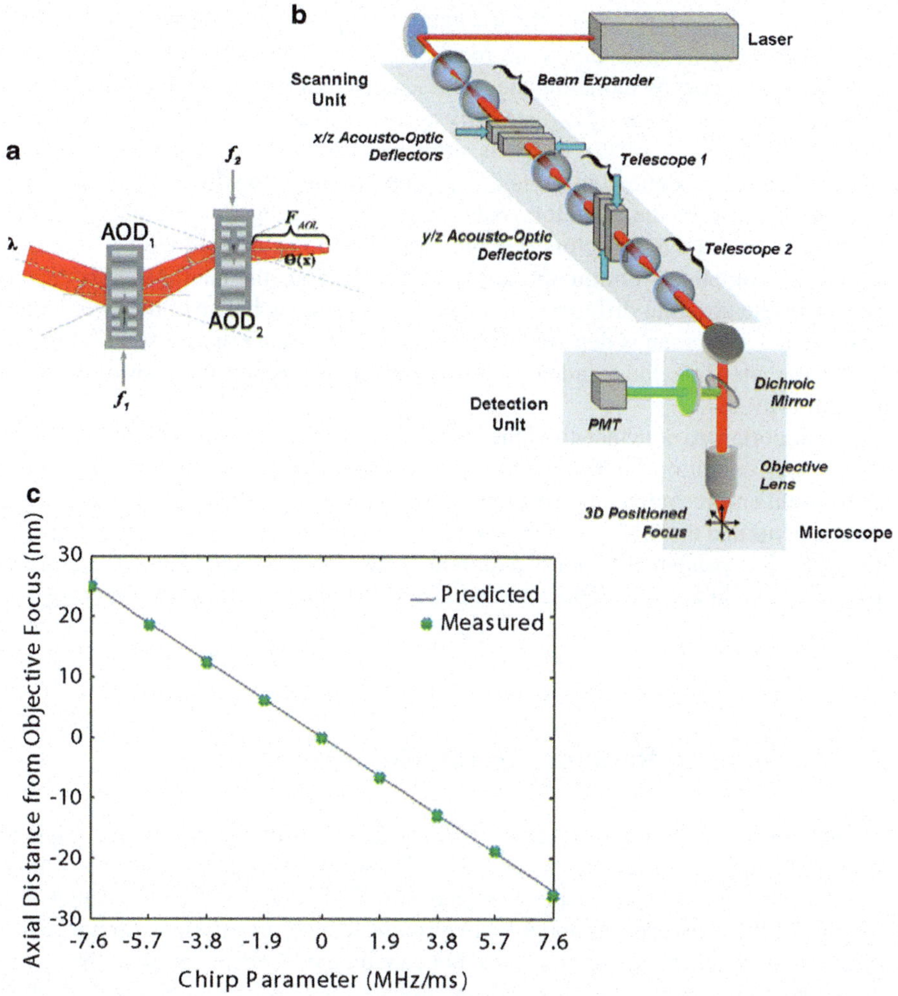

Fig. 18.3 3D AOD Scanning Microscope. (**a**) Change in collimation (focal length F_{AOL}) and angular deflection ($\Theta(x)$) by counter-propagating chirped acoustic waves (f_1, f_2) in a pair of AODs. (**b**) Two orthogonal pairs of AODs support random-access scanning within a volume. (**c**) Predicted (*solid line*) and measured (*crosses*) axial distance of the focal position from the inherent focal plane of the objective lens

$$\theta\ (x) = -\left(2\alpha\frac{\lambda}{v^2}\right)x + \Delta f\frac{\lambda}{v} \tag{18.3}$$

The new variable α in these equations is the chirp parameter and defines the rate of change in frequency. It can be seen that both the right hand side of Eq. (18.2) and the first part of the right hand side of Eq. (18.3), are position (x) dependent. In fact, the specific relationship between the angle of deflection and position is identical to that of a cylindrical lens, with a focal length of $f_{AOL} = \frac{v^2}{2\alpha\lambda}$. As shown in Fig. 18.3c, which compares the measured axial focal distances of an AOD-based microscope with those predicted by modeling the AOD scanner as a lens, this value for the focal length of the AOD system, f_{AOL}, strongly agrees with experiment. The second part of the right hand side of Eq. (18.3) is similar to Eq. (18.1) and defines a position independent lateral deflection angle that is proportional to the difference in acoustic frequencies between the deflectors. Note that by eliminating the difference between the frequencies in the deflectors (i.e, making $\Delta f = 0$), we can reduce Eq. (18.3) to Eq. (18.2), completely eliminating the lateral scan. Thus the lateral scan is dependent upon the absolute difference in acoustic frequencies between the two deflectors, and the axial scan is dependent upon the rate of change in acoustic frequency. Both parameters are independent and user-definable through the control over the acoustic waves.

It is important to note that all of the above equations are time-independent. This is a necessary condition for the random-access capabilities, since a time-dependent term would enforce a fixed pattern to the scan. However, if the random-access scan requirements and the S/N-improving advantage of increased dwell time are dropped, then time independence is not a requirement for 3D scanning. Indeed, a chirped acoustic wave in a single deflector can be used to create a cylindrical lens effect as well as a continuously changing deflection angle (Vucinic and Sejnowski 2007).

3 Multiphoton Random-Access Microscopy

As mentioned above, an AOD-based beam scanner has previously been combined with a custom light microscope and fluorescent optical indicators to study both voltage and calcium transients in neuronal cultures (Bullen et al. 1997; Bullen and Saggau 1998). However, although optically ideal, such preparations do not represent *in vivo* environments since they do not contain the 3D architecture of the neural tissue. Therefore, when studying realistic preparations, such as a brain slice or even intact cortex, optical sectioning techniques such as confocal laser scanning microscopy (CLSM) or multiphoton microscopy are essential.

However, the application of random-access beam scanning using AODs in either CLSM or multiphoton microscopy requires additional challenges to be addressed. In CLSM for example, a separate random-access method must be used for spatial filtering of the collected fluorescence, since attempting to de-scan the collected fluorescence *via* the AODs would result in significant signal loss secondary to the limited diffraction efficiency of the deflectors (Bansal et al. 2006). AOD-based multiphoton microscopy does not share this limitation, since no de-scanning of the fluorescent signal is necessary; however, it creates three distinct technical challenges: spatial dispersion, temporal dispersion, and power loss.

3.1 Spatial Dispersion

This type of dispersion is a result of the fact that ultrafast laser pulses are utilized in multiphoton microscopes to achieve the high photon flux required for useful multiphoton excitation. These laser pulses have durations of hundreds of femtoseconds and consequently significantly larger frequency (and thus wavelength) bandwidths compared to standard continuous wave (CW) emission. As a result, the standard equation for angle of deflection in an AOD is changed from a single angle θ in Eq. (18.1) to a range of angles $\Delta\theta$ in Eq. (18.4) below

$$\Delta\theta = \frac{\Delta\lambda f}{\nu} \tag{18.4}$$

As can be seen in Fig. 18.4a, this results in a diverging beam at the output of the AOD. If left uncorrected, this would drastically reduce the spatial resolution, since the spot size at the object plane would be much larger than the diffraction limit.

To deal with this issue, several methods have been developed to minimize the amount of spatial dispersion. The simplest method has been to use longer temporal pulsewidths (Iyer et al. 2006), which reduces the $\Delta\lambda$ term in Eq. (18.4). Other methods minimize the total amount of spatial dispersion by introducing a degree of spatial dispersion in the opposite direction using a dispersive element such as a diffraction grating (Iyer et al. 2003; Salome et al. 2006), a prism (Bi et al. 2006; Lechleiter et al. 2002; Zeng et al. 2006), or an additional AOD (Fig. 18.4b) (Iyer et al. 2003; Salome et al. 2006). Effectively, these methods mimic reducing the absolute value of the frequency term (f) in Eq. (18.4), while maintaining similar values for the range of frequencies (thereby keeping the same scan range). The 3D AOD-based-scanning method inherently compensates for spatial resolution by a similar method. Indeed as can be seen by comparing Eq. (18.3) with Eq. (18.2), the lateral deflection angle (and hence the amount of spatial dispersion) in this method is determined by the difference in acoustic frequencies between the two deflectors, which is optimally zero at the center. However, even at the largest scan angles, the maximum amount of spatial dispersion in the 3D method is proportional to the acoustic frequency bandwidth of the deflector, which for practical reasons is significantly smaller than the absolute value of the acoustic frequencies used in each deflector. Indeed, it has been shown that the reduction in spatial dispersion in 3D AOD scanning (Reddy and Saggau 2005) is at least a factor of 3 over the best uncompensated single AOD situation.

3.2 Temporal Dispersion

This dispersion, which is another major challenge of combining AOD-based scanning with multiphoton microscopy, is a consequence of the dependence of the

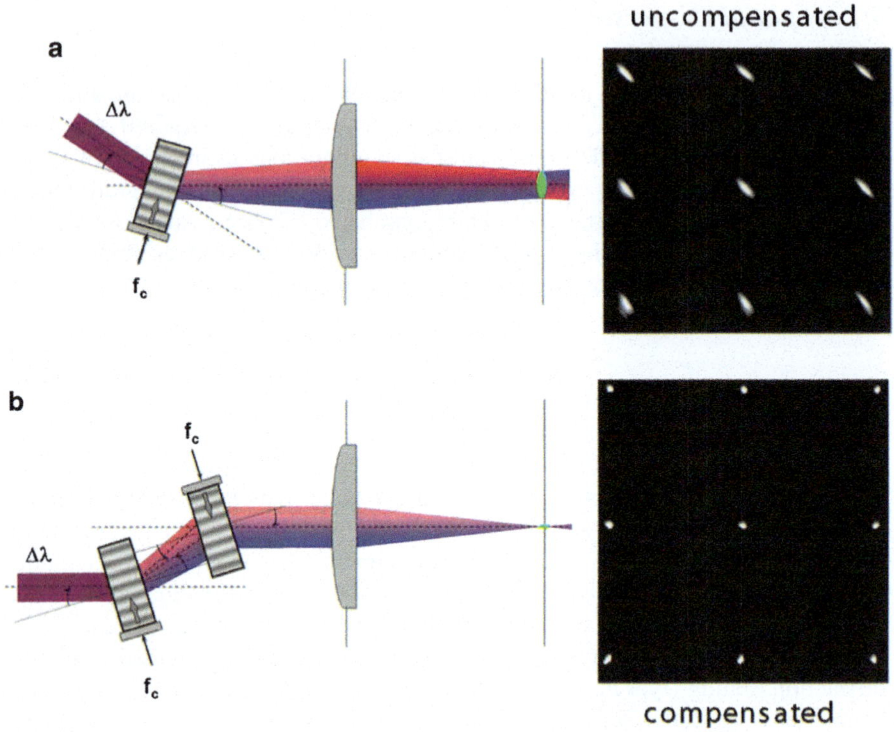

Fig. 18.4 Spatial dispersion and compensation. (**a**) Frequency components (bandwidth Δλ) of ultrafast pulses are spectrally separated by an AOD. Uncompensated spatial dispersion in a 2D imaging system results in focal spots elongated along the 45° axis (3 × 3scan pattern). (**b**) Compensation of spatial dispersion by second parallel AOD operated with counter-propagating sound wave. Compensated scan pattern showing resulting spot sizes

refractive index of the material in each deflector on the frequency of light. Most material exhibit what is known as normal dispersion in the visible and near infrared frequency ranges, wherein the refractive index increases with increasing optical frequency. As a consequence, at higher optical frequencies, light has a higher refractive index. Thus, when an ultrafast optical pulse travels through a normal dispersive material, the red portion of the pulse spectrum will travel faster than the blue portion, which in turn will lead to increased pulse duration. Since the optical pulse now has a linearly changing optical frequency relationship in time, it is referred to as *optically* "chirped" (not to be confused with the *acoustic* "chirp" needed for 3D scanning).

A method of characterizing the degree to which a pulse will be broadened by a given material is through the amount of accumulated second-order phase, which is commonly known as the group delay dispersion (GDD) of the material. This value can be normalized for the length of the material, in which case, it is called the group velocity dispersion (GVD). Mathematically, this value can be expressed as Eq. (18.5)

$$GVD = \frac{\varphi''(\omega)}{L} = \frac{d^2}{d\omega^2}\left(n(\omega)\frac{\omega}{c}\right) = \frac{\lambda^3}{2\pi c^2}n''(\lambda), \tag{18.5}$$

where φ (ω), the phase of the light wave with respect to the angular frequency (ω), is equivalent to $\frac{n(\omega)L\omega}{c}$, λ is the wavelength, L is the total distance traveled through the material, n is the refractive index of the material, and c is the speed of light in a vacuum. Note that since the wave vector k is just the accumulated phase per unit length (i.e., $k = \frac{\varphi(\omega)}{L}$), the GVD can also be expressed as the second derivative of the wavevector with respect to the angular frequency or $\frac{d^2k}{d\omega^2}$, or, since the first derivative of the wavevector with respect to the angular frequency is the inverse of the group velocity (i.e., $\frac{dk}{d\omega} = \frac{1}{v_{group}}$), as $\frac{d}{d\omega}\left(\frac{1}{v_{group}}\right)$.

The wavelength dependent value of the refractive index can be determined from the empirically derived Sellmeier equations (Weber 1986) for the transparent materials often used for AOD fabrication. These in turn can be used to calculate the amount of GVD introduced by the acousto-optic medium. Assuming a Gaussian pulse envelope, the output pulsewidth (τ_{out}) after the dispersion by the AODs can be calculated using Eq. (18.6)

$$\tau_{out} = \tau_{in}\sqrt{1 + \left(\frac{4\ln 2 \cdot GVD \cdot L}{\tau_{in}}\right)^2}, \tag{18.6}$$

where τ_{in} is the input pulsewidth and L is the total path length through the medium. Since the material used for fabricating AODs often has a substantial thickness, the degree of pulse broadening is significantly higher than in mirror-based multiphoton microscopes. Indeed, measured values for the GVD introduced by a single AOD have ranged from ~4,500 to 6,000 fs^2, depending upon wavelength (Iyer et al. 2003; Lv et al. 2006; Salome et al. 2006). As a result, 2D AOD imaging systems have reported pulse broadening ranging from 2.7 to 6.3 times the incident pulsewidth and 3D AOD imaging systems have reported pulse broadening of ~9 times the incident pulsewidth (Reddy et al. 2008).

Since temporal dispersion has a detrimental effect in many areas of ultrafast optics, several different types of compensating schemes have been developed to reduce the total amount of pulse broadening. All these methods introduce a specific amount of negative GVD to compensate the positive GVD produced by the optical system, a process known as “pre-chirping”. These methods include schemes that are prism based (Fork et al. 1984; Bi et al. 2006), grating based (Treacy 1969), coated mirror based (Steinmeyer 2006), and even AOD-based (Nakazawa et al. 1988).

3.3 Diffraction Efficiency

The last challenge in using AODs for beam scanning in a multiphoton microscope is the power loss due to the limited diffraction efficiency of each deflector. In contrast to reflection based scanning, which is almost 100 % efficient, AOD-based diffraction is practically ~70 % efficient, although specialized deflectors can achieve higher diffraction efficiencies. As a result after two deflectors, almost 50 % of the incident light is lost. Similarly in 3D AOD scanning situations, almost 75 % is lost. This loss of power, combined with the elongated pulse widths, place a strong power requirement on the excitation light source of the system. However, the increased availability of multi-watt femtosecond lasers makes it possible to achieve several hundred milliwatts of output power, which is sufficient for many biological applications.

4 Imaging

Despite the challenges listed above, several systems have been developed which successfully combine multiphoton microscopy with AOD-based scanning. These include systems for 2D scanning (Iyer et al. 2006; Salome et al. 2006; Lv et al. 2006) as well as for 3D random-access scanning (Reddy et al. 2008; Katona et al. 2012; Cotton et al. 2013).

4.1 Imaging Quality

The effective spatial resolution of the 3D AOD microscope, as determined by measuring point spread functions (PSFs) at different positions in space, is shown in Fig. 18.5. From the full width at half maximum (FWHM) of the PSFs, it can be seen that in the compensated system, due to residual spatial dispersion, there is an approximately 50 % decrease in lateral resolution at the furthest scan angles, which correspond to 83 μm with a 60× objective. Similarly, due to spherical aberration effects as well as relative axial reductions in numerical aperture (NA) (Reddy et al. 2008), there is a maximum decrease in axial resolution of 50 % over the axial scan range, which was 50 μm with a 60× objective lens.

A stronger appreciation for the degree to which these aberrations practically affect resolution is obtained by inspecting acquired structural images. In this regard, Fig. 18.6 shows a portion of a dendritic branch at higher magnification obtained by AOD imaging at three different axial locations. The image at the inherent focus of the objective lens (middle panel), which is referred to as the reference plane, was obtained with the optimal spatial resolution. The axial locations are then changed by adjusting the laser beam focus using the AODs, positioning the focus either

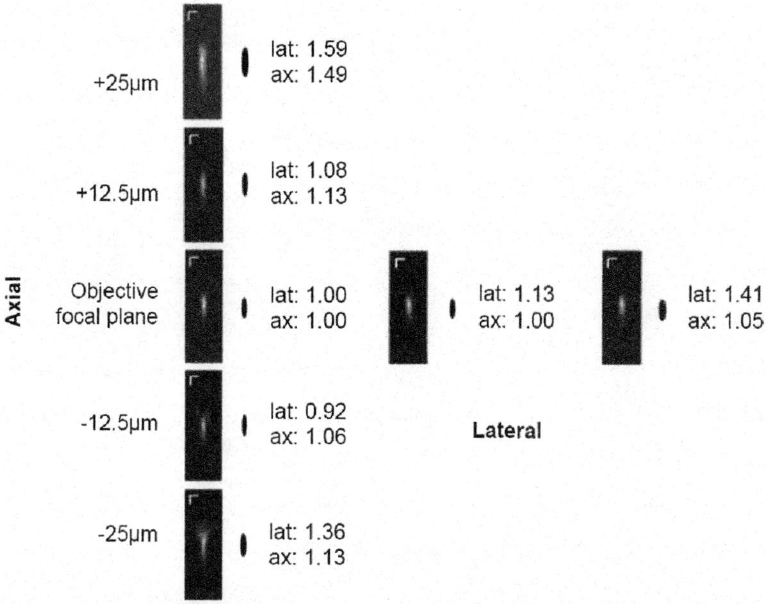

Fig. 18.5 Point spread functions in 3D scanning mode. Lateral and axial point spread functions (PSFs) at different axial and lateral positions as determined by imaging 200 nm fluorescent beads. Relative changes in the full width half maximum of the PSFs compared with inherent objective focus (490 nm × 2.3 μm) are shown to the right of each image

25 μm above (top panel) or 25 μm below (bottom panel) the reference plane. In order to maintain the field of view however, the objective lens is also moved correspondingly to adjust the effective focus back at the reference plane.

Combining the results of the PSF data with the reconstructed structural images leads to two distinct features of AOD-based 3D imaging. Since there are noticeable changes in both axial and lateral resolution, this imaging scheme is not the optimal solution for obtaining high-resolution structural images. However, the structural images using AOD-based scanning offer adequate spatial resolution to discern submicron features and identify neurons. In this regard, the scans obtained by AOD-based methods are perfectly suited to generate the initial structural image on which further random-access functional scans can be performed. This intrinsic possibility eliminates the need to register the functional scan with a structural scan obtained using another method.

4.2 *Brain Slice Imaging*

Experiments using the 3D random-access multiphoton microscope have provided dendritic multi-site measurements of localized calcium transients in response to

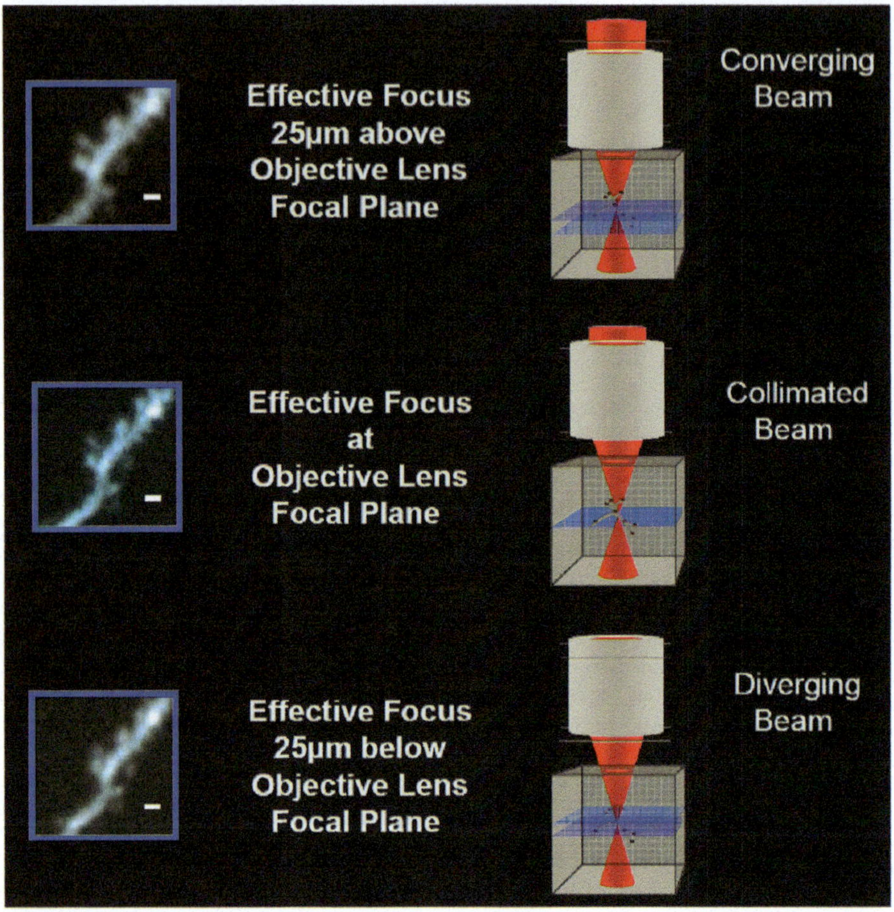

Fig. 18.6 Structural imaging in 3D scanning mode. Dendritic segment imaged with high resolution (zoom-in) using a 3D multiphoton microscope at three different AOD-controlled focal distances. Reference image (*middle panel*), taken at inherent focal distance of the objective lens (reference plane). Test images, taken at AOD-controlled effective focal planes 25 μm above (*top*) and 25 μm below (*bottom*) the reference plane with the objective lens moved in the opposite axial direction to compensate (scale bars, 1 μm)

back-propagating action potentials (bAPs) over axial ranges of up to 50 μm when using a 60× objective lens. Recorded sites include locations along the main apical dendrite as well as oblique dendrites (Fig. 18.7). Note that the sites selected for each scan were from different lateral and axial locations as is indicated by the color-coded depth scheme.

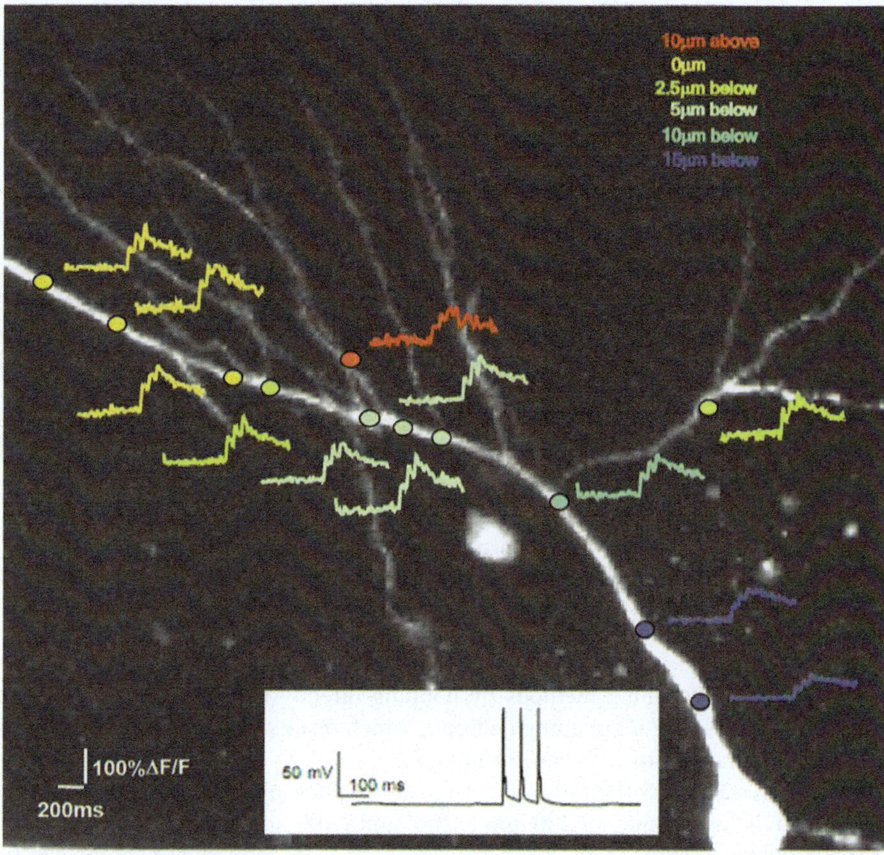

Fig. 18.7 Fast 3D functional imaging of dendritic calcium dynamics. Recording sites selected from a structural image acquired in 3D scanning mode. 3D functional image (using a random access scan) of calcium transients along main and oblique dendrites during a train of three bAPs. Acquisition frame rate ~3 kHz. The axial distances of the user-selected sites are color coded (measured in microns relative to the inherent objective lens focus). All values rounded to nearest micron

4.3 In Vivo Imaging

Three-dimensional random-access multiphoton (3D RAMP) microscopy allows dense recording from nearly all the neurons within a cube of several hundred microns in the mouse visual cortex and studying their tuning properties (Katona et al. 2012; Cotton et al. 2013; Froudarakis et al. 2014). Because of the three dimensional distribution of neurons in the brain, the ability to rapidly "hop" between neurons at different depths allows recording from significantly larger populations than methods limited to a single plane, and with substantially higher sampling rates. Using this technology *in vivo* exploits the S/N benefits of AODs

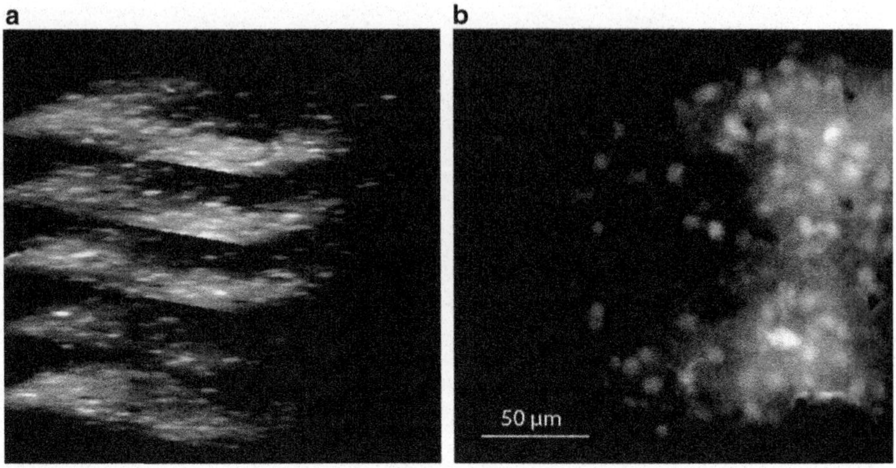

Fig. 18.8 Structural imaging. (**a**) Subset of image planes from a volume acquired with a 3D RAMP scanner. Neurons were stained with the calcium sensitive fluorophore OGB-1 which is trapped intracellularly and makes cell bodies brighter than the surrounding neuropil. The volume was acquired by sequentially hopping over points within a 200 μm × 200 μm × 100 μm volume. (**b**) Middle plane. Neurons are clearly stained and can be localized to then acquire functional fluorescent data

over continuous scanning methods by hopping directly between neurons without having to traverse the intervening neuropil, which results in imaging from neurons for a greater fraction of the acquisition time.

Neurons stained with the calcium-sensitive fluorophore OGB-1 are clearly visualized in a volume of 200 μm × 200 μm × 100 μm acquired with a 20× 1.0 NA objective (Fig. 18.8). From the structural scan, individual neurons can be localized and their functional fluorescence recorded over time. Calcium events corresponding to the influx of calcium during an action potential can be detected from hundreds of simultaneously recorded neurons and used to reconstruct the underlying firing rates (Fig. 18.9a, b). By showing gratings at different orientations to the mouse, the orientation tuning curves of many individual neurons can be characterized and produces results consistent with traditional electrophysiological recordings (Fig. 18.9c). This technique has also been utilized to study properties of population codes, such as sparseness, that cannot be measured with traditional techniques (Froudarakis et al. 2014).

4.4 Motion Tracking

A major challenge of applying random access techniques to *in vivo* imaging is that motion-induced fluorescence signals from the tissue can be confounded with functional brain activity. Because one of the main purposes of large scale *in vivo*

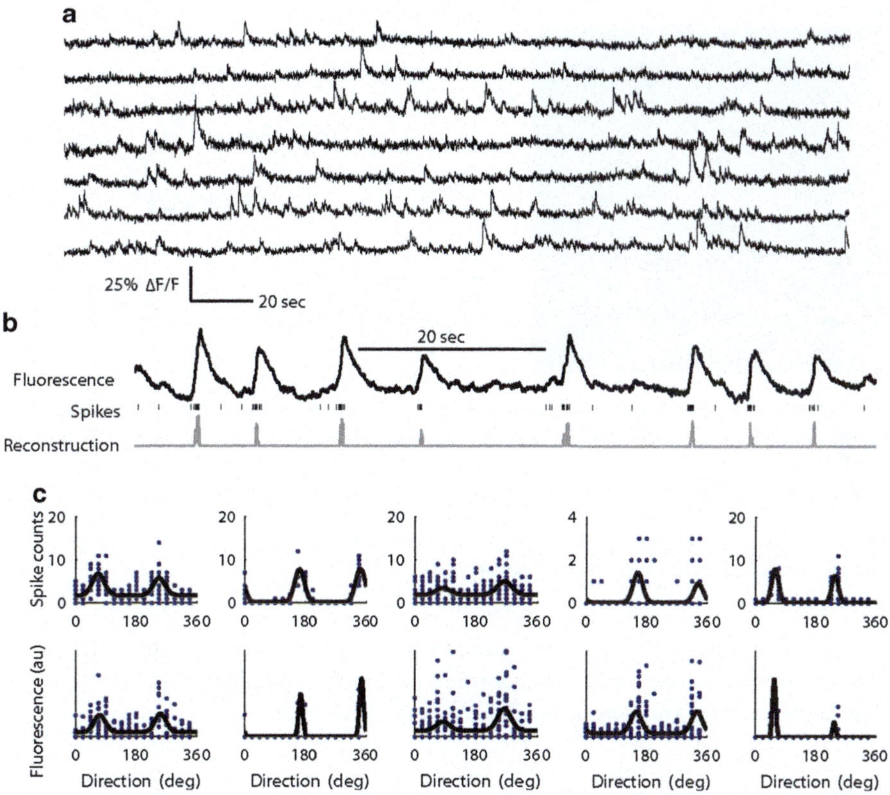

Fig. 18.9 Functional imaging. (**a**) Data from a subset of 411 neurons simultaneously imaged in vivo with a 3D RAMP scanner. Shown are neuronal calcium events, rapid rises in the fluorescence intensity followed by a slower decay. Data was acquired at 120 Hz and down-sampled to 20 Hz for visualization. (**b**) *Top*: Fluorescence activity from a single cell. Middle: Spike times recorded from simultaneous electrophysiology. The calcium events align with the timing of action potentials. Bottom: Inferred spiking activity (*gray*) reconstructed from the fluorescent trace. (**c**) Tuning curves measured for cells with electrophysiology (*top row*) and reconstructed firing rates from the fluorescent data (*bottom row*). The x-axis indicates the direction of motion of a Gabor grating presented to the animal for 1 s. *Individual dots* indicate single trials and *black curves* are von Mises fit to the data. The tuning curves measured with 3D-RAMP are very similar to those measured with electrophysiology

imaging is to study the coordinated activity of many neurons, this confound is extremely problematic. Traditional scanning methods generate a "movie" that shows the structure, and thus tissue movement is clearly visible and can be corrected (Greenberg and Kerr 2009). With random access scanning there is no such movie and thus movement cannot be compensated *post hoc*.

This problem can be addressed by interleaving small image planes over high contrast structures with the functional scanning pattern in order to simultaneously track motion with the fluorescence activity (Fig. 18.10). While this approach does

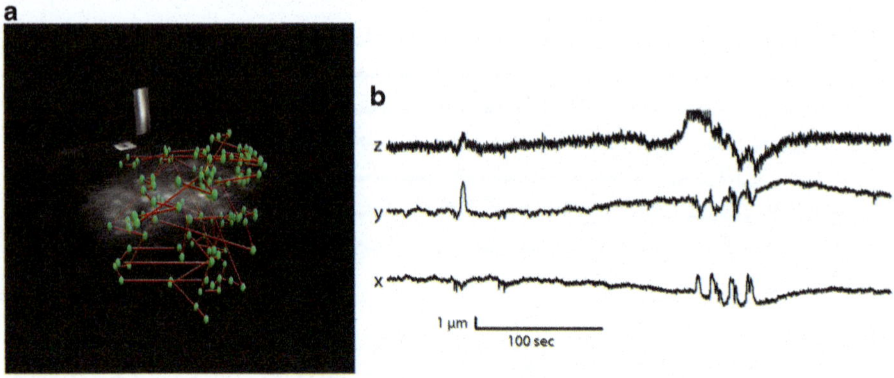

Fig. 18.10 Motion Tracking. (**a**) Visualization of interleaving motion tracking planes with functional scanning. *Green points* indicate locations of recorded neurons with hopping path (*red*) between them. Scanning path is interleaved with points placed on a horizontal and vertical plane within the volume. These planes are placed over high contrast objects allowing motion to be tracked with all three directions simultaneously. (**b**) Larger motion artifact measured motion during an in vivo experiment. Starting and ending positions are quite similar, with periods of abrupt movement measuring several microns

not allow removing the motion confound *per se*, it permits detecting motion and discarding data where motion artifacts are excessive. Ongoing efforts are targeting real-time tracking and compensation of tissue movement.

5 Conclusion

In conclusion, AODs offer a viable method for random-access scanning in both two and three dimensions. When combined with multiphoton microscopy principles, the resulting imaging system has the unique ability to selectively record from multiple noncontiguous sites of interest in brain tissue. As a result, random-access multiphoton (RAMP) microscopy has been utilized with fluorescent optical indicators to study neuronal function with increased temporal resolution and S/N. Combined with the new advances in two-photon voltage imaging described in another chapter in this volume, RAMP microscopy can be the solution best suited for resolving signals from voltage-sensitive dyes *in vivo*.

Ongoing developments in both acousto-optic devices and ultrafast optics offer ways to further improve the usefulness of RAMP microscopy. These include new deflectors with higher acceptance angles (Xu et al. 1996) that offer the possibility of extending the axial range of 3D RAMP microscopes. One microscope has already been demonstrated that expands the field of view to a tissue volume of 700 μm × 700 μm × 1400 μm, although the axial point spread function extends to 8 μm when imaging further from the reference plane (Katona et al. 2012). Also new dispersion compensating optical fibers could allow straightforward compensation of the large amount of GVD introduced by multi-AOD scanners (Mogilevtsev et al. 1998). Finally, techniques for spherical aberration correction (Escobar et al. 2006) offer the possibility of improving the overall spatial resolution for the system.

So far, random-access multiphoton microscopes have been shown to be uniquely suited for studying single neuron function in brain slices and recording the simultaneous activity of larger *in vivo* populations than is possible with other methods. Thus, together with advanced voltage- and ion-sensitive indicators, fast 3D random-access microscopes promise to become important tools in the arsenal of the experimental neurophysiologist.

Acknowledgements The development of AOD-based fast 3D random-access multi-photon microscopy is the result of the continuing efforts of many members and affiliates of the Saggau and Tolias labs. These efforts were supported by numerous grants from NIH and NSF.

References

Bansal V, Patel S, Saggau P (2006) High-speed addressable confocal microscopy for functional imaging of cellular activity. J Biomed Opt 11:34003

Bi K, Zeng S et al (2006) Position of the prism in a dispersion-compensated acousto-optic deflector for multiphoton imaging. Appl Opt 45:8560–8565

Bullen A, Patel SS, Saggau P (1997) High-speed, random-access fluorescence microscopy: I. High-resolution optical recording with voltage-sensitive dyes and ion indicators. Biophys J 73:477–491

Bullen A, Saggau P (1998) Indicators and optical configuration for simultaneous high-resolution recording of membrane potential and intracellular calcium using laser scanning microscopy. Pflugers Arch 436:788–796

Cotton RJ, Froudarakis E, Storer P, Saggau P, Tolias AS (2013) Three-dimensional mapping of microcircuit correlation structure. Front Neural Circ 7(151)

Escobar I, Saavedra G et al (2006) Reduction of the spherical aberration effect in high-numerical-aperture optical scanning instruments. J Opt Soc Am A Opt Image Sci Vis 23:3150–3155

Fork RL, Martinez OE, Gordon JP (1984) Negative dispersion using pairs of prisms. Opt Lett 9:150–152

Froudarakis E, Berens P, Ecker AS, Cotton RJ, Sinz FH, Yatsenko D et al (2014) Population code in mouse V1 facilitates readout of natural scenes through increased sparseness. Nat Neurosci 17(6):851–857

Gobel W, Kampa BM, Helmchen F (2007) Imaging cellular network dynamics in three dimensions using fast 3D laser scanning. Nat Methods 4:73–79

Greenberg DS, Kerr JN (2009) Automated correction of fast motion artifacts for two-photon imaging of awake animals. J Neurosci Methods 176(1):1–15

Iyer V, Hoogland TM, Saggau P (2006) Fast functional imaging of single neurons using random-access multiphoton (RAMP) microscopy. J Neurophysiol 95:535–545

Iyer V, Losavio BE, Saggau P (2003) Compensation of spatial and temporal dispersion for acousto-optic multiphoton laser-scanning microscopy. J Biomed Opt 8:460–471

Katona G, Szalay G, Maák P, Kaszás A, Veress M, Hillier D et al (2012) Fast two-photon in vivo imaging with three-dimensional random-access scanning in large tissue volumes. Nat Methods 9(2):201–208

Lechleiter JD, Lin DT, Sieneart I (2002) Multi-photon laser scanning microscopy using an acoustic optical deflector. Biophys J 83:2292–2299

Lv XH, Zhan C et al (2006) Construction of multiphoton laser scanning microscope based on dual-axis acousto-optic deflector. Rev Sci Instrum 77:046101

Mogilevtsev D, Birks TA, Russell PS (1998) Group-velocity dispersion in photonic crystal fibers. Opt Lett 23:1662–1664

Nakazawa M, Nakashima T, Kubota H (1988) Optical pulse-compression using a Teo2 acoustooptical light deflector. Opt Lett 13:120–122

Pawley J (1995) Handbook of biological confocal microscopy, 2nd edn. Plenum, New York

Reddy GD, Kelleher K et al (2008) Three-dimensional random access multiphoton microscopy for functional imaging of neuronal activity. Nat Neurosci 11:713–720

Reddy GD, Saggau P (2005) Fast three-dimensional laser scanning scheme using acousto-optic deflectors. J Biomed Opt 10:064038

Saggau P, Bullen A, Patel SS (1998) Acousto-optic random-access laser scanning microscopy: fundamentals and applications to optical recording of neuronal activity. Cell Mol Biol 44:827–846

Salome R, Kremer Y et al (2006) Ultrafast random-access scanning in two-photon microscopy using acousto-optic deflectors. J Neurosci Methods 154:161–174

Steinmeyer G (2006) Femtosecond dispersion compensation with multilayer coatings: toward the optical octave. Appl Opt 45:1484–1490

Treacy EB (1969) Optical pulse compression with diffraction gratings. IEEE J Quant Electron 5:454–458

Vucinic D, Sejnowski TJ (2007) A compact multiphoton 3D imaging system for recording fast neuronal activity. PLoS One 2, e699

Weber MJ (1986) Handbook of laser science and technology. CRC, Boca Raton, FL

Xu J, Weverka RT, Wagner KH (1996) Wide-angular-aperture acousto-optic devices. Proc SPIE 2754:104–114

Zeng S, Lv X et al (2006) Simultaneous compensation for spatial and temporal dispersion of acousto-optical deflectors for two-dimensional scanning with a single prism. Opt Lett 31:1091–1093

Chapter 19
Second Harmonic Imaging of Membrane Potential

Leslie M. Loew and Aaron Lewis

Abstract The non-linear optical effect known as second harmonic generation (SHG) has been recognized since the earliest days of the laser. But it has only been in the last 20 years that it has begun to emerge as a viable microscope imaging contrast mechanism for visualization of cell and tissue structure and function. This is because only small modifications are required to equip a standard laser scanning 2-photon microscope for second harmonic imaging microscopy (SHIM). SHG signals from certain membrane-bound dyes are highly sensitive to membrane potential, indicating that SHIM may become a valuable probe of cell physiology. However, for the current generation of dyes and microscopes, the small signal size limits the number of photons that can be collected during the course of a fast action potential. Better dyes and optimized microscope optics could ultimately lead to the ability to image neuronal electrical activity with SHIM.

Keywords Fluorescence • Non-linear optics • Membrane potential • Voltage sensor • Electrochromism • Chemistry • Optics • 2-Photon • Microscopy • Imaging • SHG

1 Introduction and Overview

SHG, also called "frequency doubling", is a non-linear optical process that has been known since the earliest days of laser physics. This phenomenon requires intense laser light passing through a highly polarizable material with a non-centrosymmetric molecular organization. The emerging light is at precisely half the wavelength (or twice the frequency) of the light that entered the material. The most common application of SHG in laser optics is to produce visible laser light from an infrared

L.M. Loew (✉)
Department of Cell Biology, R. D. Berlin Center for Cell Analysis and Modeling, University of Connecticut Health Center, Farmington, CT 06030-1507, USA
e-mail: les@volt.uchc.edu

A. Lewis
Department of Applied Physics, Hebrew University, Jerusalem, Israel
e-mail: LEWISU@vms.huji.ac.il

M. Canepari et al. (eds.), *Membrane Potential Imaging in the Nervous System and Heart*, Advances in Experimental Medicine and Biology 859,
DOI 10.1007/978-3-319-17641-3_19

laser; the SHG material for such an application is an inorganic crystal placed in the light path immediately following the primary laser. As in the case of two-photon excitation, the probability of producing SHG is proportional to the square of the incident light intensity. Therefore, when incorporated into a microscope, it has the same intrinsic optical sectioning attributes as two-photon excited fluorescence (2PF) microscopy.

The idea that 2PF and SHG might each be used for microscopy was first proposed by Sheppard (Sheppard et al. 1977). The first biological SHG microscopy experiments were by Freund et al. in 1986 (Freund et al. 1986). They utilized SHG to study the orientation of collagen fibers in rat tail tendon at approximately 50 μm resolution and showed the collagen fibers formed highly dipolar structures on this size scale. Our work focused from the start on the possibility that electrochromic membrane dyes could produce SHG. We first showed that the styryl dyes such as di-4-ANEPPS, which had been developed for voltage sensing via fluorescence, could produce strong SHG when arrayed in a monolayer (Huang et al. 1988). We then measured the second harmonic response of styryl dyes to membrane potential on a model membrane system (Bouevitch et al. 1993) and demonstrated the possibility of imaging live cells by SHG (Ben-Oren et al. 1996). In all this earlier work, stage scanning with a picosecond laser source focused through the microscope was used and frame rates of minutes to hours were required. The application of mode-locked lasers to 2PF microscopy was demonstrated in 1990 by Denk, Strickler and Webb (Denk et al. 1990), making non-linear optical microscopy a practical tool for cell biology and physiology. We were able to apply scanning of mode-locked lasers to produce the first high resolution SHG microscopy images of cells acquired with frame rates of seconds (Campagnola et al. 1999). The use of SHIM to investigate membrane potential is the primary focus of this chapter, but it should also be noted that the use of SHIM to investigate collagen and other endogenous biological assemblies has mushroomed in the last 10 years and is widely being investigated for clinical diagnostic applications (Bianchini and Diaspro 2008; Campagnola and Loew 2003; Mohler et al. 2003).

Since SHG is a non-linear optical phenomenon, it shares many of the features of 2PF microscopy. But it is critical to appreciate the differences. While 2PF involves the absorption of two photons to excite a fluorophore, followed by relaxation and fluorescence emission, SHG is a process in which two photons are directly converted into a single photon of twice the energy. SHG light retains the coherence of the fundamental laser beam, while the fluorescence from 2PF is no longer coherent. Furthermore, SHG should be considered a material property rather than a molecular property. It requires a distribution of "harmonophores" that lacks a center of symmetry; this constraint is readily satisfied at cellular membranes in which SHG-active constituents are unevenly distributed between the two leaflets of the lipid bilayer. However, dyes bound to highly convoluted membranous structures such as the endoplasmic reticulum do not produce SHG; this is because the orientations of the dye molecules are random on the spatial scale of the wavelength of light, thus violating the requirement for a noncentrosymmetric distribution of harmonophores. Finally, a key difference is that fluorescence intensity is linearly related to the concentration of fluorophores, while SHG is quadratically dependent on the density of harmonophores.

SHG does not involve excitation of molecules into an excited state. Thus, it might be expected that photobleaching and phototoxicity of live specimens should be eliminated. However, there can be collateral damage if the incident laser light also produces 2-photon excitation of chromophores in the specimen. Electrochromic membrane dyes can be used to produce SHG that is sensitive to membrane potential. This promises to enable new optical approaches to mapping electrical activity in complex neuronal systems. Excitation uses near infrared wavelengths, allowing excellent depth penetration, and hence this method is well-suited for studying thick tissue samples such as brain slices or in vivo brain imaging. Additionally, 2PF images can be collected in a separate data channel simultaneously with SHG, permitting the simultaneous imaging of membrane potential and the distribution of fluorescent labels. Ratiometric SHG:2PF techniques allow for the quantitation of chiral-enhancement and voltage-sensitivity, while normalising out irrelevant parameters arising from laser fluctuations, sample movement and non-uniform staining (Campagnola et al. 1999).However, a major challenge that will be discussed is the small flux of SHG photons from the current generation of membrane staining voltage-sensitive dyes.

The rest of this chapter will be devoted to more detailed discussions of many of these points. We will also provide guidance on how to set up an SHIM experiment and review some of the attempts to use SHIM for voltage imaging.

2 The Photophysics of SHG

An energy diagram (also known as a Jablonski diagram) will help us to understand what happens to the incident light during the process of fluorescence and SHG. Figure 19.1 contrasts the energetics of 1 and 2 photon fluorescence and SHG. For one photon fluorescence, the energy of the photon (blue arrow) matches the gap between the ground state (GS) and a sublevel within the excited state (ES), which is composed of a manifold of electronic and vibrational states. This initial excited state rapidly relaxes to the lowest energy state of the ES manifold, losing some energy to heat in the process. The ES has an average lifetime of Δt (on the order of ns) before relaxing back to the GS by emitting a fluorescent photon. Because there is always time for relaxation of the excited state, usually involving vibrational relaxation and solvent reorganization, the wavelength of the emitted fluorescence is always longer (i.e. lower energy) than the absorbed photon. For 2PF (center of Fig. 19.1), the wavelength of the exciting light is approximately twice that of the corresponding 1 photon absorption process; under these circumstances 2 photons (represented by two red arrows in the figure) are required to deliver sufficient energy for excitation to the ES manifold; this has a much lower probability and therefore requires light of much higher intensity (i.e. photon flux) than 1 photon absorption. Because 2 photons are involved, the probability of absorption is

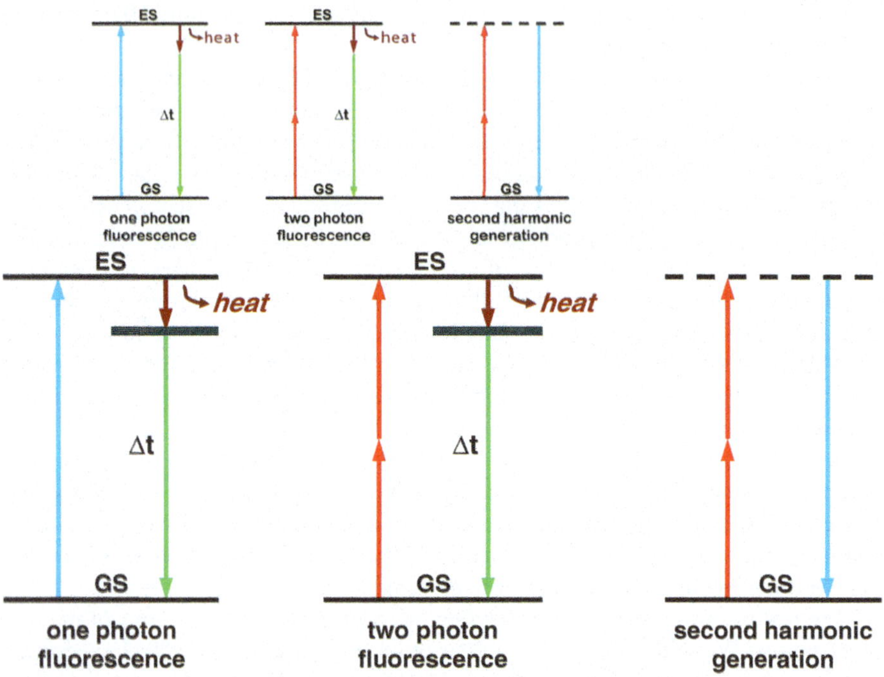

Fig. 19.1 Jablonski diagrams for 1 photon fluorescence, 2PF and SHG. GS = ground state; ES = excited state; Δt = excited state lifetime

proportional to the square of the incident light intensity. But once the molecule reaches the ES, it does not retain a memory of where it came from and it behaves exactly as described for 1 photon excitation. SHG, shown on the right of Fig. 19.1, does not involve an excited state and is an instantaneous transformation of 2 photons (the red arrows) into a single photon with precisely twice the energy (or half the wavelength), represented by the blue arrow pointing downwards. Importantly, there is no excited state involved in SHG and the light that is produced retains the coherence of the incident fundamental laser light.

To explore the origin of SHG requires consideration of how the light interacts with the non-linear optical material. A qualitative approach will provide a physical understanding of the process and why it is fundamentally so different from fluorescence. Much as when a guitar string is plucked too hard, if an intense pulse of light interacts with a material, it can interact with the electrons in that material and emerge with additional harmonics. The ability of the oscillating electric field in the light wave to interact with the material depends on its polarisability. We can understand how this might produce harmonics of the original fundamental frequency by referring to Fig. 19.2. A high amplitude electromagnetic wave impinges on an array of molecules that are oriented with their electron clouds aligned so as to reinforce the light wave for the positive component of the oscillating electric filed and oppose the negative electric field component. The emerging electromagnetic

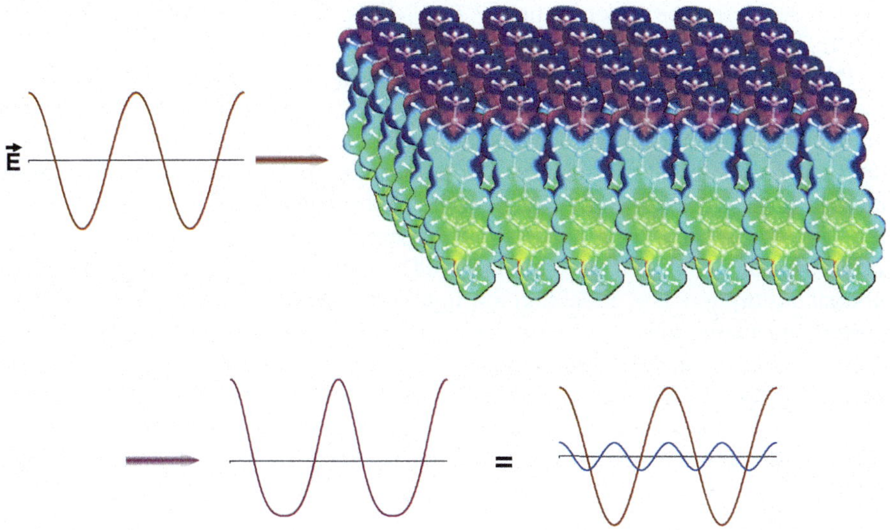

Fig. 19.2 Qualitative picture of the optical physics underlying SHG. The oscillating electric field of the intense fundamental light beam (*red* sine wave on the *top left*) impinges on an ordered array of polarizable molecules. The interaction of the oscillating wave with the material induces a distortion of the emergent light wave (*violet*, *bottom left*), that is equivalent to the mixing of the original fundamental frequency with a small contribution from twice the frequency, i.e. the second harmonic (shown in *blue* on the *bottom right*). The molecular array shown in the *upper right* was created by 2D translations of an electron density map of the aminonaphthylethenylpyridinium (ANEP) chromophore. The electron density map was rendered in *colors* from the output of molecular orbital calculations

wave is correspondingly distorted and can be decomposed into components that correspond to the fundamental frequency plus a component at double that frequency. It is apparent from this simple picture that SHG depends on the collective properties of the array; this distinguishes it from absorbance and fluorescence, which depend on the interaction of photons with individual molecules. Reference to Fig. 19.2 makes it easy to understand why a fully centrosymmetric array of molecules or randomly oriented molecules in a homogeneous solution cannot produce SHG: it is because the electric field vector of the fundamental light wave would not be differentially distorted, as it is in Fig. 19.2 for an ordered array. One can also appreciate from this picture that the SHG will depend strongly on the polarization of the incoming light and its angle of incidence with respect to the array of harmonophores in the non-linear optical material—i.e. tensor relationships. Thus while this simple picture cannot fully describe the details of SHG, it nicely explains many qualitative features of the phenomenon.

With this qualitative picture in hand, it is worthwhile to consider the photophysics at a more rigorous level in order to appreciate some of the more subtle features of SHG and how it differs from 2PF. Such a treatment necessarily begins with the general expression for polarisability, which can be expressed as a

Taylor series in the electric field: $P = \chi^{(1)} \otimes E + \chi^{(2)} \otimes E \otimes E + \chi^{(3)} \otimes E \otimes E \otimes E + \ldots$

P is the induced polarisation vector, E represents the vector electric field, $\chi^{(i)}$ is the ith-order non-linear susceptibility tensor and $\otimes$ represents a combined tensor product and integral over frequencies. It is again important to emphasize that the $\chi^{(i)}$ are bulk properties of the material. However, they parallel the interaction of an electric field with molecules based on the molecular properties, respectively, of dipole moment, and polarizability and hyperpolarisabilities (Cole and Kreiling 2002; Moreaux et al. 2000a; Williams 1984). The first term in the series describes normal absorption and reflection of light. The second- and third-order terms correspond to the phenomena of SHG and 2PF and their sensitivity to electric field. To begin to appreciate how SHG might directly depend on the electric field associated with the membrane potential in addition to the oscillating electric field of the light, one can decompose the above equation into a combination of a true second-order process and a third-order process dependent on a constant electric field:

$$P = \chi^{(2)} \otimes E(\omega) \otimes E(\omega) + \chi^{(3)} \otimes E \otimes E(\omega) \otimes E(\omega) \tag{19.1}$$

$$P = \chi^{(2)}{}_{\text{eff}}(E) \otimes E(\omega) \otimes E(\omega) \tag{19.2}$$

Here, $\chi^{(2)}{}_{\text{eff}}(E)$ is an effective second-order non-linear susceptibility (Conboy and Richmond 1997; Yan et al. 1998) that depends on a static electric field vector, E. This term could impart a membrane potential sensitivity for harmonophores embedded on one side of a cell membrane.

The SHG intensity, I(2ν), is related to the square of the second order non-linear susceptibility, $\chi^{(2)}$ (or $\chi^{(2)}{}_{\text{eff}}(E)$):

$$I(2\nu) \propto \frac{\left[\chi^{(2)} I(\nu)\right]^2}{\tau} \tag{19.3}$$

where $I(\nu)$ is the laser pulse energy, ν is the fundamental frequency and 2ν is the second harmonic frequency. As in 2PF, the signal has a quadratic dependence on laser intensity and an inverse dependence on pulsewidth, τ, but since SHG is an instantaneous process, signal will only be generated during the duration of the laser pulse.

We now turn to a consideration of the molecular origin of SHG. $\chi^{(2)}$ is related to the molecular property β, the first hyperpolarizability according to:

$$\chi^{(2)} = N_s < \beta > \tag{19.4}$$

where N_s is the density of molecules and the brackets denotes an orientational average. This further underscores the need for a non-centrosymmetric region, since <ß> would vanish for an isotropic distribution. Examination of Eqs. (19.3) and

(19.4) also reveals a major difference in the expected contrast for the 2PF and SHG signal levels: SHG depends on the square of the surface density whereas fluorescence intensity is linear with the density of fluorophores. This can lead to significant differences in these two mechanisms even when signal arises from the same dye in the membrane. For example, at high dye concentrations the fluorescence signal will become quenched due to dye aggregation and subsequent non-radiative decay, while the SHG would actually be larger due to the square dependency on the surface density. SHG can be resonance-enhanced when the second harmonic transition overlaps with an electronic absorption band. The resonance-enhanced molecular hyperpolarizability (Dirk et al. 1986; Morley 1988; Nicoud and Twieg 1987), which determines the resonance contribution, is given by:

$$\beta = \frac{4\mu_{eg}^2\left(\mu_e - \mu_g\right)}{3h^2\left(\nu_{eg}^2 - \nu^2\right)\left(\nu_{eg}^2 - 4\nu^2\right)} \tag{19.5}$$

Where h is Planck's constant, μ_g and μ_e are the dipole moments of the ground and excited states, respectively, μ_{eg} is the transition moment, ν is the frequency of the incident light, and $h\nu_{eg}$ is the energy for the electronic transition from the ground to the excited state. For the very broad absorption bands of organic dyes in solution, the resonance condition where 2ν overlaps ν_{eg} will pertain over a large range of wavelengths. Other consequences of this simple expression are that β is enhanced when there is a large difference in the ground and excited state electron distribution, i.e. a large $\mu_e - \mu_g$, and when the dye has a large extinction coefficient (which is proportional to μ_{eg}, the transition dipole). These are precisely the features that have been behind the design of our electrochromic styryl dyes and this is what led us to initially explore the possibility that they may produce large SHG signals This relationship between SHG and electrochromism manifest in Eqs. (19.2) and (19.5) was also the original motivation for us to explore whether SHG from membranes stained with our dyes could be sensitive to membrane potential.

3 SHG Response to Membrane Potential

Our first foray into the non-linear optical properties of our dyes, involved creating a monolayer of di-4-ANEPPS in a Langmuir-Blodgett trough (Huang et al. 1988). In a series of papers that culminated in the first high resolution SHG images (Ben-Oren et al. 1996; Bouevitch et al. 1993; Campagnola et al. 1999), we were able to show that the SHG signal from ANEP dye stained membranes is sensitive to membrane potential. These experiments were continued (Campagnola et al. 2001; Clark et al. 2000; Millard et al. 2003a, b, 2004, 2005a, b; Teisseyre et al. 2007; Yan et al. 2006) and were validated and extended by the laboratories of Jerome Mertz, Rafael Yuste, Kenneth Eisenthal and Watt Webb (Araya et al. 2006, 2007;

Dombeck et al. 2004, 2005; Jiang et al. 2007; Moreaux et al. 2000a, b, 2003; Nuriya et al. 2006; Pons et al. 2003).

Table 19.1, taken from a published paper (Teisseyre et al. 2007), summarizes results from a series of very promising new probes for both fluorescence and SHG. The absorbance and emission maxima in this Table are for dyes bound to lipid vesicles. The first three dyes in the Table are noteworthy because of the huge fluorescence sensitivities observed for them on the voltage clamped hemispherical bilayer (column 4; see Chap. 2 in this volume). Versions of these chromophores have been extraordinarily successful in fluorescence optical recording from brain slice and heart studies. A completely new class of hemicyanines employing an amino-thiophene donor moiety are the "PY" probes (for their inventor, Ping Yan) in the Table. We had employed thiophenes as linker moieties for many years, but this is the first set of hemicyanine chromophores that we are aware of where amino-thiophenes serve as the electron donor. The rationale for this donor moiety is that the thiophene behaves more like a rigidified diene than an aromatic heterocycle; thus through conjugation of the unshared electrons on the amino nitrogen is more effective than for the aniline and amino-naphthalene donor moieties in the earlier generation of dyes. Indeed, the spectra of these dyes are much further red shifted than comparable dyes employing aniline or amino-naphthalene donors. We also felt that these dyes should be particularly suitable for SHG imaging because the thiophenes are highly polarizable; additionally there should be a large change in dipole moment in going from ground to excited state for resonance enhanced SHG (Eq. 19.5). All of the dyes have absorbance maxima in the range of 530–560 nm in lipid membranes. This makes them ideal for SHG with our 1,064 nm femtosecond fiber laser and these dyes do indeed show strong SHG and the best SHG sensitivity to potential that we know of (PY-1278 is twice as good as FM-4-64). Also very striking is the 2-photon result shown for PY-1268 of 16.6 %/50 mV. It should be noted that all the results reported in columns 6 and 7 were for excitation with the 1,064 nm laser and emission at the 2 wavelength ranges indicated; no attempt was made to find optimal wavelengths for the 2PF sensitivity to potential, so most of these numbers are likely to be improvable. Indeed the 1-photon wavelength optima are generally quite different. A full account of the synthesis of this novel class of dyes has been published (Yan et al. 2008). A major problem that will need to be addressed for these dyes before they can be truly useful VSDs is their poor photostability (Zhou et al. 2007).

The labs of Rafael Yuste and Ken Eisenthal at Columbia U. have collaborated to perform SHG measurements of FM-4-64 applied to single neocortical pyramidal neurons through a patch pipette. FM-4-64 is a hemicyanine dye similar to our first generation styryl dyes in that it contains an aniline donor moiety (rather than an aminonaphthalene) conjugated via a trienyl linkage to a pyridinium acceptor. It has found extensive utility as a probe for synaptic vesicle cycling because it can be applied to preparations from the outside and readily becomes internalized in the vesicles. When applied from the inside, as in the Columbia experiments, the dye remains in the neuron because of its doubly positive charge; i.e. the negative resting potential of the cell will hold this relatively water soluble dye in the interior of the

Table 19.1 Linear and non-linear responses of Gen3 dyes

Dye	1PF Abs. Max (nm)	1PF Em. Max (nm)	1PF Change (ΔF/F – 100 mv); Ex/Em (nm)	SHG Change (%50 mV)	2PF Change (%, 615 – 665 nm/50 mV)	2PF Change (%, 750 – 850 nm/50 mV)	SHG Kinetics (ms)	Structure
JPW-6008 di-4-ANEQPQ	528	670	1.2E–001; 610/>665	3.84 ± 0.38	0.79 ± 0.37	5.03 ± 0.26	76 ± 13	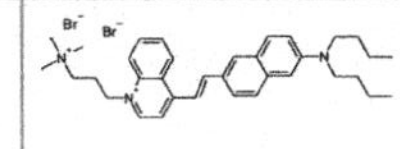
JPW-6003 di-4-ANBDQPQ	539	708	2.0E–001; 618/>715	3.55 ± 0.55	–6.61 ± 0.88	4.32 ± 0.57	<5	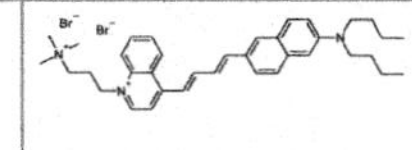
JPW-6027 di-4-ANHTQPQ	504	655	5.4E–002; 630/>715	4.15 ± 0.28	–7.51 ± 0.73	2.83 ± 0.45	34 ± 3	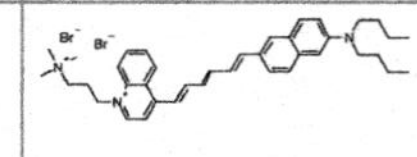
PY-1261	547	686	5.0E–002; 625/>715	9.56 ± 0.42	–2.57 ± –0.24	5.22 ± 0.28	<5	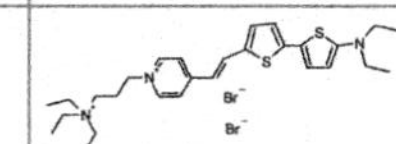
PY-1266	552	694	1.0E–001; 635/>715	3.75 ± 0.33	–4.80 ± 0.53	2.25 ± 0.47	23 ± 2	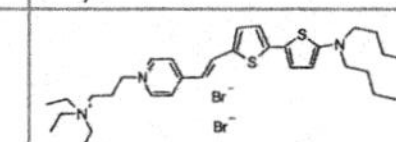
PY-1268	535	714	4.2E–002; 640/>715	1.82 ± 0.18	–16.64 ± 0.85	–4.28 ± 0.54	<5	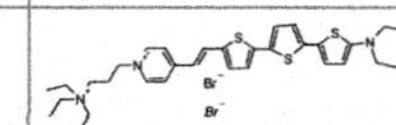
PY-1274	551	691	5.0E–002; 620/>715	5.65 ± 0.45	–4.20 ± 0.13	3.12 ± 0.18	–	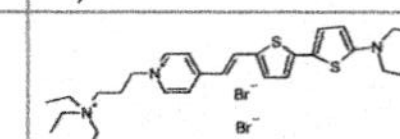
PY-1278	555	688	7.9E–002; 640/>715	10.70 ± 0.59	–3.86 ± 0.53	6.52 ± 0.76	<5	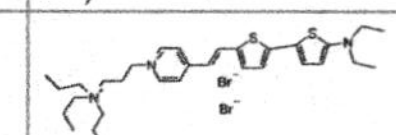
PY-1280	547	690	1.1E–001; 640/>715	10.13 ± 0.88	–6.12 ± 0.87	5.77 ± 1.22	109 ± 30	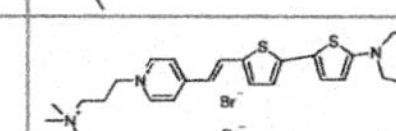
PY-1282	530	676	8.0E–002; 640/>715	11,64 ± 0.60	0.29 ± 0.39	8.78 ± 1.08	16 ± 1	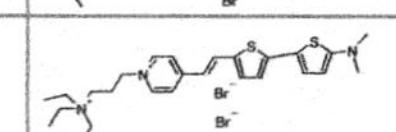

(continued)

Table 19.1 (continued)

Dye	1PF Abs. Max (nm)	1PF Em. Max (nm)	1PF Change (ΔF/F – 100 mv); Ex/Em (nm)	SHG Change (%50 mV)	2PF Change (%, 615 – 665 nm/ 50 mV)	2PF Change (%, 750 – 850 nm/ 50 mV)	SHG Kinetics (ms)	Structure
PY-1284	536	588	1.5E–002; 555/>610	6.18 ± 0.68	6.72 ± 0.58	–	–	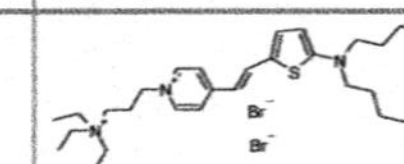
PY-1286	547	692	1.8E–001; 640/>715	8.09 ± 0.42	–5.65 ± 0.50	3.51 ± 0.66	27 ± 1	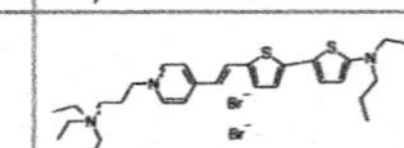

dendrite, where vesicle cycling is presumably non-existent. The Columbia group has published two papers in which they used SHG to measure electrical activity in individual spines (Araya et al. 2006; Nuriya et al. 2006) and they have used this and less direct experiments (Araya et al. 2007) to argue that the spine is indeed an electrical compartment.

Gaining an understanding of the mechanism(s) by which the SHG intensity responds to membrane potential changes will, of course, allow us to design better SHG-based voltage sensors to fully exploit this new modality for optical electrophysiology. The theory of SHG, which was outlined in the previous section, provides a basis for a purely electronic mechanism by which SHG could respond to electric field. Specifically, Eqs. (19.1) and (19.2) show how a DC electric field can cause mixing of a $\chi^{(3)}$ contribution into the second order susceptibility that is responsible for SHG; Eq. (19.5) suggests that resonance enhancement of SHG depends on a large charge redistribution between the ground and excited states; an electric field should produce a second order effect on this charge redistribution. Such a purely electronic mechanism would necessarily be instantaneous on the timescale of an action potential. Although there has been some investigation of the mechanism by which SHG from the styryl dyes is sensitive to membrane potential that suggests a purely electronic mechanism (Campagnola and Loew 2003; Jiang et al. 2007; Pons et al. 2003), kinetic data, showing non-instantaneous responses, are inconsistent with this explanation (Millard et al. 2005a; Teisseyre et al. 2007).

We have shown that several of our dyes show large potential dependent changes in SHG (as high as 38 %/100 mV for di-4-ANEPPS), with some instantaneous responders, within the limits of our 5 ms time resolution, but others with response times in the tens of milliseconds (Millard et al. 2005a; Teisseyre et al. 2007). A non-instantaneous response precludes a purely electronic mechanism and suggests some form of conformational change or reorientation is involved in the mechanism of the SHG response. What is particularly curious is that in 2PF mode, which we also measured, all of the dyes showed rapid responses (consistent with the putative electrochromic mechanism for the 1PF voltage sensitivity). However in some of the kinetic traces there is a small slow component to the 2PF response that may have the same time course as the slow SHG response; Fig. 19.3, taken from Millard et al. (2005), shows a pair of kinetic traces for di-4-ANEPPS that is the best example of this behavior. Thus, it would appear that whatever molecular motion underlies the slow SHG response does not strongly affect the 2PF. We also showed that the dye responses to potential are linear and have a shallow wavelength dependence that peaks around the optimal second harmonic resonance with the 1 photon absorbance spectrum (Millard et al. 2003b, 2004, 2005a). The Mertz lab showed that a styryl quinolinium dye, di-6-APEQBS (our lab calls the same dye di-6-ASQBS (Fluhler et al. 1985)), displays a mixture of a relatively weak (3 %/100 mV) electronic response and a stronger (−12 %/100 mV) reorientation response (Pons et al. 2003). They dissected these mechanisms by analyzing the polarization dependence of the voltage response. Interestingly, both of these responses were linear and fast, but in opposite directions. The Eiesenthal-Yuste collaboration at Columbia U. recently looked at the kinetics and polarization

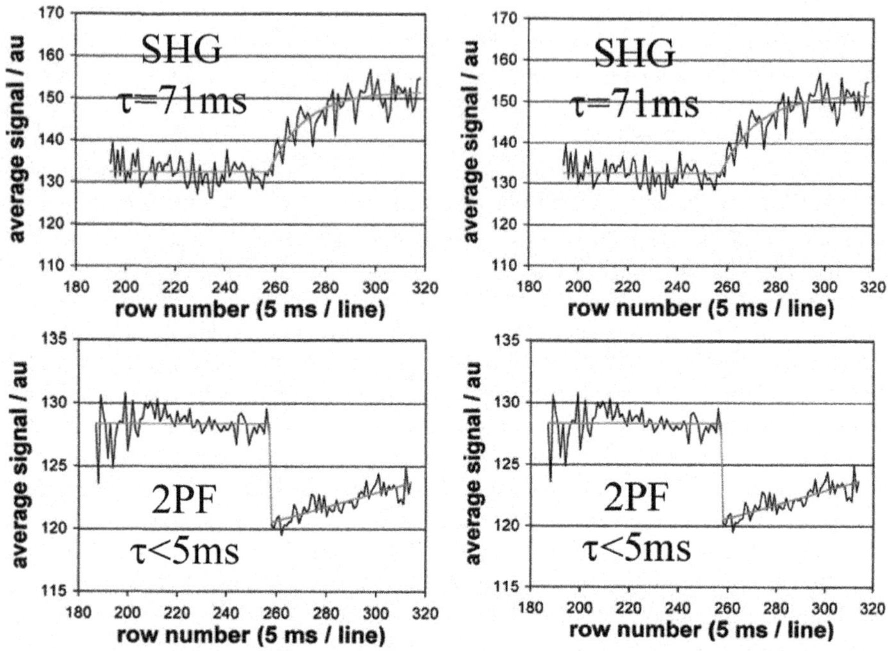

Fig. 19.3 Kinetic response to a 50 mV step of non-linear optical signals on patch clamped di-4-ANEPPS stained neuroblastoma cells. The voltage is changed at the midpoint of the trace. The 2PF signal shows an immediate decrease in fluorescence followed by a slow increase. See Millard et al. (2005), for details. Reprinted with Permission of the Journal of Membrane Biology

dependence of the 14 %/100 mV FM-4-64 SHG response (Jiang et al. 2007). They reported fast kinetics, linear voltage dependence and no change in polarization following a voltage step. By process of elimination, they concluded that the mechanism was purely electronic. However, a recent study of FM-4-64 in HEK cells (Theer et al. 2011) showed similar voltage sensitivity but a slow voltage response kinetics. The discrepancy in the kinetics between these results may be because different cell types were used, but it is clear that the voltage response mechanism of FM-4-64 can, at least in part, involve a change in its association with the plasma membrane.

It is important to emphasize that a mechanism that involves a rapid redistribution of harmonophores from a random and therefore SHG-silent population, to a well ordered non-centrosymmetric array within the membrane could produce an increase in SHG intensity without any change in its polarization. This will be the result if both states are populated both before and after the voltage change, with the voltage change effecting a change in the equilibrium between them. Based on the experiments carried out by us and by the other labs, this mechanism would be indistinguishable from what has been termed "electronic" or "electrooptic". Our hypothesis is that the mechanism that leads to the slow voltage sensitive responses could also pertain to the SHG probes that have faster kinetics. All that is required to produce the faster kinetics is that the rearrangement rate be increased by about a factor of 100 from the dye with the slowest response (PY1280, Table 19.1). In terms of the free energy barrier, this corresponds to a rather small difference of 2.8Kcal/ mol. In fact, comparing PY-1280, with a SHG time constant of 110 ms, to PY-1261, which has a time constant of <5 ms, the only difference in the structure is a trimethyammonium vs. a triethylammonium headgroup. *Despite the difference in kinetics, these 2 dyes have virtually identical SHG sensitivities to potential (20 %/ 100 mV)!* Clearly the dye with the slow kinetics cannot be using an "electoroptic" mechanism and it would be surprising if the change from methyl to ethyl substituents could switch the mechanism. Our ability to synthesize many of these dyes and then compare their voltage-dependent non-linear optical properties (e.g. Fig. 19.4 and other similar comparisons in Teisseyre et al. (2007)), will be our most important tool in uncovering the mechanism(s) of the voltage sensitivity.

Happily, as can be seen from Table 19.1, we discovered that some of the more sensitive SHG dyes also displayed fast kinetics. These fast SHG dyes are also more sensitive to potential than FM-4-64. Not included in Table 19.1 is the kinetic response of the dyes in 2PF mode; interestingly, all the dyes responded with fast kinetics (i.e. faster than the 5 ms resolution of our measurement) and some of the 2PF responses are truly huge. Considering that we have not attempted to find optimal wavelengths for these dyes, using 1,064 nm excitation and just two fixed choices of emission filters, it is likely that these dyes will produce even better sensitivities than those listed in Table 19.1. But there are some new problems as already discussed regarding the poor photostability of some of the PY dyes (Zhou et al. 2007). Efforts to synthesize and characterize new SHG dyes will ultimately lead to an understanding of the mechanism of the dye response to potential and allow us to design optimal potentiometric indicators. This approach is illustrated by several systematic structure-property relations reported in Teisseyre et al. (2007), and exemplified in Fig. 19.4.

We now consider a class of dyes other than the styryl dyes discussed above. These dyes are based on retinal, the chromophore associated with rhodopsin found in visual systems and in bacterial rhodopsins where a retinal chromophore is bound in the membrane embedded protein. In fact studies of resonant second harmonic generation in membranes began with the retinal protein bacteriorhodopsin (Huang and Lewis 1989). However, even though all trans retinal was used in a study by the group of Yuste in Columbia (Nemet et al. 2004), there has been little effort to investigate this broad class of chromophores in terms of their second harmonic membrane voltage sensitivity.

Therefore, this was the objective of a recent study (Theer et al. 2011). That compared all trans retinal and several synthetic retinals to the styryl dye FM4-64, discussed above. This study found that the retinal chromophores had a large sensitivity to membrane potential changes, which were also quantified in terms of response kinetics. The voltage sensitivity, for example, of all trans retinal was over twice that of FM4-64. Moreover, as part of the analysis it was possible to provide information about the probe's location and orientation inside the membrane.

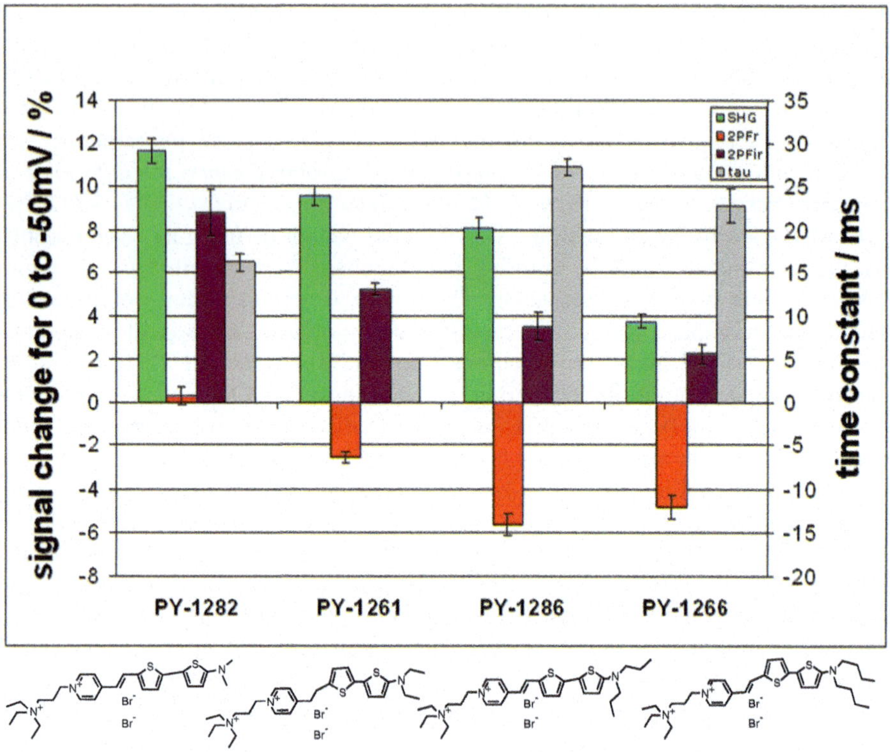

Fig. 19.4 Comparison of the effect of lipophilic tale length on the non-linear optical response of a series of amino-thiophene dyes. The incident light is from a 1,064 nm fs laser. The SHG in *green* was recorded at 532 nm. 2PF was recorded at emission ranges of 615–665 nm (*red*) and 750–850 nm (*violet*). The *right axis* refers to the SHG response time constant (*gray*). From Teisseyre et al. (2007). Reprinted with the permission of Journal of Biomedical Optics

The theoretical framework used in this paper is similar to that presented above in relation to Eqs. (19.1) and (19.2); it should be important for a full comparative analysis of other dyes employed for SHG studies of membrane potential sensitivity and provides a rational basis for fine tuning molecular structure to improve probe performance in the future. Specifically, they considered the case when $\chi_3 E_m$ is not small relative to χ_2, as would be expected for a voltage drop across a 7.5 nm membrane. Under this condition these workers showed that a parabolic expression is obtained for the change in second harmonic signal versus the voltage. This is important since χ_3 normally changes more rapidly with frequency than χ_2 and thus, in order to perform a good comparison of data in this field, the effect of frequency on the signal has to be evaluated in the resonance case especially when the signal change is larger than 10 %. The effectiveness of this analysis was shown for the retinal dyes in terms of their second harmonic sensitivity as a function of voltage. In the case of all trans retinal a full analysis was possible due to the fact that the first and second order non-linear molecular hyperpolarizabilities, β and γ are known.

Therefore, a best fit based on the derived equation could be affected by adjusting the surface (E_s), dipole (E_d) and transmembrane potential (E). This analysis yielded the probe's location in the membrane and the probe's orientation in the membrane could be deduced. It was shown that a best fit of the data was obtained with the β-ionone ring of the retinal embedded in the membrane with the saturated carbon chain oriented toward the outside of the membrane.

In addition, the study demonstrated for comparison purposes that these retinal dyes in the same cell culture environment had a much faster response than FM4-64. As noted above, even though Jiang et al. 2007 reported that FM4-64 responds to voltage changes on a submillisecond timescale in the cell types they used, the Theer et al. study measure a 1.6 ms time constant in their cells. When all dyes are compared in the same cell type the retinal dyes showed an instrumentally limited response (i.e. <100 μs) unlike FM4-64.

4 Measuring SHG Response to Membrane Potential

Imaging SHG requires a microscope equipped with a mode-locked laser excitation source similar to the setup used for a 2-photon microscope (Wilson et al. 2014). Laser scanning, as in confocal microscopy, permits rapid acquisition of high resolution images. Our original non-linear microscope allowed us to produce the first high resolution SHG images, for which we coined the acronym Second Harmonic Imaging Microscopy (SHIM) (Campagnola and Loew 2003). This was an inverted microscope that used a mode locked Ti-Sapphire laser pumped with an argon ion laser to produce 120 fs pulses in the range of 720–980 nm. We have added a fiber laser to this system that has a fixed wavelength of 1,064 nm, but is extremely robust and turnkey; we have also upgraded our laser system to use a solid state pump laser. The recent availability of computer controlled integrated solid state pumped Ti-Sapphire laser systems that allow automated choice of wavelength in the entire range from 680 nm to 1080 nm, makes the laser technology both more powerful and more accessible to experimentalists without significant optics training.

The primary difference between 2-photon laser scanning microscopy and SHIM is in the detection path. Indeed, it is quite straightforward to take advantage of this difference so as to acquire both 2PF and SHIM from the same specimen simultaneously. The light paths are shown in Fig. 19.5. The key point is that because the SHG signal is coherent and co-propagates with the fundamental laser light, it is optimally collected in the transmitted light path of the microscope. A good condenser is sufficient to capture the SHG transmission. On the other hand, 2PF emanating from the focus of the laser, propagates in all directions. It is typically and most conveniently collected in the backward direction with the same objective that was used to deliver the focused laser light. Clearly, the appropriate filters need to be employed in each light path to collect the desired wavelengths and reject the transmitted or reflected fundamental laser light (not shown in Fig. 19.5). As has

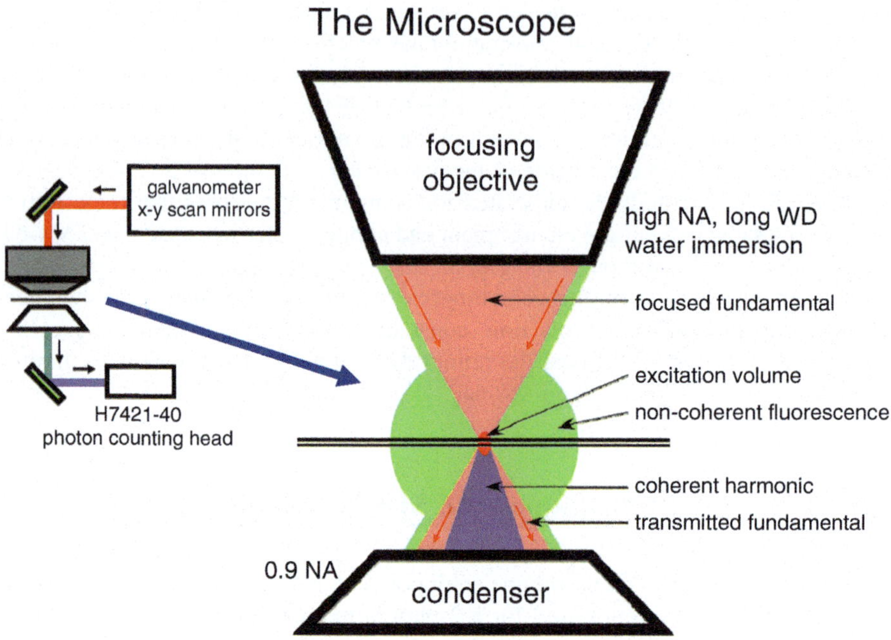

Fig. 19.5 Optical setup for non-linear optical microscopy. This arrangement is designed for simultaneous acquisition of 2PF through the backward (epi) light path and SHG through the forward (transmitted) light path. Raster scanned laser light (*red*) is focused onto the specimen with the objective and produces both 2PF (*green*) and SHG (*violet*) only at the focal plane. The 2PF emanates in all directions and a portion is collected by the objective in the backward direction, as is common in most fluorescence microscopes. For optical recording from brain slices, the objective should be water immersion and have a long working distance. It also should have good transmittance in the entire range from 700 to 1,100 nm. The SHG co-propagates with the fundamental laser light and both are collected by a high numerical aperture condenser. A short pass reflector combined with a band pass filter at half the wavelength of the fundamental laser beam is inserted before the detector to reject the fundamental and pass the SHG

been emphasized preciously, the SHG signal occurs at precisely ½ the wavelength of the fundamental; therefore the choice of filter will be dependent on the laser wavelength. This is not the case for 2PF, as the emission spectrum is independent of the exciting wavelength.

Depending on the needs of the experiment, dye can be applied either from the bathing medium or internally through the patch pipette. The former can be used to stain the entire preparation and the latter allows the experimentalist to target a single neuron. As has been discussed above, a non-centrosymmetric distribution of dye is required for SHG. Consequently if the stained membranes in a multicellular preparation are too close to each other, the SHG signal will be weak. For dyes applied to the outside, therefore, it would be important that they preferentially stain a subpopulation of cells. But a better application of SHG to membrane potential imaging might be for single cell measurements. Dye applied through a patch pipet will stain not just the plasma membranes, but also internal membranous organelles—most notably the endoplasmic reticulum—this produces large background signal when measuring fluorescence. Importantly, however, the SHG signal from the endoplasmic reticulum is very weak compared to that from the plasma membrane. This is because the convoluted structure of the endoplasmic reticulum effectively creates a centrosymmetric array of dye molecules. Thus, SHIM of cells stained internally primarily show plasma membrane intensity, while 2PF of the same dye in these cells display huge background from endoplasmic reticulum fluorescence. Indeed, the successful test cases in the Webb and Yuste labs were all measurements of single isolated cells (Dombeck et al. 2005; Nuriya et al. 2006; Sacconi et al. 2006). Most recently, several other labs have measured voltage-dependent SHG from aoxns in isolated neurons and from neurons in brain slice preparations (Nuriya and Yasui 2010; Rama et al. 2010) Fig. 19.6, taken from Dombeck et al. 2005, shows SHG from hippocampal neurons in a brain slice. In these experiments, FM4-64 is applied through a whole cell patch pipet. Very little intensity is observed from the interior of the cells, although 2PF images of the same cells (see Figure 1 of Dombeck et al. 2005) shows significant fluorescence from internal membranes. Figure 19.6 also shows optically recorded action potentials from the SHG intensity recorded on patches of the membrane using the laser scanning line scan mode. Notably, 55 scans were averaged to obtain the data in

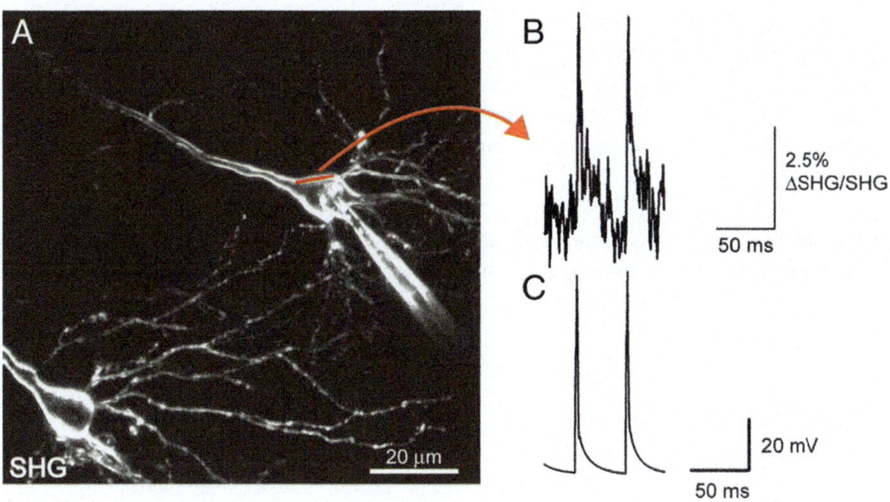

Fig. 19.6 Fast SHG line scan recording of elicited APs in brain slice. (**a**) To show recording of action potentials (APs) with SHG, this neuron was patch clamped and filled with FM4-64. Straight red line represents scanned line where elicited APs were recorded optically by integrating over the width. (**b**) SHG recording of APs with S/N of 7–8. This intensity plot of SHG emission vs. time is obtained from averaged line scans (1,200 lines/s) of the line denoted in **a**. $n = 55$ line scans were averaged. (**c**) Average current-clamp trace of elicited APs recorded optically in **b**. This figure was taken from Dombeck et al. (2005) and is reprinted by permission of the authors and the Journal of Neurophysiology

Fig. 19.6. Thus, despite the large signal to background ratio and despite the good sensitivity of the signal to membrane potential, the low intensity of the SHG signal requires significant averaging to overcome the poor signal to noise ratio of SHG measurements.

5 Conclusion

In this chapter we have reviewed the current status of second harmonic generation as a new modality for optical recording of membrane potential. The special properties of SHG were described, followed by an exposition of their origins based on both qualitative physical arguments and quantitative theory. We summarized the ways that SHG from dye-stained membranes could be modulated by membrane potential and showed how such an experiment is performed. The largest potentiometric sensitivity of SHG from a dye stained membrane is approximately 40 %/100 mV, which is significantly higher than the best fluorescence recordings. SHG also shows advantages for recording from internally stained single cells in brain slice preparations because there is no background signal from internal organelle membranes, as there is for similar fluorescence-based experiments. However, a significant impediment to the practical application of SHG for optical recording is the small number of photons that are produced compared to fluorescence. We and other will continue efforts to design and synthesize new SHG dyes with improved properties—most importantly, increased SHG efficiency. We expect that as the dyes are improved and the SHG methods for measuring membrane potential are optimized, recording of membrane potential from the membranes of internally stained cells will become the dominant application for SHG-based optical recording.

Acknowledgment We are indebted to the many talented students and colleagues who have participated in the development of SHG imaging in our laboratories over the last 20 years. This work was supported by NIH EB001963.

References

Araya R, Jiang J, Eisenthal KB, Yuste R (2006) The spine neck filters membrane potentials. Proc Natl Acad Sci U S A 103:17961–17966

Araya R, Nikolenko V, Eisenthal KB, Yuste R (2007) Sodium channels amplify spine potentials. Proc Natl Acad Sci U S A 104:12347–12352

Ben-Oren I, Peleg G, Lewis A, Minke B, Loew LM (1996) Infrared nonlinear optical measurements of membrane potential in photoreceptor cells. Biophys J 71:1616–1620

Bianchini P, Diaspro A (2008) Three-dimensional (3D) backward and forward second harmonic generation (SHG) microscopy of biological tissues. J Biophotonics 1:443–450

Bouevitch O, Lewis A, Pinevsky I, Wuskell JP, Loew LM (1993) Probing membrane potential with non-linear optics. Biophys J 65:672–679

Campagnola P, Loew LM (2003) Second-harmonic imaging microscopy for visualizing biomolecular arrays in cells, tissues and organisms. Nat Biotechnol 21:1356–1360

Campagnola PJ, Clark HA, Mohler WA, Lewis A, Loew LM (2001) Second harmonic imaging microscopy of living cells. J Biomed Opt 6:277–286

Campagnola PJ, Wei M-d, Lewis A, Loew LM (1999) High resolution optical imaging of live cells by second harmonic generation. Biophys J 77:3341–3349

Clark HA, Campagnola PJ, Wuskell JP, Lewis A, Loew LM (2000) Second harmonic generation properties of fluorescent polymer encapsulated gold nanoparticles. J Am Chem Soc 122:10234–10235

Cole JM, Kreiling S (2002) Exploiting structure/property relationships in organic non-linear optical materials: developing strategies to realize the potential of TCNQ derivatives. CrystEngComm 4:232–238

Conboy JC, Richmond GL (1997) Examination of the electrochemical interface between two immiscible electrolye solutions by second harmonic generation. J Phys Chem B 101:983–990

Denk W, Strickler JH, Webb WW (1990) Two-photon laser scanning fluorescence microscopy. Science 248:73–76

Dirk CW, Twieg RJ, Wagniere J (1986) The Contribution of Pi Electrons to Second Harmonic Generation in Organic Molecule. J Am Chem Soc 108:5387–5395

Dombeck DA, Blanchard-Desce M, Webb WW (2004) Optical recording of action potentials with second-harmonic generation microscopy. J Neurosci 24:999–1003

Dombeck DA, Sacconi L, Blanchard-Desce M, Webb WW (2005) Optical recording of fast neuronal membrane potential transients in acute mammalian brain slices by second-harmonic generation microscopy. J Neurophysiol 94:3628–3636

Fluhler E, Burnham VG, Loew LM (1985) Spectra, membrane binding and potentiometric responses of new charge shift probes. Biochemistry 24:5749–5755

Freund I, Deutsch M, Sprecher A (1986) Connective tissue polarity. Optical second-harmonic microscopy, crossed-beam summation, and small-angle scattering in rat-tail tendon. Biophys J 50:693–712

Huang J, Lewis A (1989) Determination of the absolute orientation of the retinyldiene chromophore in purple membrane by a second harmonic interference technique. Biophys J 55:835–842

Huang JY, Lewis A, Loew LM (1988) Non-linear optical properties of potential sensitive styryl dyes. Biophys J 53:665–670

Jiang J, Eisenthal KB, Yuste R (2007) Second harmonic generation in neurons: electro-optic mechanism of membrane potential sensitivity. Biophys J 93:L26–L28

Millard AC, Campagnola PJ, Mohler W, Lewis A, Loew LM (2003a) Second harmonic imaging microscopy. In: Marriott G, Parker I (eds) Methods in enzymology, vol. 361B. Academic, San Diego, pp 47–69

Millard AC, Jin L, Lewis A, Loew LM (2003b) Direct measurement of the voltage sensitivity of second-harmonic generation from a membrane dye in patch-clamped cells. Opt Lett 28:1221–1223

Millard AC, Jin L, Wei M-d, Wuskell JP, Lewis A, Loew LM (2004) Sensitivity of second harmonic generation from styryl dyes to trans-membrane potential. Biophys J 86:1169–1176

Millard AC, Jin L, Wuskell JP, Boudreau DM, Lewis A, Loew LM (2005a) Wavelength- and time-dependence of potentiometric non-linear optical signals from styryl dyes. J Membr Biol 208:103–111

Millard AC, Lewis A, Loew LM (2005b) Second harmonic imaging of membrane potential. In: Yuste R, Konnerth A (eds) Imaging in neuroscience and development. Cold Spring Harbor Laboratory Press, Cold Spring Harbor, pp 463–474

Mohler W, Millard AC, Campagnola PJ (2003) Second harmonic generation imaging of endogenous structural proteins. Methods 29:97–109

Moreaux L, Pons T, Dambrin V, Blanchard-Desce M, Mertz J (2003) Electro-optic response of second-harmonic generation membrane potential sensors. Opt Lett 28:625–627

Moreaux L, Sandre O, Blanchard-desce M, Mertz J (2000a) Membrane imaging by simultaneous second-harmonic generation and two-photo microscopy. Opt Lett 25:320–322
Moreaux L, Sandre O, Mertz J (2000b) Membrane imaging by second harmonic generation microscopy. J Opt Soc Am B 17:1685–1694
Morley JO (1988) Non-linear optical properties of organic molecules. 7. Calculated hyperpolarizabilities of azulenes and sesquifulvalene. J Am Chem Soc 110:7660–7663
Nemet BA, Nikolenko V, Yuste R (2004) Second harmonic imaging of membrane potential of neurons with retinal. J Biomed Opt 9:873–881
Nicoud JF, Twieg RJ (1987) Design and synthesis of organic molecular compounds for efficient second-harmonic generation. In: Chemla DS, Zyss J (eds) Nonlinear optical properties of organic molecules and crystals, vol. 1. Academic, Orlando, pp 227–296
Nuriya M, Jiang J, Nemet B, Eisenthal KB, Yuste R (2006) Imaging membrane potential in dendritic spines. Proc Natl Acad Sci U S A 103:786–790
Nuriya M, Yasui M (2010) Membrane potential dynamics of axons in cultured hippocampal neurons probed by second-harmonic-generation imaging. J Biomed Opt 15:020503
Pons T, Moreaux L, Mongin O, Blanchard-Desce M, Mertz J (2003) Mechanisms of membrane potential sensing with second-harmonic generation microscopy. J Biomed Opt 8:428–431
Rama S, Vetrivel L, Semyanov A (2010) Second-harmonic generation voltage imaging at subcellular resolution in rat hippocampal slices. J Biophotonics 3:784–790
Sacconi L, Dombeck DA, Webb WW (2006) Overcoming photodamage in second-harmonic generation microscopy: real-time optical recording of neuronal action potentials. Proc Natl Acad Sci U S A 103:3124–3129
Sheppard C, Kompfner R, Gannaway J, Walsh D (1977) IEEE J Quant Electron 13D:100E
Teisseyre TZ, Millard AC, Yan P, Wuskell JP, Wei M-d, Lewis A, Loew LM (2007) Non-linear optical potentiometric dyes optimized for imaging with 1064 nm light. J Biomed Opt 12:044001–044008
Theer P, Denk W, Sheves M, Lewis A, Detwiler PB (2011) Second-harmonic generation imaging of membrane potential with retinal analogues. Biophys J 100:232–242
Williams DJ (1984) Organic polymeric and non-polymeric materials with large optical non-linearities. Angew Chem Int Ed Engl 23:690–703
Wilson SA, Millard A, Lewis A, Loew LM (2014) Monitoring membrane potential with second-harmonic generation. Cold Spring Harb Protoc 2014:643–654
Yan ECY, Liu Y, Eisenthal KB (1998) New method for determination of surface potential of microscopic particles by second harmonic generation. J Phys Chem B 102:6331–6336
Yan P, Millard AC, Wei M, Loew LM (2006) Unique contrast patterns from resonance-enhanced chiral SHG of cell membranes. J Am Chem Soc 128:11030–11031
Yan P, Xie A, Wei M-d, Loew LM (2008) Amino(oligo)thiophene-based environmentally sensitive biomembrane chromophores. J Org Chem 73:6587–6594
Zhou W-L, Yan P, Wuskell JP, Loew LM, Antic SD (2007) Intracellular long wavelength voltage-sensitive dyes for studying the dynamics of action potentials in axons and thin dendrites. J Neurosci Methods 164:225–239

Chapter 20
Genetically Encoded Protein Sensors of Membrane Potential

Douglas Storace, Masoud Sepehri Rad, Zhou Han, Lei Jin, Lawrence B. Cohen, Thom Hughes, Bradley J. Baker, and Uhna Sung

Abstract Organic voltage-sensitive dyes offer very high spatial and temporal resolution for imaging neuronal function. However these dyes suffer from the drawbacks of non-specificity of cell staining and low accessibility of the dye to some cell types. Further progress in imaging activity is expected from the development of genetically encoded fluorescent sensors of membrane potential. Cell type specificity of expression of these fluorescent protein (FP) voltage sensors can be obtained via several different mechanisms. One is cell type specificity of infection by individual virus subtypes. A second mechanism is specificity of promoter expression in individual cell types. A third, depends on the offspring of transgenic animals with cell type specific expression of cre recombinase mated with an animal that has the DNA for the FP voltage sensor in all of its cells but its expression is dependent on the recombinase activity. Challenges remain. First, the response time constants of many of the new FP voltage sensors are slower (2–10 ms) than those of organic dyes. This results in a relatively small fractional fluorescence change, ΔF/F, for action potentials. Second, the largest signal presently available is only ~40 % for a 100 mV depolarization and many of the new probes have signals that are substantially smaller. Large signals are especially important when attempting to

D. Storace • Z. Han • L. Jin
Department of Cellular and Molecular Physiology, Yale University School of Medicine, New Haven, CT 06520, USA

M.S. Rad • B.J. Baker • U. Sung
Center for Functional Connectomics, Korea Institute of Science and Technology (KIST), Seoul 136-791, Korea

L.B. Cohen (✉)
Center for Functional Connectomics, Korea Institute of Science and Technology (KIST), Seoul 136-791, Korea

Department of Physiology, Yale University School of Medicine, 333 Cedar St., New Haven, CT 06510, USA
e-mail: lawrence.b.cohen@hotmail.com

T. Hughes
Department of Cell Biology and Neuroscience, Montana State University, Bozeman, MT 59717, USA

M. Canepari et al. (eds.), *Membrane Potential Imaging in the Nervous System and Heart*, Advances in Experimental Medicine and Biology 859,
DOI 10.1007/978-3-319-17641-3_20

detect fast events because the shorter measurement interval results in a relatively small number of detected photons and therefore a relatively large shot noise (see Chap. 1). Another kind of challenge has occurred when attempts were made to transition from one species to another or from one cell type to another or from cell culture to *in vivo* measurements.

Several laboratories have recently described a number of novel FP voltage sensors. Here we attempt to critically review the current status of these developments in terms of signal size, time course, and *in vivo* function.

Keywords FP voltage sensors • Voltage-sensitive dyes • Genetically encoded voltage sensors • *Ciona* • GFP • Membrane potential • Mammalian cells • Membrane expression • Thomas Knopfel • Ehud Isacoff • *Ciona* voltage sensitive domain • Mermaid2 • ArcLight

1 Introduction

Optical imaging is a flexible method for studying cellular and population activities. Organic dyes have been developed that can faithfully report many biological variables including calcium concentration, pH, or membrane potential (Brown et al. 1975; MacDonald and Jobsis 1976; Davila et al. 1973). These optical probes enable simultaneous measurements from many locations and have been used to study the physiology of individual spines, the spiking activity of many individual neurons, as well as the activity of large populations of cells (for earlier reviews see Cohen and Salzberg (1978); Miesenbock and Kevrekidis (2005); Grinvald and Hildesheim (2004); and Baker et al. (2005)).

Organic voltage sensitive dyes are outstandingly fast. Their response time constants are less than 2 μs (Loew et al. 1985; Salzberg et al. 1993), faster than all known neurobiological membrane potential changes. Two drawbacks of organic dyes are the non-specificity of cell staining and the low accessibility of the dye to some cell types. Organic dyes either stain all cells in a tissue to which they are exposed, or if delivered *via* a patch pipette they label only one patched cell at a time. When all the cells and processes are stained, optical signals of interest are often overshadowed by signals from neurons that are not the focus of interest or the signal is lost in the noise from background fluorescence from inactive cells. One solution to this problem is to make voltage sensors from proteins which can then be targeted to specific cell types.

Several protein-based voltage sensors are a "mosaic" of two parts: a "sensor" membrane protein that undergoes a potential dependent conformational rearrangement and an attached Fluorescent Protein(s) (FP) reporter(s) whose emission is modulated by changes in the "mosaic" sensor protein. A second type of sensor are microbial rhodopsins which are membrane resident. Lastly there are two component sensors one of which resides in the membrane and the other in the cytoplasm but bound to the membrane.

1.1 Kinds of Signals: Measurement Compromises

Organic voltage sensitive dyes were used to measure two kinds of signals. Signals from individual cells or portions of individual cells (e.g. data from the Dejan Zecevic laboratory) or signals that are the average of a population of cells (e.g. data from the Amiram Grinvald laboratory).

1.2 Individual Cells

In measurements from individual cells a sensor with a linear voltage dependence is an advantage because one can directly interpret the optical signal in terms of hyperpolarization, a subthreshold event, and/or an action potential. In addition the sensor should be bright, have a large signal-to-noise ratio, and be very fast. All of these characteristics are obtained with presently available organic voltage sensitive dyes. In mammalian slice preparations where cells and processes can be close to the slice surface and only one or a few cells are monitored, organic dyes remain the method of choice.

1.3 Many Individual Cells

In translucent brains (e.g., *Drosophila* and zebrafish) the cell type specificity of a protein sensor is important. Thus, protein sensors will be useful for monitoring activity from many individual cells. However, in the mammalian brain the situation is more complicated because light scattering blurs the image of individual cells in wide-field imaging. 2-photon microscopy is needed to obtain cellular resolution. But, because of the low efficiency of 2-photon emission, frames rates of 3–30 Hz are used and these are much too slow for optimal detection of millisecond events. Only line scans have the necessary time resolution. Thus, presently available FP voltage sensors are unlikely to be useful for monitoring the spike activity of individual neurons *in vivo*. The sensor would need to have a very long lasting response to a brief depolarization and even then accurate information about spike timing could not be obtained. The relatively slow calcium signals (durations >100 ms) are a much better fit for 2-photon frame rates. As a result, imaging calcium signals with protein calcium sensors is used as a surrogate for measuring action potentials in *in vivo* mammalian preparations.

1.4 Population Signals

Population signals can also be very useful for understanding brain function. For example in the mammalian olfactory bulb only one class of olfactory receptor neurons send their axons to an individual glomerulus and the signal from that population of nerve terminals provides a useful measure of information coming from the nose to that glomerulus. Similarly, the output of each glomerulus is carried by ~25 mitral/tufted cells and the population signal from those cells would be very informative about olfactory bulb output. A linear voltage relationship might not be optimal for interpreting a population signal. Signals from neurons that hyperpolarize would subtract from depolarizing signals. Subthreshold depolarizations could be confused with a train of many asynchronous action potentials. The interpretation of the optical signal would be simpler if the sensor functioned over a restricted voltage range. Many mosaic FP voltage sensors have a sigmoidal fluorescence *vs* voltage curve (see below). This sigmoid response can, inprinciple, be "tuned" along the voltage axis by modification of transmembrane domains so that the signal emphasizes either hyperpolarizing potentials, subthreshold potentials, or suprathreshold potentials. Thus, with appropriate steepness of the fluorescence *vs* voltage response and appropriate tuning, a FP voltage sensor could selectively detect specific types of neuron activity.

1.5 Calcium Sensors

The calcium sensor family of GCaMPs is frequently used as a surrogate for monitoring spike activity in many individual mammalian neurons in intact behaving organisms. However, there may be important limitations. Detection of fast-spiking activity will be difficult, subthreshold depolarizations may not be detected, and hyperpolarization is likely to be missed. Furthermore, in some neurons there are large activity related calcium signals that occur in the absence of action potentials (Charpak et al. 2001). Nonetheless, calcium sensors are likely to continue to be used in mammalian preparations because the signals are large and their time course is better suited to the frame rates available in 2-photon microscopy.

Below we summarize the development of the first and second generations of FP voltage sensors but provide more detail about recently described third generation sensors.

2 First Generation FP Voltage Sensors

The first generation FP voltage sensors were mosaics constructed by molecular fusion of a GFP-based fluorescent reporter and voltage-gated ion channels (or components thereof). The first sensor, named FlaSH, was developed in the Isacoff laboratory by inserting wtGFP after the last transmembrane segment of the *Drosophila* Shaker potassium channel (Fig. 20.1a; Siegel and Isacoff 1997). When FlaSH was expressed in oocytes, an 80 mV depolarization of the plasma membrane

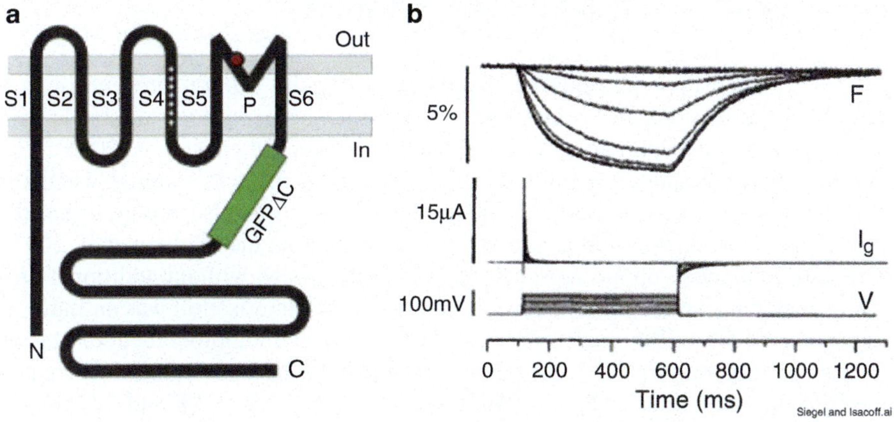

Fig. 20.1 FlaSh. (**a**) *GFPΔC* (*green*) was inserted in-frame into the Shaker K*1* channel (*black*). A point mutation *W434F* (*red circle*) was made in the pore of the channel to eliminate ionic current through the sensor. (**b**) Simultaneous two-electrode voltage-clamp recording and photometry show current and fluorescence changes in response to voltage steps (V) between −60 and 10 mV, in 10 mV increments. Holding potential was −80 mV. The mutated FlaSh exhibits no ionic current. FlaSh fluorescence (F) decreases reversibly in response to membrane depolarizations. Traces are the average of 20 sweeps. Fluorescence scale, 5 % ΔF/F (Modified from Siegel and Isacoff (1997))

resulted in a 5 % decrease in fluorescence (Fig. 20.1b). To reduce unwanted effects on the cell's physiology, FlaSh was rendered non-conducting by introducing a W434F mutation (Perozo et al. 1993), preventing ions from moving through the pore while maintaining voltage dependent rearrangements. In order to resolve action potentials, an ideal sensor would generate a robust signal on a millisecond timescale. The on and off rates of FlaSh are much too slow (tau-on ~100) for efficient spike detection.

FlaSh, Flare (a variant based on a mammalian potassium channel), and two additional first generation FP voltage sensors also based on mammalian channels (VSFP1 and SPARC (Sakai et al. 2001; Ataka and Pieribone 2002)) functioned well in frog oocytes, but they have not been useful in mammalian systems because of poor plasma membrane expression in mammalian cells. No voltage dependent optical signals could be detected in the average of 16 trials using either Flare, VSFP1, or SPARC (Baker et al. 2007). Several strategies to improve membrane expression were tried. Unfortunately, mutagenesis of potential ER retention signals, addition of ER release motifs, and expression in hippocampal neurons that might endogenously express trafficking partners all failed to improve the plasma membrane expression (K. Ray, K. Tomita, Y. Iwamoto, D. Dimitrov, A. Perron, T. Knöpfel, L.B. Cohen, and B.R. Baker, unpublished observations).

3 Second Generation FP Voltage Sensors

The voltage-sensitive phosphatase from the sea squirt, *Ciona intestinalis,* is a voltage-controlled enzyme consisting of a transmembrane voltage sensor domain and a cytosolic phosphoinositide phosphatase domain (Murata et al. 2005). The voltage sensor domain from this protein was shown to be functional when the enzyme domain was removed (Murata et al. 2005). It was also reported that the phosphatase can exist as a monomer in the plasma membrane (Kohout et al. 2007). This self-contained voltage sensor domain that functioned without additional protein components and without the need for subunit multimerisation was an interesting candidate for a FP-voltage sensor. With the hope of an improvement of plasma membrane targeting in mind, the Thomas Knopfel laboratory explored the use of the *Ciona* voltage sensor domain (Dimitrov et al. 2007). The CFP and YFP FRET pair from the first generation VSFP1 was fused to the C-terminus of the *Ciona* voltage sensitive domain. The cytosolic phosphatase domain was replaced by the FRET pair. When expressed in PC12 cells or hippocampal neurons, the second generation VSFP (VSFP2) displayed bright fluorescence and clear targeting to the plasma membrane (Fig. 20.2a) as well as fluorescence changes in response to changes in membrane potential (Dimitrov et al. 2007; Fig. 20.2b). This is a remarkable result. Sensors based on mammalian membrane proteins did not express well in mammalian cells while a sensor based on the sea quirt protein did express well.

The FRET fluorescence change could result from a change in the distance or angle between the two chromophores or from environmental sensitive spectral shifts. In some VSFP2 variants the CFP signal is substantially larger than the YFP signal and/or the two signals have differing kinetics suggesting that non-FRET mechanisms are also occurring (Dimitrov et al. 2007; Lundby et al. 2008). At present we have very limited understanding of the mechanism (s) coupling the voltage dependent structural changes in the sensor domain to the changes in fluorescence of the FPs.

Consistent with 'gating' currents measured in *Ciona* phosphatase, the fluorescence voltage curves for VSFP2 exhibited half maximal activation ($V_{1/2}$) at potentials far above the physiological range of mammalian neurons. To shift the voltage dependency to a more physiological range, a series of mutational alterations were explored in the putative S4 segment of the *Ciona* voltage sensor domain. The R217Q mutation resulted in the protein termed VSFP2.1 which had a $V_{1/2}$ value of ~−70 mV (Dimitrov et al. 2007).

The signal size, ΔF/F, of the second generation VSFP2.1 was ~3 % for a 100 mv depolarization (Dimitrov et al. 2007). Tsutsui et al. (2008) generated and tested an additional 300 *Ciona* based FP-sensors. One of these, named Mermaid, was said to have a signal size substantially larger than that of previously reported sensors. Mutoh et al. (2009) made a direct comparison of Mermaid and two other *Ciona* variants, VSFP2.3 and VSFP2.4 and reported that the signal sizes (ΔF/F ~5 %) were similar for all three constructs.

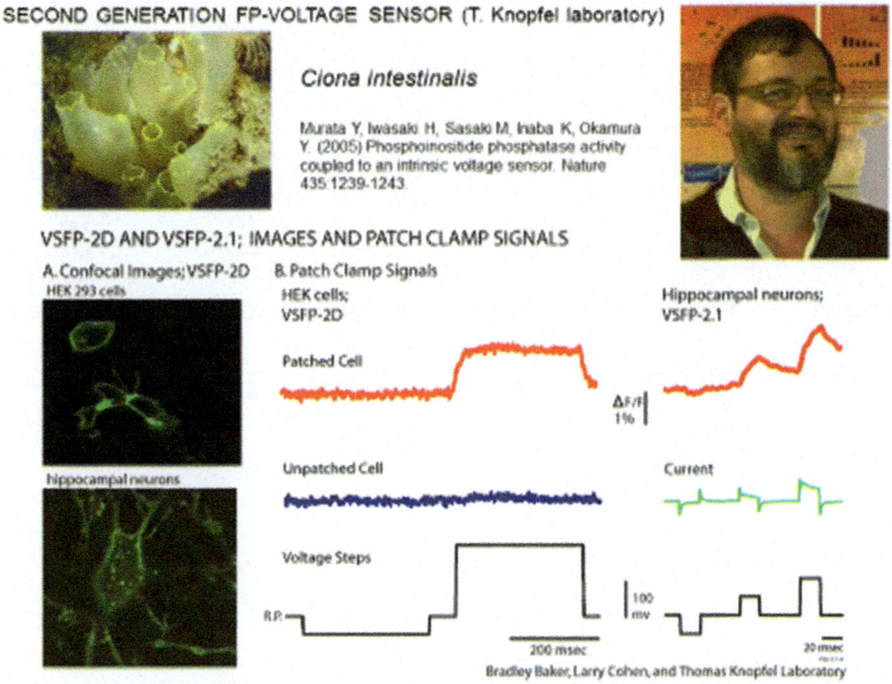

Fig. 20.2 Second generation FP-voltage sensors using the *Ciona intenstinalis* voltage sensing domain. (A) Images of transfected cultured neurons showing good trafficking to the external membrane of VSFP-2D. (B) Fluorescence signals in response to voltage clamp steps in HEK293 cells and hippocampal neurons. These signals are relatively small (~3 % per 100 mv) and not very fast

The response time courses of the *Ciona* constructs were best fit by two exponentials (Dimitrov et al. 2007; Lundby et al. 2008; Villalba-Galea et al. 2009; Tsutsui et al. 2008; Mutoh et al. 2009), with time constants of about 5 and 100 ms. The relative proportion of the faster time constant increased with larger depolarizations.

4 Third Generation FP Voltage Sensors

An accelerating effort has been undertaken to develop sensors with larger and faster signals. There have been three approaches. The first is to make changes in all three components of mosaics like VSFP2 and Mermaid: the voltage sensitive domain, the linker between the *Ciona* protein and the FP, and the FP itself (selected results are illustrated in Table 20.1). The second is the use of microbial rhodopsins as both the voltage sensor and the fluorescent reporter (Table 20.2). The third is the

Table 20.1 Selected mosaic FP voltage sensors based on the *Ciona* (or an analogue) voltage sensitive domain

	ΔF/F for a 100 mv depolarization (HEK293 cells)	Fast ON tau for a 100 mv depolarization (HEK293 cells) (ms)	Single trial signals (neurons in culture)	Single trial *in vivo* signals
Single FP sensors				
ArcLight[a]	40 %	10	Yes	Yes[b,c]
Bongwoori[d]	20 %	8	Yes	Not tested
ASAP1[e]	17 %	2	Yes	Not tested
Butterfly FRET sensors				
Nabi2.213[f]	10 %	3	Yes	Not tested
Chimeric VSFP-Butterfly YR[g]	not presented	2	Not tested	Yes
Mermaid2[h]	20 %	2	Yes	Yes

[a]Jin et al. (2012)
[b]Storace et al. (2014)
[c]Cao et al. (2013)
[d]Piao et al. (2014)
[e]St-Pierre et al. (2014)
[f]Sung et al. (2014)
[g]Mishina et al. (2014)
[h]Tsutsui et al. (2013)

Table 20.2 Selected FP Voltage sensors based on microbial rhodopsin sensors

Sensor	ΔF/F for a 100 mv depolarization (HEK293 cells)	Fast ON tau for a 100 mv depolarization (HEK293 cells) (ms)	Single trial signals (neurons in culture)	Single trial *in vivo* signals
Archer1[a]	80 %[b]	2[b]	Yes	Yes[c]
QuasAr2[d]	90 %	<1	Yes	Not tested
MacQ-mCitrine[e]	20 %	5	Yes	Yes

[a]Flytzanis et al. (2014)
[b]In cultured neurons
[c]In *C. elegans* AWC neuron, small S/N, average from two worms
[d]Hochbaum et al. (2014)
[e]Gong et al. (2014)

development of a two component FRET system: an absorbing molecule that resides in the membrane and a membrane bound FP that serves as the FRET partner (Table 20.3). As indicated in the three tables, there has been remarkable recent success; many third generation sensors have larger and faster signals.

Table 20.3 A two component FP voltage sensor

Sensor	ΔF/F for a 100 mv depolarization (PC12 cells)	ON tau (PC12 cells) (ms)	Single trial signals (neurons in culture)	Single trial *in vivo* signals
nVOS[a, b]	30 %[b]	0.5[a]	Yes[a]	Not tested

[a]Wang et al. (2012)
[b]Wang et al. (2010)

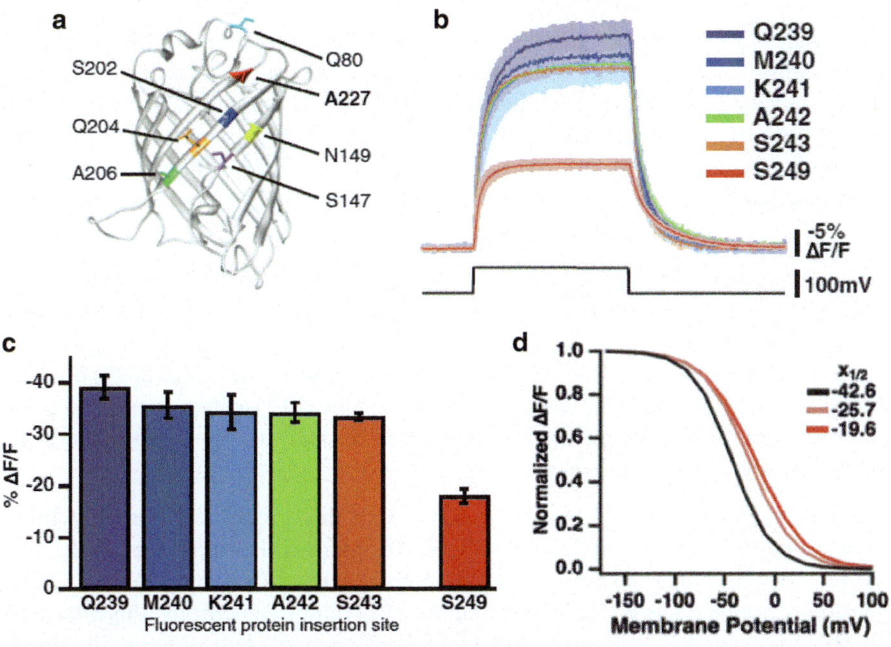

Fig. 20.3 ArcLight, a Ciona based FP voltage sensor with relatively large fluorescence changes. (**a**) The FP in ArcLight is super ecliptic pHluorin with an A227D mutation. This spontaneous mutation increased the signal size, ΔF/F, from ~1 % to 15 %. The resulting signal is shown in *red* in (**b**). (**b**, **c**) Shortening the linker resulted in a further twofold increase in the signal size. (**d**) The fluorescence vs voltage curve for ArcLight (*red*) is sigmoidal. The $V_{1/2}$ is −20 mV (Modified from Jin et al. (2012))

4.1 Modifying Mosaic Second Generation Probes Based on the Ciona Voltage Sensitive Domain

Dramatic improvements in signal size and speed have been obtained (Table 20.1). ArcLight signals are about eight times larger than those of VSFP2.3 or Mermaid. The large signals were the results of a spontaneous mutation, A227D, in the FP, super ecliptic pHluorin, together with a reduction in the linker length between the *Ciona* protein and the FP (Fig. 20.3a–c). Nabi2.213, ASAP1, Chimeric VSFP-

Butterfly YR, and Mermaid2 all have fast component time constants of ~2 ms; substantially faster than VSFP2.3 or Mermaid (Table 20.1). The functional properties of new mutations are often unpredictable. Thus the improvements in signal size and kinetics are, in large part, the result of testing many mutations in the three components of the mosaic sensors.

Voltage sensitive domain of the phosphatase from other species. Baker et al. (2012) and Han et al. (2013) tested the voltage sensitive domains from species other than *Ciona*. Signals from chicken, *Gallus gallus,* and zebrafish, *Danio rerio*, were relatively large, ΔF/F ~10 %, and fast, tau ~5 ms.

New kinds of FRET probes. In the second generation Mermaid and VSFP, the donor and acceptor FPs were inserted in tandem downstream of the S4 helix. Positioning donor and acceptor FPs at flanking regions of S1-S4 domain was introduced in VSFP Butterfly (Akemann et al. 2012). Three of the probes in Table 20.1, Nabi, Chimeric VSFP-Butterfly YR, and Mermaid2, have butterfly FP locations. Their signal size and kinetics are improved in comparison with the tandem arrangement used in the second generation sensors.

Using circularly permuted FPs. Following the example of GCaMP calcium sensors, probes that incorporated circularly permuted FPs (Baird et al. 1999) into the Ciona voltage sensitive domain were tested but the initial attempts resulted in signals that were small (Gautam et al. 2009: Barnett et al. 2012). St-Pierre et al. (2014) obtained relatively large and fast signals with ASAP1 (Table 20.1) which uses a circularly permutted FP in the extracellular loop between S3 and S4 in the voltage sensitive domain of the chicken phosphatase.

Quantum efficiency. The FP voltage sensors in this group are relatively bright because the FPs are, in part, chosen for their brightness. For example, ArcLight is 50-fold brighter than the microbial rhodopsin sensor QuasAr2 (Adam Cohen, personal communication). In some circumstances (cultured neurons) this difference in brightness can be overcome by using more intense incident illumination for the microbial rhodopsin sensor. In other situations (e.g. *in vivo* mammalian brain) the high intensity illumination needed for the rhodopsin is likely to cause damage. In addition the signals from a dim sensor can be obscured by the intrinsic fluorescence change and the "intrinsic imaging" blood flow signals.

External membrane expression. The mosaic FP voltage sensors also appear to have relatively selective external membrane expression (Fig. 20.2; Jin et al. 2012; Tsutsui et al. 2013; Storace et al. 2014; Mishina et al. 2014; St-Pierre et al. 2014).

Fluorescence vs membrane potential. All of the sensors in Table 20.1 (except ASAP1) have sigmoidal fluorescence *vs* voltage relationships (e.g., Fig. 20.3d). The $V_{1/2}$ can be shifted along the voltage axis by mutations in the transmembrane helices. Thus these FP voltage sensors could be modified to be selectively sensitive to hyperpolarizations, subthreshold depolarizations, or action potentials. Piao et al. (2014) have studied a large number of mutations of the transmembrane helical elements of Bongwoori and related *Ciona* based sensors.

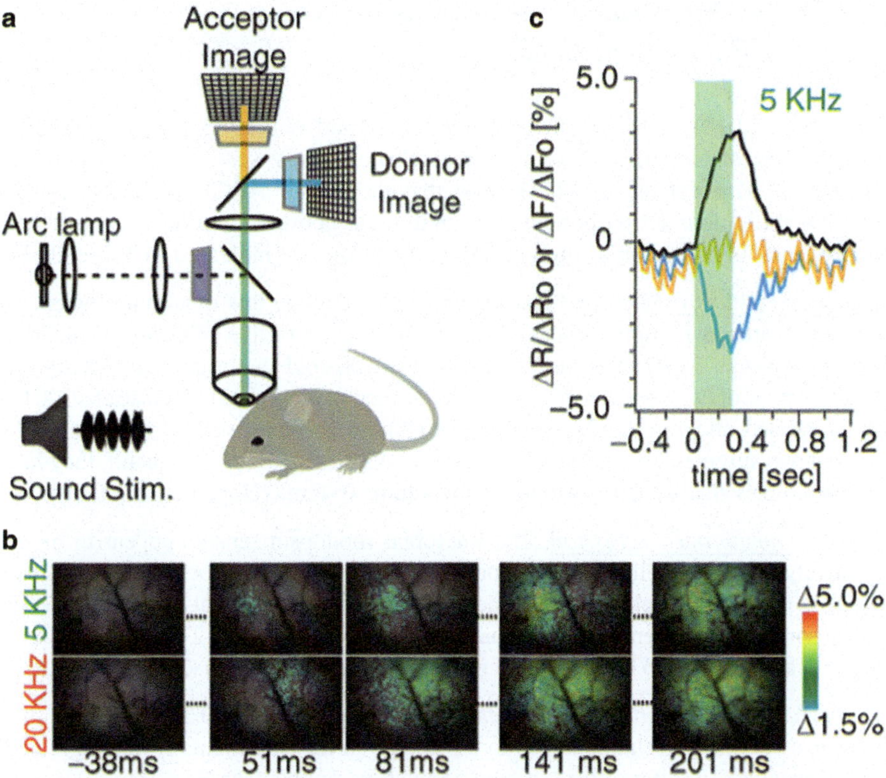

Fig. 20.4 Cortical responses to auditory stimuli detected using Mermaid2 *in vivo* (**a**), schematic diagram showing the setup for transcranial FRET imaging of cortical surface fluorescence *in vivo*. Sound stimuli (5 or 20 kHz tones) were delivered to the mice by using an electrostatic speaker, and the primary auditory cortex of the right hemisphere was simultaneously imaged. (***b***) pseudo-colored images of the ratiometric response of Mermaid2 to tonal stimuli at 5 and 20 kHz (60 dB to show the tonotopic organization noted in single-trial measurements. Scale bar = 1 mm. (**c**) Representative time courses of the changes in YFP intensity (*yellow*), CFP intensity (*cyan*) and their ratio (*black*). The stimulus period is *shaded*. The CFP and YFP signals generally contained substantial common noise, which, however, was cancelled in the ratio. The noise has frequency at ~650 cycles min^{-1} and probably comes from the motions associated with heartbeats (Modified from Tsutsui et al. (2013))

In vivo measurements. Three of the probes in Table 20.1 have been used to record single trial signals from the mammalian brain. As expected from the arguments given in the Sect. 1, all three recordings were of population signals. Figure 20.4 illustrates results from mouse auditory cortex following embryonic electroporation with Mermaid2 DNA (Tsutsui et al. 2013). Panel c shows the time courses of the donor (blue) and acceptor (orange) signals as well as the ratio signal (black). Both the donor and acceptor signals have heart beat artifacts which are greatly reduced in the ratio signal. Panel b shows a comparison of the signals resulting from a 5 and a 20 kHz acoustic input. The initial responses at 51 ms and 81 ms are in different locations on the cortex.

4.2 Using Microbial Rhodopsins as the Sensor and Fluorescent Reporter

Kralj et al. (2012) revived earlier findings (Junge and Witt 1968; Shinar et al. 1977) that the optical properties of some microbial rhodopsins were membrane potential dependent. Kralj et al. (2012) showed that the sensor ARCH could be used to monitor individual action potentials in individual neurons in culture. More recent improvements in microbial rhodopsin based sensors are indicated in Table 20.2.

Quantum efficiency. Microbial rhodopsins have relatively low quantum efficiencies. Recent efforts have improved the brightness by a factor of five (Flytzanis et al. 2014; Hochbaum et al. 2014) but the new probes are still 50-fold less bright than ArcLight. A further factor of two increase in brightness was recently reported (McIsaac et al. 2014). As discussed above, the relatively low brightness is likely to limit their utility for *in vivo* measurements in mammalian nervous systems. On the other hand, they have proven to be valuable sensors in neuronal culture systems (Hoppa et al. (2014)).

External membrane expression. The microbial rhodopsin sensors appear to have a relatively large intracellular expression (Flytzanis et al. 2014; Hochbaum et al. 2014; Gong et al. 2014) which will have a negative impact on the signal-to-noise ratio. The red absorbing FRET sensor, a combination of mRuby2 and Arch-2 (a QuasAr2 homologue, Table 20.2) also had relatively poor membrane expression (Zou et al. 2014).

Fluorescence vs membrane potential. The sensors in Table 20.2 have linear fluorescence *vs* voltage relationships; an advantage in measurements from single cells but a drawback in interpreting *in vivo* population signals (see Sect. 1).

4.3 A Two Component System

The two component probe, hVoS, couples the fluorescence of membrane-anchored FPs to the movement of a membrane-embedded hydrophobic anion, dipicrylamine (Chanda et al. 2005). The dipicrylamine is not fluorescent but its ability to quench the fluorescence of the FP depends on the membrane potential which causes the charged dipicrylamine to move within the membrane. The results from recent improved versions of in hVoS are shown in Table 20.3.

5 Discussion

5.1 Pharmacological and Photodynamic Effects

The FP voltage sensor can interfere with intrinsic biochemical and/or biophysical cell functions. Hence FP sensor-derived data require controls to demonstrate the absence of probe-induced undesired effects. Unlike organic voltage sensitive dyes, which exhibit phototoxicity in some situations (Grinvald et al. 1982; Kalyanaraman et al. 1987), FP-based sensors are not known to form toxic photo byproducts.

5.2 *Capacitative Loading*

The charges in the voltage sensor domain move across the membrane in response to changes in membrane potential. This charge movement is manifested as a trans-membrane capacitive current. Expression of FP-voltage sensors therefore will add an extra capacitance to the membrane which may influence the spiking properties of the target cells. However, Jin et al. (2012) reported that ArcLight expression in cultured neurons did not affect the action potential time course. Thus the effect of capacitative loading by ArcLight appears to be small. The effect of capacitative load associated with VSFP2 variants has been characterized by computational modeling (Akemann et al. 2009). Thomas Knopfel (personal communication) pointed out that the use of the *Ciona* voltage sensitive domain where the FP to S4 ratio is 1:1 might result in a relatively reduced capacitative load compared to FlaSh where the ratio is 1:4.

Extra capacitance is a known problem from the dipicrylamine used in the hVoS probe. The dipicrylamine concentration has to be limited to avoid suppressing action potential activity (Wang et al. 2010).

5.3 *Future Directions*

Our understanding of FP voltage sensor function is primitive; it rarely happens that a engineered variation in the sensor will have an expected effect. The parameter space is very large and the ten laboratories trying to improve the FP voltage sensors are not close to a complete exploration of even one of the parameters. Nonetheless, many different ideas are being tested and as a result there have been substantial improvements in signal size and response kinetics. This improvement is obvious comparing the third generation probes (Tables 20.1, 20.2, and 20.3) with the second generation sensors.

Further development is underway on several lines: linker optimization, alternative FP variants, the search for additional locations for inserting FPs into the voltage sensitive domain, and the use of alternative sensor proteins.

Linker Optimized Variants. The sensitivity and kinetics of FP voltage sensors depend on the length and amino acid composition of the linker between the sensor protein and the FP (Dimitrov et al. 2007; Mutoh et al. 2009; Villalba-Galea et al. 2009; Jin et al. 2012; Jung et al. 2015). An example is shown in Fig. 20.3b, c. However, a full exploration of linker length and composition would require a very large effort.

Alternative FP Colors. The cyan- and yellow-emitting variants (CFP and YFP) of GFP were the most widely used components for the second generation FRET based mosaic biosensors. However, red shifted versions are more desirable because: (1) red variants would be more suitable for deep tissue imaging, (2) reduced interference from tissue autofluorescence and hemoglobin absorption, and (3) spectral separation when used in combination with green fluorescent molecules. Lam et al. (2012) have introduced a new FRET pair, Clover and mRuby2, that includes a red FP. This pair was used successfully in the Nabi probes.

Modify the voltage sensor. In the voltage sensor domain, it is not only the S4 segment that responds to voltage shifts, but rather the whole domain undergoes conformational changes in a complex manner (Li et al. 2014). It follows that finding alternative sites for insertions of FPs along the voltage sensor domain may lead to improved function. ASAP1 (St-Pierre et al. 2014) and the Nabi probes (Sung et al. 2014) are an initial step in this direction.

6 Conclusion

Fluorescent protein sensors will open paths to yet unexplored territories of functional neuroimaging. In addition to the FP-voltage sensors on which this review focuses, there are now many different fluorescent protein sensors for monitoring biochemical processes and transmitter dynamics (Tsien 2005) at the level of neuronal circuits.

Both the signal size and the speed of FP voltage sensors need to be improved. It is hoped that voltage probe development community will be able to make progress toward meeting these goals.

Acknowledgements Supported by the World Class Institute program of the National Research Foundation of Korea, grant Number WCI 2009-003, the Center for Functional Connectomics at the Korea Institute of Science and Technology, and U.S. NIH grants DC005259 and NS054270, a James Hudson Brown – Alexander Brown Coxe fellowship from Yale University, and a Ruth L. Kirschstein National Research Service Award DC012981.

References

Akemann W, Mutoh H, Perron A, Park YK, Iwamoto Y, Knöpfel T (2012) Imaging neural circuit dynamics with a voltage-sensitive fluorescent protein. J Neurophysiol 108:2323–2337. doi:10.1152/jn.00452.2012

Akemann W, Lundby A, Mutoh H, Knöpfel T (2009) Effect of voltage sensitive fluorescent proteins on neuronal excitability. Biophys J 96:3959–3976

Ataka K, Pieribone VA (2002) A genetically targetable fluorescent probe of channel gating with rapid kinetics. Biophys J 82:509–516

Baird GS, Zacharias DA, Tsien RY (1999) Circular permutation and receptor insertion within green fluorescent proteins. Proc Natl Acad Sci U S A 96:11241–11246

Baker BJ, Kosmidis EK, Vucinic D, Falk C, Cohen L, Djurisic M, Zecevic D (2005) Imaging brain activity with voltage- and calcium-sensitive dyes. Cell Mol Neurobiol 25:245–282

Baker BJ, Lee H, Pieribone VA, Cohen LB, Isacoff EY, Knopfel T, Kosmidis EK (2007) Three fluorescent protein voltage sensors exhibit low plasma membrane expression in mammalian cells. J Neurosci Methods 161:32–38

Baker BJ, Jin L, Zhou H, Cohen LB, Popovic M, Platisa J, Pieribone V (2012) Genetically encoded voltage sensors based on the *Nematostella* and *Dario* voltage sensing phosphatases exhibit fast kinetics, J. Neurosci Methods 208:190–196, NIHMSID #390003

Barnett L, Platisa J, Popovic M, Pieribone VA, Hughes T (2012) A fluorescent, genetically-encoded voltage probe capable of resolving action potentials. PLoS One 7(9), e43454

Brown JE, Cohen LB, De Weer P, Pinto LH, Ross WN, Salzberg BM (1975) Rapid changes in intracellular free calcium concentration. Detection by metallochromic indicator dyes in squid giant axon. Biophys J 15:1155–1160

Cao G, Platisa J, Pieribone VA, Raccuglia D, Kunst M, Nitabach MN (2013) Genetically targeted optical electrophysiology in intact neural circuits. Cell 154:904–913

Chanda B, Blunck R, Faria LC, Schweizer FE, Mody I, Bezanilla F (2005) A hybrid approach to measuring electrical activity in genetically specified neurons. Nat Neurosci 8:1619–1626

Charpak S, Mertz J, Beaurepaire E, Moreaux L, Delaney K (2001) Odor-evoked calcium signals in dendrites of rat mitral cells. Proc Natl Acad Sci U S A 98:1230–1234

Cohen LB, Salzberg BM (1978) Optical measurement of membrane potential. Rev Physiol Biochem Pharmacol 83:35–88

Davila HV, Salzberg BM, Cohen LB, Waggoner AS (1973) A large change in axon fluorescence that provides a promising method for measuring membrane potential. Nat New Biol 241:159–160

Dimitrov D, He Y, Mutoh H, Baker BJ, Cohen L, Akemann W, Knöpfel T (2007) Engineering and characterization of an enhanced fluorescent protein voltage sensor. PLoS One 2(5), e440. doi:10.1371/journal.pone.0000440

Flytzanis NC, Bedbrook CN, Chiu H, Engqvist MKM, Xiao C, Chan KY, Sternberg PW, Arnold FH, Gradinaru V (2014) Archaerhodopsin variants with enhanced voltage-sensitive fluorescence in mammalian and *Caenorhabditis elegans* neurons. Nat Commun 5:4894. doi:10.1038/ncomms5894

Gautam SG, Perron A, Mutoh H, Knöpfel T (2009) Exploration of fluorescent protein voltage probes based on circularly permuted fluorescent proteins. Front Neuroeng 2:14. doi:10.3389/neuro.16.014.2009

Grinvald A, Hildesheim R, Farber IC, Anglister L (1982) Improved fluorescent probes for the measurement of rapid changes in membrane potential. Biophys J 39:301–308

Grinvald A, Hildesheim R (2004) VSDI: a new era in functional imaging of cortical dynamics. Nat Rev Neurosci 5:874–885

Gong Y, Wagner MJ, Li JZ, Schnitzer MJ (2014) Imaging neural spiking in brain tissue using FRET-opsin protein voltage sensors. Nat Commun 5:3674. doi:10.1038/ncomms4674

Han Z, Jin L, Platisa J, Cohen LB, Baker BJ, Pieribone VA (2013) Fluorescent protein voltage probes derived from ArcLight that respond to membrane voltage changes with fast kinetics. PLoS One 10:1371, journal.pone.0081295

Hochbaum DR, Zhao Y, Farhi SL, Klapoetke N, Werley CA, Kapoor V, Zou P, Kralj JM, Maclaurin D, Smedemark-Margulies N, Saulnier JL, Boulting GL, Straub C, Cho YK, Melkonian M, Wong GK, Harrison DJ, Murthy VN, Sabatini BL, Boyden ES, Campbell RE, Cohen AE (2014) All-optical electrophysiology in mammalian neurons using engineered microbial rhodopsins. Nat Methods 11:825–833. doi:10.1038/nmeth.3000

Hoppa MB, Gouzer G, Armbruster M, Ryan TA (2014) Control and plasticity of the presynaptic action potential waveform at small CNS nerve terminals. Neuron 84(4):778–789

Jin L, Han Z, Platisa J, Wooltorton JRA, Cohen LB, Pieribone VA (2012) Single action potentials and subthreshold electrical events visualized in neurons using a novel fluorescent protein voltage sensor. Neuron 75:779–785

Jung A, Garcia JE, Kim E, Yoon B-J, Baker BJ (2015) Linker length and fusion site composition improve the optical signal of genetically-encoded fluorescent voltage sensors. J Neurophoton 2 (2):021012

Junge W, Witt HT (1968) On the ion transport system of photosynthesis—investigations on a molecular level. Z Naturforsch B 23:244–254

Kalyanaraman B, Feix JB, Sieber F, Thomas JP, Girotti AW (1987) Photodynamic action of merocyanine 540 on artificial and natural cell membranes: involvement of singlet molecular oxygen. Proc Natl Acad Sci U S A 84:2999–3003

Kohout SC, Ulbrich MH, Bell SC, Isacoff EY (2007) Subunit organization and functional transitions in Ci-VSP. Nat Struct Mol Biol 15:106–108

Kralj JM, Douglass AD, Hochbaum DR, Maclaurin D, Cohen AE (2012) Optical recording of action potentials in mammalian neurons using a microbial rhodopsin. Nat Methods 9:90–95

Lam AJ, St-Pierre F, Gong Y, Marshall JD, Cranfill PJ, Baird MA, McKeown MR, Wiedenmann J, Davidson MW, Schnitzer MJ, Tsien RY, Lin MZ (2012) Improving FRET dynamic range with bright green and red fluorescent proteins. Nat Methods 9:1005–1012

Li Q, Wanderling S, Paduch M, Medovoy D, Singharoy A, McGreevy R, Villalba-Galea, CA, Hulse RE, Roux B, Schulten K, Kossiakoff A & Perozo E (2014) Structural mechanism of voltage-dependent gating in an isolated voltage-sensing domain. Nat Struct Mol Biol 21:244–252

Loew LM, Cohen LB, Salzberg BM, Obaid AL, Bezanilla F (1985) Charge-shift probes of membrane potential. Characterization of aminostyrylpyridinium dyes on the squid giant axon. Biophys J 47:71–77

Lundby A, Mutoh H, Dimitrov D, Akemann W, Knöpfel T (2008) Engineering of a genetically encodable fluorescent voltage sensor exploiting fast Ci-VSP voltage-sensing movements. PLoS One 3(6), e2514

MacDonald VW, Jobsis FF (1976) Spectrophotometric studies on the pH of frog skeletal muscle. pH change during and after contractile activity. J Gen Physiol 68:179–195

McIsaac RS, Engqvist MKM, Wannier T, Rosenthal AZ, Herwig L, Flytzanis NC, Imasheva ES, Lanyi JK, Balashov SP, Gradinaru V, Arnold FH (2014) Directed evolution of a far-red fluorescent rhodopsin. Proc Natl Acad Sci U S A 111:13034–13039

Miesenbock G, Kevrekidis IG (2005) Optical imaging and control of genetically designated neurons in functioning circuits. Annu Rev Neurosci 28:533–563

Mishina Y, Mutoh H, Song C, Knöpfel T (2014) Exploration of genetically encoded voltage indicators based on a chimeric voltage sensing domain. Front Mol Neurosci 7:78. doi:10.3389/fnmol.2014.00078

Murata Y, Iwasaki H, Sasaki M, Inaba K, Okamura Y (2005) Phosphoinositide phosphatase activity coupled to an intrinsic voltage sensor. Nature 435:1239–1243

Mutoh H, Perron A, Dimitrov D, Iwamoto Y, Akemann W, Chudakov DM, Knöpfel T (2009) Spectrally-resolved response properties of the three most advanced FRET Based fluorescent protein voltage probes. PLoS One 4(2), e4555, PMCID: PMC2641041

Perozo E, MacKinnon R, Bezanilla F, Stefani E (1993) Gating currents from a nonconducting mutant reveal open-closed conformations in Shaker K+ channels. Neuron 11:353–358

Piao HH, Rajakumar D, Kang BE, Kim EH, Baker BJ (2014) Combinatorial mutagenesis of the voltage-sensing domain enables the optical resolution of action potentials firing at 60 Hz by a genetically-encoded fluorescent sensor of membrane potential. J Neurosci 35(1):372–385

Sakai R, Repunte-Canonigo V, Raj CD, Knopfel T (2001) Design and characterization of a DNA-encoded, voltage-sensitive fluorescent protein. Eur J Neurosci 13:2314–2318

Salzberg BM, Obaid AL, Bezanilla F (1993) Microsecond response of a voltage-sensitive merocyanine dye: fast voltage-clamp measurements on squid giant axon. Jpn J Physiol 43 (Suppl 1):S37–S41

Shinar R, Druckmann S, Ottolenghi M, Korenstein R (1977) Electric field effects in bacteriorhodopsin. Biophys J 19:1–5

Siegel MS, Isacoff EY (1997) A genetically encoded optical probe of membrane voltage. Neuron 19:735–741

Storace DA, Braubach OR, Jin L., Cohen LB, Sung U (2015) Monitoring brain activity with protein voltage and calcium sensors. Nature Scientific Reports, in press

Sung U, Sepehri-Rad M, Jin L, Hughes T, Cohen LB, Baker BJ (2014) Improving signal dynamics of fluorescent protein voltage sensors by optimizing FRET interactions. Biophys J 108(2): p152a

St-Pierre F, Marshall JD, Yang Y, Gong Y, Schnitzer MJ, Lin MZ (2014) High-fidelity optical reporting of neuronal electrical activity with an ultrafast fluorescent voltage sensor. Nat Neurosci 17:884–889

Tsien RY (2005) Building and breeding molecules to spy on cells and tumors. FEBS Lett 579:927–932

Tsutsui H, Karasawa S, Okamura Y, Miyawaki A (2008) Improving membrane voltage measurements using FRET with new fluorescent proteins. Nat Methods 8:683–685

Tsutsui H, Jinno Y, Tomita A, Niino Y, Yamada Y (2013) Improved detection of electrical activity with a voltage probe based on a voltage-sensing phosphatase. J Physiol 591:4427–4437

Villalba-Galea CA, Sandtner W, Dimitrov D, Mutoh H, Knopfel T, Bezanilla F (2009) Charge movement of the voltage sensitive fluorescent protein. Biophys J 96:L19–L21

Wang D, McMahon S, Zhang Z, Jackson MB (2012) Hybrid voltage sensor imaging of electrical activity from neurons in hippocampal slices from transgenic mice. J Neurophysiol 108:3147–3160

Wang D, Zhang Z, Chanda B, Jackson MB (2010) Improved probes for hybrid voltage sensor imaging. Biophys J 99:2355–2365

Zou P, Zhao Y, Douglass AD, Hochbaum DR, Brinks D, Werley CA, Harrison DJ, Campbell RE, Cohen AE (2014) Bright and fast multicoloured voltage reporters via electrochromic FRET. Nat Commun 5:4625. doi:10.1038/ncomms5625

Index

M. Canepari et al. (eds.), *Membrane Potential Imaging in the Nervous System and Heart*, Advances in Experimental Medicine and Biology 859,
DOI 10.1007/978-3-319-17641-3

T

V

The manufacturer's authorised representative in the EU is Springer Nature Customer Service Centre GmbH, Europaplatz 3, 69115 Heidelberg, Germany. If you have any concerns regarding our products, please contact ProductSafety@springernature.com

Printed and bound by CPI Group (UK) Ltd, Croydon, CR0 4YY
30/09/2026
02225435-0015